Compliant Mechanisms

Compliant Mechanisms
Design of Flexure Hinges

Second Edition

Nicolae Lobontiu

CRC Press
Taylor & Francis Group
Boca Raton London New York

CRC Press is an imprint of the
Taylor & Francis Group, an **informa** business

Second edition published 2021
by CRC Press
6000 Broken Sound Parkway NW, Suite 300, Boca Raton, FL 33487-2742

and by CRC Press
2 Park Square, Milton Park, Abingdon, Oxon, OX14 4RN

© 2021 Taylor & Francis Group, LLC

First edition published by CRC Press 2002

CRC Press is an imprint of Taylor & Francis Group, LLC

Library of Congress Cataloging-in-Publication Data
Names: Lobontiu, Nicolae, author.
Title: Compliant mechanisms : design of flexure hinges / Nicolae Lobontiu.
Description: Second edition. | Boca Raton : CRC Press, 2021. | Includes
bibliographical references and index.
Identifiers: LCCN 2020022285 (print) | LCCN 2020022286 (ebook) |
ISBN 9781439893692 (hbk) | ISBN 9780429184666 (ebk) | ISBN 9781439893708
(adobe pdf) | ISBN 9780429542107 (mobi) | ISBN 9780429527401 (epub)
Subjects: LCSH: Mechanical movements—Design and construction. |
Hinges—Design and construction.
Classification: LCC TJ181.L63 2021 (print) | LCC TJ181 (ebook) |
DDC 621.8/2—dc23
LC record available at https://lccn.loc.gov/2020022285
LC ebook record available at https://lccn.loc.gov/2020022286

ISBN: 978-1-4398-9369-2 (hbk)
ISBN: 978-0-429-18466-6 (ebk)

Typeset in Times
by codeMantra

To my parents, Ana and Nicolae

Contents

Preface to the Second Edition

Since the first edition publication of this book, substantial developments have occurred in the field of flexure-based compliant mechanisms, both at the modeling (analysis) and at the design (synthesis) levels, as reflected by the impressive number of publications in scientific journals and monographs. While the major topics of the first edition, such as the analytical description of flexure hinges in terms of compliances, the static, dynamic and finite element modeling of hinge-based compliant mechanisms, have solidified their worth in time, an updating of that information became necessary in order to better capture the advances in this field.

The second edition is a substantial update of the first edition in terms of topical coverage, modeling approaches and application pool. New straight-axis flexible-hinge geometries are included here, alongside curvilinear-axis designs, which are dedicated each a separate chapter. Both hinge categories realize their final, often-times complex, configuration by combining basic (or primitive) segments in either series or parallel, and by utilizing a compliance matrix approach to formulate their elastic characteristics. The quasi-static response of flexible-hinge mechanisms is studied in two chapters: one describes the serial mechanisms and the other discusses the parallel mechanism architectures. The chapter formulating the dynamic model of flexible-hinge mechanisms has a comprehensive new section on hinge inertia that are derived by means of compliance-based shape functions, similar to finite element models. The finite element chapter incorporates a new section dedicated to circular-axis line elements. A miscellaneous chapter proposes a new section on straight- and circular-axis hinge precision of rotation, as well as new sections detailing flexible-hinge stress concentration, and actuation/sensing by means of piezoelectric multilayer active hinges and blocks.

The matrix approach is applied throughout this book to model all problems either by compliance-based methods or by finite element modeling.

The sense that a more application-oriented approach in this type of book would benefit the reader was materialized by numerous newly solved examples of flexible-hinge mechanism applications.

My sincere hope is that this new edition could be of real assistance to the researcher interested in designing flexible hinges and flexible-hinge mechanisms that operate in the small-displacement/deformation domain and utilize linear models.

I would like to express my profound gratitude to all my colleagues at University of Alaska Anchorage, Technical University of Cluj-Napoca (Romania) and Cornell University who have helped this research book shape up with their direct contribution and advice on the research projects we have been collaborating since the publication of the first edition of this book. I was also very fortunate to be able to exchange ideas, discover new perspectives and hands-on solutions while exploring flexible-hinge mechanisms with my University of Alaska Anchorage (former) students Paul Bilodeau, Raphael Wunderle, Tim Kirk, Josh Lazaro,

Jesse Wight-Crask, Collette Kawagley, Kaitie McCloud, Beth Steele, Dennis Kudryn and Jeff Leath.

My sincere thanks go to Jonathan Plant, former Mechanical Engineering and Applied Mechanics Executive Editor at Taylor & Francis/CRC Press, for patiently working with me over an extended period of time and gracefully offering me almost-endless deadline extensions to allow me to complete this manuscript. I am also indebted to Nicola Sharpe, Mechanical Engineering Editor at Taylor & Francis/CRC Press, for her prompt and professional support with finalizing this project.

Last but not least, I am beyond grateful to my wife Simona for her constant encouragement, patience and unwavering faith that I will somehow, sometime get to the end of this effort and fully return to more mundane activities – I would not have done it without her.

April 2020, Anchorage, AK

Author

Nicolae Lobontiu is Professor of Mechanical Engineering at the University of Alaska Anchorage, USA. Dr. Lobontiu's research interests for the past two decades have focused on flexure/flexible hinges and macro-/micro-scale hinge-based compliant mechanisms. He has published several journal papers on these research topics and is the author of four other books: *System Dynamics for Engineering Students – Second Edition, Dynamics of Microelectromechanical Systems, Mechanical Design of Microresonators* and *Mechanics of Microelectromechanical Systems* (with E. Garcia). Professor Lobontiu's teaching background includes courses in dynamics of systems, controls, mechanical vibrations, finite element analysis, dynamics and mechanics of materials.

Author

Nicolás...

1 Introduction

This introductory chapter gives a brief description of flexible hinges and flexure-based compliant mechanisms for macro- and micro-scale applications by highlighting the main traits of these mechanical members/devices. An outline of the subject matter and the associated approach in this book is also sketched here in order to identify and possibly locate the work in the context of similar dedicated information that has already been published.

Derived from the term *compliance*, utilized by Paros and Weisbord [1] to identify the reciprocal of stiffness, the *compliant* qualifier is utilized in this book to identify any mechanical device that comprises at least one elastically deformable joint, called *flexible* (or *flexure*) *hinge*. Such a mechanism usually includes other rigid links, which ensure the necessary rigidity and serve as input/output ports. The flexible-hinge *mechanisms* employ their mobility to perform conversions between electric, thermal, magnetic, etc. and mechanical energies. Many of these mechanisms, which are typically of monolithic architecture, are designed to undergo relatively low levels of rotation/deformation in order to generate accurate, quasi-continuous output motion. This work covers the flexure-based compliant mechanisms possessing all these traits.

Figure 1.1 shows the schematic representation of two microaccelerometers – sensing microdevices used to evaluate external acceleration. Both devices have a planar structure and comprise rigid links (fixed or movable) that are connected by

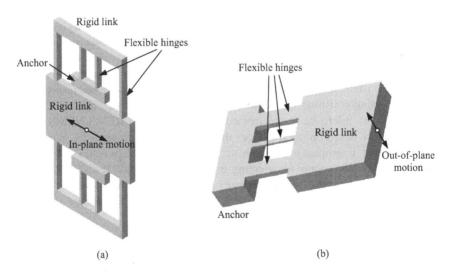

(a) (b)

FIGURE 1.1 Flexible-hinge microaccelerometers with: (a) in-plane motion; (b) out-of-the-plane motion.

flexible hinges of (constant) rectangular cross-section. The thinner in-plane cross-sectional dimension of the eight hinges of the microaccelerometer, as shown in Figure 1.1a, also *known as* folded-beam accelerometer (Bhushan [2]), enables the central rigid link (the proof mass) to undergo in-plane, back-and-forth translatory motions when its anchor links are attached to an external vibrating source along the motion direction. Alternatively, in order to sense acceleration along a direction perpendicular to the device plane, the design of Figure 1.1b, named resonant-beam accelerometer (see Tudor et al. [3]), utilizes flexible hinges whose out-of-plane dimension is smaller than the in-plane cross-sectional thickness; consequently, these hinges can bend along an in-plane axis and enable the mobile rigid link (also known as seismic mass) to displace out of the plane. Both devices use additional sensing (capacitive, for instance) to convert the mechanical acceleration into electrical voltage.

Flexible hinges are largely utilized in the auto/aviation industry in applications such as acceleration/speed/position sensors, adaptable seats, airbags, fault-tolerant connectors, single-surface-independent aircraft control devices, actuators for configurable-geometry foils, steering columns, antifriction bearings, suspension systems, satellite small-angle tilting mechanisms, laser-beam communication systems between spacecraft, and flexible couplings. The biomedical industry is also a beneficiary of mechanisms that are based on flexible hinges, and applications in this category include devices for vascular catheters, urethral compression devices, intravascular endoprostheses, cardiac massage apparatuses, orthotic devices and biopsy devices. The computers and fiber optics industries have also applications that incorporate flexible hinges, and the examples in this category include disk drive suspensions, laser systems, optical mirrors, optical disks, microscopes, cameras, print heads, optical scanning equipment, vibrating beam accelerometers, keyboard assemblies, kinematic lens mountings and rotary actuators for disk drives. Various other fields utilize flexible-hinge mechanisms, and a few application examples are coin packaging systems, systems for remotely playing percussion musical instruments, collapsible fishing net mechanisms, table tennis ball-retrieving systems, snow blade attachments, foot propulsion devices for float tube users, bicycle seats, steerable wheels for skate/roller ski, respiratory masks, grinding/polishing machines, fluid jet cutting machines and flywheels. The compliant micro-/nano-electromechanical systems (MEMS/NEMS) are almost entirely based on microdevices that generate their motions by means of flexible hinges. Examples in this industry comprise optical switches, miniature load cells, flexible mounts for imaging masks, load-sensitive resonators, gyroscopes, gravity gradiometers, disk memory head positioners, wire bonding heads, microfluidic devices, accelerometers, scan modules for bar code readers and cantilevers for microscopy.

A flexible hinge is a relatively thin member that provides the relative rotation between two adjacent rigid links mainly through flexing (bending), as sketched in Figure 1.2, where a conventional rotational joint is shown next to a flexible hinge.

In terms of this rotary function, a flexible hinge can be considered the structural correspondent of a bearing with limited rotation capability, as illustrated in

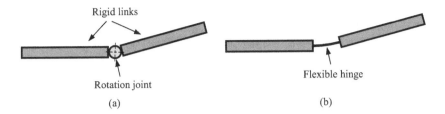

FIGURE 1.2 Joints enabling relative rotation in mechanisms: (a) conventional rotation joint; (b) skeleton representation of a generic flexible hinge.

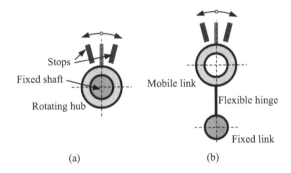

FIGURE 1.3 Functional similarity between rotary bearings and flexible hinges: (a) collocated (concentric) rotation produced by a rotary bearing; (b) non-collocated rotation produced by a flexible hinge.

Figure 1.3. In a classical rotary bearing, the relative rotation takes place between a shaft and its housing, these mating parts are concentrically located, and the rotation can be limited to a specific angular sector, as indicated in Figure 1.3a. A flexible hinge can provide a similar limited rotary output, and the only difference is attributable to the fact that the "centers" of the two adjacent members undergoing the relative rotation are no longer collocated, as shown in Figure 1.3b.

Physically, a flexible hinge can mainly be achieved in two different ways:

- Use an independently fabricated member (like a strip or shim in two-dimensional applications or a cylinder-like part in three-dimensional applications) to connect two rigid members that are designed to undergo a relative rotation;
- Generate the flexible hinge by either subtractive or additive fabrication procedures. In this case, the flexure is integral (or monolithic) to the rigid links it connects.

Among the benefits provided by the flexible hinges, the most notable are:

- No friction losses;
- No need of lubrication;

- No hysteresis;
- Compactness;
- Capacity to be utilized in small-scale applications;
- Ease of fabrication;
- Virtually no maintenance needed;
- No need of repair since the mechanism will operate until something will fail (usually the hinges) because of fatigue or overloading;
- Producing small-increment, quasi-continuous and very precise motion.

Flexible hinges do have limitations, however, and a few examples of such drawbacks are:

- The hinges, particularly the notch ones, provide relatively low levels of rotations due to stress limitations;
- The rotation is not pure because the deformation of a flexible hinge is complex as it is produced by axial loading, shear and torsion in addition to bending;
- The flexible hinge is usually sensitive to temperature variations, and therefore, its dimensions change as a result of thermal expansion/contraction, which leads to modifications in the original hinge elastic properties.

For two-dimensional applications where the flexible hinge is fabricated by removing material from a blank piece, the manufacturing processes that are being utilized for this purpose include end milling, electrodischarge machining (EDM), laser cutting, water/liquid jetting, metal stamping, three-dimensional printing and photolithographic/microfabrication techniques for MEMS/NEMS.

In two-dimensional, planar flexible mechanisms, the hinge is designed to be compliant about a single axis only (the input, compliant or sensitive axis), along which the relative rotation between the adjacent rigid links is taking place, and stiff (as much as possible) about all other axes and motions – see the designs of Figure 1.4a and c. These configurations are also known as *single-axis* flexible hinges. For other hinge architectures that have rotational symmetry (they are revolute and their cross-section is circular), like the one pictured in Figure 1.4b, bending is indiscriminately possible about any axis that is perpendicular to the axial direction, and therefore, the compliant axis can instantaneously be set by the loading/boundary conditions of the three-dimensional compliant mechanism. Consistent with the terminology that has already been utilized such flexible hinges are also called *multi-axis*. Single-axis flexible hinges are specifically intended for two-dimensional compliant mechanisms that have in-plane and out-of-plane motion, whereas multi-axis hinges are implemented in three-dimensional applications in order to take advantage of their capacity to produce relative rotation about two or more compliant axes.

While the designs of Figure 1.4a and b have straight longitudinal axes, flexible hinges with curvilinear longitudinal axis, like the one shown in Figure 1.4c, have also been designed and utilized in various applications.

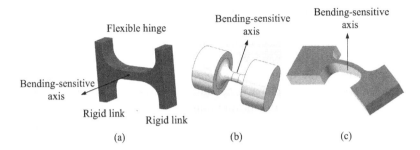

FIGURE 1.4 (a) Straight-axis flexible hinge of rectangular cross-section; (b) straight-axis flexible hinge of circular cross-section; (c) curvilinear-axis flexible hinge of rectangular cross-section.

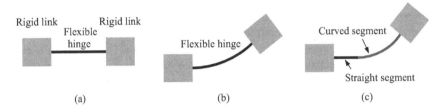

FIGURE 1.5 Skeleton representation of: (a) straight-axis flexible hinge; (b) curvilinear-axis flexible hinge; (c) flexible hinge with straight- and curvilinear-axis segments.

Oftentimes, it is convenient to use a simplified, skeleton representation of flexible hinges by just drawing the longitudinal axis of the hinge, as illustrated in Figure 1.5.

A considerable number of compliant mechanisms utilize a reduced pool of flexible-hinge configurations. Apart from the constant cross-section (either rectangular or circular) designs that present the inconvenience of high stress concentration at the areas joining the rigid links, two other geometries – the circular design (Paros and Weisbord [1]) and the right circularly corner-filleted configuration (Lobontiu et al. [4]) (these basic configurations are sketched in Figure 1.6) – occupy an ample space of choices for flexible-hinge applications. The simplicity of their geometry and the relative ease of fabrication are two main reasons that prompted the widespread use of these two types of hinges. In

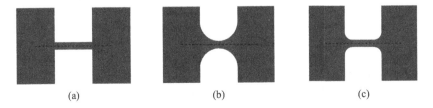

FIGURE 1.6 Commonly utilized flexible-hinge configurations with straight longitudinal axis: (a) constant cross-section; (b) right circular; (c) right circularly corner-filleted.

addition, both configurations come with the advantage of being able to reduce the stress concentration levels through the filleted regions at their corners (ends).

A thorough analytical presentation of single-axis and two-axis circular flexure hinges is provided in Paros and Weisbord [1]. The authors of this paper interchangeably utilized the terms *flexure hinge* and *flexure* to denote a mechanical member, which is "compliant in bending about one axis but rigid about the cross-axes", and mentioned that flexure hinges are incorporated in applications where angular motion is required about an axis – the compliant axis. The key merit of this seminal work consisted in the clear mathematical definition of a flexure hinge as a spring element, which displays two distinct behaviors: it is compliant about one axis, in order to produce the desired rotation, and stiff (as much as possible) in all other motions about the other axes in order to prevent/ minimize the respective parasitic motions. This paper provided analytical equations, both exact and approximate, for the spring rates (reciprocal to compliances) of single-axis and two-axis circular symmetric flexures in terms of the motions generated through the consideration of bending and axial effects. The alternative terminology of *flexural pivot* has also been utilized over the years in referring to the same mechanical member. The *notch hinge* has also been employed by several authors; see Smith [5], for instance, especially for single-axis flexible hinges. While the term *flexure* aptly depicts the primary function and operation mode of an elastic joint to flex (bend), there are also hinge configurations whose main intended/functional deformation is torsional. In order to cover all possible deformation modes, the terminology *flexible hinge* is used throughout this book. The term *compliance* (also known as *flexibility*) is utilized here in the original definition given in Paros and Weisbord [1] such that the term refers to the quantity which is the reciprocal of one flexure's spring rate (or stiffness). As a terminology extension from *compliance* – see also previous mention at the beginning of this chapter – the qualifier *compliant* is used in this book to denote the behavior of a mechanical device/mechanism that moves by the elastic deformation of its flexible hinges in order to realize a design objective (actuation or sensing).

Expanding the flexible-hinge design domain with new geometries and configurations is prompted by practical need. The compliant behavior of flexible hinges and flexure-based mechanisms largely depend on the specific geometry of the hinge for a given material. Slight alterations in geometry and dimensions can sensibly modify the elastic response. This aspect is particularly important in mechanisms where precision is a key performance parameter or where a finely tuned, quasi-continuous output is expected in terms of displacement, force or frequency (resonant) response. This book provides compliance equations for a large variety of flexible-hinge configurations that have been developed and reported in the specialty literature over the past years or which are new. It also qualifies and compares the flexible hinges in terms of level and precision of the output motion, sensitivity to parasitic (undesired) loading and stress levels.

The book essentially develops not only a process of discretization through which the compliance (stiffness) is primarily derived, but also studies the inertia and damping characteristics that are derived for a large variety of flexible hinges

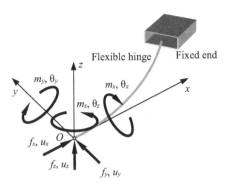

FIGURE 1.7   Skeleton fixed-free flexible hinge with six DOF at the free end.

both analytically and by means of the finite element technique. With the analytical approach, a distributed-parameter flexible hinge is equivalently transformed into a discrete one with properties defined individually and independently in terms of its degrees of freedom (DOF). For a three-dimensional flexible hinge, one of its ends can move with respect to the other one (presumed fixed) by three translations and three rotations, respectively, as sketched in Figure 1.7. Consequently, a flexible hinge can be modeled as a six-DOF member in the three-dimensional space when only the position of one end relative to the opposite one is of interest.

Planar flexible hinges of rectangular cross-sections can be designed to operate in their original plane, as the microaccelerometer of Figure 1.1a does, or out of their plane, as the case is with the mechanism of Figure 1.1b.

For in-plane operation, a planar flexible hinge, like the curvilinear-axis one shown in Figure 1.8a, has its in-plane dimension thinner than its (constant) out-of-plane dimension w. As a result, the free end O can translate along the x and y axes and rotate around the z axis. The quasi-static behavior of such a hinge, as shown in Chapters 2–4, is characterized by an *in-plane compliance matrix* connecting the load vector formed of the forces f_x and f_y and the moment m_z to the corresponding

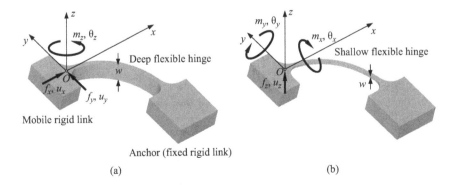

FIGURE 1.8   Curvilinear-axis flexible hinges for: (a) in-plane operation; (b) out-of-plane operation.

displacement vector defined by the translations u_x and u_y and the rotation θ_z. The hinge deformation includes bending around the z axis, axial deformation along the x axis and shear along the y axis. Out-of-plane deformations produced by the other three loads, namely, m_x, m_y and f_z, are undesired and therefore parasitic.

Conversely, the flexible hinge shown in Figure 1.8b, with an out-of-plane dimension w smaller than the in-plane thickness, can easily deform out of its original structural plane; as such, an out-of-plane compliance matrix can be derived to connect the load vector defined by the moments m_x and m_y and the force f_z to the resulting displacement vector formed of the rotations θ_x and θ_y and displacement u_z. This particular hinge mainly deforms in bending around the y axis, torsion around the x axis and shear along the z axis. All other deformations resulting from the in-plane loads f_x, f_y and m_z are undesired (parasitic).

It should be noted again that flexible hinges are approached in this book as members that are subjected to small deformations. This assumption stems from the reality that in numerous engineering applications the flexible hinges actually behave according to this model, which enables utilization of the linear (or first-order) bending theory and derivation of the spring rates for various hinge configurations.

In addition to the compliance-based approach utilized to model the static and dynamic behaviors of flexible-hinge mechanisms, which is a variant of the *force* (or *flexibility matrix*) *method*, this book also studies the flexible hinges and their mechanisms by means of the finite element technique using the classical *displacement* (*stiffness matrix*) *method*. A flexible hinge of either straight or curvilinear/circular axis is modeled as a two- or three-node line element with six DOF per node. A generic finite element formulation is developed by defining the elemental stiffness, mass and damping matrices. Explicit forms of these elemental matrices are derived for a few flexible-hinge configurations. These equations are subsequently utilized to characterize the static or dynamic response of various flexible-hinge mechanism examples.

Several other related topics, including stress concentration, combined stresses and theories of failure, flexible-hinge precision of rotation, layered flexible hinges, temperature effects and piezoelectric actuation/sensing, are also studied in this book.

REFERENCES

1. Paros, J.M. and Weisbord, L., How to design flexure hinges, *Machine Design*, November, 151, 1965.
2. Bhushan, B. (editor), *Springer Handbook of Nanotechnology*, 3rd ed., Springer, Berlin, 2010.
3. Tudor, M.J. et al., Silicon resonator sensors: interrogation techniques and characteristics, *IEEE Proceedings on Control Theory and Applications*, 135 (5), 364, 1988.
4. Lobontiu, N. et al., Corner-filleted flexure hinges, *ASME Journal of Mechanical Design*, 123, 346, 2001.
5. Smith, S.T., *Flexures – Elements of Elastic Mechanisms*, Gordon and Breach Science Publishers, Amsterdam, 2000.

2 Compliances of Basic Flexible-Hinge Segments

Using the strain-energy Castigliano's second theorem, this chapter derives and studies the compliances (flexibilities) of various basic planar segments with either straight or curvilinear longitudinal (centroidal) axis, which are used to build flexible hinges of more complex geometry in Chapters 3 and 4. Each basic (or primitive) segment has its longitudinal profile defined by a planar analytical curve, such as a straight line or another curve. Individual compliances are subsequently arranged in compliance matrices that model the in-plane and out-of-plane elastic behavior of the basic flexible-hinge segments. The Appendix at this chapter's end offers a compliance database for some of the common straight-axis and curvilinear-axis hinge segments.

2.1 STRAIGHT-AXIS FLEXIBLE-HINGE SEGMENTS

Two categories of flexible-hinge segments with straight longitudinal axis are discussed here: one category is formed of configurations of rectangular cross-section with either one dimension variable (and the other one constant) or both dimensions variable, and the other category comprises geometries with circular cross-section of variable diameter.

The generic hinge segment longitudinal profile of Figure 2.1 has a length l and is defined by the curves $y(x)$ and $-y(x)$, which are symmetric with respect to the centroidal x axis. This is a necessary feature of any configuration whose centroidal (longitudinal) axis is a straight line irrespective of the shape of $y(x)$.

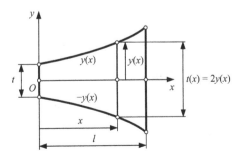

FIGURE 2.1 Front view of generic axially symmetric hinge segment whose variable thickness $t(x)$ is defined by an analytical curve $y(x)$.

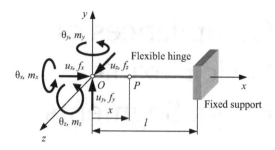

FIGURE 2.2 Schematic representation of generic fixed-free, straight-axis flexible hinge with free-end loads and deformations.

For a rectangular cross-section hinge, the end thickness is $t = 2y$ and the variable thickness $t(x)$ is equal to $2y(x)$ at an arbitrary distance x. A circular cross-section hinge segment with the same longitudinal profile of Figure 2.1 has an end diameter $d = 2y$ and a variable-position diameter $d(x) = 2y(x)$. In order to keep the notation simple, this section utilizes t and $t(x)$ for both rectangular and circular cross-section hinge segments.

2.1.1 GENERIC COMPLIANCES

Consider the generic straight-axis flexible hinge, which is shown in skeleton form in Figure 2.2 as a fixed-free member in the reference frame $Oxyz$. Under the action of the free-end point forces f_x, f_y and f_z and moments m_x, m_y and m_z, the corresponding deflections u_x, u_y and u_z and rotations (slopes) θ_x, θ_y and θ_z are produced.

The *Castigliano's second theorem* (also known as *Castigliano's displacement theorem*) is a very useful technique that enables calculating the deformations of elastic bodies under the action of external loading and support reactions, as detailed in Timoshenko [1], Richards [2], Muvdi and McNabb [3], Harker [4], Langhaar [5] or Cook and Young [6]. The theorem offers a simple and elegant mathematical tool to calculate a local deformation (either linear or angular) that is produced by a corresponding external load/support reaction (either force or moment) acting at that particular location. Specifically, the local deformation of an elastic body with linear elastic material properties is expressed as the partial derivative of the total strain energy stored in that body in terms of the force/moment acting at that location and along the direction of the specified deformation. A prerequisite of the theorem is that the elastic body under study is sufficiently supported, meaning that any rigid-body motions (motions allowed by insufficient external support and not produced by elastic deformation) are prohibited. A second requirement is that the strain energy be expressed in terms of loads and, consequently, it should contain no displacements. As shown in the following, this requirement is easy to comply with, by the very nature of formulating the strain energy.

The elastic body represented in Figure 2.3 undergoes both linear and angular deformations under the action of external loading and support reactions.

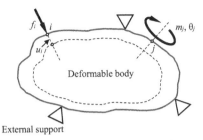

FIGURE 2.3 Spatially constrained, elastically deformable body under external load.

According to Castigliano's second theorem, the linear displacement u_i at point i is expressed in terms of the force f_i acting at that location; similarly, the angular deformation θ_j is calculated in terms of the moment m_j acting at point j as:

$$u_i = \frac{\partial U}{\partial f_i}; \quad \theta_j = \frac{\partial U}{\partial m_j}, \tag{2.1}$$

where U is the elastic strain energy of the deformed body. Note that f_i and u_i need to be aligned at point i; similarly, m_j and θ_j are aligned at point j, as illustrated in Figure 2.3.

Another formulation, known as the *Castigliano's first* (or *force*) *theorem*, reverses the causality of the second theorem by expressing the load that is necessary to produce a specific deformation of an elastic body based on the strain energy. This theorem is applied in Chapter 8 to derive flexible hinge stiffnesses by the finite element method. It is worth noting that while applicable to linear elastic materials, the Castigliano's theorems are also valid for nonlinear elastic materials when the complementary energy needs to be used instead of the strain energy.

The displacements u_x, u_y, u_z, θ_x, θ_y and θ_z at the free end of the generic flexible segment of Figure 2.2 are evaluated by means of Castigliano's second theorem according to which:

$$u_x = \frac{\partial U}{\partial f_x}; \quad u_y = \frac{\partial U}{\partial f_y}; \quad u_z = \frac{\partial U}{\partial f_z}; \quad \theta_x = \frac{\partial U}{\partial m_x}; \quad \theta_y = \frac{\partial U}{\partial m_y}; \quad \theta_z = \frac{\partial U}{\partial m_z}. \tag{2.2}$$

The fixed-free elastic flexible hinge is mainly deformed by the normal (axial) force, torsion moment and bending moments about the y and z axes. As detailed in Richards [2], Cook and Young [6] or Den Hartog [7], the total strain energy is expressed as:

$$U = U_a + U_t + U_{b,z} + U_{b,y} = \frac{1}{2E} \cdot \int_0^l \frac{N(x)^2}{A(x)} dx + \frac{1}{2G} \cdot \int_0^l \frac{M_t(x)^2}{I_t(x)} dx$$

$$+ \frac{1}{2E} \cdot \int_0^l \frac{M_{b,z}(x)^2}{I_z(x)} dx + \frac{1}{2E} \cdot \int_0^l \frac{M_{b,y}(x)^2}{I_y(x)} dx, \tag{2.3}$$

where "*a*", "*t*" and "*b*" stand for axial, torsion and bending, respectively. The axial force $N(x)$, torsion moment $M_t(x)$, z-axis bending moment $M_{b,z}(x)$ and y-axis bending moment $M_{b,y}(x)$ are defined at the generic point P of Figure 2.2 as:

$$N(x) = f_x; \quad M_t(x) = m_x; \quad M_{b,z}(x) = -f_y \cdot x + m_z; \quad M_{b,y}(x) = f_z \cdot x + m_y. \quad (2.4)$$

The flexible-hinge cross-section, which is assumed variable along the length, is defined in Eq. (2.3) by the area $A(x)$, moments of area in torsion $I_t(x)$, and bending $I_y(x)$ and $I_z(x)$. In the same equation, E and G are the material Young's and shear moduli.

Combining Eqs. (2.2)–(2.4) yields:

$$\begin{cases} u_x = \dfrac{1}{E} \cdot \displaystyle\int_0^l \dfrac{N(x)}{A(x)} \cdot \dfrac{\partial N(x)}{\partial f_x} dx = C_{u_x - f_x} \cdot f_x \\[3mm] u_y = \dfrac{1}{E} \cdot \displaystyle\int_0^l \dfrac{M_{b,z}(x)}{I_z(x)} \cdot \dfrac{\partial M_{b,z}(x)}{\partial f_y} dx = C_{u_y - f_y} \cdot f_y + C_{u_y - m_z} \cdot m_z \\[3mm] u_z = \dfrac{1}{E} \cdot \displaystyle\int_0^l \dfrac{M_{b,y}(x)}{I_y(x)} \cdot \dfrac{\partial M_{b,y}(x)}{\partial f_z} dx = C_{u_z - f_z} \cdot f_z + C_{u_z - m_y} \cdot m_y \\[3mm] \theta_x = \dfrac{1}{G} \cdot \displaystyle\int_0^l \dfrac{M_t(x)}{I_t(x)} \cdot \dfrac{\partial M_t(x)}{\partial m_x} dx = C_{\theta_x - m_x} \cdot m_x \\[3mm] \theta_y = \dfrac{1}{E} \cdot \displaystyle\int_0^l \dfrac{M_{b,y}(x)}{I_y(x)} \cdot \dfrac{\partial M_{b,y}(x)}{\partial m_y} dx = C_{\theta_y - f_z} \cdot f_z + C_{\theta_y - m_y} \cdot m_y \\[3mm] \theta_z = \dfrac{1}{E} \cdot \displaystyle\int_0^l \dfrac{M_{b,z}(x)}{I_z(x)} \cdot \dfrac{\partial M_{b,z}(x)}{\partial m_z} dx = C_{\theta_z - f_y} \cdot f_y + C_{\theta_z - m_z} \cdot m_z \end{cases} \quad (2.5)$$

The necessary partial derivatives are calculated based on Eqs. (2.4). The following *compliances* (denoted by C) are used in Eqs. (2.5):

$$\begin{cases} C_{u_x - f_x} = \dfrac{1}{E} \cdot \displaystyle\int_0^l \dfrac{dx}{A(x)}; \quad C_{\theta_x - m_x} = \dfrac{1}{G} \cdot \displaystyle\int_0^l \dfrac{dx}{I_t(x)} \\[4mm] C_{u_y - f_y} = \dfrac{1}{E} \cdot \displaystyle\int_0^l \dfrac{x^2 dx}{I_z(x)}; \quad C_{u_y - m_z} = C_{\theta_z - f_y} = -\dfrac{1}{E} \cdot \displaystyle\int_0^l \dfrac{x dx}{I_z(x)}; \quad C_{\theta_z - m_z} = \dfrac{1}{E} \cdot \displaystyle\int_0^l \dfrac{dx}{I_z(x)} \\[4mm] C_{u_z - f_z} = \dfrac{1}{E} \cdot \displaystyle\int_0^l \dfrac{x^2 dx}{I_y(x)}; \quad C_{u_z - m_y} = C_{\theta_y - f_z} = \dfrac{1}{E} \cdot \displaystyle\int_0^l \dfrac{x dx}{I_y(x)}; \quad C_{\theta_y - m_y} = \dfrac{1}{E} \cdot \displaystyle\int_0^l \dfrac{dx}{I_y(x)} \end{cases}$$

$$(2.6)$$

to define the generic, variable cross-section flexible-hinge segment. The first part of a compliance's subscript indicates any of the six free-end displacements, whereas the second subscript portion denotes the load component generating the particular displacement. The first Eq. (2.5) provides the x-axis displacement at the free end in terms of solely f_x as a result of axial loading, and similarly, the fourth Eq. (2.5) expresses the x-axis rotation of the free end as a function of m_x through torsion deformation. The remaining four equations of Eq. (2.5), which characterize bending about the y and z axes, express the end-point slopes and deflections in terms of both a force and a moment; therefore, they reflect coupling effects. The second Eq. (2.5), for instance, which gives u_y, shows that contributions are made by both the end-point force f_y and the moment m_z. Similarly, coupling occurs for f_z and m_y in producing the deformations u_z and θ_y.

The compliance formulation so far was concerned with flexible hinges that are relatively long compared to their cross-sectional dimensions, and that are treated as Euler–Bernoulli-beam-type members. For such a model, the planar cross-section remains perpendicular to the neutral axis after the external bending has been applied. This model ignores the shear stresses and associated deformations. However, for relatively short beams, the shear effects need to be considered together with their corresponding additional deformation. A model that accounts for such additional shear effects is the Timoshenko short-beam model, which also incorporates rotary inertia effects to better describe the dynamic response of such members.

The extra shear-produced strain energy of an elastic beam-like component along the y axis, as shown in Young et al. [8] and Cowper [9], for instance, is expressed as:

$$U_{s,y} = \frac{\alpha_s}{2G} \cdot \int_0^l \frac{S_y(x)^2}{A(x)} dx \quad \text{with} \quad S_y(x) = f_y, \tag{2.7}$$

where $S_y(x)$ is the shear force along the y direction at the generic point P of the flexible hinge shown in Figure 2.2 and α_s is a shear correction factor that depends on the shape of the cross-section. For rectangular cross-sections, Cowper [9] suggests using $\alpha_s = (12 + 11\mu)/10/(1 + \mu)$, where μ is Poisson's ratio. By applying Castigliano's second theorem, the shear-generated extra deflection is calculated as:

$$
\begin{aligned}
u_{y,s} &= \frac{\partial U_{s,y}}{\partial f_y} = \frac{\alpha_s}{G} \cdot \int_0^l \frac{S_y(x)}{A(x)} \cdot \frac{\partial S_y(x)}{\partial f_y} dx = \frac{\alpha_s}{G} \cdot \left(\int_0^l \frac{dx}{A(x)} \right) \cdot f_y \\
&= \frac{\alpha_s \cdot E}{G} \cdot C_{u_x - f_x} \cdot f_y = 2\alpha_s \cdot (1 + \mu) \cdot C_{u_x - f_x} \cdot f_y
\end{aligned}
\tag{2.8}
$$

where the first Eq. (2.6) has been used. As a consequence, the total deflection about the y axis (that sums up bending and shear effects) is:

$$u_y = C_{u_y-f_y} \cdot f_y + C_{u_y-m_z} \cdot m_z + \frac{\alpha_s \cdot E}{G} \cdot C_{u_x-f_x} \cdot f_y$$

$$= \left(C_{u_y-f_y} + 2\alpha_s \cdot (1+\mu) \cdot C_{u_x-f_x} \right) \cdot f_y + C_{u_y-m_z} \cdot m_z, \tag{2.9}$$

which yields the modified compliance:

$$C_{u_y-f_y,s} = C_{u_y-f_y} + \frac{\alpha_s \cdot E}{G} \cdot C_{u_x-f_x} = C_{u_y-f_y} + 2\alpha_s \cdot (1+\mu) \cdot C_{u_x-f_x}. \tag{2.10}$$

Equations (2.8)–(2.10) used the relationship between the shear modulus G and Young's modulus E: $E = 2(1 + \mu) \cdot G$.

Considering now the shear effects along the z axis, the direct-bending compliance corresponding to this axis, similar to Eq. (2.10), changes to:

$$C_{u_z-f_z,s} = C_{u_z-f_z} + \frac{\alpha_s \cdot E}{G} \cdot C_{u_x-f_x} = C_{u_z-f_z} + 2\alpha_s \cdot (1+\mu) \cdot C_{u_x-f_x}. \tag{2.11}$$

2.1.1.1 In-Plane Load, Displacement Vectors and Compliance Matrix

With respect to the generic hinge segment of Figure 2.2, the following loads and displacements are formally assembled into the *in-plane vectors*:

$$\left[f_{ip} \right] = \begin{bmatrix} f_x & f_y & m_z \end{bmatrix}^T; \quad \left[u_{ip} \right] = \begin{bmatrix} u_x & u_y & \theta_z \end{bmatrix}^T. \tag{2.12}$$

The first, second and sixth Eqs. (2.5) can therefore be written as:

$$\left[u_{ip} \right] = \left[C_{ip} \right]\left[f_{ip} \right]; \quad \left[C_{ip} \right] = \begin{bmatrix} C_{u_x-f_x} & 0 & 0 \\ 0 & C_{u_y-f_y} & C_{u_y-m_z} \\ 0 & C_{u_y-m_z} & C_{\theta_z-m_z} \end{bmatrix}, \tag{2.13}$$

where $[C_{ip}]$ is the *in-plane compliance matrix*. For relatively short hinges, when shear effects are considered, the compliance matrix of Eqs. (2.13) uses Eqs. (2.10) and becomes:

$$\left[C_{ip} \right] = \begin{bmatrix} C_{u_x-f_x} & 0 & 0 \\ 0 & C_{u_y-f_y} + 2\alpha_s \cdot (1+\mu) \cdot C_{u_x-f_x} & C_{u_y-m_z} \\ 0 & C_{u_y-m_z} & C_{\theta_z-m_z} \end{bmatrix}. \tag{2.14}$$

2.1.1.2 Out-of-Plane Load, Displacement Vectors and Compliance Matrix

The following *out-of-plane* load and displacement *vectors* are formed with the remaining components identified in Figure 2.2:

$$\left[f_{op} \right] = \left[\begin{array}{ccc} m_x & m_y & f_z \end{array} \right]^T; \quad \left[u_{op} \right] = \left[\begin{array}{ccc} \theta_x & \theta_y & u_z \end{array} \right]^T. \quad (2.15)$$

These vectors enable reformulating the corresponding Eqs. (2.5) as:

$$\left[u_{op} \right] = \left[C_{op} \right] \left[f_{op} \right]; \quad \left[C_{op} \right] = \left[\begin{array}{ccc} C_{\theta_x - m_x} & 0 & 0 \\ 0 & C_{\theta_y - m_y} & C_{\theta_y - f_z} \\ 0 & C_{\theta_y - f_z} & C_{u_z - f_z} \end{array} \right], \quad (2.16)$$

where $[C_{op}]$ is the *out-of-plane compliance matrix*. When shear effects need to be considered in the case of short hinges, the compliance matrix of Eq. (2.16) combines with Eq. (2.11) and changes to:

$$\left[C_{op} \right] = \left[\begin{array}{ccc} C_{\theta_x - m_x} & 0 & 0 \\ 0 & C_{\theta_y - m_y} & C_{\theta_y - f_z} \\ 0 & C_{\theta_y - f_z} & C_{u_z - f_z} + 2\alpha_s \cdot \left(1 + \mu\right) \cdot C_{u_x - f_x} \end{array} \right]. \quad (2.17)$$

Notes on cross-sectional properties

Equations (2.6) utilize the following cross-section parameters: area $A(x)$, moments of area $I_y(x)$ and $I_z(x)$, and torsional moment of area $I_t(x)$.

A generic, variable cross-section hinge segment of length l and possessing axial symmetry is illustrated in front view in Figure 2.4a; a rectangular cross-section is shown in Figure 2.4b, while a circular cross-section is sketched in

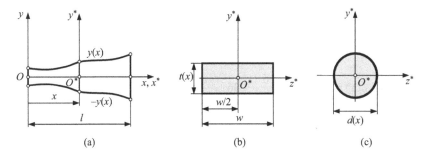

(a)   (b)   (c)

FIGURE 2.4 Planar geometry of variable cross-section flexible-hinge segment: (a) front view; (b) rectangular cross-section of constant width w and variable thickness $t(x)$ at an arbitrary position; (c) circular cross-section of variable diameter $d(x)$ at an arbitrary position.

Figure 2.4c, both at an arbitrary abscissa x. The variable thickness of a rectangular cross-section is $t(x) = 2y(x)$, while the variable diameter is $d(x) = 2y(x)$ for the circular cross-section. As a consequence, the variable cross-section properties need to be evaluated with respect to the reference frame located at point O^* in the same Figure 2.4a, which is a point on the centroidal axis of this generic flexible segment.

Circular cross-section

For a circular cross-section defined by a variable diameter $d(x)$, the parameters are:

$$A(x) = \frac{\pi d(x)^2}{4}; \quad I_z(x) = I_y(x) = I(x) = \frac{\pi d(x)^4}{64}; \quad I_t(x) = I_p(x) = \frac{\pi d(x)^4}{32},$$

$$(2.18)$$

where the torsional moment of area $I_t(x)$ becomes the polar moment of area $I_p(x)$. Because of that, the following compliance relationships result from Eqs. (2.6):

$$\begin{cases} C_{\theta_x - m_x} = \dfrac{E}{2G} \cdot C_{\theta_y - m_y} = (1 + \mu) \cdot C_{\theta_y - m_y} \\[2ex] C_{u_y - f_y} = C_{u_z - f_z}; \, C_{u_y - m_z} = -C_{\theta_y - f_z}; \, C_{\theta_y - m_y} = C_{\theta_z - m_z} \end{cases}.$$

$$(2.19)$$

Rectangular cross-section

For a rectangular cross-section hinge segment with w constant and $t(x)$ variable, the area and the moments of area are: $A(x) = w \cdot t(x)$, $I_z(x) = w \cdot t(x)^3/12$ and $I_y(x) = w^3 \cdot t(x)/12$. Note that the two moments of area are calculated in the reference frame $O^* x^* y^* z^*$ in Figure 2.4b. The torsion moment of area of this cross-section can be expressed in three different ways depending on the dimensional relationship between w and $t(x)$ as discussed here. One equation for the torsion moment of area in Eq. (2.16) is suggested in Young et al. [8] as:

$$I_{t,l}(x) = \begin{cases} I_{t,lw}(x) = \dfrac{w^3 \cdot t(x)}{16} \cdot \left[\dfrac{16}{3} - 3.36 \cdot \dfrac{w}{t(x)} \cdot \left(1 - \dfrac{1}{12} \cdot \dfrac{w^4}{t(x)^4} \right) \right] & \text{for } w < t(x) \\[3ex] I_{t,lt}(x) = \dfrac{w \cdot t(x)^3}{16} \cdot \left[\dfrac{16}{3} - 3.36 \cdot \dfrac{t(x)}{w} \cdot \left(1 - \dfrac{1}{12} \cdot \dfrac{t(x)^4}{w^4} \right) \right] & \text{for } w > t(x) \end{cases}.$$

$$(2.20)$$

For either $w < t(x)$ or $w > t(x)$, the ratio formed of the two cross-section variables ($t(x)$ and w) to the power of 5 is a very small quantity, which can be ignored and results in the simplified torsion moment of area:

$$I_{t,2}(x) = \begin{cases} I_{t,2w}(x) = \dfrac{w^3 \cdot t(x)}{16} \cdot \left(\dfrac{16}{3} - 3.36 \cdot \dfrac{w}{t(x)} \right) & \text{for} \quad w < t(x) \\[4mm] I_{t,2t}(x) = \dfrac{w \cdot t(x)^3}{16} \cdot \left(\dfrac{16}{3} - 3.36 \cdot \dfrac{t(x)}{w} \right) & \text{for} \quad w > t(x) \end{cases} \tag{2.21}$$

When the differences between the two cross-section dimensions w and $t(x)$ are very large, the torsion moments of area provided in Eqs. (2.21) can be further simplified to:

$$I_{t,3}(x) = \begin{cases} I_{t,3w}(x) = \dfrac{w^3 \cdot t(x)}{3} & \text{for} \quad w \ll t(x) \\[4mm] I_{t,3t}(x) = \dfrac{w \cdot t(x)^3}{3} & \text{for} \quad w \gg t(x) \end{cases} \tag{2.22}$$

There are situations where $t(x) < w$ for a length portion $[0, x_1]$, but as a result of increasing $t(x)$, $t(x) > w$ for the remaining length interval $[x_1, l]$. The torsional compliance in this case needs to be evaluated as:

$$C_{\theta_x - m_x} = \frac{1}{G} \cdot \left(\int_0^{x_1} \frac{dx}{I_{t,it}} + \int_{x_1}^l \frac{dx}{I_{t,iw}} \right), \quad i = 1,2 \text{ or } 3. \tag{2.23}$$

The threshold value x_1 is calculated by solving the equation $2y(x_1) = w$, where $y(x)$ is the curve defining the geometric profile of the axially symmetric hinge segment.

Using Eqs. (2.6) and (2.22), the following connections result between the torsion and the bending compliances for hinge configurations with either $w \ll t(x)$ or $w \gg t(x)$:

$$C_{\theta_x - m_x, 3} = \begin{cases} \dfrac{E}{4G} \cdot C_{\theta_y - m_y} = \dfrac{1 + \mu}{2} \cdot C_{\theta_y - m_y} & \text{for} \quad w \ll t(x) \\[4mm] \dfrac{E}{4G} \cdot C_{\theta_z - m_z} = \dfrac{1 + \mu}{2} \cdot C_{\theta_z - m_z} & \text{for} \quad w \gg t(x) \end{cases} \tag{2.24}$$

where "3" has been added to the torsional compliance to indicate that model 3 of the torsional moment of area – Eq. (2.22) – has been utilized.

Because the torsion moment of area can be expressed by one of the three expressions given in Eqs. (2.20)–(2.22), the torsion compliances can be expressed in three different ways by means of Eq. (2.6). A simple method identifying the proper torsional compliance (and corresponding moment of area) studies the following relative compliance ratios:

$$\begin{cases} r_{12} = \dfrac{\left| C_{\theta_x - m_x, 1} - C_{\theta_x - m_x, 2} \right|}{C_{\theta_x - m_x, 1}} \\[4mm] r_{13} = \dfrac{\left| C_{\theta_x - m_x, 1} - C_{\theta_x - m_x, 3} \right|}{C_{\theta_x - m_x, 1}} \end{cases} \tag{2.25}$$

Should, for instance, the compliance ratio r_{13} (which compares the torsional compliances resulting from the torsion moment $C_{\theta_x - m_x, 1}$ of Eq. (2.20), to the most simplified moment $C_{\theta_x - m_x, 3}$ of Eq. (2.22)) be smaller than an acceptable error level, then the most simplified compliance $C_{\theta_x - m_x, 3}$ can be used with sufficient accuracy. If the errors resulting from $C_{\theta_x - m_x, 3}$ are not acceptable, then the ratio r_{12} is calculated (which compares the compliance $C_{\theta_x - m_x, 1}$ with the intermediate-complexity compliance $C_{\theta_x - m_x, 2}$ of Eq. (2.21)) and its value is compared to the tolerable error. If that error is acceptable, then $C_{\theta_x - m_x, 2}$ will be used. If the error is too large, then $C_{\theta_x - m_x, 1}$ needs to be utilized to evaluate the torsion compliance. *Example 2.1* studies these options in the next section.

2.1.2 GEOMETRIC CONFIGURATIONS

This section derives in-plane and out-of-plane compliances of basic (primitive) straight-axis hinge segments whose cross-section dimensions are defined by a single analytical curve $y(x)$. Various curves $y(x)$ are proposed to define the axial profile of hinge segments, including the circle, ellipse, hyperbola, parabola and Bézier curves. While this section only provides equations of the profile curve (explicitly, parametrically or in polar form) and the corresponding equations for the cross-section dimensions, the Appendix at the end of this chapter includes closed-form compliance equations for these hinge segments with rectangular and circular cross-sections.

2.1.2.1 Straight-Line Profile Segments

In this section, the hinge segments have longitudinal profiles that are defined by line segments and are obtained when the curve $y(x)$ defining the upper edge of the generic hinge segment in Figure 2.1 is a straight line. The generating line segment can be either parallel to the centroidal axis – case where the hinge has a constant cross-section – or can be inclined with respect to the centroidal axis, and the resulting longitudinal profile of the hinge is trapezoid.

2.1.2.1.1 Constant Cross-Sectional Segment

The in-plane and out-of-plane compliances of the constant cross-section segment of Figure 2.5, defined by the planar geometric parameters l and t (or d), are calculated with respect to the frame $Oxyz$.

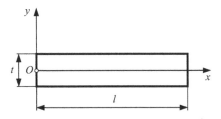

FIGURE 2.5 Front view with planar geometry of constant cross-section flexible segment.

Rectangular cross-section:
A rectangular cross-section member has the following in-plane compliances:

$$C_{u_x-f_x} = \frac{l}{E \cdot w \cdot t}; \quad C_{u_y-f_y} = \frac{4l^3}{E \cdot w \cdot t^3};$$

$$C_{u_y-m_z} = -\frac{6l^2}{E \cdot w \cdot t^3}; \quad C_{\theta_z-m_z} = \frac{12l}{E \cdot w \cdot t^3} \tag{2.26}$$

whereas the out-of-plane bending compliances are:

$$C_{u_z-f_z} = \frac{4l^3}{E \cdot w^3 \cdot t}; \quad C_{u_z-m_y} = \frac{6l^2}{E \cdot w^3 \cdot t}; \quad C_{\theta_y-m_y} = \frac{12l}{E \cdot w^3 \cdot t}; \quad C_{\theta_x-m_x} = \frac{l}{G \cdot I_t}. \tag{2.27}$$

Both Eqs. (2.26) and (2.27) result from the generic Eqs. (2.6) by using $t(x) = t$.
When $t(x) = t$, the torsion moment of area given in Eq. (2.20) becomes:

$$I_{t,1} = \begin{cases} \dfrac{w^3 \cdot t}{16} \cdot \left[\dfrac{16}{3} - 3.36 \cdot \dfrac{w}{t} \cdot \left(1 - \dfrac{1}{12} \cdot \dfrac{w^4}{t^4} \right) \right] & \text{for} \quad w < t \\[4mm] \dfrac{w \cdot t^3}{16} \cdot \left[\dfrac{16}{3} - 3.36 \cdot \dfrac{t}{w} \cdot \left(1 - \dfrac{1}{12} \cdot \dfrac{t^4}{w^4} \right) \right] & \text{for} \quad w > t \end{cases} \tag{2.28}$$

Similarly, the simplified torsion moment of area of Eq. (2.21) is:

$$I_{t,2} = \begin{cases} \dfrac{w^3 \cdot t}{16} \cdot \left(\dfrac{16}{3} - 3.36 \cdot \dfrac{w}{t} \right) & \text{for} \quad w < t \\[4mm] \dfrac{w \cdot t^3}{16} \cdot \left(\dfrac{16}{3} - 3.36 \cdot \dfrac{t}{w} \right) & \text{for} \quad w > t \end{cases} \tag{2.29}$$

Eventually, the torsion moments of areas formulated in Eqs. (2.22) simplify to:

$$I_{t,3} = \begin{cases} \dfrac{w^3 \cdot t}{3} & \text{for} \quad w \ll t \\[2em] \dfrac{w \cdot t^3}{3} & \text{for} \quad w \gg t \end{cases}. \tag{2.30}$$

Example 2.1

Given a constant rectangular cross-section (dimensions are $w < t$), straight-axis hinge, study the relative variation of the torsional compliance in terms of the w/t ratio when considering the possibility and applicability of using the three torsional moments of area of Eqs. (2.28)–(2.30).

Solution:

The second compliance ratio of Eq. (2.25) becomes for this design:

$$r_{13}(w/t) = \frac{\left| C_{\theta_x - m_x,1} - C_{\theta_x - m_x,3} \right|}{C_{\theta_x - m_x,1}} = \frac{I_{t,3} - I_{t,1}}{I_{t,1}}. \tag{2.31}$$

The plot of r_{13} in terms of the cross-sectional ratio w/t is shown in Figure 2.6.

For w/t ratios smaller than 0.7, the errors in the torsional compliance are less than 10% when using the most simplified torsional moment of inertia instead of the most complex one, as seen in Figure 2.6. Imposing certain threshold values to this relative difference in compliance, the limit values of w/t given in Table 2.1 are obtained, which indicate that the simplified compliances based on I_{t3} of Eq. (2.30) can be used safely up to that limit with the corresponding error listed in the same row of the table.

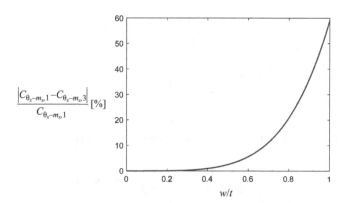

FIGURE 2.6 Plot of the relative errors between the torsional compliances defined by I_{t1} and I_{t3}.

TABLE 2.1

Upper Limits of the *w/t* Ratio for Allowable Percentage Relative Compliance Errors Based on I_{t1} and I_{t3}

$\dfrac{C_{\theta_x-m_x,1} - C_{\theta_x-m_x,3}}{C_{\theta_x-m_x,1}}(\%)$	*w/t*
1	0.4
2	0.47
5	0.58
10	0.68

If the errors resulting from using the most simplified torsional moment of inertia are unacceptable, the first compliance ratio introduced in Eq. (2.25) needs to be utilized instead; for the constant cross-section, this ratio is:

$$r_{12}(w/t) = \frac{\left| C_{\theta_x-m_x,1} - C_{\theta_x-m_x,2} \right|}{C_{\theta_x-m_x,1}} = \frac{I_{t,2} - I_{t,1}}{I_{t,2}}, \qquad (2.32)$$

which used the torsion moments of areas of Eqs. (2.28) and (2.29) corresponding to $w < t$.

As seen in the plot of Figure 2.7, the percentage of the relative compliance difference of Eq. (2.32) increases with the increasing ratio *w/t*; for an upper limit of *w/t* = 1, the percentage is slightly more than 12%, which might be unacceptable. However, when a maximum (acceptable) value is associated with the relative compliance difference of Eq. (2.32), the same equation can be solved for *w/t* and Table 2.2 includes a few pairs of acceptable percentage differences and their related dimensional ratios. Provided a relative compliance difference is selected, values of the *w/t* ratio up to the threshold value of Table 2.2 can be utilized in conjunction with the simplified torsion compliances given in Eqs. (2.29).

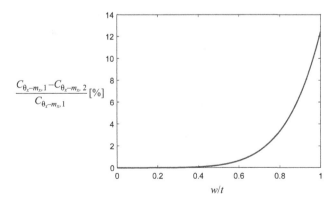

FIGURE 2.7 Plot of the relative errors between the torsional compliances defined by I_{t1} and I_{t2}.

TABLE 2.2

Upper Limits of the *w/t* Ratio for Allowable Percentage Relative Compliance Errors Based on I_{t1} and I_{t2}

$\dfrac{C_{\theta_x-m_x,1} - C_{\theta_x-m_x,2}}{C_{\theta_x-m_x,1}}(\%)$	w/t
1	0.65
2	0.73
5	0.86
10	0.96

Circular cross-section:

For a circular cross-section of constant diameter d, the compliances that result from Eqs. (2.6) are:

$$C_{u_x-f_x} = \frac{4l}{\pi E d^2}; \quad C_{u_y-f_y} = C_{u_z-f_z} = \frac{64l^3}{3\pi E d^4}; \quad C_{u_y-m_z} = -C_{u_z-m_y} = -\frac{32l^2}{\pi E d^4};$$

$$C_{\theta_z-m_z} = C_{\theta_y-m_y} = \frac{64l}{\pi G d^4}; \quad C_{\theta_x-m_x} = \frac{32l}{\pi G d^4}. \tag{2.33}$$

In regard to shear deflections with respect to long (Euler–Bernoulli) vs. short (Timoshenko) beam theory, it has become somewhat of an undisputed truism that a long beam is one whose length is "sufficiently long" compared to its cross-sectional dimensions. The threshold length-to-thickness ratio that separates long from short beams is assigned values ranging from 3 to 5 in most of the dedicated literature (a value of 3 is given in Young et al. [8], while Cowper [9] assumes a limit value of 5). It is also recognized that the deflection produced by shear in a short beam becomes comparable to the regular deflection produced by bending. A method enabling to discriminate between short and long beams evaluates the shear-to-bending deflection ratio and compares it to a limit value or error. The situation where the ratio exceeds the error limit for a particular geometry will place that particular beam into the "short" category; otherwise, it will fall in the "long" class.

Example 2.2

A straight-axis, constant rectangular cross-section hinge of length l and section dimensions w and t (assume $w<t$) is constructed of mild steel with Poisson's ratio $\mu = 0.3$. Study the variation of the l/t ratio in terms of the permissible deflection ratio $u_{y,s}/u_y$. Repeat the problem for a hinge that has a constant circular cross-section of diameter d by analyzing the ratio l/d.

Solution:

The shear deflection $u_{y,s}$ and the bending deflection u_y for a cantilever beam under the action of a tip force f_y are obtained from Eqs. (2.8) and (2.26):

$$u_{y,s} = \frac{\alpha_s \cdot l}{G \cdot A} \cdot f_y; \quad u_y = \frac{l^3}{3E \cdot I_z} \cdot f_y. \tag{2.34}$$

By considering the relationship between the shear modulus G and the Young's modulus E: $G = E/[2 \cdot (1 + \mu)]$, the ratio of the shear deflection to the bending deflection of Eq. (2.34) becomes:

$$\frac{u_{y,s}}{u_y} = \frac{6\alpha_s \cdot (1+\mu) \cdot I_z}{A \cdot l^2}. \tag{2.35}$$

The cross-sectional properties of Eq. (2.35) are the following ones for the rectangular and circular cross-sections:

$$\begin{cases} I_z = \dfrac{w \cdot t^3}{12}; \quad A = w \cdot t \\[3mm] I_z = \dfrac{\pi \cdot d^4}{64}; \quad A = \dfrac{\pi \cdot d^2}{4} \end{cases} \tag{2.36}$$

Substitution of the rectangular cross-sectional properties – first row in Eq. (2.36) – into the deflection ratio of Eq. (2.35) enables to express the following length-to-thickness ratio:

$$\frac{l}{t} = \sqrt{\frac{\alpha_s \cdot (1+\mu)}{2}} \,/\, \sqrt{\frac{u_{y,s}}{u_y}}. \tag{2.37}$$

The deflection ratio for a circular cross-section can be obtained by substituting the cross-sectional properties of the second row of Eq. (2.36) into Eq. (2.35), which results in the length-to-diameter ratio:

$$\frac{l}{d} = \sqrt{\frac{3\alpha_s \cdot (1+\mu)}{8}} \,/\, \sqrt{\frac{u_{y,s}}{u_y}}. \tag{2.38}$$

Figure 2.8 plots the l/t and l/d ratios in terms of the threshold error $u_{y,s}/u_y$ for the rectangular and circular cross-sectional hinges. The shear coefficient is $\alpha_s = 6/5$ for the rectangular cross-section and $\alpha_s = 10/9$ for the circular cross-section. These are conservative, average values provided in Young et al. [8] – the interested reader can find a more nuanced approach and refined values of α_s in Cowper [9], Rentor [10] or Stephen [11], for instance. The plot shows that as the shear deflection increases relative to the bending deflection, the length-to-thickness (or length-to-diameter) ratio that marks the long/short beam separation decreases. For a deflection ratio of 0.1, for instance, one gets $l/t = 2.793$ and $l/d = 2.327$. As also shown in Figure 2.8, circular cross-sectional cantilevers can be considered long beams for length-to-diameter ratios that are slightly smaller than the length-to-thickness ratios of the rectangular cross-sectional beams at the same error limit. This is also demonstrated by combining Eqs. (2.37) and (2.38) to get the following ratio $(l/d)/(l/t) = 5/6$.

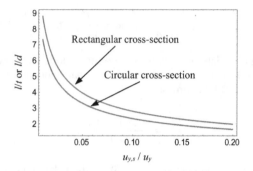

FIGURE 2.8   Plots of the l/t and l/d ratios in terms of admissible error ratios $u_{y,s}/u_y$ for constant rectangular and circular cross-section flexible hinges.

It should be noted that for variable cross-section flexible-hinge segments, Eqs. (2.34) and (2.35) are no longer valid and one must use the more generic equation:

$$\frac{u_{y,s}}{u_y} = \frac{C_{u_y-f_y} + 2\alpha_s \cdot (1+\mu) \cdot C_{u_x-f_x}}{C_{u_y-f_y}} = 1 + 2\alpha_s \cdot (1+\mu) \cdot \frac{C_{u_x-f_x}}{C_{u_y-f_y}}. \qquad (2.39)$$

2.1.2.1.2   Linearly Varying-Thickness/Diameter (Trapezoid) Segment

A trapezoid segment is sketched in Figure 2.9. Its geometry is defined by the minimum thickness t (for rectangular cross-sections) or minimum diameter d (for circular cross-sections), the length l and the longitudinal inclination angle α.

The equations of the upper profile $y(x)$ and the variable thickness $t(x)$ – or diameter $d(x)$ – are:

$$\begin{cases} y(x) = t/2 + m \cdot x \\ t(x) = 2y(x) = t + 2m \cdot x \qquad \text{with} \quad m = \tan\alpha. \\ d(x) = 2y(x) = d + 2m \cdot x \end{cases} \qquad (2.40)$$

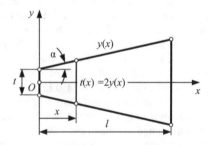

FIGURE 2.9   Front view of trapezoid flexible segment with planar geometry.

The in-plane and out-of-plane compliances are calculated based on their definition of Eqs. (2.6) and are provided in the Appendix at the end of this chapter.

2.1.2.2 Curvilinear-Profile Segments

Hinge segments that are defined by a curvilinear profile are studied in this section. Figure 2.10a shows the upper portion of an axially symmetric segment defined by a planar curve $y(x)$. The variable thickness of a rectangular cross-sectional hinge is $t(x) = 2y(x)$, and the two tangents at the curve's ends have arbitrary directions defined by the angles α_0 and α_e (where "e" stands for end). This general segment is denoted as (α_0, α_e) to indicate the two end tangents' angles. Three particular cases stemming from the general design of Figure 2.10a are possible, namely, segments $(0, \alpha_e)$, as the one of Figure 2.10b whose left tangent is horizontal (and forms an angle of 0 with the x axis); segments $(\alpha_0, \pi/2)$, as shown in Figure 2.10c where the right-end tangent is perpendicular to the x axis; and segments $(0, \pi/2)$ with the left-end tangent parallel to the x axis and the right-end tangent perpendicular to the same axis, as shown in Figure 2.10d.

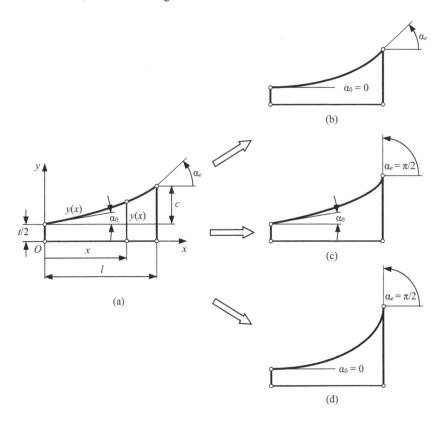

FIGURE 2.10 Base half-segments of variable thickness with: (a) generic geometry – (α_0, α_e) type; (b) horizontal left tangent – $(0, \alpha_e)$ type; (c) vertical right tangent – $(\alpha_0, \pi/2)$ type; (d) horizontal left tangent and vertical right segments – $(0, \pi/2)$ type.

The choice of a specific curve to define the upper half of a flexible-hinge segment can produce major shape changes that translate directly in large differences between similar compliances. Figure 2.11, for instance, illustrates a few curves that can be used to form a $(0, \alpha_e)$ segment. The specific curve equations and corresponding compliances are discussed in more detail in the following section of this chapter. As it can be seen in Figure 2.11, although all curves are $(0, \alpha_e)$ curves, there are major differences between a hyperbolic profile and an inverse parabolic one, as an example.

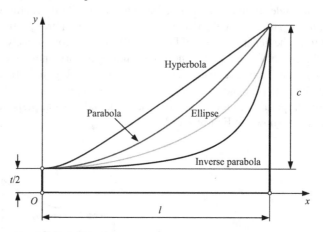

FIGURE 2.11   Front view with planar geometry of generic half-segments defined by a few curves resulting in $(0, \alpha_e)$ hinge segments.

There are multiple (α_0, α_e) curves – of the type sketched in Figure 2.10a – that can be used to generate hinge segments to be placed either as intermediate (connecting) portions or as end segments (in corner-filleted configurations whose end tangent angle is smaller than $\pi/2$). In addition to the specified end-point slopes, the thickness values are also specified at the same points – as seen in the same Figure 2.10a – such that the four end conditions are expressed as:

$$\begin{cases} t(0)/2 = y(0) = t/2; \\ \dfrac{dy(x)}{dx}\bigg|_{x=0} = \tan\alpha_0 = m_0; \\ t(l)/2 = y(l) = c + t/2; \\ \dfrac{dy(x)}{dx}\bigg|_{x=l} = \tan\alpha_e = m_e \end{cases} \tag{2.41}$$

In essence, any curve determined by four parameters can be used to satisfy the four end conditions of Eq. (2.41). Out of the multitude of curves that can be used as upper edges for the generic hinge segment of Figure 2.10a, typical curves such as circles, ellipses, hyperbolas, parabolas or polynomials are studied in this section. While all these curves result in (α_0, α_e) and $(0, \alpha_e)$ segments – as depicted in Figure 2.10a and b, the circular and standard elliptical configurations (with its semi-axes parallel to the local frame Oxy) can also generate $(\alpha_0, \pi/2)$ and $(0, \pi/2)$ segments – as shown in Figure 2.10c and d. The last part of this section focuses entirely on $(0, \pi/2)$-type segments (which are also known as *right* segments) by extending the geometry domain to include all conic-section profiles (circular, elliptic, hyperbolic, parabolic), as well as polynomial and rational Bézier curves.

2.1.2.2.1 Circular-Profile Segment

Consider the (α_0, α_e) circular half-segment of Figure 2.12a, which is defined by the radius r, the angles φ_0 and φ_e, and the minimum thickness $t/2$. By selecting an arbitrary abscissa x with respect to the reference frame $Oxyz$, the variable thickness $t(x) = t(\varphi)$ can be defined. Using the polar coordinate φ, the following relationships are established based on Figure 2.12a:

$$\begin{cases} x = r \cdot (\sin\varphi - \sin\varphi_0) \\ t(x) = t(\varphi) = t + 2r(1 - \cos\varphi) \\ y(x) = \dfrac{t(x)}{2} = \dfrac{t}{2} + r(1 - \cos\varphi) \end{cases} \quad . \tag{2.42}$$

It should also be noticed that $\alpha_0 = \varphi_0$ and $\alpha_e = \varphi_e$ due to the fact that the tangents to the circular profile are perpendicular to the respective radii at the curve end points.

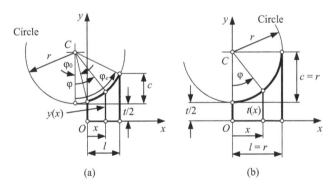

FIGURE 2.12 Front view with planar geometry of circular half-segments: (a) generic (α_0, α_e) segment, (b) right $(0, \pi/2)$ segment.

Rectangular cross-section

The cross-sectional area and area moments with respect to the z axis and the y axis of the full, axially symmetric rectangular cross-section at the arbitrary position x are calculated as:

$$A(x) = w \cdot t(x) = w \cdot t(\varphi) = A(\varphi);$$

$$I_z(x) = \frac{w \cdot t(x)^3}{3} = \frac{w \cdot t(\varphi)^3}{3} = I_z(\varphi); \quad I_y(x) = \frac{w^3 \cdot t(x)}{12} = \frac{w^3 \cdot t(\varphi)}{12} = I_y(\varphi)$$

$$(2.43)$$

The in-plane compliances are expressed using the transformation Eqs. (2.43) in their definition of Eqs. (2.6):

$$\begin{cases} C_{u_x-f_x} = \dfrac{r}{E \cdot w} \cdot \displaystyle\int_{\varphi_0}^{\varphi_e} \dfrac{\cos\varphi}{t(\varphi)} \cdot d\varphi; \\[3mm] C_{u_y-f_y} = \dfrac{12 \cdot r^3}{E \cdot w} \cdot \displaystyle\int_{\varphi_0}^{\varphi_e} \dfrac{(\sin\varphi - \sin\varphi_0)^2 \cdot \cos\varphi}{t(\varphi)^3} \cdot d\varphi; \\[3mm] C_{u_y-m_z} = -\dfrac{12 \cdot r^2}{E \cdot w} \cdot \displaystyle\int_{\varphi_0}^{\varphi_e} \dfrac{(\sin\varphi - \sin\varphi_0) \cdot \cos\varphi}{t(\varphi)^3} \cdot d\varphi; \\[3mm] C_{\theta_z-m_z} = \dfrac{12 \cdot r}{E \cdot w} \cdot \displaystyle\int_{\varphi_0}^{\varphi_e} \dfrac{\cos\varphi}{t(\varphi)^3} \cdot d\varphi; \end{cases} \qquad (2.44)$$

where the differential substitution: $dx = r \cdot \cos\varphi \cdot d\varphi$ has been used.

The out-of-plane compliances are defined similarly to Eqs. (2.44) based on Eqs. (2.6) as:

$$\begin{cases} C_{u_z-f_z} = \dfrac{12 \cdot r^3}{E \cdot w^3} \cdot \displaystyle\int_{\varphi_0}^{\varphi_e} \dfrac{(\sin\varphi - \sin\varphi_0)^2 \cdot \cos\varphi}{t(\varphi)} \cdot d\varphi; \\[3mm] C_{u_z-m_y} = \dfrac{12 \cdot r^2}{E \cdot w^3} \cdot \displaystyle\int_{\varphi_0}^{\varphi_e} \dfrac{(\sin\varphi - \sin\varphi_0) \cdot \cos\varphi}{t(\varphi)} \cdot d\varphi; \\[3mm] C_{\theta_y-m_y} = \dfrac{12 \cdot r}{E \cdot w^3} \cdot \displaystyle\int_{\varphi_0}^{\varphi_e} \dfrac{\cos\varphi}{t(\varphi)} \cdot d\varphi; \\[3mm] C_{\theta_x-m_x} = \dfrac{r}{G} \cdot \displaystyle\int \dfrac{\cos\varphi}{I_{t,i}(\varphi)} \cdot d\varphi; \end{cases} \qquad (2.45)$$

where $I_{t,i}$ ($i = 1, 2, 3$) is the torsional moment of inertia, as provided in Eqs. (2.20), (2.21) or (2.22). The final expressions of these compliances as they result from Eqs. (2.44) and (2.45) are too complex and are not included here.

For cases where $w > t$, it is possible that $t(\varphi)$ becomes larger than w starting from an angle $\varphi = \varphi_1$, where $t(\varphi_1) = w$. Using this condition, it follows that:

$$w = t + 2r\left(1 - \cos\varphi_1\right) \rightarrow \varphi_1 = \cos^{-1}\frac{t + 2r - w}{2r}, \tag{2.46}$$

and the torsional compliance is calculated as:

$$C_{\theta_x - m_x} = \frac{r}{G} \cdot \left(\int_{\varphi_0}^{\varphi_1} \frac{\cos\varphi}{I_{t,it}(\varphi)} \cdot d\varphi + \int_{\varphi_1}^{\varphi_e} \frac{\cos\varphi}{I_{t,iw}(\varphi)} \cdot d\varphi \right) \tag{2.47}$$

due to the two different torsional moments of area on the two subintervals $[\varphi_0, \varphi_1]$ and $[\varphi_1, \varphi_e]$.

The compliances corresponding to $(0, \alpha_e)$ segments are calculated by using $\varphi_0 = 0$ in Eqs. (2.44) and (2.45) since $\alpha_0 = \varphi_0$. Similarly, the compliances of $(\alpha_0, \pi/2)$ segments are determined from the same Eqs. (2.44) and (2.45) with $\varphi_e = \pi/2$ because $\alpha_e = \varphi_e$. Their equations are complicated and are not given explicitly here.

Right segments of the $(0, \pi/2)$ type are obtained by using $\varphi_0 = 0$ and $\varphi_e = \pi/2$ in Eqs. (2.44) and (2.45) when the arbitrary circular segment becomes the right quarter-circle segment of Figure 2.12b. The closed-form in-plane and out-of-plane compliances are provided in the Appendix at the end of this chapter.

Circular cross-section

The planar profile of the (α_0, α_e) circular half-segment is shown in Figure 2.12a, which is defined by the radius r, the angles φ_0 and φ_e, and the minimum diameter $d/2$. The variable diameter is:

$$d(x) = d(\varphi) = d + 2r\left(1 - \cos\varphi\right). \tag{2.48}$$

The compliances are:

$$\left| \begin{array}{l} C_{u_x - f_x} = \frac{4r}{\pi E} \int_{\varphi_0}^{\varphi_e} \frac{\cos\varphi}{d(\varphi)^2} \cdot d\varphi; \quad C_{u_y - f_y} = \frac{64r^3}{\pi E} \int_{\varphi_0}^{\varphi_e} \frac{\left(\sin\varphi - \sin\varphi_0\right)^2 \cdot \cos\varphi}{d(\varphi)^4} \cdot d\varphi; \\[4mm] C_{u_y - m_z} = -\frac{64r^2}{\pi E} \int_{\varphi_0}^{\varphi_e} \frac{\left(\sin\varphi - \sin\varphi_0\right) \cdot \cos\varphi}{d(\varphi)^4} \cdot d\varphi; \quad C_{\theta_z - m_z} = \frac{64r}{\pi E} \int_{\varphi_0}^{\varphi_e} \frac{\cos\varphi}{d(\varphi)^4} \cdot d\varphi \end{array} \right. \tag{2.49}$$

Right segments of the $(0, \pi/2)$ type are obtained by using $\varphi_0 = 0$ and $\varphi_e = \pi/2$ in the compliances of Eqs. (2.49) and are included in the end-of-chapter Appendix.

2.1.2.2.2 Elliptical-Profile Segment

A generic elliptical-profile segment of the (α_0, α_e) type is represented in Figure 2.13a, which is defined by a minimum thickness t and the elliptical

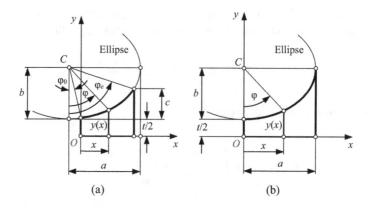

FIGURE 2.13 Front view with planar geometry of elliptical segments: (a) generic (α_0, α_e) segment, (b) right $(0, \pi/2)$ segment.

semi-axes lengths a and b. The axial dimension of the segment is set by the angles φ_0 and φ_e. The polar coordinate φ is used to express the abscissa x and the corresponding variable thickness $t(x)$ as:

$$\begin{cases} x = a \cdot \left(\sin\varphi - \sin\varphi_0 \right) \\ t(x) = t(\varphi) = t + 2b\left(1 - \cos\varphi\right), \\ y(x) = \dfrac{t(x)}{2} = \dfrac{t}{2} + b\left(1 - \cos\varphi\right) \end{cases} \qquad (2.50)$$

with φ varying between φ_0 and φ_e.

It should be remarked that, unlike the circle, the two slopes at the extremities of the elliptical upper edge do not coincide with the polar angles corresponding to those points. Using Eq. (2.50), the slope of the ellipse is calculated as:

$$\tan\alpha = \frac{dy}{dx} = \frac{dy/d\varphi}{dx/d\varphi} = \frac{b}{a} \cdot \tan\varphi \quad \text{or} \quad \tan\varphi = \frac{a}{b} \cdot \tan\alpha. \qquad (2.51)$$

As a consequence, the relationships between the slope angles and the polar angles corresponding to the two end points of the elliptical segment of Figure 2.13a are:

$$\varphi_0 = \tan^{-1}\left(\frac{a}{b} \cdot \tan\alpha_0 \right); \quad \varphi_e = \tan^{-1}\left(\frac{a}{b} \cdot \tan\alpha_e \right). \qquad (2.52)$$

Rectangular cross-section

The in-plane compliances are calculated based on their definition Eqs. (2.6) as:

$$\begin{cases} C_{u_x-f_x} = \dfrac{a}{E \cdot w} \cdot \displaystyle\int_{\varphi_0}^{\varphi_e} \dfrac{\cos\varphi}{t(\varphi)} \cdot d\varphi; \\[4mm] C_{u_y-f_y} = \dfrac{12 \cdot a^3}{E \cdot w} \cdot \displaystyle\int_{\varphi_0}^{\varphi_e} \dfrac{(\sin\varphi - \sin\varphi_0)^2 \cdot \cos\varphi}{t(\varphi)^3} \cdot d\varphi; \\[4mm] C_{u_y-m_z} = -\dfrac{12 \cdot a^2}{E \cdot w} \cdot \displaystyle\int_{\varphi_0}^{\varphi_e} \dfrac{(\sin\varphi - \sin\varphi_0) \cdot \cos\varphi}{t(\varphi)^3} \cdot d\varphi; \\[4mm] C_{\theta_z-m_z} = \dfrac{12 \cdot a}{E \cdot w} \cdot \displaystyle\int_{\varphi_0}^{\varphi_e} \dfrac{\cos\varphi}{t(\varphi)^3} \cdot d\varphi \end{cases} \tag{2.53}$$

where Eqs. (2.50) have been used, as well as the differential substitution: $dx = a \cdot \cos\varphi \cdot d\varphi$. Note that for $a = b = r$ (when the ellipse becomes a circle), the variable thickness of the elliptical segment in Eq. (2.50) is identical to the one of Eq. (2.42), which corresponds to a circular segment; as a consequence, the elliptical-segment compliances of Eqs. (2.53) become identical to the compliances of Eqs. (2.44), which correspond to a circular-thickness segment.

The out-of-plane compliances are calculated from Eqs. (2.6) as:

$$\begin{cases} C_{u_z-f_z} = \dfrac{12 \cdot a^3}{E \cdot w^3} \cdot \displaystyle\int_{\varphi_0}^{\varphi_e} \dfrac{(\sin\varphi - \sin\varphi_0)^2 \cdot \cos\varphi}{t(\varphi)} \cdot d\varphi; \\[4mm] C_{u_z-m_y} = \dfrac{12 \cdot a^2}{E \cdot w^3} \cdot \displaystyle\int_{\varphi_0}^{\varphi_e} \dfrac{(\sin\varphi - \sin\varphi_0) \cdot \cos\varphi}{t(\varphi)} \cdot d\varphi; \\[4mm] C_{\theta_y-m_y} = \dfrac{12 \cdot a}{E \cdot w^3} \cdot \displaystyle\int_{\varphi_0}^{\varphi_e} \dfrac{\cos\varphi}{t(\varphi)} \cdot d\varphi; \\[4mm] C_{\theta_x-m_x} = \dfrac{a}{G} \cdot \displaystyle\int \dfrac{\cos\varphi}{I_{t,i}(\varphi)} \cdot d\varphi; \end{cases} \tag{2.54}$$

The closed-form equations resulting after the integrations necessary in Eqs. (2.53) and (2.54) are quite complex and are not included here.

For cases where $w > t$, $t(\varphi)$ may become larger than w from an angle $\varphi = \varphi_1$, where $t(\varphi_1) = w$. This angle is:

$$w = t + 2b\left(1 - \cos\varphi_1\right) \to \varphi_1 = \cos^{-1}\dfrac{t + 2b - w}{2b}, \tag{2.55}$$

and the torsional compliance is calculated as:

$$C_{\theta_x-m_x} = \frac{a}{G} \cdot \left(\int_{\varphi_0}^{\varphi_1} \frac{\cos\varphi}{I_{t,it}(\varphi)} \cdot d\varphi + \int_{\varphi_1}^{\varphi_e} \frac{\cos\varphi}{I_{t,iw}(\varphi)} \cdot d\varphi \right) \tag{2.56}$$

due to the two different torsional moments of area on the two subintervals $[\varphi_0, \varphi_1]$ and $[\varphi_1, \varphi_e]$.

The compliances of $(0, \alpha_e)$ elliptic segments are determined by using $\alpha_0 = 0$ in Eqs. (2.53) and (2.54). Similarly, the compliances corresponding to $(\alpha_0, \pi/2)$ segments are determined from the same Eqs. (2.53) and (2.54) by using $\alpha_e = \pi/2$ at the end point of the elliptical edge. Their equations are, again, quite involved and are not provided explicitly here.

The particular design $(0, \pi/2)$ is the right quarter-ellipse segment, which is sketched in Figure 2.13b and whose angular limits are $\alpha_0 = 0$ and $\alpha_e = \pi/2$. Equations (2.53) and (2.54) yield the closed-form compliance equations for the in-plane and out-of-plane compliances – they can be found in the Appendix at the end of this chapter.

Circular cross-section:

The variable diameter is:

$$d(x) = d(\varphi) = d + 2b(1 - \cos\varphi), \tag{2.57}$$

and the hinge segment compliances are:

$$\begin{cases} C_{u_x-f_x} = \dfrac{4a}{\pi E} \displaystyle\int_{\varphi_0}^{\varphi_e} \frac{\cos\varphi}{d(\varphi)^2} \cdot d\varphi; \quad C_{u_y-f_y} = \dfrac{64a^3}{\pi E} \displaystyle\int_{\varphi_0}^{\varphi_e} \frac{(\sin\varphi - \sin\varphi_0)^2 \cdot \cos\varphi}{d(\varphi)^4} \cdot d\varphi; \\[4mm] C_{u_y-m_z} = -\dfrac{64a^2}{\pi E} \displaystyle\int_{\varphi_0}^{\varphi_e} \frac{(\sin\varphi - \sin\varphi_0) \cdot \cos\varphi}{d(\varphi)^4} \cdot d\varphi; \quad C_{\theta_z-m_z} = \dfrac{64a}{\pi E} \displaystyle\int_{\varphi_0}^{\varphi_e} \frac{\cos\varphi}{d(\varphi)^4} \cdot d\varphi \end{cases}$$

$$\tag{2.58}$$

The compliances of a right segment $(0, \pi/2)$ are obtained by substituting $\alpha_0 = 0$ and $\alpha_e = \pi/2$ in Eqs. (2.58) – they are explicitly given in the Appendix at the end of this chapter.

2.1.2.2.3 Hyperbolic-Profile Segment

A hyperbolic curve is illustrated in Figure 2.14a, which defines the upper half of an axially symmetric generic (α_0, α_e) flexible segment. The hyperbola possesses a symmetry axis, and its equation in the reference frame $O_1 x_1 y_1 z_1$ (where the y_1 axis coincides with the symmetry axis) is:

$$\frac{y_1(x_1)^2}{c_1^2} - \frac{x_1^2}{c_0^2} = 1, \tag{2.59}$$

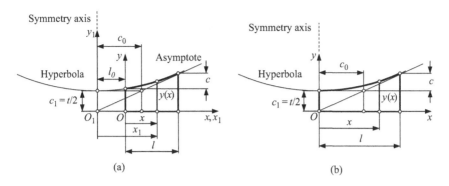

FIGURE 2.14 Front view with planar geometry of hyperbolic segments: (a) generic (α_0, α_e) segment; (b) (0, α_e) segment.

with the slope at the origin being zero.

Rectangular cross-section:

Using the general boundary conditions of Eq. (2.41), the constants c_0 and c_1 of Eq. (2.59) are determined (note that $c_1 = t/2$). Taking into account that $x_1 = x + l_0$ (as seen in Figure 2.14a), the variable thickness (which is related to $y_1(x_1)$ of Eq. (2.59)) becomes:

$$t(x) = 2y(x) = \frac{1}{l_0 + l} \cdot \sqrt{4c(c+t) \cdot (x + l_0)^2 + (l_0 + l)^2 \cdot t^2}. \qquad (2.60)$$

With the thickness of Eq. (2.60), the in-plane and out-of-plane compliances of the general (α_0, α_e) hyperbolic segment can be determined by means of their definition Eqs. (2.6).

The (0, α_e) hyperbolic segment is illustrated in Figure 2.14b, and its variable thickness is found by using $l_0 = 0$ in Eq. (2.60), namely:

$$t(x) = 2y(x) = \frac{1}{l} \cdot \sqrt{4c \cdot (c+t) \cdot x^2 + l^2 \cdot t^2}. \qquad (2.61)$$

The in-plane and out-of-plane compliance equations are calculated for the (0, α_e) configuration by means of the definition of Eqs. (2.6) and the variable thickness of Eq. (2.61); they are included in the Appendix at the end of this chapter.

Circular cross-section:

The variable diameter $d(x)$ is expressed similarly to the variable thickness $t(x)$ of Eq. (2.60) with the help of Figure 2.14a for a generic (α_0, α_e) hyperbolic segment as:

$$d(x) = \frac{1}{l_0 + l} \cdot \sqrt{4c \cdot (c+d) \cdot (x + l_0)^2 + (l_0 + l)^2 \cdot d^2}. \qquad (2.62)$$

For the $(0, \varphi_e)$ segment, whose planar profile is illustrated in Figure 2.14b, the variable diameter is obtained from Eq. (2.62) with $l_0 = 0$:

$$d(x) = \frac{1}{l} \cdot \sqrt{4c \cdot (c+d) \cdot x^2 + l^2 \cdot d^2} . \tag{2.63}$$

The compliances of this segment are expressed in the Appendix at the end of this chapter.

2.1.2.2.4 Parabolic-Profile Segment

A generic (α_0, α_e) parabolic flexible-hinge segment is illustrated in Figure 2.15a in its half-configuration.

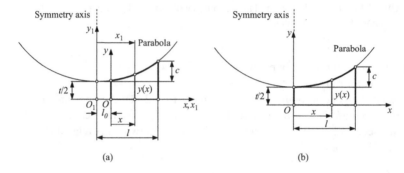

(a)   (b)

FIGURE 2.15 Front view with planar geometry of parabolic segments: (a) generic (α_0, α_e) segment; (b) $(0, \alpha_e)$ segment.

Rectangular cross-section:

The hinge segment of Figure 2.15a has a length l, and its full minimum thickness is t. The upper curve of this profile is a parabola, which is symmetric with respect to the y_1 axis and which defines a variable thickness of the generic segment that is expressed in the $O_1x_1 y_1z_1$ frame as:

$$y_1(x_1) = c_2 x_1^2 + c_0 = \frac{t(x_1)}{2}, \tag{2.64}$$

with the slope at the origin being zero. Using the boundary conditions:

$$y_1(0) = t/2; \quad y_1(l_0 + l) = t/2 + c \tag{2.65}$$

in conjunction with the change of variable $x_1 = x + l_0$, the variable thickness of Eq. (2.64) is expressed in the local frame $Oxyz$ as:

$$\begin{cases} y(x) = \dfrac{c}{(l_0+l)^2} \cdot (l_0 + x)^2 + t/2; \\[3mm] t(x) = 2y(x) = \dfrac{2c}{(l_0+l)^2} \cdot (l_0 + x)^2 + t \end{cases} . \tag{2.66}$$

Using the definition Eqs. (2.6) and the thickness of Eq. (2.66) enables expressing the in-plane and out-of-plane compliances of the generic (α_0, α_e) parabolic segment, which are too complex and are not included here.

A (0, α_e) parabolic segment is depicted in Figure 2.15b whose variable thickness results from the one of Eq. (2.66) by using $l_0 = 0$, namely:

$$y(x) = \frac{c}{l^2} \cdot x^2 + t/2; \quad t(x) = 2y(x) = \frac{2c}{l^2} \cdot x^2 + t. \tag{2.67}$$

The compliances of this segment are calculated with this thickness substituted in the definition Eqs. (2.6); they are explicitly provided in the Appendix at the end of this chapter.

Circular cross-section:

The variable diameter $d(x)$ is expressed similarly to the variable thickness $t(x)$ of Eq. (2.66) with the help of Figure 2.15a for a generic (α_0, α_e) parabolic segment as:

$$d(x) = \frac{2c}{(l_0 + l)^2} \cdot (x + l_0)^2 + d. \tag{2.68}$$

For the (0, α_e) segment, whose planar profile is illustrated in Figure 2.15b, the variable diameter is obtained from Eq. (2.68) by using $l_0 = 0$:

$$d(x) = \frac{2c}{l^2} \cdot x^2 + d. \tag{2.69}$$

The compliances of this segment are included in the Appendix at the end of this chapter.

2.1.2.2.5 Inverse Parabolic-Profile Segment

A generic (α_0, α_e) inverse parabolic-profile segment is similar in shape and parameters to the parabolic segment pictured in Figure 2.15a. The upper-edge curve is symmetric with respect to the y_1 axis and defines the variable thickness of the segment in the $O_1 x_1 y_1 z_1$ frame as:

$$y_1(x_1) = \frac{a}{b^2 - x_1^2}. \tag{2.70}$$

Equation (2.70) can also be expressed by means of the parameter θ as:

$$x_1(\theta) = b \cdot \sin\theta; \quad y_1(\theta) = \frac{a}{b^2 \cdot \cos^2\theta}. \tag{2.71}$$

The slope of this curve is found by means of Eqs. (2.71) as:

$$\frac{dy_1}{dx_1} = \frac{dy_1(\theta)/d\theta}{dx_1(\theta)/d\theta} = \frac{2a \cdot \sin\theta}{b^3 \cdot \cos^4\theta}. \tag{2.72}$$

As indicated in Eq. (2.72), the slope at the origin (with $x_1 = 0$ and $\theta = 0$) is zero, and therefore, an inverse-parabola curve can produce a $(0, \alpha_e)$ flexible segment, like the one of Figure 2.15b.

Rectangular cross-section:

The constants a and b are determined with the boundary conditions of Eqs. (2.41) as:

$$a = \frac{(l_0 + l)^2 \cdot (t/2 + c)t}{2c}; \quad b^2 = \frac{(l_0 + l)^2 \cdot (t/2 + c)}{c}. \tag{2.73}$$

The in- and out-of-plane compliances can now be calculated with Eqs. (2.6) by using the following variable thickness:

$$y(x) = \frac{a}{b^2 - (x + l_0)^2}; \quad t(x) = 2y(x). \tag{2.74}$$

They are evaluated in the hinge reference frame $Oxyz$, as indicated in Figure 2.15a.

The $(0, \alpha_e)$ segment of Figure 2.15b has a variable thickness of:

$$y(x) = \frac{a}{b^2 - x^2}; \quad t(x) = 2y(x) \quad \text{with} \quad \begin{cases} a = \dfrac{l^2 \cdot (t/2 + c) \cdot t}{2c} \\ b^2 = \dfrac{l^2 \cdot (t/2 + c)}{c} \end{cases}, \tag{2.75}$$

resulting from Eq. (2.74) for $l_0 = 0$. The closed-form compliances for this particular segment are provided in the Appendix at the end of this chapter.

Circular cross-section:

For a generic (α_0, α_e) inverse-parabola segment, the variable diameter $d(x)$ is expressed similarly to the variable thickness $t(x)$ of Eq. (2.74) as:

$$d(x) = \frac{2a}{b^2 - (x + l_0)^2} \quad \text{with} \quad \begin{cases} a = \dfrac{(l_0 + l)^2 \cdot \left(\dfrac{d}{2} + c\right) \cdot d}{2c} \\ b^2 = \dfrac{(l_0 + l)^2 \cdot \left(\dfrac{d}{2} + c\right)}{c} \end{cases}. \tag{2.76}$$

For the $(0, \varphi_e)$ segment, the variable diameter is obtained from Eq. (2.76) for $l_0 = 0$:

$$d(x) = \frac{2a}{b^2 - x^2} \quad \text{with} \quad \begin{cases} a = \dfrac{l^2 \left(\dfrac{d}{2} + c\right) d}{2c} \\ b^2 = \dfrac{l^2 \left(\dfrac{d}{2} + c\right)}{c} \end{cases}. \tag{2.77}$$

The closed-form compliances of this segment are included in the Appendix at the end of this chapter.

2.1.2.2.6 Right Corner-Filleted (0, π/2) Segments by Rational Bézier Functions

A right corner-filleted flexible hinge is designed by means of a curve whose end-point tangents coincide with the perpendicular edges of two adjacent segments, as sketched in Figure 2.16a. These segments were introduced as $(0, \pi/2)$ sections in Figure 2.10d, where the tangent at the start point to the curve makes an angle $\alpha_0 = 0$ with the horizontal, whereas the tangent to the end (terminal) point of the curve makes an angle $\alpha_e = \pi/2$ with the horizontal. The resulting region that is shown in Figure 2.16b is a right corner-filleted half-segment that can be used to form more complex hinge configurations. Obviously, the right corner-filleted design is a particular case of the filleted hinge where the two edges of the adjacent segments form an angle different from $\pi/2$.

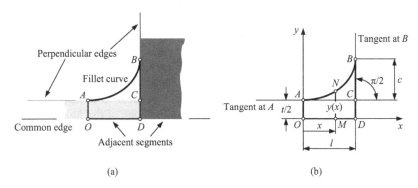

(a)   (b)

FIGURE 2.16 (a) Fillet curve tangent to two perpendicular edges of adjacent segments; (b) resulting right corner-filleted half-hinge segment.

Filleted hinge configurations are mainly utilized to reduce stress concentration at the connection region between a non-filleted hinge segment and a rigid link. Besides this primary usage, filleted hinges can substantially alter the flexible properties of non-filleted designs, particularly their compliances. Corner-filleted flexible hinges bring about another advantage over non-filleted hinge configurations. Compare, for instance, the two designs of Figure 2.17 where the segments to the right of the hinges are (supposed to be) rigid segments. While for the non-filleted configuration of Figure 2.17a, the supposedly rigid segments have an in-plane thickness that might not be sufficient to render them truly rigid, the filleted design of Figure 2.17b allows us to increase the thickness of the segments adjacent to the hinge to values that will ensure the deformations of these segments are negligible.

Moreover, a right corner-filleted design would usually be a choice over a corner-filleted configuration with non-perpendicular edges due to both design simplicity and the fact that compliance equations are simpler than those of the non-right variant.

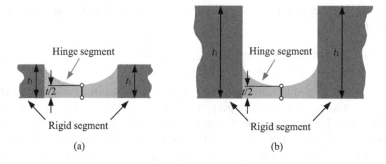

FIGURE 2.17 (a) Flexible hinge without fillet; (b) right corner-filleted flexible hinge.

Two right corner-filleted segments that have already been discussed in a previous section are the circular one of Figure 2.12b, which utilizes a quarter-circle arc, and the elliptic shape of Figure 2.13b, which uses a quarter ellipse to realize the respective fillet regions.

The circle and the ellipse are not the only planar curves resulting in right corner-filleted flexure segments. Rational Bézier curves can also be used to generate right corner-filleted flexible hinges, and these curves are actually conic-section curves, as shown here. Conic-section curves include the parabola, hyperbola and ellipse (or circle, as a limit case of an ellipse with equal semi-axes). They result by intersecting a conic surface with a plane that is in one of the following positions: at an angle with respect to the cone axis and intersecting both the lateral surface and the circular base of the cone – this results in a parabola, parallel to the cone axis – producing a hyperbola, and at an angle with respect to the cone axis and intersecting the lateral cone surface but not the circular base of the cone– this results in an ellipse, perpendicular to the cone axis – producing a circle; the parabola, hyperbola and ellipse are illustrated in Figure 2.18.

The general equation of a conic is:

$$A \cdot x^2 + B \cdot x \cdot y + C \cdot y^2 + D \cdot x + E \cdot y + F = 0, \tag{2.78}$$

with x and y being the coordinates of an arbitrary point in the planar reference frame Oxy. The theory of conics – see Hall [12], for instance – shows that the discriminant $\Delta = B^2 - 4A \cdot C$ determines whether a conic is a hyperbola, a parabola or an ellipse as follows:

$$B^2 - 4A \cdot C > 0 \rightarrow \text{hyperbola}$$

$$B^2 - 4A \cdot C = 0 \rightarrow \text{parabola} \quad . \tag{2.79}$$

$$B^2 - 4A \cdot C < 0 \rightarrow \text{ellipse}$$

Rational (or polynomial) Bézier curves are part of the larger family of splines and are defined by the parametric equations:

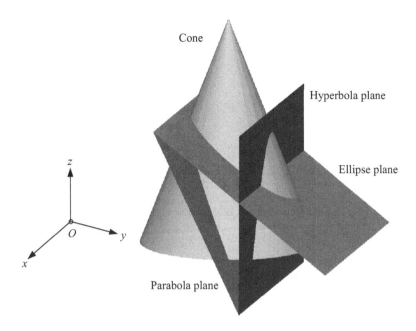

FIGURE 2.18 Parabola, hyperbola and ellipse resulting from particular planes intersecting a cone.

$$
\begin{cases}
x = \dfrac{\displaystyle\sum_{i=0}^{n} B_i(p) \cdot w_i \cdot x_{Pi}}{\displaystyle\sum_{i=0}^{n} B_i(p) \cdot w_i} \\[2em]
y = \dfrac{\displaystyle\sum_{i=0}^{n} B_i(p) \cdot w_i \cdot y_{Pi}}{\displaystyle\sum_{i=0}^{n} B_i(p) \cdot w_i}
\end{cases}
\quad \text{with} \quad B_i(p) = \sum_{i=0}^{n} \frac{n!}{i!(n-i)!} \cdot p^i (1-p)^{n-i}, \quad (2.80)
$$

where the points P_i of coordinates x_{Pi} and y_{Pi} are called control points, $B_i(p)$ are known as the Bernstein polynomials and w_i are weight coefficients – more details can be found in Cohen et al. [13]. The parameter p varies between 0 and 1. Two important properties of polynomial Bézier curves are that they pass through the extreme control points P_0 and P_n and are tangent to the extreme lines P_0P_1 and $P_{n-1}P_n$, as illustrated in Figure 2.19.

The curves defined as in Eqs. (2.80) will pass through the extreme points when the weights of those points are $w_0 = w_n = 1$.

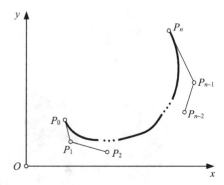

FIGURE 2.19 Plot of a generic planar Bézier curve of degree n.

Rational quadratic Bézier curves are a frequent choice and their parametric equations are obtained from Eqs. (2.80) for $n = 2$ as:

$$\begin{cases} x(p) = \dfrac{(1-p)^2 \cdot w_0 \cdot x_{P_0} + 2(1-p) \cdot p \cdot w_1 \cdot x_{P_1} + p^2 \cdot w_2 \cdot x_{P_2}}{(1-p)^2 \cdot w_0 + 2(1-p) \cdot p \cdot w_1 + p^2 \cdot w_2} \\[4mm] y(p) = \dfrac{(1-p)^2 \cdot w_0 \cdot y_{P_0} + 2(1-p) \cdot p \cdot w_1 \cdot y_{P_1} + p^2 \cdot w_2 \cdot y_{P_2}}{(1-p)^2 \cdot w_0 + 2(1-p) \cdot p \cdot w_1 + p^2 \cdot w_2} \end{cases} \tag{2.81}$$

A rational quadratic Bézier curve requires three control points to define it, which are P_0, P_1 and P_2. The following coordinates of the control points: $x_{P0} = 0$, $x_{P1} = x_{P2} = l$, $y_{P0} = y_{P1} = t/2$, and $y_{P2} = t/2 + c$, as well as $w_0 = w_2 = 1$, are substituted in Eqs. (2.81), which become:

$$\begin{cases} x(p) = \dfrac{2(1-p) \cdot p \cdot w_1 \cdot l + p^2 l}{(1-p)^2 + 2(1-p) \cdot p \cdot w_1 + p^2} = \dfrac{h_1(p)}{h_0(p)} \\[4mm] y(p) = \dfrac{(1-p)^2 \cdot t/2 + (1-p) \cdot p \cdot w_1 \cdot t + p^2 \cdot (t/2+c)}{(1-p)^2 + 2(1-p) \cdot p \cdot w_1 + p^2} = \dfrac{h_2(p)}{h_0(p)} \end{cases} \tag{2.82}$$

To implicitize the parametric Eqs. (2.82), the auxiliary functions f and g are built as:

$$\begin{cases} f = x \cdot h_0(p) - h_1(p) = f_2 \cdot p^2 + f_1 \cdot p + f_0 \\ g = y \cdot h_0(p) - h_2(p) = g_2 \cdot p^2 + g_1 \cdot p + g_0 \end{cases} \tag{2.83}$$

where the coefficients of Eqs. (2.83) are:

$$\begin{cases} f_2 = 2(1-w_1)x + l \cdot (2w_1 - 1) \\ f_1 = 2(w_1 - 1)x - 2l \cdot w_1 \\ f_0 = x \\ g_2 = 2(1-w_1)y + tw_1 - t - c \\ g_1 = 2(w_1 - 1)y + t - tw_1 \\ g_0 = y - t/2 \end{cases} \quad (2.84)$$

The implicit equation of the curve defined parametrically in Eqs. (2.84) is found by equating the Sylvester determinant (a polynomial using the coefficients of f and g) to zero – see more details in Sederberg and Anderson [14] or Bareiss [15], which is:

$$S(f,g) = \begin{vmatrix} f_2 & 0 & g_2 & 0 \\ f_1 & f_2 & g_1 & g_2 \\ f_0 & f_1 & g_0 & g_1 \\ 0 & f_0 & 0 & g_0 \end{vmatrix} = 0. \quad (2.85)$$

Using the coefficients given in Eq. (2.84), Eq. (2.85) yields:

$$c^2 \cdot x^2 + 2c \cdot l \cdot (2w_1^2 - 1) \cdot x \cdot y + l^2 \cdot y^2 + c \cdot l \cdot t \cdot (1 - 2w_1^2) \cdot$$
$$x - l^2 \cdot (t + 4c \cdot w_1^2) \cdot y + l^2 \cdot t \cdot (t/4 + 2c \cdot w_1^2) = 0, \quad (2.86)$$

which is the equation of a conic section. The discriminant of this equation is obtained by comparison to the generic Eq. (2.78) as:

$$\Delta = B^2 - 4A \cdot C = 4\left[c \cdot l \cdot (2w_1^2 - 1)\right]^2 - 4 \cdot c^2 \cdot l^2 = 16 \cdot c^2 \cdot l^2 \cdot w_1^2 \cdot (w_1^2 - 1) \quad (2.87)$$

and therefore, as defined in Eq. (2.79), the conic section is:

$$\Delta > 0 \quad \text{or} \quad w_1^2 > 1 \rightarrow \text{hyperbola}$$

$$\Delta = 0 \quad \text{or} \quad w_1^2 = 1 \rightarrow \text{parabola} \quad (2.88)$$

$$\Delta < 0 \quad \text{or} \quad w_1^2 < 1 \rightarrow \text{ellipse}.$$

The weight w_1 corresponding to the middle point of the three points defining the conic section is therefore instrumental in generating one of the three types of conic sections, as well as in altering the curve shape. Figure 2.20 plots an ellipse for $w_1 = 1/3$, a parabola that is drawn for $w_1 = 1$ and a hyperbola corresponding to $w_1 = 3$.

As the graphs of Figure 2.20 indicate, the two perpendicular lines connecting the three control points P_0, P_1 and P_2 and passing through the end points P_0 and P_1 are also tangent to the curve at these points. This latter trait is always valid as seen from calculating the slope of the spline from Eqs. (2.82):

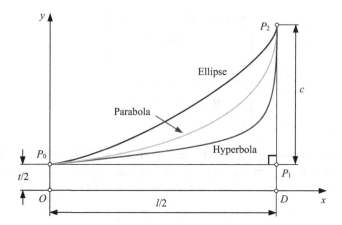

FIGURE 2.20 Right corner-filleted hinge segment shapes generated by means of various conic-section curves expressed as Bézier quadratic rational functions.

$$\frac{dy(p)}{dx(p)} = \frac{\partial y(p)/\partial p}{\partial x(p)/\partial p} = -\frac{c \cdot p \cdot \left[1 + (w_1 - 1) \cdot p\right]}{l \cdot (p - 1) \cdot \left[w_1 - (w_1 - 1) \cdot p\right]}.$$
(2.89)

The slopes at the curve's ends are therefore:

$$m_{P_0} = \left.\frac{dy(p)}{dx(p)}\right|_{p=0} = 0; \quad m_{P_2} = \left.\frac{dy(p)}{dx(p)}\right|_{p=1} = \infty,$$
(2.90)

which demonstrate that indeed, the two tangent lines are horizontal and vertical, respectively, with respect to the reference frame Oxy.

While for a fixed group of points P_0, P_1, P_2 (as in Figure 2.20, where the parameters t, c and l clearly define these points), there is one single parabola passing through the two end points because the parameter w_1 assumes the value of 1, there is an infinite number of hyperbolae and ellipses that are all passing through the end points – Figure 2.21 graphs several plot examples. With $w_1 > 1$, the hyperbolae

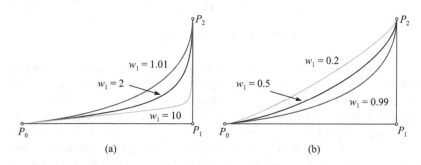

FIGURE 2.21 Right corner-filleted hinge of: (a) hyperbolic profile; (b) elliptic profile.

move closer to the perpendicular tangent lines P_0P_1 and P_1P_2 as w_1 increases. Similarly, the ellipses move further away from the tangent lines and become (apparently) more deformed from their regular elliptic aspect as $w_1 < 1$ decreases.

Rectangular cross-section:

For rectangular cross-section hinges of constant width w, the following relationships need to be used in Eqs. (2.6):

$$dx = \frac{\partial x(p)}{\partial p} \cdot dx; \quad t(p) = 2y(p), \tag{2.91}$$

with $x(p)$ and $y(p)$ of Eq. (2.82). The in-plane compliances of this right corner-filleted segment are calculated as:

$$\begin{cases} C_{u_x-f_x} = \frac{1}{E \cdot w} \cdot \int_0^1 \frac{\frac{\partial x(p)}{\partial p}}{t(p)} \cdot dp; \quad C_{u_y-f_y} = \frac{12}{E \cdot w} \cdot \int_0^1 \frac{x(p)^2 \cdot \frac{\partial x(p)}{\partial p}}{t(p)^3} \cdot dp; \\ \\ C_{u_y-m_z} = -\frac{12}{E \cdot w} \cdot \int_0^1 \frac{x(p) \cdot \frac{\partial x(p)}{\partial p}}{t(p)^3} \cdot dp; \quad C_{\theta_z-m_z} = \frac{12}{E \cdot w} \cdot \int_0^1 \frac{\frac{\partial x(p)}{\partial p}}{t(p)^3} \cdot dp \end{cases} \tag{2.92}$$

The out-of-plane bending-related compliances are evaluated as:

$$C_{u_z-f_z} = \frac{12}{E \cdot w^3} \cdot \int_0^1 \frac{x(p)^2 \cdot \frac{\partial x(p)}{\partial p}}{t(p)} \cdot dp; \quad C_{u_z-m_y} = \frac{12}{E \cdot w^3} \cdot \int_0^1 \frac{x(p) \cdot \frac{\partial x(p)}{\partial p}}{t(p)} \cdot dp;$$

$$\tag{2.93}$$

$$C_{\theta_y-m_y} = \frac{12}{E \cdot w^3} \cdot \int_0^1 \frac{\frac{\partial x(p)}{\partial p}}{t(p)} \cdot dp.$$

The torsion compliances are calculated based on the following equation:

$$C_{\theta_x-m_x} = \begin{cases} \dfrac{3}{G \cdot w^3} \cdot \displaystyle\int_0^1 \frac{\frac{\partial x(p)}{\partial p}}{t(p)} \cdot dp, & w \ll t(p) \\ \\ \dfrac{3}{G \cdot w} \cdot \displaystyle\int_0^1 \frac{\frac{\partial x(p)}{\partial p}}{t(p)^3} \cdot dp, & w \gg t(p) \end{cases} \tag{2.94}$$

Flexible hinges of rectangular cross-section that are obtained using Bézier curves have also been studied by Vallance et al. [16].

Circular cross-section:

The variable diameter in this case is:

$$d(x) = 2y(p), \qquad (2.95)$$

and the compliances of this segment are calculated as:

$$
\begin{cases}
C_{u_x - f_x} = \dfrac{4}{\pi E} \cdot \displaystyle\int_0^1 \dfrac{\dfrac{\partial x(p)}{\partial p}}{d(p)^2} \cdot dp; \quad
C_{u_y - f_y} = \dfrac{64}{\pi E} \cdot \displaystyle\int_0^1 \dfrac{x(p)^2 \cdot \dfrac{\partial x(p)}{\partial p}}{d(p)^4} \cdot dp; \\[4mm]
C_{u_y - m_z} = -\dfrac{64}{\pi E} \cdot \displaystyle\int_0^1 \dfrac{x(p) \cdot \dfrac{\partial x(p)}{\partial p}}{d(p)^4} \cdot dp; \quad
C_{\theta_z - m_z} = \dfrac{64}{\pi E} \cdot \displaystyle\int_0^1 \dfrac{\dfrac{\partial x(p)}{\partial p}}{d(p)^4} \cdot dp
\end{cases}
\qquad (2.96)
$$

2.1.2.2.7 Segment of Rectangular Cross-Section with Variable Width and Thickness

The segment of Figure 2.22 has a rectangular cross-section where both dimensions are variable over the hinge length. There are two pairs of identical and mirrored notches that form this configuration, and each pair is symmetric with respect to the longitudinal axis – this results in a hinge with axial symmetry. One notch pair is in the *xy* plane, while the other one is located in the *xz* plane. In the *xy* plane, the defining outline curve is $y(x)$, while in the *xz* plane, the similar curve is $z(x)$. At the abscissa x, the variable dimensions of the rectangular cross-sections are:

$$t(x) = 2y(x); \quad w(x) = 2z(x). \qquad (2.97)$$

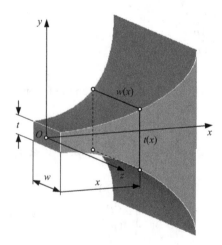

FIGURE 2.22 Rectangular cross-section hinge segment with variable thickness $t(x)$ and width $w(x)$.

It is assumed that the minimum cross-sectional dimensions are t and w, and they correspond to $x = 0$, as illustrated in Figure 2.22.

The eight compliances that characterize the generic segment of Figure 2.22 are calculated as per Eqs. (2.6) where the cross-sectional parameters are: $A(x) = w(x) \cdot t(x)$, $I_z(x) = w(x) \cdot t(x)^3/12$ and $I_y(x) = w(x)^3 \cdot t(x)/12$. The torsional moment of area $I_t(x)$ is calculated as in Eqs. (2.20), (2.21) or (2.22). In all these equations, $w(x)$ should be used instead of w.

Example 2.3

Express the compliances of a two-axis notch segment like the one of Figure 2.22 where the two defining curves are the parabolas shown in Figure 2.23.

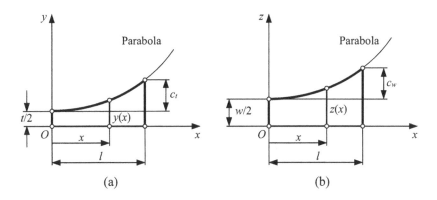

(a)   (b)

FIGURE 2.23 Planar geometry of two-axis flexible segment defined by parabolas in (a) Oxy plane; (b) Oxz plane.

Solution:

The variable cross-sectional dimensions are:

$$t(x) = t + \frac{2c_t}{l^2} \cdot x^2; \quad w(x) = w + \frac{2c_w}{l^2} \cdot x^2. \tag{2.98}$$

The torsional moment of area is calculated with one of the Eqs. (2.20), (2.21) or (2.22) depending on the relationship between the two cross-sectional dimensions and using $w(x)$ instead of w. While the explicit algebraic compliances of the parabolic segment are not included here, *Example 3.4* in Chapter 3 studies a notch hinge with transverse symmetry whose half-segment is the one discussed in this example.

2.2 CURVILINEAR-AXIS FLEXIBLE-HINGE SEGMENTS

This section studies flexible-hinge segments whose centroidal axis is a curve, such as circle, ellipse, parabola or hyperbola. Similar to the straight-axis segments, planar configurations are considered of either rectangular cross-section (with the out-of-plane dimension w constant) or circular. These segments can be combined with other segments to form full flexible hinges such as those studied in Chapter 4.

2.2.1 GENERIC COMPLIANCES

The small-displacement analytic compliances of generic flexible hinges with curvilinear centroidal (longitudinal) axis are formulated here to characterize their elastic response to in-plane and out-of-plane loading. Figure 2.24a shows the skeleton representation of a generic curvilinear-axis flexible-hinge segment, while Figure 2.24b plots the inner and outer curves $y_i(x)$ and $y_o(x)$ that define the in-plane profile of the same hinge segment.

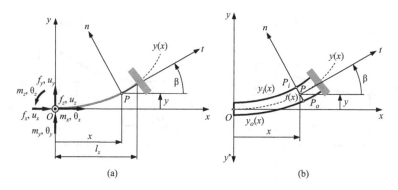

(a)   (b)

FIGURE 2.24 Curvilinear-axis flexible hinge with fixed-free ends: (a) skeleton representation with free-end loads and displacements; (b) front-view representation with planar geometry.

The forces f_x, f_y and f_z and the moments m_x, m_y and m_z are applied at the free end O of a flexible-hinge segment whose longitudinal axis is a planar curve defined as $y(x)$ in the Cartesian frame $Oxyz$, as illustrated in Figure 2.24a. The aim is to relate the displacements u_x, u_y, u_z, θ_x, θ_y and θ_z at the same point O to the loads in terms of the particular geometry and material parameters of the flexible hinge. The displacements are determined using Castigliano's second theorem based on the total strain energy U stored by the flexible hinge under load as per Eqs. (2.2).

The total strain energy is expressed in relation to a generic point P on the centroidal axis in Figure 2.24a. At P, the tangent (denoted by t) is defined, as well as the normal (perpendicular to the tangent and identified as n) and the binormal axis b (perpendicular to the hinge plane, which is also the plane formed by t and n; b is parallel to z and has the same direction as z). The six free-end point loads are reduced to a group of three loads at P that deform the flexible hinge in its plane and that are formed of N (the normal force along t), the shear force S_n (along the normal axis n) and the bending moment $M_{b,z}$ (acting along the binormal). The three other loads, namely, the torsion moment M_t (along the t axis), the bending moment $M_{b,n}$ (along the n axis) and the shear force S_z (perpendicular to the hinge plane and oriented along the binormal direction), cause the flexible hinge to deform out of its original plane. Correspondingly, the in-plane displacements at O are u_x, u_y and θ_z, whereas the out-of-plane displacements are θ_x, θ_y and u_z. The loads at P are obtained from the original free-end loads based on the geometry of Figure 2.24a as:

$$
\begin{cases}
N = f_x \cdot \cos\beta + f_y \cdot \sin\beta; \\[4pt]
S_n = -f_x \cdot \sin\beta + f_y \cdot \cos\beta; \\[4pt]
M_{b,z} = f_x \cdot y - f_y \cdot x + m_z; \\[4pt]
M_t = m_x \cdot \cos\beta + m_y \cdot \sin\beta + f_z \cdot (x \cdot \sin\beta - y \cdot \cos\beta); \\[4pt]
M_{b,n} = -m_x \cdot \sin\beta + m_y \cdot \cos\beta + f_z \cdot (x \cdot \cos\beta + y \cdot \sin\beta); \\[4pt]
S_z = f_z
\end{cases}
\tag{2.99}
$$

with $\beta = \tan^{-1}(dy(x)/dx)$. For a homogeneous and isotropic material, the total strain energy produced by the six loads of Eqs. (2.99) is:

$$
\begin{aligned}
U = {} & \frac{1}{2E} \cdot \int_0^l \frac{N^2}{A(s)} \cdot ds + \frac{\alpha_s}{2G} \cdot \int_0^l \frac{(S_n)^2}{A(s)} \cdot ds + \frac{1}{2E} \cdot \int_0^l \frac{(M_{b,z})^2}{I_z(s)} \cdot ds \\[6pt]
& + \frac{1}{2G} \cdot \int_0^l \frac{(M_t)^2}{I_t(s)} \cdot ds + \frac{1}{2E} \cdot \int_0^l \frac{(M_{b,n})^2}{I_n(s)} \cdot ds + \frac{\alpha_s}{2G} \cdot \int_0^l \frac{(S_z)^2}{A(s)} \cdot ds
\end{aligned}
\tag{2.100}
$$

$A(s)$, $I_t(s)$, $I_z(s)$ and $I_n(s)$ are the arbitrary-position cross-section area and moments of area (in either torsion – along the tangential direction or bending – along the z or along the normal direction). The parameter l in Eq. (2.100) is the length of the curvilinear element measured along the centroidal axis. The elementary length ds along the curvilinear centroidal axis and the trigonometric functions of Eqs. (2.99) are expressed as:

$$
\begin{cases}
ds = \sqrt{1 + \left(dy(x)/dx\right)^2} \cdot dx; \quad \tan\beta = dy(x)/dx; \\[10pt]
\sin\beta = \dfrac{dy(x)/dx}{\sqrt{1 + \left(dy(x)/dx\right)^2}}; \quad \cos\beta = \dfrac{1}{\sqrt{1 + \left(dy(x)/dx\right)^2}}.
\end{cases}
\tag{2.101}
$$

When substituting Eqs. (2.101) into Eqs. (2.99) and (2.100), the variable x substitutes s, and the upper limit of the integrals of Eqs. (2.100) is l_x, the x-axis projection of the curvilinear length l.

2.2.1.1 In-Plane Compliances

Combining Eqs. (2.99)–(2.101) generates the following relationship between the in-plane loads f_x, f_y and m_z and the corresponding displacements u_x, u_y and θ_z by means of a symmetric in-plane compliance matrix $[C_{ip}]$:

$$
\begin{bmatrix} u_x \\ u_y \\ \theta_z \end{bmatrix} = [C_{ip}] \begin{bmatrix} f_x \\ f_y \\ m_z \end{bmatrix}; \quad [C_{ip}] = \begin{bmatrix} C_{u_x-f_x} & C_{u_x-f_y} & C_{u_x-m_z} \\ C_{u_x-f_y} & C_{u_y-f_y} & C_{u_y-m_z} \\ C_{u_x-m_z} & C_{u_y-m_z} & C_{\theta_z-m_z} \end{bmatrix}, \quad (2.102)
$$

with the individual compliances:

$$
\begin{cases}
C_{u_x-f_x} = C_{u_x-f_x}^{(b)} + C_{u_x-f_x}^{(n)} + C_{u_x-f_x}^{(s)} \\[2mm]
C_{u_x-f_y} = C_{u_x-f_y}^{(b)} + C_{u_x-f_y}^{(n)} + C_{u_x-f_y}^{(s)} \\[2mm]
C_{u_x-m_z} = C_{u_x-m_z}^{(b)} = \dfrac{1}{E} \cdot \displaystyle\int_0^{l_x} \dfrac{y(x) \cdot \sqrt{1 + (dy(x)/dx)^2}}{I_z(x)} \cdot dx \\[4mm]
C_{u_y-f_y} = C_{u_y-f_y}^{(b)} + C_{u_y-f_y}^{(n)} + C_{u_y-f_y}^{(s)} \\[2mm]
C_{u_y-m_z} = C_{u_y-m_z}^{(b)} = -\dfrac{1}{E} \cdot \displaystyle\int_0^{l_x} \dfrac{x \cdot \sqrt{1 + (dy(x)/dx)^2}}{I_z(x)} \cdot dx \\[4mm]
C_{\theta_z-m_z} = C_{\theta_z-m_z}^{(b)} = \dfrac{1}{E} \cdot \displaystyle\int_0^{l_x} \dfrac{\sqrt{1 + (dy(x)/dx)^2}}{I_z(x)} \cdot dx
\end{cases}
\quad (2.103)
$$

The superscripts "b", "n" and "s" in Eqs. (2.103) denote the compliances that are associated with bending, normal force and shear force, respectively. These compliances are calculated as:

$$
\begin{cases}
C_{u_x-f_x}^{(b)} = \dfrac{1}{E} \cdot \displaystyle\int_0^{l_x} \dfrac{y(x)^2 \cdot \sqrt{1 + (dy(x)/dx)^2}}{I_z(x)} \cdot dx \\[4mm]
C_{u_x-f_x}^{(n)} = \dfrac{1}{E} \cdot \displaystyle\int_0^{l_x} \dfrac{dx}{A(x) \cdot \sqrt{1 + (dy(x)/dx)^2}} \\[4mm]
C_{u_x-f_x}^{(s)} = \dfrac{\alpha_s}{G} \cdot \displaystyle\int_0^{l_x} \dfrac{(dy(x)/dx)^2}{A(x) \cdot \sqrt{1 + (dy(x)/dx)^2}} \cdot dx
\end{cases}
\quad (2.104)
$$

$$\begin{cases} C_{u_x-f_y}^{(b)} = -\frac{1}{E} \cdot \int_0^{l_x} \frac{xy(x) \cdot \sqrt{1+(dy(x)/dx)^2}}{I_z(x)} \cdot dx \\ \\ C_{u_x-f_y}^{(n)} = \frac{1}{E} \cdot \int_0^{l_x} \frac{(dy(x)/dx)}{A(x) \cdot \sqrt{1+(dy(x)/dx)^2}} \cdot dx \\ \\ C_{u_x-f_y}^{(s)} = -2\alpha_s \cdot (1+\mu) \cdot C_{u_x-f_y}^{(n)} \end{cases} \qquad (2.105)$$

$$\begin{cases} C_{u_y-f_y}^{(b)} = \frac{1}{E} \cdot \int_0^{l_x} \frac{x^2 \cdot \sqrt{1+(dy(x)/dx)^2}}{I_z(x)} \cdot dx \\ \\ C_{u_y-f_y}^{(n)} = \frac{1}{E} \cdot \int_0^{l_x} \frac{(dy(x)/dx)^2}{A(x) \cdot \sqrt{1+(dy(x)/dx)^2}} \cdot dx. \\ \\ C_{u_y-f_y}^{(s)} = 2\alpha_s \cdot (1+\mu) \cdot C_{u_x-f_x}^{(n)} \end{cases} \qquad (2.106)$$

While the compliances of Eqs. (2.102) were defined with respect to the $Oxyz$ reference frame shown in Figure 2.24b, there are cases when it is necessary to use the reference frame Oxy^*z^* (also pictured in Figure 2.24b) whose y^* and z^* axes are oriented in directions opposite to the y and z axes, respectively. When referenced to the Oxy^*z^* frame, the in-plane compliance matrix of Eq. (2.102) becomes:

$$\left[C_{ip}^* \right] = \begin{bmatrix} C_{u_x-f_x} & -C_{u_x-f_y} & -C_{u_x-m_z} \\ -C_{u_x-f_y} & C_{u_y-f_y} & C_{u_y-m_z} \\ -C_{u_x-m_z} & C_{u_y-m_z} & C_{\theta_z-m_z} \end{bmatrix}, \qquad (2.107)$$

with the individual compliances provided in Eqs. (2.103) through (2.106). As Eq. (2.107) indicates and compared to the compliance matrix of Eq. (2.102), the compliances involving the axes pairs x-y and x-z changed their signs, but all other compliances kept their original plus sign. This modification is due to the fact that when formulating compliances by means of Eqs. (2.99) and (2.100), a sign change occurs in any of the six loads of Eqs. (2.99) when either y or z is flipped since f_y becomes $-f_y$ and m_z changes to $-m_z$.

The straight-axis hinge compliances as a particular case:

Assuming the curvilinear-axis hinge becomes a straight-axis one, both the slope $dy(x)/dx$ and $y(x)$ become zero – see Figure 2.24a. In this particular case, the x-axis length projection l_x is simply the length l. As a consequence, as it can be checked easily from Eqs. (2.103) – (2.106), these compliances become:

$$
\begin{cases}
C_{u_x-f_x} = C^{(n)}_{u_x-f_x} = \frac{1}{E} \cdot \int_0^l \frac{dx}{A(x)}; \quad C_{u_x-f_y} = 0; \quad C_{u_x-m_z} = 0; \\[3mm]
C_{u_y-f_y} = C^{(b)}_{u_y-f_y} + C^{(s)}_{u_y-f_y} = \frac{1}{E} \cdot \int_0^l \frac{x^2}{I_z(x)} \cdot dx + 2\alpha_s \cdot (1+\mu) \cdot C_{u_x-f_x}; \\[3mm]
C_{u_y-m_z} = C^{(b)}_{u_y-m_z} = -\frac{1}{E} \cdot \int_0^l \frac{x}{I_z(x)} \cdot dx; \\[3mm]
C_{\theta_z-m_z} = C^{(b)}_{\theta_z-m_z} = \frac{1}{E} \cdot \int_0^l \frac{dx}{I_z(x)}
\end{cases}
\tag{2.108}
$$

which are the compliance equations of a straight axis with variable cross-section as per the definition Eqs. (2.6).

2.2.1.2 Out-of-Plane Compliances

A relationship similar to the in-plane Eq. (2.102) is obtained by combining Eqs. (2.99)–(2.101) that connect the out-of-plane loads m_x, m_y and f_z to the corresponding displacements θ_x, θ_y and u_z by means of a symmetric out-of-plane compliance matrix $[C_{op}]$ as:

$$
\begin{bmatrix} \theta_x \\ \theta_y \\ u_z \end{bmatrix} = [C_{op}] \begin{bmatrix} m_x \\ m_y \\ f_z \end{bmatrix}; \quad [C_{op}] = \begin{bmatrix} C_{\theta_x-m_x} & C_{\theta_x-m_y} & C_{\theta_x-f_z} \\ C_{\theta_x-m_y} & C_{\theta_y-m_y} & C_{\theta_y-f_z} \\ C_{\theta_x-f_z} & C_{\theta_y-f_z} & C_{u_z-f_z} \end{bmatrix}, \tag{2.109}
$$

with the individual compliances being calculated as:

$$
\begin{cases}
C_{\theta_x-m_x} = C^{(b)}_{\theta_x-m_x} + C^{(t)}_{\theta_x-m_x} \\[2mm]
C_{\theta_x-m_y} = C^{(b)}_{\theta_x-m_y} + C^{(t)}_{\theta_x-m_y} \\[2mm]
C_{\theta_x-f_z} = C^{(b)}_{\theta_x-f_z} + C^{(t)}_{\theta_x-f_z} \\[2mm]
C_{\theta_y-m_y} = C^{(b)}_{\theta_y-m_y} + C^{(t)}_{\theta_y-m_y} \\[2mm]
C_{\theta_y-f_z} = C^{(b)}_{\theta_y-f_z} + C^{(t)}_{\theta_y-f_z} \\[2mm]
C_{u_z-f_z} = C^{(b)}_{u_z-f_z} + C^{(t)}_{u_z-f_z} + C^{(s)}_{u_z-f_z}
\end{cases}
, \tag{2.110}
$$

where the new superscript "t" indicates torsion. The compliances in the right-hand side of Eqs. (2.110) are:

$$\begin{cases} C^{(b)}_{\theta_x-m_x} = \dfrac{1}{E} \cdot \displaystyle\int_0^{l_x} \dfrac{\left(dy(x)/dx\right)^2}{I_n(x) \cdot \sqrt{1+\left(dy(x)/dx\right)^2}} \cdot dx \\[20pt] C^{(t)}_{\theta_x-m_x} = \dfrac{1}{G} \cdot \displaystyle\int_0^{l_x} \dfrac{dx}{I_t(x) \cdot \sqrt{1+\left(dy(x)/dx\right)^2}} \end{cases} \tag{2.111}$$

$$\begin{cases} C^{(b)}_{\theta_x-m_y} = -\dfrac{1}{E} \cdot \displaystyle\int_0^{l_x} \dfrac{\left(dy(x)/dx\right)}{I_n(x) \cdot \sqrt{1+\left(dy(x)/dx\right)^2}} \cdot dx \\[20pt] C^{(t)}_{\theta_x-m_y} = \dfrac{1}{G} \cdot \displaystyle\int_0^{l_x} \dfrac{\left(dy(x)/dx\right)}{I_t(x) \cdot \sqrt{1+\left(dy(x)/dx\right)^2}} \cdot dx \end{cases} \tag{2.112}$$

$$\begin{cases} C^{(b)}_{\theta_x-f_z} = -\dfrac{1}{E} \cdot \displaystyle\int_0^{l_x} \dfrac{x \cdot \left[dy(x)/dx\right] + y(x) \cdot \left[dy(x)/dx\right]^2}{I_n(x) \cdot \sqrt{1+\left(dy(x)/dx\right)^2}} \cdot dx \\[20pt] C^{(t)}_{\theta_x-f_z} = \dfrac{1}{G} \cdot \displaystyle\int_0^{l_x} \dfrac{x \cdot \left[dy(x)/dx\right] - y(x)}{I_t(x) \cdot \sqrt{1+\left(dy(x)/dx\right)^2}} \cdot dx \end{cases} \tag{2.113}$$

$$\begin{cases} C^{(b)}_{\theta_y-m_y} = \dfrac{1}{E} \cdot \displaystyle\int_0^{l_x} \dfrac{dx}{I_n(x) \cdot \sqrt{1+\left(dy(x)/dx\right)^2}} \\[20pt] C^{(t)}_{\theta_y-m_y} = \dfrac{1}{G} \cdot \displaystyle\int_0^{l_x} \dfrac{\left[dy(x)/dx\right]^2}{I_t(x) \cdot \sqrt{1+\left(dy(x)/dx\right)^2}} \cdot dx \end{cases} \tag{2.114}$$

$$\begin{cases} C^{(b)}_{\theta_y-f_z} = \dfrac{1}{E} \cdot \displaystyle\int_0^{l_x} \dfrac{x + y(x) \cdot \left[dy(x)/dx\right]}{I_n(x) \cdot \sqrt{1+\left(dy(x)/dx\right)^2}} \cdot dx \\[20pt] C^{(t)}_{\theta_y-f_z} = \dfrac{1}{G} \cdot \displaystyle\int_0^{l_x} \dfrac{x \cdot \left[dy(x)/dx\right]^2 - y(x) \cdot \left[dy(x)/dx\right]}{I_t(x) \cdot \sqrt{1+\left(dy(x)/dx\right)^2}} \cdot dx \end{cases} \tag{2.115}$$

$$
\begin{cases}
C_{u_z-f_z}^{(b)} = \dfrac{1}{E} \cdot \displaystyle\int_0^{l_x} \dfrac{\left\{x + y(x) \cdot [dy(x)/dx]\right\}^2}{I_n(x) \cdot \sqrt{1 + (dy(x)/dx)^2}} \cdot dx \\[18pt]
C_{u_z-f_z}^{(t)} = \dfrac{1}{G} \cdot \displaystyle\int_0^{l_x} \dfrac{\left\{x \cdot [dy(x)/dx] - y(x)\right\}^2}{I_t(x) \cdot \sqrt{1 + (dy(x)/dx)^2}} \cdot dx \\[18pt]
C_{u_z-f_z}^{(s)} = \dfrac{\alpha_s}{G} \cdot \displaystyle\int_0^{l_x} \dfrac{\sqrt{1 + (dy(x)/dx)^2}}{A(x)} \cdot dx
\end{cases}
\tag{2.116}
$$

When the y and z axes are replaced by their opposite-direction y^* and z^* axes, the out-of-plane compliance matrix of Eq. (2.109) becomes:

$$
\left[C_{op}^* \right] =
\begin{bmatrix}
C_{\theta_x-m_x} & -C_{\theta_x-m_y} & -C_{\theta_x-f_z} \\
-C_{\theta_x-m_y} & C_{\theta_y-m_y} & C_{\theta_y-f_z} \\
-C_{\theta_x-f_z} & C_{\theta_y-f_z} & C_{u_z-f_z}
\end{bmatrix},
\tag{2.117}
$$

with the individual compliances calculated as in Eqs. (2.110) – (2.116) by following a reasoning similar to the one applied in deriving $\left[C_{ip}^* \right]$ from $\left[C_{ip} \right]$.

The straight-axis hinge compliances as a particular case:

When the curved-axis hinge becomes a straight-axis one and similar to the in-plane case, the out-of-plane compliances of Eqs. (2.110) – (2.116) become:

$$
\begin{cases}
C_{\theta_x-m_x} = C_{\theta_x-m_x}^{(t)} = \dfrac{1}{G} \cdot \displaystyle\int_0^l \dfrac{dx}{I_t(x)}; \quad C_{\theta_x-m_y} = 0; \quad C_{\theta_x-f_z} = 0; \\[18pt]
C_{\theta_y-m_y} = C_{\theta_y-m_y}^{(b)} = \dfrac{1}{E} \cdot \displaystyle\int_0^l \dfrac{dx}{I_y(x)}; \\[18pt]
C_{\theta_y-f_z} = C_{\theta_y-f_z}^{(b)} = \dfrac{1}{E} \cdot \displaystyle\int_0^l \dfrac{x}{I_y(x)} \cdot dx; \\[18pt]
C_{u_z-f_z} = C_{u_z-f_z}^{(b)} + C_{u_z-f_z}^{(s)} = \dfrac{1}{E} \cdot \displaystyle\int_0^l \dfrac{x^2}{I_y(x)} \cdot dx + 2\alpha_s \cdot (1+\mu) \cdot C_{u_x-f_x}
\end{cases}
\tag{2.118}
$$

To derive Eqs. (2.118), which are identical to the definition Eqs. (2.6), the following particular values have been used in Eqs. (2.110) through (2.116): $dy(x)/dx = 0$, $y(x) = 0$ $l_x = l$ and $I_n(x) = I_y(x)$.

2.2.2 GEOMETRIC CONFIGURATIONS

This section formulates in- and out-of-plane compliances of flexible segments defined by various longitudinal curves, including circular, elliptic and parabolic profiles. The cross-section is assumed variable.

2.2.2.1 Circular-Axis Hinge Segment

Consider the particular design of Figure 2.25 where the curve $y(x)$ is a circular segment. The in-plane and out-of-plane compliances that have been defined for a generic curve $y(x)$ will be expressed here.

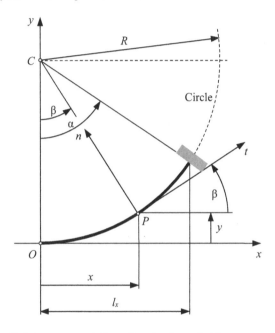

FIGURE 2.25 Skeleton representation of a circular-axis hinge segment with defining geometry.

The Cartesian coordinates x and y are expressed in terms of the polar coordinate β as:

$$x = R \cdot \sin\beta; \quad dx = R \cdot \cos\beta \cdot d\beta; \quad y = R \cdot (1 - \cos\beta), \qquad (2.119)$$

which results in:

$$\frac{dy}{dx} = \frac{\dfrac{dy(\beta)}{d\beta}}{\dfrac{dx(\beta)}{d\beta}} = \tan\beta; \quad \sqrt{1 + (dy/dx)^2} = \frac{1}{\cos\beta}. \qquad (2.120)$$

Substituting Eqs. (2.119) and (2.120) into Eqs. (2.104)–(2.106), the following in-plane compliances are obtained for a generic circular-axis hinge segment:

$$
\begin{cases}
C_{u_x-f_x}^{(b)} = \dfrac{R^3}{E} \cdot \displaystyle\int_0^\alpha \dfrac{(1-\cos\beta)^2}{I_z(\beta)} \cdot d\beta \\[2ex]
C_{u_x-f_x}^{(n)} = \dfrac{R}{E} \cdot \displaystyle\int_0^\alpha \dfrac{\cos^2\beta}{A(\beta)} \cdot d\beta \\[2ex]
C_{u_x-f_x}^{(s)} = \dfrac{\alpha_s \cdot R}{G} \cdot \displaystyle\int_0^\alpha \dfrac{\sin^2\beta\, d\beta}{A(\beta)}
\end{cases}
\tag{2.121}
$$

$$
\begin{cases}
C_{u_x-f_y}^{(b)} = -\dfrac{R^3}{E} \cdot \displaystyle\int_0^\alpha \dfrac{\sin\beta \cdot (1-\cos\beta)}{I_z(\beta)} \cdot d\beta \\[2ex]
C_{u_x-f_y}^{(n)} = \dfrac{R}{E} \cdot \displaystyle\int_0^\alpha \dfrac{\sin\beta \cdot \cos\beta}{A(\beta)\cdot} \cdot d\beta
\end{cases}
\tag{2.122}
$$

$$
C_{u_x-m_z} = C_{u_x-m_z}^{(b)} = \dfrac{R^2}{E} \cdot \int_0^\alpha \dfrac{1-\cos\beta}{I_z(\beta)} \cdot d\beta
\tag{2.123}
$$

$$
\begin{cases}
C_{u_y-f_y}^{(b)} = \dfrac{R^3}{E} \cdot \displaystyle\int_0^\alpha \dfrac{\sin^2\beta}{I_z(\beta)} \cdot d\beta \\[2ex]
C_{u_y-f_y}^{(n)} = \dfrac{R}{E} \cdot \displaystyle\int_0^\alpha \dfrac{\sin^2\beta}{A(\beta)} \cdot d\beta
\end{cases}
\tag{2.124}
$$

$$
\begin{cases}
C_{u_y-m_z} = C_{u_y-m_z}^{(b)} = -\dfrac{R^2}{E} \cdot \displaystyle\int_0^\alpha \dfrac{\sin\beta}{I_z(\beta)} \cdot d\beta \\[2ex]
C_{\theta_z-m_z} = C_{\theta_z-m_z}^{(b)} = \dfrac{R}{E} \cdot \displaystyle\int_0^\alpha \dfrac{d\beta}{I_z(\beta)}
\end{cases}
\tag{2.125}
$$

Combination of Eqs. (2.119) and (2.120) with Eqs. (2.111) – (2.116) yields the out-of-plane compliances of a generic circular-axis flexible-hinge segment:

$$
\begin{cases}
C^{(b)}_{\theta_x-m_x} = \dfrac{R}{E} \cdot \displaystyle\int_0^\alpha \dfrac{\sin^2\beta}{I_n(\beta)} \cdot d\beta \\[4mm]
C^{(t)}_{\theta_x-m_x} = \dfrac{R}{G} \cdot \displaystyle\int_0^\alpha \dfrac{\cos^2\beta}{I_t(\beta)} \cdot d\beta
\end{cases}
\tag{2.126}
$$

$$
\begin{cases}
C^{(b)}_{\theta_x-m_y} = -\dfrac{R}{E} \cdot \displaystyle\int_0^\alpha \dfrac{\sin\beta \cdot \cos\beta}{I_n(\beta)} \cdot d\beta \\[4mm]
C^{(t)}_{\theta_x-m_y} = \dfrac{R}{G} \cdot \displaystyle\int_0^\alpha \dfrac{\sin\beta \cdot \cos\beta}{I_t(\beta)} \cdot d\beta
\end{cases}
\tag{2.127}
$$

$$
\begin{cases}
C^{(b)}_{\theta_x-f_z} = -\dfrac{R^2}{E} \cdot \displaystyle\int_0^\alpha \dfrac{\sin^2\beta}{I_n(\beta)} \cdot d\beta \\[4mm]
C^{(t)}_{\theta_x-f_z} = \dfrac{R^2}{G} \cdot \displaystyle\int_0^\alpha \dfrac{\cos\beta \cdot (1-\cos\beta)}{I_t(\beta)} \cdot d\beta
\end{cases}
\tag{2.128}
$$

$$
\begin{cases}
C^{(b)}_{\theta_y-m_y} = \dfrac{R}{E} \cdot \displaystyle\int_0^\alpha \dfrac{\cos^2\beta}{I_n(\beta)} \cdot d\beta \\[4mm]
C^{(t)}_{\theta_y-m_y} = \dfrac{R}{G} \cdot \displaystyle\int_0^\alpha \dfrac{\sin^2\beta}{I_t(\beta)} \cdot d\beta
\end{cases}
\tag{2.129}
$$

$$
\begin{cases}
C^{(b)}_{\theta_y-f_z} = \dfrac{R^2}{E} \cdot \displaystyle\int_0^\alpha \dfrac{\sin\beta \cdot \cos\beta}{I_n(\beta)} \cdot d\beta \\[4mm]
C^{(t)}_{\theta_y-f_z} = \dfrac{R^2}{G} \cdot \displaystyle\int_0^\alpha \dfrac{\sin\beta \cdot (1-\cos\beta)}{I_t(\beta)} \cdot d\beta
\end{cases}
\tag{2.130}
$$

$$
\begin{cases}
C^{(b)}_{u_z-f_z} = \dfrac{R^3}{E} \cdot \displaystyle\int_0^\alpha \dfrac{\sin^2\beta}{I_n(\beta)} \cdot d\beta \\[4mm]
C^{(t)}_{u_z-f_z} = \dfrac{R^3}{G} \cdot \displaystyle\int_0^\alpha \dfrac{(1-\cos\beta)^2}{I_t(\beta)} \cdot d\beta \\[4mm]
C^{(s)}_{u_z-f_z} = \dfrac{\alpha_s R}{G} \cdot \displaystyle\int_0^\alpha \dfrac{d\beta}{A(\beta)}
\end{cases}
\tag{2.131}
$$

The Appendix at the end of this chapter includes the compliances of a circular-axis hinge segment of constant cross-section.

2.2.2.2   Elliptical-Axis Hinge Segment

This section formulates the in-plane and out-of-plane compliances of a hinge segment whose centroidal axis is an ellipse – Figure 2.26 shows its skeleton representation. The compliance derivation is limited to the general case of a variable
cross-section hinge segment.

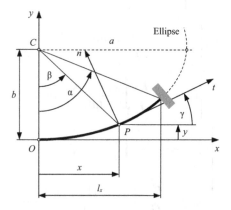

FIGURE 2.26   Skeleton representation of an elliptical-axis hinge segment with defining
geometry.

The following relationships connecting the Cartesian coordinates x and y to
the polar coordinate β of Figure 2.26 are needed in the compliance integrations:

$$x = a \cdot \sin\beta; \quad dx = a \cdot \cos\beta \cdot d\beta; \quad y = b \cdot (1 - \cos\beta);$$

$$\tan\gamma = \frac{dy}{dx} = \frac{\dfrac{dy(\beta)}{d\beta}}{\dfrac{dx(\beta)}{d\beta}} = \frac{b}{a} \cdot \tan\beta; \quad \sqrt{1 + (dy/dx)^2} = \frac{\sqrt{a^2 \cdot \cos^2\beta + b^2 \cdot \sin^2\beta} \cdot}{a \cdot \cos\beta}$$

$$(2.132)$$

Equations (2.104)–(2.106) are used in conjunction with Eq. (2.132), to yield the
following in-plane compliances for a generic elliptical-axis hinge segment of
variable cross-section:

$$\begin{cases} C_{u_x-f_x}^{(b)} = \dfrac{b^2}{E} \cdot \displaystyle\int_0^\alpha \dfrac{(1-\cos\beta)^2 \cdot \sqrt{a^2\cos^2\beta + b^2 \cdot \sin^2\beta}}{I_z(\beta)} \cdot d\beta \\[4mm] C_{u_x-f_x}^{(n)} = \dfrac{a^2}{E} \cdot \displaystyle\int_0^\alpha \dfrac{\cos^2\beta}{A(\beta) \cdot \sqrt{a^2\cos^2\beta + b^2 \cdot \sin^2\beta}} \cdot d\beta \\[4mm] C_{u_x-f_x}^{(s)} = \dfrac{\alpha_s \cdot b^2}{G} \cdot \displaystyle\int_0^\alpha \dfrac{\sin^2\beta\, d\beta}{A(\beta) \cdot \sqrt{a^2\cos^2\beta + b^2 \cdot \sin^2\beta}} \end{cases} \qquad (2.133)$$

$$\begin{cases} C^{(b)}_{u_x-f_y} = -\dfrac{a \cdot b}{E} \cdot \int_0^\alpha \dfrac{\sin\beta \cdot (1-\cos\beta) \cdot \sqrt{a^2\cos^2\beta + b^2 \cdot \sin^2\beta}}{I_z(\beta)} \cdot d\beta \\[4mm] C^{(n)}_{u_x-f_y} = \dfrac{a \cdot b}{E} \cdot \int_0^\alpha \dfrac{\sin\beta \cdot \cos\beta}{A(\beta) \cdot \sqrt{a^2\cos^2\beta + b^2 \cdot \sin^2\beta}} \cdot d\beta \end{cases}$$

(2.134)

$$C_{u_x-m_z} = C^{(b)}_{u_x-m_z} = \dfrac{b}{E} \cdot \int_0^\alpha \dfrac{(1-\cos\beta) \cdot \sqrt{a^2\cos^2\beta + b^2 \cdot \sin^2\beta}}{I_z(\beta)} \cdot d\beta \quad (2.135)$$

$$\begin{cases} C^{(b)}_{u_y-f_y} = \dfrac{a^2}{E} \cdot \int_0^\alpha \dfrac{\sin^2\beta \cdot \sqrt{a^2\cos^2\beta + b^2 \cdot \sin^2\beta}}{I_z(\beta)} \cdot d\beta \\[4mm] C^{(n)}_{u_y-f_y} = \dfrac{b^2}{E} \cdot \int_0^\alpha \dfrac{\sin^2\beta}{A(\beta) \cdot \sqrt{a^2\cos^2\beta + b^2 \cdot \sin^2\beta}} \cdot d\beta \\[4mm] C^{(s)}_{u_y-f_y} = \dfrac{\alpha_s \cdot a^2}{G} \cdot \int_0^\alpha \dfrac{\cos^2\beta}{A(\beta) \cdot \sqrt{a^2\cos^2\beta + b^2 \cdot \sin^2\beta}} \cdot d\beta \end{cases}$$

(2.136)

$$\begin{cases} C_{u_y-m_z} = C^{(b)}_{u_y-m_z} = -\dfrac{a}{E} \cdot \int_0^\alpha \dfrac{\sin\beta \cdot \sqrt{a^2\cos^2\beta + b^2 \cdot \sin^2\beta}}{I_z(\beta)} \cdot d\beta \\[4mm] C_{\theta_z-m_z} = C^{(b)}_{\theta_z-m_z} = \dfrac{1}{E} \cdot \int_0^\alpha \dfrac{\sqrt{a^2\cos^2\beta + b^2 \cdot \sin^2\beta}}{I_z(\beta)} \cdot d\beta \end{cases}$$

. (2.137)

The parameters of Eqs. (2.132) are substituted into Eqs. (2.111) – (2.116), which results in the out-of-plane compliances of a generic elliptical-axis flexible-hinge segment:

$$\begin{cases} C^{(b)}_{\theta_x-m_x} = \dfrac{b^2}{E} \cdot \int_0^\alpha \dfrac{\sin^2\beta}{I_n(\beta) \cdot \sqrt{a^2\cos^2\beta + b^2 \cdot \sin^2\beta}} \cdot d\beta \\[4mm] C^{(t)}_{\theta_x-m_x} = \dfrac{a^2}{G} \cdot \int_0^\alpha \dfrac{\cos^2\beta}{I_t(\beta) \cdot \sqrt{a^2\cos^2\beta + b^2 \cdot \sin^2\beta}} \cdot d\beta \end{cases}$$

(2.138)

$$\begin{cases} C^{(b)}_{\theta_x-m_y} = -\dfrac{a \cdot b}{E} \cdot \int_0^\alpha \dfrac{\sin\beta \cdot \cos\beta}{I_n(\beta) \cdot \sqrt{a^2\cos^2\beta + b^2 \cdot \sin^2\beta}} \cdot d\beta \\[4mm] C^{(t)}_{\theta_x-m_y} = \dfrac{a \cdot b}{G} \cdot \int_0^\alpha \dfrac{\sin\beta \cdot \cos\beta}{I_t(\beta) \cdot \sqrt{a^2\cos^2\beta + b^2 \cdot \sin^2\beta}} \cdot d\beta \end{cases}$$

(2.139)

$$
\begin{cases}
C^{(b)}_{\theta_x - f_z} = -\dfrac{b}{E} \cdot \displaystyle\int_0^\alpha \dfrac{\left[b^2 + \left(a^2 - b^2\right) \cdot \cos\beta\right] \cdot \sin^2\beta}{I_n(\beta) \cdot \sqrt{a^2 \cos^2\beta + b^2 \cdot \sin^2\beta}} \cdot d\beta \\[4mm]
C^{(t)}_{\theta_x - f_z} = \dfrac{a^2 \cdot b}{G} \cdot \displaystyle\int_0^\alpha \dfrac{\cos\beta \cdot \left(1 - \cos\beta\right)}{I_t(\beta) \cdot \sqrt{a^2 \cos^2\beta + b^2 \cdot \sin^2\beta}} \cdot d\beta
\end{cases}
\tag{2.140}
$$

$$
\begin{cases}
C^{(b)}_{\theta_y - m_y} = \dfrac{a^2}{E} \cdot \displaystyle\int_0^\alpha \dfrac{\cos^2\beta}{I_n(\beta) \cdot \sqrt{a^2 \cos^2\beta + b^2 \cdot \sin^2\beta}} \cdot d\beta \\[4mm]
C^{(t)}_{\theta_y - m_y} = \dfrac{b^2}{G} \cdot \displaystyle\int_0^\alpha \dfrac{\sin^2\beta}{I_t(\beta) \cdot \sqrt{a^2 \cos^2\beta + b^2 \cdot \sin^2\beta}} \cdot d\beta
\end{cases}
\tag{2.141}
$$

$$
\begin{cases}
C^{(b)}_{\theta_y - f_z} = \dfrac{a}{E} \cdot \displaystyle\int_0^\alpha \dfrac{\left[b^2 + \left(a^2 - b^2\right) \cdot \cos\beta\right]\sin\beta}{I_n(\beta) \cdot \sqrt{a^2 \cos^2\beta + b^2 \cdot \sin^2\beta}} \cdot d\beta \\[4mm]
C^{(t)}_{\theta_y - f_z} = \dfrac{a \cdot b^2}{G} \cdot \displaystyle\int_0^\alpha \dfrac{\sin\beta \cdot \left(1 - \cos\beta\right)}{I_t(\beta) \cdot \sqrt{a^2 \cos^2\beta + b^2 \cdot \sin^2\beta}} \cdot d\beta
\end{cases}
\tag{2.142}
$$

$$
\begin{cases}
C^{(b)}_{u_z - f_z} = \dfrac{1}{E} \cdot \displaystyle\int_0^\alpha \dfrac{\left[b^2 + \left(a^2 - b^2\right) \cdot \cos\beta\right]^2 \cdot \sin^2\beta}{I_n(\beta) \cdot \sqrt{a^2 \cos^2\beta + b^2 \cdot \sin^2\beta}} \cdot d\beta \\[4mm]
C^{(t)}_{u_z - f_z} = \dfrac{a^2 \cdot b^2}{G} \cdot \displaystyle\int_0^\alpha \dfrac{\left(1 - \cos\beta\right)^2}{I_t(\beta) \cdot \sqrt{a^2 \cos^2\beta + b^2 \cdot \sin^2\beta}} \cdot d\beta. \\[4mm]
C^{(s)}_{u_z - f_z} = \dfrac{\alpha_s}{G} \cdot \displaystyle\int_0^\alpha \dfrac{\sqrt{a^2 \cos^2\beta + b^2 \cdot \sin^2\beta}}{A(\beta)} \cdot d\beta
\end{cases}
\tag{2.143}
$$

2.2.2.3 Parabolic-Axis Hinge Segment

Consider the centroidal axis of the hinge segment shown in Figure 2.24 is a parabola whose equation and associated parameters are:

$$
y(x) = a_p \cdot x^2; \quad dy(x)/dx = 2a_p \cdot x; \quad \sqrt{1 + \left(dy(x)/dx\right)^2} = \sqrt{1 + \left(2a_p \cdot x\right)^2}. \tag{2.144}
$$

The in-plane compliances of a variable cross-section hinge result from Eqs. (2.103) – (2.106), while the out-of-plane compliances of the same segment are obtained from Eqs. (2.110) – (2.116); they are not included here, but the Appendix at the end of this chapter provides the equations of the in-plane and out-of-plane compliances of a constant cross-section parabolic-axis hinge segment.

2.2.2.4 Corner-Filleted Hinge Segments

Similar to straight-axis corner-filleted segments that can be combined with other flexible segments to join (usually) rigid links and reduce stress concentration in those areas, corner-filleted segments of curved axis can also be considered for similar purposes. The fillet area is obtained by means of a curve arc that is tangent to both the hinge profile curve and the rectilinear segment that intersects that hinge profile curve and marks one edge of the rigid link adjoining the hinge. As schematically shown in Figure 2.27a, the rigid-link edge segment can coincide with the normal to the centroidal curve (which is perpendicular to the tangent to the centroidal curve), case where the filleted corner is a right one. For angles between the terminal edge of the adjacent rigid link and the tangent other than 90° – as illustrated in Figure 2.27b – the resulting hinge is a regular corner-filleted one.

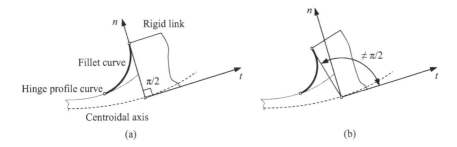

FIGURE 2.27 Connection areas between curvilinear-axis hinge segment and rigid link resulting in: (a) right corner fillet; (b) regular corner fillet.

The particular design of Figure 2.28 (more details on this configuration can be found in Lobontiu and Cullin [17]) shows a planar hinge segment that is constructed around a circular centroidal axis of radius R. The two end segments AB and DE are directed along two radial directions spaced at a center angle α. The two circular fillets of radii r_1 and r_2 are both tangent to the radial line CA, and the minimum segment in-plane thickness is t. Assuming this segment connects to a rigid link to the right of segment AB, the segment $ABDE$ of Figure 2.28 is a right circularly corner-filleted one. This fillet segment can connect to another flexible segment to the left of DE in order to form a flexible-hinge configuration.

The two fillet radii r_1 and r_2 are expressed in terms of R, t and α as:

$$r_1 = \frac{(R-t/2)\sin\alpha}{1+\sin\alpha}; \quad r_2 = \frac{(R+t/2)\sin\alpha}{1-\sin\alpha}. \tag{2.145}$$

The length of segment MN in Figure 2.28 represents the variable thickness $t(\beta)$ that is evaluated at an arbitrary angle β. This length is calculated based on the angles $\delta = \angle EC_2M$; $\varepsilon = \angle C_1NC$:

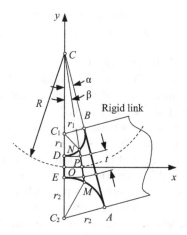

FIGURE 2.28 Right circular-axis fillet hinge segment of variable thickness $t(\beta)$.

$$\delta = \sin^{-1}\left(\frac{R+t/2+r_2}{r_2}\cdot\sin\beta\right)-\beta; \quad \varepsilon = \sin^{-1}\left(\frac{R-t/2-r_1}{r_1}\cdot\sin\beta\right) \quad (2.146)$$

as:

$$t(\beta) = \frac{r_2\sin\delta - r_1\sin(\beta+\varepsilon)}{\sin\beta}. \quad (2.147)$$

The maximum thickness of the root segment AB, t_{max}, is calculated as:

$$t_{max} = t(\alpha) = \frac{r_2 - r_1}{\tan\alpha} \quad (2.148)$$

from Eqs. (2.145) – (2.147) for $\beta = \alpha$.

In the design of Figure 2.28, the circular longitudinal axis of radius R is not a centroidal axis because the two thickness portions determined between the circle and the inner and outer curves, MP and PN, are not equal; it can simply be shown that:

$$\begin{cases} MP = r_2\cdot\dfrac{\sin\delta}{\sin\beta} - R \\[2ex] PN = R - r_1\cdot\dfrac{\sin(\varepsilon+\beta)}{\sin\beta} \end{cases} \quad (2.149)$$

It can be checked that $MP > PN$ for a generic angular position in Eq. (2.149), which would mean that the compliances that are calculated using the circular axis as the centroidal axis (specifically those using the moments of area involved in bending and torsion) need to be corrected. One possible solution is to determine the centroidal axis from the design of Figure 2.28, in other words, to find the equation of a new centroidal axis that would ensure that $MP = PN$ when P lies on this centroidal axis. However, the new centroidal axis to be determined will cause the radial-direction thickness not to be perpendicular to the tangent, which will require projecting the radial thickness on the direction perpendicular to the new centroidal axis – this would lead to a complex mathematical model, which is unnecessary provided the errors by the circular-centroid model are relatively small, as discussed in a related application in Chapter 4.

APPENDIX A2: CLOSED-FORM COMPLIANCES

A2.1 STRAIGHT-AXIS SEGMENTS

A2.1.1 Rectangular Cross-Section of Constant Out-of-Plane Width w and Minimum In-Plane Thickness t

A2.1.1.1 Trapezoid Segment

$$
\begin{cases}
C_{u_x-f_x} = \dfrac{1}{2E \cdot w \cdot m} \cdot \ln\left(1 + \dfrac{2l \cdot m}{t}\right); \\[2em]
C_{u_y-f_y} = \dfrac{3 \cdot \left[(2l \cdot m + t)^2 \cdot \ln\left(1 + \dfrac{2l \cdot m}{t}\right) - 2l \cdot m \cdot (3l \cdot m + t)\right]}{2E \cdot w \cdot (2l \cdot m + t)^2 \cdot m^3}; \\[2em]
C_{u_y-m_z} = -\dfrac{6l^2}{E \cdot w \cdot t \cdot (2l \cdot m + t)^2}; \\[2em]
C_{\theta_z-m_z} = \dfrac{12l \cdot (l \cdot m + t)}{E \cdot w \cdot t^2 \cdot (2l \cdot m + t)^2}
\end{cases}
\tag{A2.1}
$$

$$
\begin{cases}
C_{u_z-f_z} = \dfrac{3 \cdot \left[t^2 \cdot \ln\left(1 + \dfrac{2l \cdot m}{t}\right) + 2l \cdot m \cdot (l \cdot m - t)\right]}{2E \cdot w^3 \cdot m^3}; \\[2em]
C_{u_z-m_y} = \dfrac{3 \cdot \left[2l \cdot m - t \cdot \ln\left(1 + \dfrac{2l \cdot m}{t}\right)\right]}{E \cdot w^3 \cdot m^2}; \\[2em]
C_{\theta_y-m_y} = \dfrac{6 \cdot \ln\left(1 + \dfrac{2l \cdot m}{t}\right)}{E \cdot w^3 \cdot m}
\end{cases}
\tag{A2.2}
$$

A2.1.1.2   Right Circular Segment

$$
\begin{cases}
C_{u_x - f_x} = \dfrac{1}{Ew} \cdot \left[\dfrac{(2r+t)}{\sqrt{t \cdot (4r+t)}} \cdot \tan^{-1} \sqrt{1 + \dfrac{4r}{t}} - \dfrac{\pi}{4} \right]; \\[4mm]
C_{u_y - f_y} = \dfrac{3}{4Ewt \cdot (4r+t)} \cdot \left[\dfrac{8r^3 + 8(\pi+2)r^2 t + 2(3\pi+2)rt^2 + \pi t^3}{2r+t} \right. \\[4mm]
\qquad\qquad \left. + \dfrac{4(2r+t)\left(2r^2 - 4rt - t^2\right)}{\sqrt{t(4r+t)}} \cdot \tan^{-1} \sqrt{1 + \dfrac{4r}{t}} \right]; \\[4mm]
C_{u_y - m_z} = -\dfrac{6r^2}{Ewt^2 \cdot (2r+t)}; \\[4mm]
C_{\theta_z - m_z} = \dfrac{12r}{Ewt^2 \cdot (4r+t)^2} \cdot \left[\dfrac{6r^2 + 4rt + t^2}{2r+t} + \dfrac{6r(2r+t)}{\sqrt{t(4r+t)}} \cdot \tan^{-1} \sqrt{1 + \dfrac{4r}{t}} \right]
\end{cases}
\tag{A2.3}
$$

$$
\begin{cases}
C_{u_z - f_z} = \dfrac{3}{4E \cdot w^3} \cdot \left[2(4+\pi)r^2 + 4(1+\pi)rt + \pi t^2 \right. \\[4mm]
\qquad\qquad \left. - 4(2r+t)\sqrt{t(4r+t)} \cdot \tan^{-1} \sqrt{1 + \dfrac{4r}{t}} \right] \\[4mm]
C_{u_z - m_y} = \dfrac{3}{E \cdot w^3} \cdot \left[(2r+t) \cdot \ln\left(1 + \dfrac{2r}{t}\right) - 2r \right] \\[4mm]
C_{\theta_y - m_y} = \dfrac{3}{E \cdot w^3} \cdot \left[\dfrac{4(2r+t)}{\sqrt{t(4r+t)}} \cdot \tan^{-1} \sqrt{1 + \dfrac{4r}{t}} - \pi \right]
\end{cases}
\tag{A2.4}
$$

A2.1.1.3   Right Elliptic Segment

$$
\begin{cases}
C_{u_x - f_x} = \dfrac{a}{Ewb} \cdot \left[\dfrac{2b+t}{\sqrt{t(4b+t)}} \cdot \tan^{-1} \sqrt{1 + \dfrac{4b}{t}} - \dfrac{\pi}{4} \right]; \\[4mm]
C_{u_y - f_y} = \dfrac{3a^3}{4Ewb^3 t \cdot (4b+t)} \cdot \left[\dfrac{8b^3 + 8(2+\pi)b^2 t + 2(2+3\pi)bt^2 + \pi t^3}{2b+t} \right. \\[4mm]
\qquad\qquad \left. + \dfrac{4(2b+t)\left(2b^2 - 4bt - t^2\right)}{\sqrt{t(4b+t)}} \tan^{-1} \sqrt{1 + \dfrac{4b}{t}} \right]; \\[4mm]
C_{u_y - m_z} = -\dfrac{6a^2}{Ewt^2 \cdot (2b+t)}; \\[4mm]
C_{\theta_z - m_z} = \dfrac{12a}{Ewt^2 \cdot (4b+t)^2} \cdot \left[\dfrac{6b^2 + 4bt + t^2}{2b+t} + \dfrac{6b(2b+t)}{\sqrt{t(4b+t)}} \cdot \tan^{-1} \sqrt{1 + \dfrac{4b}{t}} \right]
\end{cases}
\tag{A2.5}
$$

$$\begin{cases} C_{u_z-f_z} = \dfrac{3a^3}{4E \cdot w^3 \cdot b^3} \cdot \Bigg[2(4+\pi)b^2 + 4(1+\pi)bt + \pi t^2 \\[2em] \qquad\qquad - 4(2b+t)\sqrt{t(4b+t)} \cdot \tan^{-1}\sqrt{1+\dfrac{4b}{t}} \Bigg]; \\[3em] C_{u_z-m_y} = \dfrac{3a^2 \cdot}{E \cdot w^3 \cdot b^2} \cdot \Bigg[(2b+t)\cdot \ln\left(1+\dfrac{2b}{t}\right) - 2b \Bigg]; \\[3em] C_{\theta_y-m_y} = \dfrac{3a}{E \cdot w^3 \cdot b} \cdot \Bigg[\dfrac{4(2b+t)}{\sqrt{t(4b+t)}} \cdot \tan^{-1}\sqrt{1+\dfrac{4b}{t}} - \pi \Bigg] \end{cases} \tag{A2.6}$$

A2.1.1.4 Hyperbolic Segment

$$\begin{cases} C_{u_x-f_x} = \dfrac{l}{2Ew\sqrt{c(c+t)}} \cdot \ln\Bigg[1+\dfrac{2\left(\sqrt{c(c+t)}+c\right)}{t} \Bigg]; \\[3em] C_{u_y-f_y} = \dfrac{3l^3}{2Ew} \cdot \Bigg[\dfrac{1}{\sqrt{c^3(c+t)^3}} \cdot \ln\left(1+\dfrac{2\left(\sqrt{c(c+t)}+c\right)}{t}\right) - \dfrac{2}{c(c+t)(2c+t)} \Bigg]; \\[3em] C_{u_y-m_z} = -\dfrac{6l^2}{Ewt(c+t)(2c+t)}; \\[3em] C_{\theta_z-m_z} = \dfrac{12l}{Ewt^2(2c+t)} \end{cases}$$

$$\tag{A2.7}$$

$$\begin{cases} C_{u_z-f_z} = \dfrac{3l^3}{4Ew^3c(c+t)} \cdot \Bigg\{ 2(2c+t) - \dfrac{t^2}{\sqrt{c(c+t)}} \cdot \ln\Bigg[1+\dfrac{2\left(\sqrt{c(c+t)}+c\right)}{t} \Bigg] \Bigg\}; \\[3em] C_{u_z-m_y} = \dfrac{6l^2}{Ew^3(c+t)}; \\[3em] C_{\theta_y-m_y} = \dfrac{6l}{Ew^3\sqrt{c(c+t)}} \cdot \ln\Bigg[1+\dfrac{2\left(\sqrt{c(c+t)}+c\right)}{t} \Bigg] \end{cases}$$

$$\tag{A2.8}$$

A2.1.1.5 Parabolic Segment

$$\begin{cases}
C_{u_x-f_x} = \dfrac{l}{\sqrt{2}Ew\sqrt{c\cdot t}} \cdot \tan^{-1}\sqrt{2c/t}; \\[3mm]
C_{u_y-f_y} = \dfrac{3\cdot l^3}{8Ew\cdot c\cdot t}\left[\dfrac{2(2c-t)}{(2c+t)^2} + \dfrac{\sqrt{2}}{\sqrt{c\cdot t}}\cdot\tan^{-1}\sqrt{2c/t}\right]; \\[3mm]
C_{u_y-m_z} = -\dfrac{6l^2(c+t)}{Ewt^2(2c+t)^2}; \\[3mm]
C_{\theta_z-m_z} = \dfrac{3l}{4Ewt^2}\cdot\left[\dfrac{2(6c+5t)}{(2c+t)^2} + \dfrac{3\sqrt{2}}{\sqrt{c\cdot t}}\cdot\tan^{-1}\sqrt{2c/t}\right]
\end{cases}$$

(A2.9)

$$\begin{cases}
C_{u_z-f_z} = \dfrac{3l^3}{Ew^3c}\cdot\left(2 - \sqrt{\dfrac{2t}{c}}\cdot\tan^{-1}\sqrt{2c/t}\right); \\[3mm]
C_{u_z-m_y} = \dfrac{3l^2}{Ew^3c}\cdot\ln\left(1+\dfrac{2c}{t}\right); \\[3mm]
C_{\theta_y-m_y} = \dfrac{6\sqrt{2}l}{Ew^3\sqrt{ct}}\cdot\tan^{-1}\sqrt{2c/t}
\end{cases}$$

(A2.10)

A2.1.1.6 Inverse Parabolic Segment

$$\begin{cases}
C_{u_x-f_x} = \dfrac{l\cdot(4c+3t)}{3Ewt(2c+t)}; \\[3mm]
C_{u_y-f_y} = \dfrac{4l^3\cdot\left(128c^3+288c^2\cdot t+252c\cdot t^2+105t^3\right)}{105\cdot E\cdot w\cdot t^3\cdot(2c+t)^3}; \\[3mm]
C_{u_y-m_z} = -\dfrac{6l^2\cdot(c+t)\cdot\left(2c^2+2c\cdot t+t^2\right)}{E\cdot w\cdot t^3(2c+t)^3}; \\[3mm]
C_{\theta_z-m_z} = \dfrac{12l\cdot\left(128c^3+224c^2\cdot t+140c\cdot t^2+35t^3\right)}{35E\cdot w\cdot t^3(2c+t)^3}
\end{cases}$$

(A2.11)

$$\begin{cases}
C_{u_z-f_z} = \dfrac{4l^3\cdot(4c+5t)}{5E\cdot w^3\cdot t\cdot(2c+t)}; \\[3mm]
C_{u_z-m_y} = \dfrac{6l^2\cdot(c+t)}{E\cdot w^3\cdot t\cdot(2c+t)}; \\[3mm]
C_{\theta_y-m_y} = \dfrac{4l\cdot(4c+3t)}{E\cdot w^3\cdot t\cdot(2c+t)}.
\end{cases}$$

(A2.12)

A2.1.1.7   Quadratic Polynomial Bézier Segment

$$
\left\{
\begin{aligned}
C_{u_x-f_x} &= \frac{l}{2E \cdot w \cdot c} \cdot \left[2\sqrt{\frac{2c}{t}} \cdot \tan^{-1}\sqrt{\frac{2c}{t}} - \ln\left(1 + \frac{2c}{t}\right) \right]; \\[2ex]
C_{u_y-f_y} &= \frac{3 \cdot l^3 \cdot \left\{ (2c+t) \cdot \left[\sqrt{2c} \cdot (8c+15t) \cdot \tan^{-1}\sqrt{\frac{2c}{t}} - 4\sqrt{t^3} \cdot \ln\left(1+\frac{2c}{t}\right) \right] - 2c\sqrt{t} \cdot (24c+11t) \right\}}{8E \cdot w \cdot c^3 \cdot \sqrt{t^3} \cdot (2c+t)}; \\[2ex]
C_{u_y-m_z} &= -\frac{3l^2 \cdot \left[2\sqrt{c} \cdot (8c+3t) - 3\sqrt{2t} \cdot (2c+t) \cdot \tan^{-1}\sqrt{\frac{2c}{t}} \right]}{4E \cdot w \cdot t^2 \cdot \sqrt{c^3} \cdot (2c+t)}; \\[2ex]
C_{\theta_z-m_z} &= \frac{3l \cdot \left[\frac{2\sqrt{t}}{2c+t} + 3\sqrt{\frac{2}{c}} \cdot \tan^{-1}\sqrt{\frac{2c}{t}} \right]}{2E \cdot w \cdot \sqrt{t^5}} \, .
\end{aligned}
\right.
\tag{A2.13}
$$

$$
\left\{
\begin{aligned}
C_{u_z-f_z} &= \frac{l^3 \cdot \left[2c \cdot (17c-27t) + 6\sqrt{2c \cdot t} \cdot (5t-8c) \cdot \tan^{-1}\sqrt{\frac{2c}{t}} + 3t \cdot (16c-t) \cdot \ln\left(1+\frac{2c}{t}\right) \right]}{2E \cdot w^3 \cdot c^3}; \\[2ex]
C_{u_z-m_y} &= \frac{3l^2 \cdot \left[6\sqrt{2c \cdot t} \cdot \tan^{-1}\sqrt{\frac{2c}{t}} + (4c-t) \cdot \ln\left(1+\frac{2c}{t}\right) - 10c \right]}{E \cdot w^3 \cdot c^2}; \\[2ex]
C_{\theta_y-m_y} &= \frac{6l \cdot \left[2\sqrt{\frac{2c}{t}} \cdot \tan^{-1}\sqrt{\frac{2c}{t}} - \ln\left(1+\frac{2c}{t}\right) \right]}{E \cdot w^3 \cdot c}
\end{aligned}
\right.
$$

$$\tag{A2.14}$$

A2.1.2   Circular Cross-Section

For all configurations: $C_{u_y-m_z} = -C_{u_z-m_y}$.

A2.1.2.1   Linearly Varying-Diameter Segment

$$
C_{u_x-f_x} = \frac{4l}{\pi E d (d+2lm)}; \quad C_{u_y-f_y} = \frac{64l^3}{3\pi E d (d+2lm)^3};
$$

$$
C_{u_y-m_z} = -\frac{32l^2 (3d+2lm)}{3\pi E d^2 (d+2lm)^3}; \quad C_{\theta_z-m_z} = \frac{64l (3d^2 + 6dlm + 4l^2 m^2)}{3\pi E d^3 (d+2lm)^3};
$$

$$\tag{A2.15}$$

A2.1.2.2 Right Circular Segment

$$
\left\{
\begin{aligned}
C_{u_x-f_x} &= \frac{4r}{\pi E d(d+4r)}\left[1+\frac{4r}{\sqrt{d(4r+d)}}\cdot\tan^{-1}\sqrt{1+\frac{4r}{d}}\right]; \\[2mm]
C_{u_y-f_y} &= \frac{64r^3}{3\pi E(2r+d)^2\sqrt{d^5(4r+d)^5}}\left[\sqrt{d(4r+d)}\left(d^2+4dr+6r^2\right)\right. \\[2mm]
&\quad\left. +6r(2r+d)\cdot\tan^{-1}\sqrt{1+\frac{4r}{d}}\right]; \quad C_{u_y-m_z}=-\frac{32r^2(4r+3d)}{3\pi E d^3(2r+d)^2}; \\[2mm]
C_{\theta_z-m_z} &= \frac{64r}{3\pi E(2r+d)^2\sqrt{d^7(4r+d)^7}}\cdot \\[2mm]
&\quad\left[\begin{aligned}
&\sqrt{d(4r+d)}\left(3d^4+24d^3r+92d^2r^2+176dr^3+120r^4\right) \\
&+24r(2r+d)^2\left(d^2+4dr+5r^2\right)\cdot\tan^{-1}\sqrt{1+\frac{4r}{t}}
\end{aligned}\right]
\end{aligned}
\right.
\qquad . \text{ (A2.16)}
$$

A2.1.2.3 Right Elliptical Segment

$$
\left\{
\begin{aligned}
C_{u_x-f_x} &= \frac{4a}{\pi E d(d+4b)}\cdot\left[1+\frac{4b}{\sqrt{d(4b+d)}}\cdot\tan^{-1}\sqrt{1+\frac{4b}{d}}\right]; \\[2mm]
C_{u_y-f_y} &= \frac{64a^3}{3\pi E(2b+d)^2\sqrt{d^5(4b+d)^5}}\cdot\left[\sqrt{d(4b+d)}\left(d^2+4db+6b^2\right)\right. \\[2mm]
&\quad\left. +6b(2b+d)\cdot\tan^{-1}\sqrt{1+\frac{4b}{d}}\right]; \\[2mm]
C_{u_y-m_z} &= -\frac{32a^2(4b+3d)}{3\pi E d^3(2b+d)^2}; \\[2mm]
C_{\theta_z-m_z} &= \frac{64a}{3\pi E(2b+d)^2\sqrt{d^7(4b+d)^7}}\cdot \\[2mm]
&\quad\left[\begin{aligned}
&\sqrt{d(4b+d)}\left(3d^4+24d^3b+92d^2b^2+176db^3+120b^4\right) \\
&+24b(2b+d)^2\left(d^2+4db+5b^2\right)\cdot\tan^{-1}\sqrt{1+\frac{4b}{t}}
\end{aligned}\right]
\end{aligned}
\right.
\qquad . \text{ (A2.17)}
$$

A2.1.2.4   Hyperbolic Segment

$$
\left\{
\begin{aligned}
C_{u_x-f_x} &= \frac{2l}{\pi E d \sqrt{c(c+d)}} \cdot \tan^{-1}\!\left(\frac{2\sqrt{c(c+d)}}{d}\right); \\[2mm]
C_{u_y-f_y} &= \frac{4l^3}{\pi E d (2c+d)^2 \sqrt{c^3(c+d)^3}} \cdot \left[(2c+d)^2 \tan^{-1}\!\left(\frac{2\sqrt{c(c+d)}}{d}\right) - 2d\sqrt{c(c+d)}\right]; \\[2mm]
C_{u_y-m_z} &= -\frac{32l^2}{\pi E d^2 (2c+d)^2}; \\[2mm]
C_{\theta_z-m_z} &= \frac{16l}{\pi E d^3} \cdot \left[\frac{2d}{(2c+d)^2} + \frac{1}{\sqrt{c(c+d)}} \cdot \tan^{-1}\!\left(\frac{2\sqrt{c(c+d)}}{d}\right)\right]
\end{aligned}
\right.
\tag{A2.18}
$$

A2.1.2.5   Parabolic Segment

$$
\left\{
\begin{aligned}
C_{u_x-f_x} &= \frac{l}{\pi E d} \cdot \left[\frac{2}{2c+d} + \sqrt{\frac{2}{cd}} \cdot \tan^{-1}\!\left(\frac{2c}{d}\right)\right]; \\[2mm]
C_{u_y-f_y} &= \frac{l^3}{3\pi E \sqrt{c^3 d^5}} \cdot \left[\frac{2\sqrt{cd}\,(6c-d)(2c+3d)}{(2c+3d)^3} + 3\sqrt{2}\,\tan^{-1}\!\left(\frac{2c}{d}\right)\right]; \\[2mm]
C_{u_y-m_z} &= -\frac{32l^2\left(4c^2+6cd+3d^2\right)^2}{\pi E d^3 (2c+d)^3}; \\[2mm]
C_{\theta_z-m_z} &= \frac{2l}{3\pi E \sqrt{d^7}} \cdot \left[\frac{2\sqrt{d}\,(60c^2+80cd+33d^2)}{(2c+d)^3} + 15\sqrt{\frac{2}{c}} \cdot \tan^{-1}\!\left(\frac{2c}{d}\right)\right]
\end{aligned}
\right.
\tag{A2.19}
$$

A2.1.2.6   Inverse-Parabolic Segment

$$
\left\{
\begin{aligned}
C_{u_x-f_x} &= \frac{4l\left(32c^2+40cd+15d^2\right)}{15\pi E d^2 (2c+d)^2}; \\[2mm]
C_{u_y-f_y} &= \frac{64l^3\left(2048c^4+5632c^3d+6336c^2d^2+3696cd^3+1155d^4\right)}{3465\pi E d^4 (2c+d)^4}; \\[2mm]
C_{u_y-m_z} &= -\frac{32l^2\left(16c^4+40c^3d+40c^2d^2+20cd^3+5d^4\right)}{5\pi E d^4 (2c+d)^4}; \\[2mm]
C_{\theta_z-m_z} &= \frac{64l\left(2048c^4+4608c^3d+4032c^2d^2+1680cd^3+315d^4\right)}{315\pi E d^4 (2c+d)^4}
\end{aligned}
\right.
\tag{A2.20}
$$

A2.1.2.7　Quadratic Polynomial Bézier Segment

$$
\left\{
\begin{aligned}
C_{u_x-f_x} &= \frac{2\sqrt{2}l}{\pi E\sqrt{cd^3}} \cdot \tan^{-1}\sqrt{\frac{2c}{d}}; \\[2mm]
C_{u_y-f_y} &= \frac{(8c+5d)l^3}{3\pi E(2c+d)^2\sqrt{c^5d^5}} \cdot \left[3\sqrt{2}(2c+d)^2\cdot\tan^{-1}\sqrt{\frac{2c}{d}} - 2\sqrt{cd}(10c+3d)\right]; \\[2mm]
C_{u_y-m_z} &= -\frac{2l^2}{3\pi E(2c+d)^2\sqrt{c^3d^3}} \cdot \left[2\sqrt{c}\left(64c^2+54cd+9d^2\right)-9\sqrt{2d}(2c+d)^2\tan^{-1}\sqrt{\frac{2c}{d}}\right]; \\[2mm]
C_{\theta_z-m_z} &= \frac{4l}{3\pi E\sqrt{d^7}} \cdot \left[\frac{2\sqrt{d}(14c+9d)}{(2c+d)^2}+15\sqrt{\frac{2}{c}}\tan^{-1}\sqrt{\frac{2c}{d}}\right]
\end{aligned}
\right.
$$

$$(A2.21)$$

A2.2　Curvilinear-Axis Segments of Constant Cross-Section

A2.2.1　Circular-Axis Segment

$$
C_{u_x-f_x}^{(b)} = \frac{R^3\cdot\left[6\alpha-8\cdot\sin\alpha+\sin(2\alpha)\right]}{4EI_z}; \quad C_{u_x-f_x}^{(n)} = \frac{R\cdot\left[\alpha+\dfrac{\sin(2\alpha)}{2}\right]}{2EA};
$$

$$(A2.22)$$

$$
C_{u_x-f_x}^{(s)} = \frac{\alpha_s\cdot R\cdot\left[\alpha-\dfrac{\sin(2\alpha)}{2}\right]}{2GA}
$$

$$
C_{u_x-f_y}^{(b)} = -\frac{2R^3\cdot\sin^4(\alpha/2)}{E\cdot I_z}; \quad C_{u_x-f_y}^{(n)} = \frac{R\cdot\sin^2\alpha}{2E\cdot A} \tag{A2.23}
$$

$$
C_{u_x-m_z} = C_{u_x-m_z}^{(b)} = \frac{R^2\cdot(\alpha-\sin\alpha)}{E\cdot I_z} \tag{A2.24}
$$

$$
C_{u_y-f_y}^{(b)} = \frac{R^3\cdot\left[\alpha-\dfrac{\sin(2\alpha)}{2}\right]}{2E\cdot I_z}; \quad C_{u_y-f_y}^{(n)} = \frac{R\cdot\left[\alpha-\dfrac{\sin(2\alpha)}{2}\right]}{2E\cdot A};
$$

$$(A2.25)$$

$$
C_{u_y-f_y}^{(s)} = \frac{\alpha_s\cdot R\cdot\left[\alpha+\dfrac{\sin(2\alpha)}{2}\right]}{2G\cdot A}
$$

$$
C_{u_y-m_z} = C_{u_y-m_z}^{(b)} = -\frac{R^2\cdot(1-\cos\alpha)}{E\cdot I_z}; \quad C_{\theta_z-m_z} = C_{\theta_z-m_z}^{(b)} = \frac{R\cdot\alpha}{E\cdot I_z} \tag{A2.26}
$$

$$C_{\theta_x-m_x}^{(b)} = \frac{R \cdot \left[\alpha - \frac{\sin(2\alpha)}{2}\right]}{2E \cdot I_n}; \quad C_{\theta_x-m_x}^{(t)} = \frac{R \cdot \left[\alpha + \frac{\sin(2\alpha)}{2}\right]}{2G \cdot I_t} \quad (A2.27)$$

$$C_{\theta_x-m_y}^{(b)} = -\frac{R \cdot \sin^2\alpha}{2E \cdot I_n}; \quad C_{\theta_x-m_y}^{(t)} = \frac{R \cdot \sin^2\alpha}{2G \cdot I_t} \quad (A2.28)$$

$$C_{\theta_x-f_z}^{(b)} = -\frac{R^2 \cdot \left[\alpha - \frac{\sin(2\alpha)}{2}\right]}{2E \cdot I_n}; \quad C_{\theta_x-f_z}^{(t)} = -\frac{R^2 \cdot \left[(2-\cos\alpha) \cdot \sin\alpha - \alpha\right]}{2G \cdot I_t} \quad (A2.29)$$

$$C_{\theta_y-m_y}^{(b)} = \frac{R \cdot \left[\alpha + \frac{\sin(2\alpha)}{2}\right]}{2E \cdot I_n}; \quad C_{\theta_y-m_y}^{(t)} = \frac{R \cdot \left[\alpha - \frac{\sin(2\alpha)}{2}\right]}{2G \cdot I_t} \quad (A2.30)$$

$$C_{\theta_y-f_z}^{(b)} = \frac{R^2 \cdot \sin^2\alpha}{2E \cdot I_n}; \quad C_{\theta_y-f_z}^{(t)} = \frac{2R^2 \cdot \sin^4(\alpha/2)}{G \cdot I_t} \quad (A2.31)$$

$$C_{u_z-f_z}^{(b)} = \frac{R^3 \cdot \left[\alpha - \frac{\sin(2\alpha)}{2}\right]}{2E \cdot I_n}; \quad C_{u_z-f_z}^{(t)} = \frac{R^3 \cdot \left[6\alpha - 8 \cdot \sin\alpha + \sin(2\alpha)\right]}{4G \cdot I_t};$$

$$C_{u_x-f_x}^{(s)} = \frac{\alpha_s \cdot R \cdot \alpha}{G \cdot A} \quad . \quad (A2.32)$$

A2.2.2 Parabolic-Axis Segment

$$\begin{cases} C_{u_x-f_x}^{(b)} = \dfrac{3 \cdot \sinh^{-1}(2 \cdot a_p \cdot l_x) + 2 \cdot a_p \cdot l_x \cdot \left(128 \cdot a_p^4 \cdot l_x^4 + 8 \cdot a_p^2 \cdot l_x^2 - 3\right) \cdot \sqrt{1 + 4 \cdot a_p^2 \cdot l_x^2}}{1536 \cdot E \cdot a_p^3 \cdot I_z}; \\[4mm] C_{u_x-f_x}^{(n)} = \dfrac{\sinh^{-1}(2 \cdot a_p \cdot l_x)}{2 \cdot E \cdot a_p \cdot A}; \quad C_{u_x-f_x}^{(s)} = \dfrac{\alpha_s \left[2 \cdot a_p \cdot l_x \cdot \sqrt{1 + 4 \cdot a_p^2 \cdot l_x^2} - \sinh^{-1}(2 \cdot a_p \cdot l_x)\right]}{4 \cdot G \cdot a_p \cdot A} \end{cases}$$

$$(A2.33)$$

$$C_{u_x-f_y}^{(b)} = -\frac{1 + \left(6a_p^2 \cdot l_x^2 - 1\right)\sqrt{\left(1 + 4a_p^2 \cdot l_x^2\right)^3}}{120E \cdot a_p^3 \cdot I_z}; \quad C_{u_x-f_y}^{(n)} = -\frac{1 - \sqrt{1 + 4a_p^2 \cdot l_x^2}}{2E \cdot a_p \cdot A} \quad (A2.34)$$

$$C_{u_x-m_z}^{(b)} = \frac{-\sinh^{-1}(2a_p \cdot l_x) + 2a_p \cdot l_x \cdot \left(8a_p^2 \cdot l_x^2 + 1\right)\sqrt{1 + 4a_p^2 \cdot l_x^2}}{64E \cdot a_p^2 \cdot I_z} \quad (A2.35)$$

$$\begin{cases} C_{u_y-f_y}^{(b)} = \dfrac{-\sinh^{-1}\left(2a_p \cdot l_x\right) + 2a_p \cdot l_x \cdot \left(8a_p^2 \cdot l_x^2 + 1\right)\sqrt{1 + 4a_p^2 \cdot l_x^2}}{64E \cdot a_p^3 \cdot I_z}; \\[4mm] C_{u_y-f_y}^{(n)} = \dfrac{1}{2\alpha_s \cdot \left(1+\mu\right)} \cdot C_{u_x-f_x}^{(s)} \end{cases} \qquad (A2.36)$$

$$C_{u_y-m_z}^{(b)} = \frac{1 - \sqrt{\left(1 + 4 \cdot a_p^2 \cdot l_x^2\right)^3}}{12 \cdot E \cdot a_p^2 \cdot I_z} \qquad (A2.37)$$

$$C_{\theta_z-m_z}^{(b)} = \frac{\sinh^{-1}\left(2 \cdot a_p \cdot l_x\right) + 2 \cdot a_p \cdot l_x \cdot \sqrt{1 + 4 \cdot a_p^2 \cdot l_x^2}}{4 \cdot E \cdot a_p \cdot I_z} \qquad (A2.38)$$

$$C_{\theta_x-m_x}^{(b)} = \frac{A}{2\alpha_s \cdot \left(1+\mu\right) \cdot I_n} \cdot C_{u_x-f_x}^{(s)}; \quad C_{\theta_x-m_x}^{(t)} = 2\alpha_s \cdot \left(1+\mu\right) \cdot \frac{A}{I_t} \cdot C_{u_x-f_x}^{(n)} \qquad (A2.39)$$

$$C_{\theta_x-m_y}^{(b)} = -\frac{A}{I_n} \cdot C_{u_x-f_y}^{(n)}; \quad C_{\theta_x-m_y}^{(t)} = 2\alpha_s \cdot \left(1+\mu\right) \cdot \frac{A}{I_t} \cdot C_{u_x-f_y}^{(n)} \qquad (A2.40)$$

$$\begin{cases} C_{\theta_x-f_z}^{(b)} = \dfrac{5\sinh^{-1}\left(2a_p \cdot l_x\right) - 2a_p \cdot l_x \cdot \left(8a_p^2 \cdot l_x^2 + 5\right)\sqrt{1 + 4a_p^2 \cdot l_x^2}}{64E \cdot a_p^2 \cdot I_n}; \\[4mm] C_{\theta_x-f_z}^{(t)} = \dfrac{\sinh^{-1}\left(2a_p \cdot l_x\right) - 2a_p \cdot l_x \cdot \sqrt{1 + 4a_p^2 \cdot l_x^2}}{16G \cdot a_p^2 \cdot I_t} \end{cases} \qquad (A2.41)$$

$$C_{\theta_y-m_y}^{(b)} = \frac{A}{I_n} \cdot C_{u_x-f_x}^{(n)}; \quad C_{\theta_y-m_y}^{(t)} = \frac{A}{\alpha_s \cdot I_t} \cdot C_{u_x-f_x}^{(s)} \qquad (A2.42)$$

$$C_{\theta_y-f_z}^{(b)} = \frac{\left(1 + a_p^2 \cdot l_x^2\right)\sqrt{1 + 4 \cdot a_p^2 \cdot l_x^2} - 1}{6 \cdot E \cdot a_p^2 \cdot I_n}; \quad C_{\theta_y-f_z}^{(t)} = \frac{1 + \left(2a_p^2 \cdot l_x^2 - 1\right) \cdot \sqrt{1 + 4 \cdot a_p^2 \cdot l_x^2}}{12 \cdot G \cdot a_p^2 \cdot I_t}$$
$$(A2.43)$$

$$\begin{cases} C_{u_z-f_z}^{(b)} = \dfrac{2 \cdot a_p \cdot l_x \cdot \left(16 \cdot a_p^2 \cdot l_x^2 + 13\right) \cdot \left(8 \cdot a_p^2 \cdot l_x^2 + 3\right)\sqrt{1 + 4 \cdot a_p^2 \cdot l_x^2} - 39 \cdot \sinh^{-1}\left(2 \cdot a_p \cdot l_x\right)}{1536 \cdot E \cdot a_p^3 \cdot I_n}; \\[4mm] C_{u_z-f_z}^{(n)} = \dfrac{2 \cdot a_p \cdot l_x \cdot \left(8 \cdot a_p^2 \cdot l_x^2 - 3\right)\sqrt{1 + 4 \cdot a_p^2 \cdot l_x^2} + 3 \cdot \sinh^{-1}\left(2 \cdot a_p \cdot l_x\right)}{256 \cdot G \cdot a_p^3 \cdot I_t}; \\[4mm] C_{u_z-f_z}^{(s)} = 2\alpha_s \cdot \left(1+\mu\right) \cdot \dfrac{I_z}{A} \cdot C_{\theta_z-m_z}^{(b)} \end{cases}$$
$$(A2.44)$$

REFERENCES

1. Timoshenko, S.P., *History of Strength of Materials*, Dover, New York, 1983.
2. Richards, T.H., *Energy Methods in Stress Analysis with an Introduction to Finite Element Techniques*, Ellis Horwood, Chichester, 1977.
3. Muvdi, B.B. and McNabb, J.W., *Engineering Mechanics of Materials*, 2nd ed., Macmillan, New York, 1984.
4. Harker, R.J., *Elastic Energy Methods of Design Analysis*, Chapman & Hall, London, 1986.
5. Langhaar, H.L., *Energy Methods in Applied Mechanics*, Krieger, Melbourne, FL, 1989.
6. Cook, R.D. and Young, W.C., *Advanced Mechanics of Materials*, Macmillan, New York, 1985.
7. Den Hartog, J.P., *Strength of Materials*, Dover, New York, 1977.
8. Young, W.C., Budynas, R.G., and Sadegh, A.M., *Roark's Formulas for Stress and Strain*, 8th ed., McGraw-Hill, New York, 2011.
9. Cowper, G.R., The shear coefficient in Timoshenko beam theory, *Journal of Applied Mechanics*, 33 (2), 335, 1966.
10. Rentor, J.D., A note on the form of the shear coefficient, *International Journal of Solids and Structures*, 39 (14), 1681, 1997.
11. Stephen, N.G., Discussion: shear coefficients for Timoshenko beam theory, *Journal of Applied Mechanics*, 68 (6), 959, 2001.
12. Hall, J.E., *Analytic Geometry*, Brooks/Cole Publishing Company, Belmont, CA, 1968.
13. Cohen, E., Riesenfeld, R.F., and Elber, G., *Geometric Modeling with Splines: An Introduction*, AK Peters, Natick, MA, 2001.
14. Sederberg, T.W. and Anderson, D.C., Implicit representation of parametric curves and surfaces, *Computer Vision and Graphical Image Processing*, 9 (6), 52–9, 1986.
15. Bareiss, E.H., Sylvester's identity and multistep integer-preserving Gaussian elimination, *Mathematical Computing*, 22, 565, 1968.
16. Vallance, R.R., Haghighian, B., and Marsh, E.R., A unified geometric model for designing elastic pivots, *Precision Engineering: Journal of the International Societies for Precision Engineering and Nanotechnology*, 32, 278, 2008.
17. Lobontiu, N. and Cullin, M., In-plane elastic response of two-segment circular-axis symmetric notch flexure hinges: the right circular design, *Precision Engineering*, 37, 542, 2013.

3 Compliances of Straight-Axis Flexible Hinges

The straight-axis basic elastic segments introduced in Chapter 2 can be connected in series to form flexible hinges of more complex geometry and functional versatility, as discussed in this chapter. Hinges resulting from connecting two or more individual flexible hinges or segments in parallel are studied here as well. This chapter also analyzes the folded, spatially periodic designs that use several straight-axis, non-collinear and serially connected hinge segments. In-plane and out-of-plane compliance matrices are formulated for all the final hinge designs by using matrix translation, addition, rotation, symmetry and mirroring. Finite element analysis is used to confirm the analytical compliances for several hinge configurations. The analytical compliances are utilized to compare the elastic deformation capabilities of various hinge designs.

3.1 COMPLIANCE MATRIX TRANSFORMATIONS

Various straight-axis hinge segments, such as those analyzed in Chapter 2, can be combined in series or in parallel to form more complex segments to be used as building blocks for subsequent final-design hinges (through more serial or parallel connections) or to result directly in final hinge configurations. To obtain the in-plane and out-of-plane compliance matrices of the final straight-axis flexible hinges in a global reference frame (placed at one end of the hinge), the segment compliance matrices, which are defined in their local (native) reference frames, need to be transferred to the global frame by using matrix operations, such as translation, addition, rotation and mirroring.

3.1.1 COMPLIANCE MATRIX TRANSLATION

Flexible hinges can be formed by connecting two or more basic (primitive) straight-axis hinge segments in either series or parallel, as sketched in Figure 3.1a; note that the basic segments of the functionally parallel connection are also placed with their longitudinal axes in parallel, but that is not the general configuration, as seen later in this chapter. In both configurations, the local reference frame $O_i x_i y_i z_i$ of segment i and the global reference frame $Oxyz$ are identified. It is assumed that the in-plane and out-of-plane compliance matrices of a generic segment i are specified in its local frame. Known is also the position of segment i with respect to the global reference frame. Calculating the in-plane and out-of-plane compliances of serial

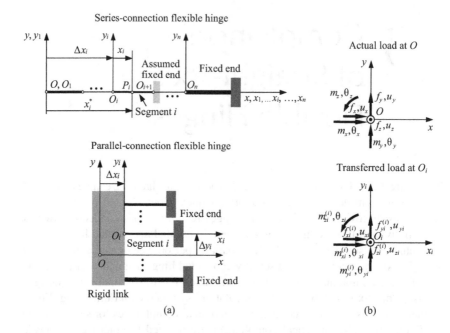

FIGURE 3.1 (a) Flexible hinge with segments connected in series or in parallel; (b) actual loads/displacements at the global-frame origin O and loads/displacements transferred at the local-frame origin O_i of flexible-hinge segment i.

flexible hinges like the one of Figure 3.1a involves transferring the local-frame hinge segment compliance matrices into the global frame, which is mathematically achieved through matrix translation.

This section derives the relationships for translating the in-plane and out-of-plane compliance matrices of hinge segments between reference frames with parallel axes.

3.1.1.1 Series Connection

3.1.1.1.1 In-Plane Compliance Matrix

The in-plane loads f_x, f_y and m_z are statically transferred from their application point O to the point O_i – the origin of flexible segment i in Figure 3.1b – as f_{xi}, f_{yi} and m_{zi}:

$$\begin{cases} f_{xi} = f_x; \\ f_{yi} = f_y; \\ m_{zi} = m_z - f_y \cdot \Delta x_i \end{cases} \quad \text{or} \quad \begin{bmatrix} f_{xi} \\ f_{yi} \\ m_{zi} \end{bmatrix} = \begin{bmatrix} 1 & 0 & 0 \\ 0 & 1 & 0 \\ 0 & -\Delta x_i & 1 \end{bmatrix} \cdot \begin{bmatrix} f_x \\ f_y \\ m_z \end{bmatrix}. \quad (3.1)$$

Equation (3.1) is written in a condensed form as:

$$\left[f_{O_i,ip}^{(i)} \right] = \left[T_{OO_i,ip}^{(i)} \right]\left[f_{O,ip}^{(i)} \right] \quad \text{with:}$$

$$\left[f_{O_i,ip}^{(i)} \right] = \begin{bmatrix} f_{xi} & f_{yi} & m_{zi} \end{bmatrix}^T$$

$$\left[f_{O,ip}^{(i)} \right] = \left[f_{O,ip} \right] = \begin{bmatrix} f_x & f_y & m_z \end{bmatrix}^T, \qquad (3.2)$$

$$\left[T_{OO_i,ip}^{(i)} \right] = \begin{bmatrix} 1 & 0 & 0 \\ 0 & 1 & 0 \\ 0 & -\Delta x_i & 1 \end{bmatrix}.$$

where the subscript "*ip*" stands for in-plane and $\left[T_{OO_i,ip}^{(i)} \right]$ is the in-plane *transla-tion matrix* that relocates the load vector from O to O_i. Equation (3.2) also enables the load transformation in the reversed manner as:

$$\left[f_{O,ip}^{(i)} \right] = \left[T_{OO_i,ip}^{(i)} \right]^{-1} \cdot \left[f_{O_i,ip}^{(i)} \right]. \qquad (3.3)$$

In the local frame $O_i x_i y_i z_i$, the load vector and the displacement vector corresponding to the flexible segment i are related by means of the native in-plane compliance matrix as:

$$\left[u_{O_i,ip}^{(i)} \right] = \left[C_{O_i,ip}^{(i)} \right] \cdot \left[f_{O_i,ip} \right] \quad \text{with} \quad \left[u_{O_i,ip}^{(i)} \right] = \begin{bmatrix} u_{xi} & u_{yi} & \theta_{zi} \end{bmatrix}^T. \qquad (3.4)$$

The displacement vector $\left[u_{O_i,ip}^{(i)} \right]$ can be statically transformed into $\left[u_{O,ip}^{(i)} \right]$ by transferring it from O_i to O as:

$$\begin{cases} u_x = u_{xi}; \\ u_y = u_{yi} - \theta_{zi} \cdot \Delta x_i; \quad \text{or} \\ \theta_z = \theta_{zi} \end{cases} \begin{bmatrix} u_x \\ u_y \\ \theta_z \end{bmatrix} = \begin{bmatrix} 1 & 0 & 0 \\ 0 & 1 & -\Delta x_i \\ 0 & 0 & 1 \end{bmatrix} \cdot \begin{bmatrix} u_{xi} \\ u_{yi} \\ \theta_{zi} \end{bmatrix}. \qquad (3.5)$$

which can be written as:

$$\left[u_{O,ip}^{(i)} \right] = \left[T_{OO_i,ip}^{(i)} \right]^T \cdot \left[u_{O_i,ip}^{(i)} \right] \quad \text{or} \quad \left[u_{O_i,ip}^{(i)} \right] = \left(\left[T_{OO_i,ip}^{(i)} \right]^T \right)^{-1} \cdot \left[u_{O,ip}^{(i)} \right]. \qquad (3.6)$$

Combining now Eqs. (3.2), (3.4) and (3.6) yields:

$$\left[u_{O,ip}^{(i)}\right]=\left[C_{O,ip}^{(i)}\right]\cdot\left[f_{O,ip}^{(i)}\right] \quad \text{with} \quad \left[C_{O,ip}^{(i)}\right]=\left[T_{OO_i,ip}^{(i)}\right]^{T}\cdot\left[C_{O_i,ip}^{(i)}\right]\cdot\left[T_{OO_i,ip}^{(i)}\right]. \quad (3.7)$$

Equation (3.7) enables transferring (translating) the native in-plane compliance matrix $\left[C_{O_i,ip}^{(i)}\right]$ from O_i to O by transforming it into $\left[C_{O,ip}^{(i)}\right]$.

3.1.1.1.2   Out-of-Plane Compliance Matrix

A procedure similar to the one applied to determine the translated in-plane compliance matrix of a hinge segment is used to evaluate the out-of-plane compliance matrix of the same segment. The out-of-plane loads m_x, m_y and f_z that are applied at the free end of the hinge O in Figure 3.1a are transferred to point O_i – the origin of flexible segment i – as m_{xi}, m_{yi} and f_{zi} in Figure 3.1b, namely:

$$\begin{cases} m_{xi} = m_x; \\ m_{yi} = m_y + f_z \cdot \Delta x_i; \quad \text{or} \\ f_{zi} = f_z \end{cases} \quad \begin{bmatrix} m_{xi} \\ m_{yi} \\ f_{zi} \end{bmatrix} = \begin{bmatrix} 1 & 0 & 0 \\ 0 & 1 & \Delta x_i \\ 0 & 0 & 1 \end{bmatrix} \cdot \begin{bmatrix} m_x \\ m_y \\ f_z \end{bmatrix}. \quad (3.8)$$

Equation (3.8) can also be formulated as:

$$\left[f_{O_i,op}^{(i)}\right]=\left[T_{OO_i,op}^{(i)}\right]\left[f_{O,op}^{(i)}\right] \quad \text{with}:$$

$$\left[f_{O_i,op}^{(i)}\right]=\begin{bmatrix} m_{xi} & m_{yi} & f_{zi} \end{bmatrix}^{T}$$

$$\left[f_{O,op}^{(i)}\right]=\left[f_{O,op}\right]=\begin{bmatrix} m_x & m_y & f_z \end{bmatrix}^{T} \quad (3.9)$$

$$\left[T_{OO_i,op}^{(i)}\right]=\begin{bmatrix} 1 & 0 & 0 \\ 0 & 1 & \Delta x_i \\ 0 & 0 & 1 \end{bmatrix},$$

where "op" means out-of-plane and $\left[T_{OO_i,op}^{(i)}\right]$ is the out-of-plane translation matrix transferring the force vector from O to O_i. The load transformation of Eq. (3.9) can also be expressed as:

$$\left[f_{O,op}^{(i)}\right]=\left[T_{OO_i,op}^{(i)}\right]^{-1}\cdot\left[f_{O_i,op}^{(i)}\right]. \quad (3.10)$$

The load and displacement vectors applied at point O_i of the flexible segment i are connected in the local reference frame by means of the native out-of-plane compliance matrix as:

$$\left[u_{O_i,op}^{(i)}\right]=\left[C_{O_i,op}^{(i)}\right]\cdot\left[f_{O_i,op}^{(i)}\right] \quad \text{with} \quad \left[u_{O_i,op}^{(i)}\right]=\left[\begin{array}{ccc} \theta_{xi} & \theta_{yi} & u_{zi} \end{array}\right]^T. \quad (3.11)$$

The displacement vector $\left[u_{O_i,op}^{(i)}\right]$ changes into $\left[u_{O,op}^{(i)}\right]$ through static transfer from O_i to O by means of the transformation:

$$\begin{cases} \theta_x = \theta_{xi}; \\ \theta_y = \theta_{yi}; \\ u_z = u_{zi} + \Delta x_i \cdot \theta_{yi} \end{cases} \quad \text{or} \quad \left[\begin{array}{c} \theta_x \\ \theta_y \\ u_z \end{array}\right] = \left[\begin{array}{ccc} 1 & 0 & 0 \\ 0 & 1 & 0 \\ 0 & \Delta x_i & 1 \end{array}\right] \cdot \left[\begin{array}{c} \theta_{xi} \\ \theta_{yi} \\ u_{zi} \end{array}\right]. \quad (3.12)$$

which is concisely written as:

$$\left[u_{O,op}^{(i)}\right]=\left[T_{OO_i,op}^{(i)}\right]^T\cdot\left[u_{O_i,op}^{(i)}\right] \quad \text{or} \quad \left[u_{O_i,op}^{(i)}\right]=\left(\left[T_{OO_i,op}^{(i)}\right]^T\right)^{-1}\cdot\left[u_{O,op}^{(i)}\right]. \quad (3.13)$$

Combining now Eqs. (3.9), (3.11) and (3.13) yields:

$$\left[u_{O,op}^{(i)}\right]=\left[C_{O,op}^{(i)}\right]\cdot\left[f_{O,op}^{(i)}\right] \quad \text{with} \quad \left[C_{O,op}^{(i)}\right]=\left[T_{OO_i,op}^{(i)}\right]^T\cdot\left[C_{O_i,op}^{(i)}\right]\cdot\left[T_{OO_i,op}^{(i)}\right]. \quad (3.14)$$

Equation (3.14) transfers (translates) the native out-of-plane compliance matrix $\left[C_{O_i,op}^{(i)}\right]$ of segment i from O_i to O by transforming it into $\left[C_{O,op}^{(i)}\right]$.

3.1.1.2 Parallel Connection

3.1.1.2.1 In-Plane Compliance Matrix

The end O_i of segment i of the parallel connection illustrated in Figure 3.1a is offset by Δx_i and Δy_i from the global frame $Oxyz$. Therefore, the z-axis moments are connected as shown in the following equation, which also changes the matrix Eq. (3.1) and the corresponding in-plane translation matrix, namely:

$$\begin{cases} f_{xi} = f_x; \\ f_{yi} = f_y; \\ m_{zi} = m_z - f_y \cdot \Delta x_i + f_x \cdot \Delta y_i \end{cases} \quad \text{or} \quad \left[\begin{array}{c} f_{xi} \\ f_{yi} \\ m_{zi} \end{array}\right] = \left[\begin{array}{ccc} 1 & 0 & 0 \\ 0 & 1 & 0 \\ \Delta y_i & -\Delta x_i & 1 \end{array}\right] \cdot \left[\begin{array}{c} f_x \\ f_y \\ m_z \end{array}\right]$$

$$\rightarrow \left[T_{OO_i,ip}^{(i)}\right] = \left[\begin{array}{ccc} 1 & 0 & 0 \\ 0 & 1 & 0 \\ \Delta y_i & -\Delta x_i & 1 \end{array}\right].$$

$$(3.15)$$

Similarly, the displacement relationships of Eqs. (3.5) change to:

$$
\begin{cases}
u_x = u_{xi} + \theta_{zi} \cdot \Delta y_i; \\
u_y = u_{yi} - \theta_{zi} \cdot \Delta x_i; \quad \text{or} \\
\theta_z = \theta_{zi}
\end{cases}
\quad
\begin{bmatrix} u_x \\ u_y \\ \theta_z \end{bmatrix} = \left[T^{(i)}_{OO_i,ip} \right]^T \cdot \begin{bmatrix} u_{xi} \\ u_{yi} \\ \theta_{zi} \end{bmatrix}.
\tag{3.16}
$$

Using the translation matrix of Eq. (3.15), the global-frame, in-plane compliance matrix can be calculated as per Eq. (3.7).

3.1.1.2.2 Out-of-Plane Compliance Matrix

The out-of-plane load vector is translated to O from O_i based on the following relationships, which stem from Eqs. (3.8):

$$
\begin{cases}
m_{xi} = m_x - f_z \cdot \Delta y_i; \\
m_{yi} = m_y + f_z \cdot \Delta x_i; \quad \text{or} \\
f_{zi} = f_z
\end{cases}
\quad
\begin{bmatrix} m_{xi} \\ m_{yi} \\ f_{zi} \end{bmatrix} =
\begin{bmatrix} 1 & 0 & -\Delta y_i \\ 0 & 1 & \Delta x_i \\ 0 & 0 & 1 \end{bmatrix} \cdot
\begin{bmatrix} m_x \\ m_y \\ f_z \end{bmatrix}
\tag{3.17}
$$

$$
\rightarrow \left[T^{(i)}_{OO_i,op} \right] =
\begin{bmatrix} 1 & 0 & -\Delta y_i \\ 0 & 1 & \Delta x_i \\ 0 & 0 & 1 \end{bmatrix}.
$$

Similar to Eqs. (3.12), the displacements are statically translated from O_i to O as:

$$
\begin{cases}
\theta_x = \theta_{xi}; \\
\theta_y = \theta_{yi}; \quad\quad\quad\quad\quad\quad \text{or} \\
u_z = u_{zi} - \Delta y_i \cdot \theta_{xi} + \Delta x_i \cdot \theta_{yi}
\end{cases}
\quad
\begin{bmatrix} \theta_x \\ \theta_y \\ u_z \end{bmatrix} = \left[T^{(i)}_{OO_i,op} \right]^T \cdot \begin{bmatrix} \theta_{xi} \\ \theta_{yi} \\ u_{zi} \end{bmatrix}.
\tag{3.18}
$$

which shows that the global-frame displacements are obtained from the local-frame ones left-multiplied by the transposed translation matrix of Eq. (3.17). The out-of-plane compliance matrix of segment i with respect to the global frame is calculated as in Eq. (3.14) with the translation matrix of Eq. (3.17).

3.1.2 COMPLIANCE MATRIX ROTATION

A basic tool in formulating compliances of hinges that display transverse symmetry is expressing the native compliance matrices of a hinge segment in a reference frame that is rotated with respect to the segment local frame. Figure 3.2 is the skeleton representation of a hinge segment referred to its local frame $Oxyz$ and a rotated frame $Ox_ry_rz_r$.

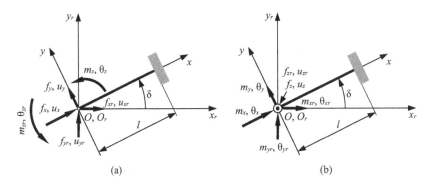

FIGURE 3.2 Flexible-hinge segment in rotated local and global reference frames with: (a) in-plane loads and displacements; (b) out-of-plane loads and displacements.

3.1.2.1 In-Plane Rotation

Assume an in-plane load formed of f_{xr}, f_{yr} and m_{zr} is applied at the free end $O \equiv O_r$ of a hinge segment (as shown in Figure 3.2a). The aim is to find the in-plane compliance matrix that connects this load to the displacements u_{xr}, u_{yr} and θ_{zr} recorded in the rotated frame as:

$$\left[u_{ip}^r \right] = \left[C_{ip}^r \right]\left[f_{ip}^r \right] \quad \text{with:}$$

$$\left[u_{ip}^r \right] = \left[\begin{array}{ccc} u_{xr} & u_{yr} & \theta_{zr} \end{array} \right]^T \qquad (3.19)$$

$$\left[f_{ip}^r \right] = \left[\begin{array}{ccc} f_{xr} & f_{yr} & m_{zr} \end{array} \right]^T .$$

by means of a rotated compliance matrix $\left[C_{ip}^r \right]$.

The in-plane loads f_{xr}, f_{yr} and m_{zr} are projected from their axes to the corresponding axes of the reference frame $Oxyz$, which is rotated in the plane $x_r y_r$ by an angle δ – see Figure 3.2a, so that the loads f_x, f_y and m_z are obtained:

$$\begin{cases} f_x = f_{xr} \cdot \cos\delta + f_{yr} \cdot \sin\delta; \\ f_y = -f_{xr} \cdot \sin\delta + f_{yr} \cdot \cos\delta; \text{ or} \\ m_z = m_{zr} \end{cases} \begin{bmatrix} f_x \\ f_y \\ m_z \end{bmatrix} = \begin{bmatrix} \cos\delta & \sin\delta & 0 \\ -\sin\delta & \cos\delta & 0 \\ 0 & 0 & 1 \end{bmatrix} \cdot \begin{bmatrix} f_{xr} \\ f_{yr} \\ m_{zr} \end{bmatrix}.$$

$$(3.20)$$

Equation (3.20) can be written in the form:

$$[f_{ip}] = [R][f_{ip}^r] \quad \text{with}:$$

$$[f_{ip}] = \begin{bmatrix} f_x & f_y & m_z \end{bmatrix}^T \tag{3.21}$$

$$[R] = \begin{bmatrix} \cos\delta & \sin\delta & 0 \\ -\sin\delta & \cos\delta & 0 \\ 0 & 0 & 1 \end{bmatrix}.$$

where $[R]$ is the *rotation matrix* at $O \equiv O_r$.

In the native frame $Oxyz$, the in-plane load and displacement vectors are related by means of the compliance matrix as:

$$[u_{ip}] = [C_{ip}] \cdot [f_{ip}] \quad \text{with} \quad [u_{ip}] = \begin{bmatrix} u_x & u_y & \theta_z \end{bmatrix}^T. \tag{3.22}$$

The displacements in the $Oxyz$ frame can be projected back to the frame $O_r x_r y_r z_r$ corresponding axes as:

$$\begin{cases} u_{xr} = u_x \cdot \cos\delta - u_y \cdot \sin\delta; \\ u_{yr} = u_x \cdot \sin\delta + u_y \cdot \cos\delta; \quad \text{or} \\ \theta_{zr} = \theta_z \end{cases} \begin{bmatrix} u_{xr} \\ u_{yr} \\ \theta_{zr} \end{bmatrix} = \begin{bmatrix} \cos\delta & -\sin\delta & 0 \\ \sin\delta & \cos\delta & 0 \\ 0 & 0 & 1 \end{bmatrix} \cdot \begin{bmatrix} u_x \\ u_y \\ \theta_z \end{bmatrix},$$

$$\tag{3.23}$$

which is written in matrix form:

$$[u_{ip}^r] = [R]^T \cdot [u_{ip}] \quad \text{or} \quad [u_{ip}] = ([R]^T)^{-1} \cdot [u_{ip}^r]. \tag{3.24}$$

Combining now Eqs. (3.21), (3.22) and (3.24) yields Eq. (3.19) with the rotated in-plane compliance matrix:

$$[C_{ip}^r] = [R]^T \cdot [C_{ip}] \cdot [R]. \tag{3.25}$$

3.1.2.2 Out-of-Plane Rotation

A similar approach is applied to evaluate the compliance matrix relating the out-of-plane loads m_{xr}, m_{yr} and f_{zr} to the displacements θ_{xr}, θ_{yr} and u_{zr} with respect to the rotated frame $O_r x_r y_r$ – see Figure 3.2b:

$$\left[u_{op}^{r} \right] = \left[C_{op}^{r} \right]\left[f_{op}^{r} \right] \quad \text{with :}$$

$$\left[u_{op}^{r} \right] = \left[\begin{array}{ccc} \theta_{xr} & \theta_{yr} & u_{zr} \end{array} \right]^{T} \qquad (3.26)$$

$$\left[f_{op}^{r} \right] = \left[\begin{array}{ccc} m_{xr} & m_{yr} & f_{zr} \end{array} \right]^{T}.$$

The out-of-plane, local-frame loads m_x, m_y, and f_z are expressed in terms of the global-frame loads m_{xr}, m_{yr} and f_{zr} as:

$$\begin{cases} m_x = m_{xr} \cdot \cos\delta + m_{yr} \cdot \sin\delta; \\ m_y = -m_{xr} \cdot \sin\delta + m_{yr} \cdot \cos\delta; \text{ or} \\ f_z = f_{zr} \end{cases} \begin{bmatrix} m_x \\ m_y \\ f_z \end{bmatrix} = \begin{bmatrix} \cos\delta & \sin\delta & 0 \\ -\sin\delta & \cos\delta & 0 \\ 0 & 0 & 1 \end{bmatrix} \cdot \begin{bmatrix} m_{xr} \\ m_{yr} \\ f_{zr} \end{bmatrix},$$

$$(3.27)$$

which can be expressed in the form:

$$\left[f_{op} \right] = [R]\left[f_{op}^{r} \right] \quad \text{with :}$$

$$\qquad (3.28)$$

$$\left[f_{op} \right] = \left[\begin{array}{ccc} m_x & m_y & f_z \end{array} \right]^{T}.$$

by means of the rotation matrix $[R]$ defined in Eq. (3.21).

The out-of-plane load and displacement vectors are connected by the compliance matrix in the native frame $Oxyz$ as:

$$\left[u_{op} \right] = \left[C_{op} \right] \cdot \left[f_{op} \right] \quad \text{with} \quad \left[u_{op} \right] = \left[\begin{array}{ccc} \theta_x & \theta_y & u_z \end{array} \right]^{T}. \qquad (3.29)$$

The displacements in the $O_r x_r y_r z_r$ frame are obtained from the ones in the native $Oxyz$ frame based on the equations:

$$\begin{cases} \theta_{xr} = \theta_x \cdot \cos\delta - \theta_y \cdot \sin\delta; \\ \theta_{yr} = \theta_x \cdot \sin\delta + \theta_y \cdot \cos\delta; \quad \text{or} \\ u_{zr} = u_z \end{cases} \begin{bmatrix} \theta_{xr} \\ \theta_{yr} \\ u_{zr} \end{bmatrix} = \begin{bmatrix} \cos\delta & -\sin\delta & 0 \\ \sin\delta & \cos\delta & 0 \\ 0 & 0 & 1 \end{bmatrix} \cdot \begin{bmatrix} \theta_x \\ \theta_y \\ u_z \end{bmatrix},$$

$$(3.30)$$

which is also:

$$\left[u_{op}^{r} \right] = [R]^{T} \cdot \left[u_{op} \right] \quad \text{or} \quad \left[u_{op} \right] = \left([R]^{T} \right)^{-1} \cdot \left[u_{op}^{r} \right]. \tag{3.31}$$

The rotated out-of-plane compliance matrix defined in Eq. (3.26) results from Eqs. (3.28), (3.29) and (3.31):

$$\left[C_{op}^{r} \right] = [R]^{T} \cdot \left[C_{op} \right] \cdot [R]. \tag{3.32}$$

3.2 SERIES-CONNECTION FLEXIBLE HINGES

Primitive straight-axis elastic segments, such as those presented in Chapter 2, can form standalone, single-profile flexible hinges. An example list (not at all exhaustive), which studies the elastic behavior of flexible hinges by analytical methods, includes work on hinges of (right) circular profile by Paros and Weisbord [1], Lobontiu and Paine [2], Yong and Lu [3], Wu and Zhou [4], Zelenika et al. [5] and Schotborg et al. [6]; conic-section (ellipse, parabola, hyperbola) geometry – as discussed in Smith et al. [7], Smith [8], Lobontiu et al. [9–11] and Chen et al. [12,13]; polynomial configuration – see Linβ et al. [14], Li et al. [15]; or defined by spline (Bézier) curves, as in Vallance et al. [16]. These hinge configurations have either rectangular cross-section with one dimension constant and the other variable, or circular cross-section of variable diameter. A flexible-hinge design of rectangular cross-section with both dimensions variable and consisting of two pairs of collocated notches is presented in Lobontiu and Garcia [17].

To enhance the design space and the elastic response performance, flexible hinges can also be built by serially connecting several primitive segments to result in a wide variety of multiple-profile configurations, of which the corner-filleted family is the most popular – examples can be found in Lobontiu et al. [18], Meng et al. [19], Lobontiu et al. [20], Lin et al. [21] and Wu et al. [22], among others. The V-shaped flexure hinges – see Tian et al. [23], Tseytlin [24] – also belong to the same category of multiple-profile hinges. While the compliance (or, conversely, stiffness) properties of multiple-profile flexible hinges can be evaluated through direct integration, the matrix approach has also been utilized to combine the simpler compliance matrices of primitive segments into the full-hinge compliance matrices by means of matrix operations such as translation, rotation, mirroring or addition – see Lobontiu et al. [25], Lobontiu [26] or [27]. Optimization of the various hinges' geometry (shape) and dimensions – as detailed in De Bona and Munteanu [28], Liu and Shih [29] or Li and Gou [30] – has also been a focal research direction in the area of flexible hinges.

Studied here are flexible hinges that are formed by serially connecting several straight-axis, compliant segments. One configuration category includes designs with similar segments whose axial directions are aligned. These hinges can be without or with transverse symmetry. The other category of hinges comprises designs that are formed of straight-axis segments that are coupled in a folded manner and whose axial directions are not collinear. The *compliance matrix approach* is used, which is similar to the finite element *force* (or *flexibility matrix*) *method*. The compliance matrix modeling procedure, which is applied in this chapter, as well as in the other chapters of this book that study series connection of elastic members, considers that all independent flexible segments/hinges are free-fixed members, which is consistent with the basic assumption used in Chapter 2 to derive segment compliance matrices.

3.2.1 Flexible Hinges without Transverse Symmetry

The majority of straight-axis flexible hinges possess transverse symmetry; however, hinge configurations without transverse symmetry can also be designed as detailed in this section. These hinges can produce deformations that are not obtainable through transversely symmetric hinge designs.

3.2.1.1 Compliance Matrices through Addition and Rotation

The overall compliance matrices of flexible hinges that are generated by serially connecting several primitive segments (whose compliances are known) can be found by adding and rotating the global-frame compliance matrices of all components' basic segments.

3.2.1.1.1 In-Plane Compliance Matrix

Returning to the flexure hinge of Figure 3.1a, which is formed of n serial segments, the aim is to determine its overall in-plane compliance matrix connecting the free-end load and displacement vectors as:

$$\left[u_{O,ip} \right] = \left[C_{O,ip} \right] \cdot \left[f_{O,ip} \right]. \tag{3.33}$$

Because the load-displacement relationships are linear for the assumed small deformations, the displacement vector of Eq. (3.33) is the sum of all displacement vectors pertaining to the individual segments, which results in:

$$\left[u_{O,ip} \right] = \sum_{i=1}^{n} \left[u_{O,ip}^{(i)} \right] = \left(\sum_{i=1}^{n} \left[T_{OO_i,ip}^{(i)} \right]^{T} \cdot \left[C_{O_i,ip}^{(i)} \right] \cdot \left[T_{OO_i,ip}^{(i)} \right] \right) \cdot \left[f_{O,ip} \right], \tag{3.34}$$

where Eq. (3.7) was used. Comparison of Eqs. (3.33) and (3.34) indicates that the overall in-plane compliance matrix is:

$$\left[C_{O,ip}\right] = \sum_{i=1}^{n}\left[C_{O,ip}^{(i)}\right] = \sum_{i=1}^{n}\left[T_{OO_i,ip}^{(i)}\right]^{T}\cdot\left[C_{O_i,ip}^{(i)}\right]\cdot\left[T_{OO_i,ip}^{(i)}\right]$$

$$= \begin{bmatrix} \displaystyle\sum_{i=1}^{n}C_{u_{xi}-f_{xi}}^{(i)} & 0 & 0 \\[4mm] 0 & \begin{aligned}\Bigg(\sum_{i=1}^{n}C_{u_{yi}-f_{yi}}^{(i)} - 2\sum_{i=1}^{n}\Delta x_i\cdot C_{u_{yi}-m_{zi}}^{(i)} \\ + \sum_{i=1}^{n}(\Delta x_i)^2\cdot C_{\theta_{zi}-m_{zi}}^{(i)}\Bigg)\end{aligned} & \displaystyle\sum_{i=1}^{n}C_{u_{yi}-f_{yi}}^{(i)} - \sum_{i=1}^{n}\Delta x_i\cdot C_{\theta_{zi}-m_{zi}}^{(i)} \\[4mm] 0 & \displaystyle\sum_{i=1}^{n}C_{u_{yi}-m_{zi}}^{(i)} - \sum_{i=1}^{n}\Delta x_i\cdot C_{\theta_{zi}-m_{zi}}^{(i)} & \displaystyle\sum_{i=1}^{n}C_{\theta_{zi}-m_{zi}}^{(i)} \end{bmatrix}$$

$$= \begin{bmatrix} C_{u_x-f_x} & 0 & 0 \\ 0 & C_{u_y-f_y} & C_{u_y-m_z} \\ 0 & C_{u_y-m_z} & C_{\theta_z-m_z} \end{bmatrix}, \tag{3.35}$$

where the native compliance matrix of segment i is:

$$\left[C_{O_i,ip}^{(i)}\right] = \begin{bmatrix} C_{u_{xi}-f_{xi}}^{(i)} & 0 & 0 \\ 0 & C_{u_{yi}-f_{yi}}^{(i)} & C_{u_{yi}-m_{zi}}^{(i)} \\ 0 & C_{u_{yi}-m_{zi}}^{(i)} & C_{\theta_{zi}-m_{zi}}^{(i)} \end{bmatrix}. \tag{3.36}$$

Equation (3.35) also used the in-plane translation matrix of Eq. (3.2).

3.2.1.1.2 Out-of-Plane Compliance Matrix

The out-of-plane compliance matrix of the full hinge sketched in Figure 3.1a connects the free-end load and displacement vectors as:

$$\left[u_{O,op}\right] = \left[C_{O,op}\right]\cdot\left[f_{o,op}\right]. \tag{3.37}$$

Due to the linear load-displacement relationship and the assumption of small deformations, the displacement vector of Eq. (3.37) is the sum of the displacement vectors resulting from all the individual segments:

$$\left[u_{O,op}\right] = \sum_{i=1}^{n}\left[u_{O,op}^{(i)}\right] = \left(\sum_{i=1}^{n}\left[T_{OO_i,op}^{(i)}\right]^{T}\cdot\left[C_{O_i,op}^{(i)}\right]\cdot\left[T_{OO_i,op}^{(i)}\right]\right)\cdot\left[f_{o,op}\right], \tag{3.38}$$

where Eq. (3.14) was used. Comparing Eqs. (3.37) and (3.38) shows that the out-of-plane compliance matrix of the full hinge is:

$$\left[C_{O,op}\right]=\sum_{i=1}^{n}\left[C_{O,op}^{(i)}\right]=\sum_{i=1}^{n}\left[T_{OO_i,op}^{(i)}\right]^{T}\cdot\left[C_{O_i,op}^{(i)}\right]\cdot\left[T_{OO_i,op}^{(i)}\right]$$

$$=\begin{bmatrix} \sum_{i=1}^{n}C_{\theta_{xi}-m_{xi}}^{(i)} & 0 & 0 \\ 0 & \sum_{i=1}^{n}C_{\theta_{yi}-m_{yi}}^{(i)} & \sum_{i=1}^{n}\Delta x_i \cdot C_{\theta_{yi}-m_{yi}}^{(i)}+\sum_{i=1}^{n}C_{\theta_{yi}-f_{zi}}^{(i)} \\ 0 & \sum_{i=1}^{n}\Delta x_i \cdot C_{\theta_{yi}-m_{yi}}^{(i)}+\sum_{i=1}^{n}C_{\theta_{yi}-f_{zi}}^{(i)} & \left(\sum_{i=1}^{n}C_{u_{zi}-f_{zi}}^{(i)}+2\sum_{i=1}^{n}\Delta x_i \cdot C_{\theta_{yi}-f_{zi}}^{(i)} + \sum_{i=1}^{n}\left(\Delta x_i\right)^2 \cdot C_{\theta_{yi}-m_{yi}}^{(i)}\right) \end{bmatrix}$$

$$=\begin{bmatrix} C_{\theta_x-m_x} & 0 & 0 \\ 0 & C_{\theta_y-m_y} & C_{\theta_y-f_z} \\ 0 & C_{\theta_y-f_z} & C_{u_z-f_z} \end{bmatrix}. \tag{3.39}$$

with the native out-of-plane compliance matrix of segment i being:

$$\left[C_{O_i,op}^{(i)}\right]=\begin{bmatrix} C_{\theta_{xi}-m_{xi}}^{(i)} & 0 & 0 \\ 0 & C_{\theta_{yi}-m_{yi}}^{(i)} & C_{\theta_{yi}-f_{zi}}^{(i)} \\ 0 & C_{\theta_{yi}-f_{zi}}^{(i)} & C_{u_{zi}-f_{zi}}^{(i)} \end{bmatrix}. \tag{3.40}$$

The expanded-form compliance matrix of Eq. (3.39) utilized the out-of-plane translation matrix of Eq. (3.9).

Example 3.1

Express the in-plane and out-of-plane components of the compliance matrices in the $Oxyz$ reference frame for the flexible hinge sketched in Figure 3.3. The hinge is formed of two serially connected generic segments whose compliances are known.

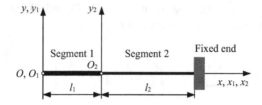

FIGURE 3.3 Skeleton representation of flexible hinge formed of two serially connected segments.

Solution:

Equation (3.35) becomes:

$$\left[C_{O,ip}\right]=\left[C_{O_1,ip}^{(1)}\right]+\left[T_{OO_2,ip}^{(2)}\right]^T\cdot\left[C_{O_2,ip}^{(2)}\right]\cdot\left[T_{OO_2,ip}^{(2)}\right]=\begin{bmatrix} C_{u_x-f_x} & 0 & 0 \\ 0 & C_{u_y-f_y} & C_{u_y-m_z} \\ 0 & C_{u_y-m_z} & C_{\theta_z-m_z} \end{bmatrix},$$

$$(3.41)$$

with $\left[T_{OO_2,ip}^{(2)}\right]$ expressed in Eq. (3.2), where $\Delta x_2 = l_1$. The individual compliances of Eq. (3.41) are:

$$\begin{cases} C_{u_x-f_x} = C_{u_{x1}-f_{x1}}^{(1)} + C_{u_{x2}-f_{x2}}^{(2)} \\ C_{u_y-f_y} = C_{u_{y1}-f_{y1}}^{(1)} + C_{u_{y2}-f_{y2}}^{(2)} - 2l_1 C_{u_{y2}-m_{z2}}^{(2)} + l_1^2 C_{\theta_{z2}-m_{z2}}^{(2)} \\ C_{u_y-m_z} = C_{u_{y1}-m_{z1}}^{(1)} + C_{u_{y2}-m_{z2}}^{(2)} - l_1 C_{\theta_{z2}-m_{z2}}^{(2)} \\ C_{\theta_z-m_z} = C_{\theta_{z1}-m_{z1}}^{(1)} + C_{\theta_{z2}-m_{z2}}^{(2)} \end{cases}. \qquad (3.42)$$

Similarly, the out-of-plane compliance matrix results from Eq. (3.39) as:

$$\left[C_{O,op}\right]=\left[C_{O_1,op}^{(1)}\right]+\left[T_{OO_2,op}^{(2)}\right]^T\cdot\left[C_{O_2,op}^{(2)}\right]\cdot\left[T_{OO_2,op}^{(2)}\right]=\begin{bmatrix} C_{\theta_x-m_x} & 0 & 0 \\ 0 & C_{\theta_y-m_y} & C_{\theta_y-f_z} \\ 0 & C_{\theta_y-f_z} & C_{u_z-f_z} \end{bmatrix}.$$

$$(3.43)$$

$\left[T_{OO_2,ip}^{(2)}\right]$ is given in Eq. (3.9). The full-hinge compliances of Eq. (3.43) are:

$$\begin{cases} C_{\theta_x-m_x} = C_{\theta_{x1}-m_{x1}}^{(1)} + C_{\theta_{x2}-m_{x2}}^{(2)} \\ C_{\theta_y-m_y}^{\cdot} = C_{\theta_{y1}-m_{y1}}^{(1)} + C_{\theta_{y2}-m_{y2}}^{(2)} \\ C_{\theta_y-f_z} = l_1 C_{\theta_{y2}-m_{y2}}^{(2)} + C_{\theta_{y1}-f_{z1}}^{(1)} + C_{\theta_{y2}-f_{z2}}^{(2)} \\ C_{u_z-f_z} = C_{u_{z1}-f_{z1}}^{(1)} + C_{u_{z2}-f_{z2}}^{(2)} + 2l_1 C_{\theta_{y2}-f_{z2}}^{(2)} + l_1^2 C_{\theta_{y2}-m_{y2}}^{(2)} \end{cases}. \qquad (3.44)$$

Two-segment hinges with one rotated segment:

Consider the two-segment flexible hinge whose front (longitudinal) view is sketched in Figure 3.4a. Let us calculate the in-plane and out-of-plane compliance matrices of the hinge with respect to the end reference frame $Oxyz$ when knowing the compliances of the two segments with respect to their corresponding reference frames $O_1x_1y_1z_1$ and $O_2x_2y_2z_2$ – see Figure 3.4b.

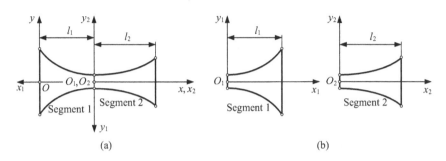

FIGURE 3.4 Front view of (a) two-segment serial flexible hinge; (b) hinge segments in their local frames.

The following operations are needed to express the in-plane and out-of-plane compliances of the two-segment hinge with respect to the end reference frame $Oxyz$:

- Rotation of segment 1 by 180° in order to align its local frame with the $Oxyz$ frame and to express its compliance matrices in terms of this frame, namely:

$$\left[C_{O_2,ip}^{(1r)}\right] = \left[R^{(1)}\right]^T \left[C_{O_2,ip}^{(1)}\right]\left[R^{(1)}\right] = \begin{bmatrix} \cos\pi & \sin\pi & 0 \\ -\sin\pi & \cos\pi & 0 \\ 0 & 0 & 1 \end{bmatrix}^T \cdot$$

$$\begin{bmatrix} C_{u_x-f_x}^{(1)} & 0 & 0 \\ 0 & C_{u_y-f_y}^{(1)} & C_{u_y-m_z}^{(1)} \\ 0 & C_{u_y-m_z}^{(1)} & C_{\theta_z-m_z}^{(1)} \end{bmatrix} \cdot \begin{bmatrix} \cos\pi & \sin\pi & 0 \\ -\sin\pi & \cos\pi & 0 \\ 0 & 0 & 1 \end{bmatrix} \qquad (3.45)$$

$$= \begin{bmatrix} C_{u_x-f_x}^{(1)} & 0 & 0 \\ 0 & C_{u_y-f_y}^{(1)} & -C_{u_y-m_z}^{(1)} \\ 0 & -C_{u_y-m_z}^{(1)} & C_{\theta_z-m_z}^{(1)} \end{bmatrix}$$

$$\left[C_{O_2,op}^{(1r)}\right]=\left[R^{(1)}\right]^T\left[C_{O_2,op}^{(1)}\right]\left[R^{(1)}\right]=\left[R^{(1)}\right]^T\cdot\begin{bmatrix} C_{\theta_x-m_x}^{(1)} & 0 & 0 \\ 0 & C_{\theta_y-m_y}^{(1)} & C_{\theta_y-f_z}^{(1)} \\ 0 & C_{\theta_y-f_z}^{(1)} & C_{u_z-f_z}^{(1)} \end{bmatrix}\cdot$$

$$\left[R^{(1)}\right]=\begin{bmatrix} C_{\theta_x-m_x}^{(1)} & 0 & 0 \\ 0 & C_{\theta_y-m_y}^{(1)} & -C_{\theta_y-f_z}^{(1)} \\ 0 & -C_{\theta_y-f_z}^{(1)} & C_{u_z-f_z}^{(1)} \end{bmatrix}\cdot$$

$$(3.45)$$

- Translation of the two-segment compliance matrices (which are now both formulated in the $O_2x_2\,y_2z_2$ frame) to express them in the end reference frame $Oxyz$, and addition of the individual segment compliances because the segments are connected in series:

$$\left[C_{ip}\right]=\left[T_{OO_2,ip}\right]^T\left(\left[C_{O_2,ip}^{(1r)}\right]+\left[C_{O_2,ip}^{(2)}\right]\right)\left[T_{OO_2,ip}\right]=\begin{bmatrix} 1 & 0 & 0 \\ 0 & 1 & 0 \\ 0 & -l_1 & 1 \end{bmatrix}^T\cdot$$

$$\begin{bmatrix} C_{u_x-f_x}^{(1)}+C_{u_x-f_x}^{(2)} & 0 & 0 \\ 0 & C_{u_y-f_y}^{(1)}+C_{u_y-f_y}^{(2)} & -C_{u_y-m_z}^{(1)}+C_{u_y-m_z}^{(2)} \\ 0 & -C_{u_y-m_z}^{(1)}+C_{u_y-m_z}^{(2)} & C_{\theta_z-m_z}^{(1)}+C_{\theta_z-m_z}^{(1)} \end{bmatrix}\cdot\begin{bmatrix} 1 & 0 & 0 \\ 0 & 1 & 0 \\ 0 & -l_1 & 1 \end{bmatrix}$$

$$\left[C_{op}\right]=\left[T_{OO_2,op}\right]^T\left(\left[C_{O_2,op}^{(1r)}\right]+\left[C_{O_2,op}^{(2)}\right]\right)\left[T_{OO_2,op}\right]=\begin{bmatrix} 1 & 0 & 0 \\ 0 & 1 & l_1 \\ 0 & 0 & 1 \end{bmatrix}^T\cdot$$

$$\begin{bmatrix} C_{\theta_x-m_x}^{(1)}+C_{\theta_x-m_x}^{(2)} & 0 & 0 \\ 0 & C_{\theta_y-m_y}^{(1)}+C_{\theta_y-m_y}^{(2)} & -C_{\theta_y-f_z}^{(1)}+C_{\theta_y-f_z}^{(2)} \\ 0 & -C_{\theta_y-f_z}^{(1)}+C_{\theta_y-f_z}^{(2)} & C_{u_z-f_z}^{(1)}+C_{u_z-f_z}^{(1)} \end{bmatrix}\cdot\begin{bmatrix} 1 & 0 & 0 \\ 0 & 1 & l_1 \\ 0 & 0 & 1 \end{bmatrix}\cdot$$

$$(3.46)$$

The final equations resulting from Eqs. (3.46) are:

$$[C_{ip}] = \begin{bmatrix} \left(C^{(1)}_{u_x-f_x} + C^{(2)}_{u_x-f_x}\right) & 0 & 0 \\[2em] 0 & \left(C^{(1)}_{u_y-f_y} + C^{(2)}_{u_y-f_y} + l_1\left[2\left(C^{(1)}_{u_y-m_z} - C^{(2)}_{u_y-m_z}\right) + l_1\left(C^{(1)}_{\theta_z-m_z} + C^{(1)}_{\theta_z-m_z}\right)\right]\right) & \left(-C^{(1)}_{u_y-m_z} + C^{(2)}_{u_y-m_z} - l_1\left(C^{(1)}_{\theta_z-m_z} + C^{(1)}_{\theta_z-m_z}\right)\right) \\[2em] 0 & -C^{(1)}_{u_y-m_z} + C^{(2)}_{u_y-m_z} - l_1\left(C^{(1)}_{\theta_z-m_z} + C^{(1)}_{\theta_z-m_z}\right) & C^{(1)}_{\theta_z-m_z} + C^{(1)}_{\theta_z-m_z} \end{bmatrix}$$

$$[C_{op}] = \begin{bmatrix} \left(C^{(1)}_{\theta_x-m_x} + C^{(2)}_{\theta_x-m_x}\right) & 0 & 0 \\[2em] 0 & C^{(1)}_{\theta_y-m_y} + C^{(1)}_{\theta_y-m_y} & \left(-C^{(1)}_{\theta_y-f_z} + C^{(2)}_{\theta_y-f_z} + l_1\left(C^{(1)}_{\theta_y-m_y} + C^{(1)}_{\theta_y-m_y}\right)\right) \\[2em] 0 & -C^{(1)}_{\theta_y-f_z} + C^{(2)}_{\theta_y-f_z} + l_1\left(C^{(1)}_{\theta_y-m_y} + C^{(1)}_{\theta_y-m_y}\right) & \left(C^{(1)}_{u_z-f_z} + C^{(2)}_{u_z-f_z} + l_1\left[2\left(C^{(2)}_{\theta_y-f_z} - C^{(1)}_{\theta_y-f_z}\right) + \left[l_1\left(C^{(1)}_{\theta_y-m_y} + C^{(1)}_{\theta_y-m_y}\right)\right]\right]\right) \end{bmatrix}.$$

(3.47)

3.2.1.2 Geometric Configurations

The generic hinge design of Figure 3.4a can result in numerous configurations that combine various straight-axis primitive segments. Figure 3.5 shows two right corner-filleted hinge designs from this category.

Example 3.2

A right circular cross-sectional flexible hinge with the structure of Figure 3.5a has its segment 1 defined by a circular profile of radius r_1 and segment 2 defined by a different circular curve of radius r_2. The flexible hinge has a circular cross-section. Analyze the variation of the compliance ratio $C_{u_y-f_y}/C_{u_x-f_x}$ in terms of the geometric parameters d (d is the minimum cross-sectional diameter), r_1 and r_2. The hinge is fabricated from mild steel with Young's modulus $E = 2.1 \cdot 10^{11}\,\text{N/m}^2$ and Poisson's ratio $\mu = 0.33$.

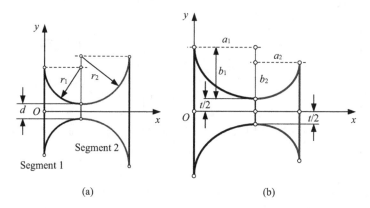

(a)　　　　　　　　　　　　　　　　　　　(b)

FIGURE 3.5　Two-segment serial right corner-filleted flexible hinges formed of dissimilar segments of: (a) circular profiles; (b) elliptic profiles.

Solution:

For the particular design of Figure 3.5a, the generic length l_1 of Figure 3.4 is $l_1 = r_1$. The first Eq. (3.47) provides the compliances needed in the example's ratio:

$$\begin{cases} C_{u_x-f_x} = C^{(1)}_{u_x-f_x} + C^{(2)}_{u_x-f_x}; \\ C_{u_y-f_y} = C^{(1)}_{u_y-f_y,s} + C^{(2)}_{u_y-f_y,s} + r_1\left[2\left(C^{(1)}_{u_y-m_z} - C^{(2)}_{u_y-m_z}\right) + r_1\left(C^{(1)}_{\theta_z-m_z} + C^{(1)}_{\theta_z-m_z}\right)\right] \end{cases} \tag{3.48}$$

The individual compliances of Eqs. (3.48) are calculated with Eqs. (A2.16) of Chapter 2. The compliances $C^{(1)}_{u_y-f_y,s}$ and $C^{(2)}_{u_y-f_y,s}$ included the shear effects as per Eqs. (2.10), with $\alpha_s = 10/9$ for the circular cross-section. Figures 3.6 and 3.7 illustrate the variation of the compliance ratio with d, r_1 and r_2. The following constant parameters have been used: $d = 0.004\,\text{m}$, $r_1 = 0.01\,\text{m}$ and $r_2 = 0.015\,\text{m}$.

The values of the ratio illustrate the active bending capability of this hinge in relation to the parasitic axial deformation capacity. Large ratio values indicate, simultaneously, high bending capacity and relatively low axial deformation potential, which reflect good performance by the flexible hinge. As the plots of Figures 3.6 and 3.7 indicate, the bending compliance along the y axis is substantially larger than that of the x-axis compliance – the ratio of the two compliances can be as high as 50,000. The ratio increases with r_1 and r_2 increasing and with the minimum diameter d decreasing.

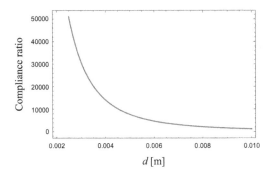

FIGURE 3.6 Plot of the compliance ratio $C_{u_y-f_y} / C_{u_x-f_x}$ in terms of the minimum diameter d.

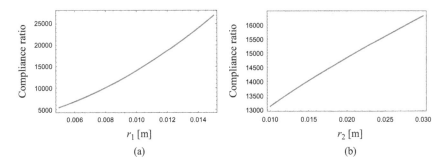

FIGURE 3.7 Plots of the compliance ratio $C_{u_y-f_y} / C_{u_x-f_x}$ in terms of the: (a) radius r_1; (b) radius r_2.

3.2.2 TRANSVERSELY SYMMETRIC FLEXIBLE HINGES

This section studies flexible hinges that are transversely symmetric, in addition to being axially symmetric. A generic mathematical model is derived that calculates the in-plane and out-of-plane compliance matrices in terms of the corresponding compliance matrices of half the transversely symmetric hinge. Several configurations are subsequently studied based on the generic compliance model.

3.2.2.1 Generic Compliance Matrices

Transverse symmetry in a flexible hinge enables expressing the full-hinge compliance matrices in terms of half-hinge compliance matrices. Moreover, when the reference frame is placed at the symmetry center (which is at the intersection of the axial and transverse symmetry axes), the in- and out-of-plane compliance matrices of the full hinge are diagonal, and therefore, the load action is fully decoupled, as discussed in this section. The generic flexible and axially symmetric segment of Figure 3.8 is mirrored (flipped) with respect to its y_h axis, and the two resulting segments are joined to obtain the transversely symmetric hinge

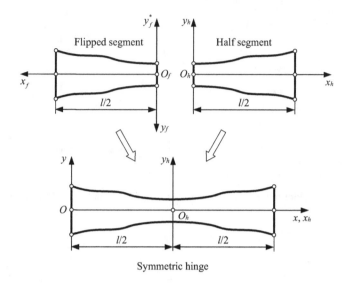

FIGURE 3.8 Transversely symmetric flexible hinge obtained by mirroring (flipping) one segment about an axis and joining the two segments.

illustrated in the same figure. Consider that known are the in- and out-of-plane compliance matrices of the original segment (identified by the superscript "h" as the half-segment), namely:

$$\left[C_{ip}^{(h)} \right] = \begin{bmatrix} C_{u_x-f_x}^{(h)} & 0 & 0 \\ 0 & C_{u_y-f_y}^{(h)} & C_{u_y-m_z}^{(h)} \\ 0 & C_{u_y-m_z}^{(h)} & C_{\theta_z-m_z}^{(h)} \end{bmatrix};$$

$$\left[C_{op}^{(h)} \right] = \begin{bmatrix} C_{\theta_x-m_x}^{(h)} & 0 & 0 \\ 0 & C_{\theta_y-m_y}^{(h)} & C_{\theta_y-f_z}^{(h)} \\ 0 & C_{\theta_y-f_z}^{(h)} & C_{u_z-f_z}^{(h)} \end{bmatrix}.$$

(3.49)

The full symmetric hinge structurally results by serially combining the flipped half-segment to the original half-segment. In terms of the central reference frame $O_h x_h y_h z_h$, the full-hinge compliance matrices are calculated by adding the transformed compliance matrices of the flipped segment to the unaltered compliance matrices of the original half-segment. When expressed in the $O_f x_f y_f z_f$ frame, the compliance matrices of the flipped segment are identical to the ones of the original half-segment in its local frame $O_h x_h y_h z_h$. At the same time, due to axial symmetry, the flipped segment has the same compliance matrices in terms of $O_f x_f y_f^* z_f^*$ and the $O_f x_f y_f z_f$ frames of Figure 3.8. As a consequence, the flipped segment compliance matrices are transferred to the central frame $O_h x_h y_h z_h$ by an 180° rotation of the original half-segment compliance matrices, and the in-plane compliance matrix of the full hinge is expressed as:

$$\left[C_{O_h,ip}\right] = [R]^T\left[C_{ip}^{(h)}\right][R] + \left[C_{ip}^{(h)}\right] \quad \text{with}$$

$$[R] = \begin{bmatrix} \cos\pi & \sin\pi & 0 \\ -\sin\pi & \cos\pi & 0 \\ 0 & 0 & 1 \end{bmatrix} = \begin{bmatrix} -1 & 0 & 0 \\ 0 & -1 & 0 \\ 0 & 0 & 1 \end{bmatrix}. \tag{3.50}$$

Combining Eqs. (3.49) and (3.50) yields:

$$\left[C_{O_h,ip}\right] = 2\cdot\begin{bmatrix} C_{u_x-f_x}^{(h)} & 0 & 0 \\ 0 & C_{u_y-f_y}^{(h)} & 0 \\ 0 & 0 & C_{\theta_z-m_z}^{(h)} \end{bmatrix}. \tag{3.51}$$

Equation (3.51) shows that the in-plane compliance matrix of a transversely and axially symmetric hinge is diagonal with respect to a frame located on the transverse symmetry axis (and at the symmetry center), and therefore, there is no coupling (cross-connections) between loads and deformations. This means that a force applied along y_h at O_h only produces a deflection along that direction (and not a rotation). Moreover, only the three direct compliances of the half-hinge are needed to evaluate the full-hinge in-plane compliance matrix. The full-hinge, in-plane compliance matrix is expressed in the end reference frame $Oxyz$ in terms of the one written in the central frame $O_h x_h y_h z_h$ by means of an x-axis translation offset $\Delta x = l/2$ accounting for the distance from O (considered now the global-frame origin) to O_h, namely:

$$\left[C_{O,ip}\right] = \left[T_{OO_h,ip}\right]^T\left[C_{O_h,ip}\right]\left[T_{OO_h,ip}\right] \quad \text{with}$$

$$\left[T_{OO_h,ip}\right] = \begin{bmatrix} 1 & 0 & 0 \\ 0 & 1 & 0 \\ 0 & -\Delta x & 1 \end{bmatrix} = \begin{bmatrix} 1 & 0 & 0 \\ 0 & 1 & 0 \\ 0 & -l/2 & 1 \end{bmatrix}. \tag{3.52}$$

Combining now Eqs. (3.51) and (3.52) results in:

$$\left[C_{O,ip}\right] = \begin{bmatrix} C_{u_x-f_x} & 0 & 0 \\ 0 & C_{u_y-f_y} & C_{u_y-m_z} \\ 0 & C_{u_y-m_z} & C_{\theta_z-m_z} \end{bmatrix} \text{with} \begin{cases} C_{u_x-f_x} = 2C_{u_x-f_x}^{(h)}; \\ C_{u_y-f_y} = 2C_{u_y-f_y}^{(h)} + \dfrac{l^2}{2}\cdot C_{\theta_z-m_z}^{(h)}; \\ C_{u_y-m_z} = -l\cdot C_{\theta_z-m_z}^{(h)}; \\ C_{\theta_z-m_z} = 2C_{\theta_z-m_z}^{(h)}. \end{cases}$$

$$\tag{3.53}$$

The non-diagonal terms in the compliance matrix of Eq. (3.53) indicate the presence of bending coupling when the reference frame is located at the end O or at any other point on the axial line that is not the symmetry center O_h. It should also be noted in Eq. (3.53) that only three direct compliances of the half-hinge, namely, $C^{(h)}_{u_x-f_x}, C^{(h)}_{u_y-f_y}$ and $C^{(h)}_{\theta_z-m_z}$, are needed to formulate the in-plane compliance matrix of the full hinge in the end-point reference frame.

A similar approach is followed to calculate the out-of-plane compliance matrices of a transversely symmetric flexible hinge. In the $O_h x_h y_h z_h$ reference frame, the out-of-plane compliance matrix is obtained in terms of the similar matrix of the half-hinge as:

$$\left[C_{O_h,op} \right] = [R]^T \left[C^{(h)}_{op} \right][R] + \left[C^{(h)}_{op} \right] = 2 \cdot \begin{bmatrix} C^{(h)}_{\theta_x-m_x} & 0 & 0 \\ 0 & C^{(h)}_{\theta_y-m_y} & 0 \\ 0 & 0 & C^{(h)}_{u_z-f_z} \end{bmatrix}. \quad (3.54)$$

with the rotation matrix of Eq. (3.50) and the half-hinge out-of-plane compliance matrix given in Eq. (3.49). Again, there is full decoupling between axes as illustrated by the diagonal form of the compliance matrix in Eq. (3.54). The compliance matrix of Eq. (3.54) is translated from O_h to the end point O as:

$$\left[C_{O,op} \right] = \left[T_{OO_h,op} \right]^T \left[C_{O_h,op} \right] \left[T_{OO_h,op} \right] \quad \text{with}$$

$$\left[T_{OO_h,op} \right] = \begin{bmatrix} 1 & 0 & 0 \\ 0 & 1 & \Delta x \\ 0 & 0 & 1 \end{bmatrix} = \begin{bmatrix} 1 & 0 & 0 \\ 0 & 1 & l/2 \\ 0 & 0 & 1 \end{bmatrix}. \quad (3.55)$$

which yields:

$$\left[C_{O,op} \right] = \begin{bmatrix} C_{\theta_x-m_x} & 0 & 0 \\ 0 & C_{\theta_y-m_y} & C_{\theta_y-f_z} \\ 0 & C_{\theta_y-f_z} & C_{u_z-f_z} \end{bmatrix} \text{ with} \begin{cases} C_{\theta_x-m_x} = 2C^{(h)}_{\theta_x-m_x}; \\ C_{\theta_y-m_y} = 2C^{(h)}_{\theta_y-m_y}; \\ C_{\theta_y-f_z} = l \cdot C^{(h)}_{\theta_y-m_y}; \\ C_{u_z-f_z} = 2C^{(h)}_{u_z-f_z} + \dfrac{l^2}{2} \cdot C^{(h)}_{\theta_y-m_y}. \end{cases}$$

$$(3.56)$$

after substituting the center-point compliance matrix of Eq. (3.54) into Eq. (3.55). Note that there is coupling between the y and z axes, as indicated by the two identical non-diagonal terms. Also note in Eq. (3.56) that only three compliances of the half-hinge, namely, $C^{(h)}_{\theta_x-m_x}, C^{(h)}_{\theta_y-m_y}$ and $C^{(h)}_{u_z-f_z}$, are necessary to express the out-of-plane compliance matrix of the full hinge with respect to the end-point reference frame.

Notes:

Rectangular cross-section:

Equations (3.53) and (3.56) show that the following compliances of the full, symmetric hinge are connected as:

$$C_{\theta_z - m_z} = -\frac{2}{l} \cdot C_{u_y - m_z}; \quad C_{\theta_y - m_y} = \frac{2}{l} \cdot C_{\theta_y - f_z}. \tag{3.57}$$

Assuming that the half-hinge is formed by serially connecting n different segments, the following compliances can be calculated:

$$C_{\theta_x - m_x}^{(h)} = \sum_{i=1}^{n} C_{\theta_{xi} - m_{xi}}^{(i)}; \quad C_{\theta_y - m_y}^{(h)} = \sum_{i=1}^{n} C_{\theta_{yi} - m_{yi}}^{(i)}; \quad C_{\theta_z - m_z}^{(h)} = \sum_{i=1}^{n} C_{\theta_{zi} - m_{zi}}^{(i)}, \tag{3.58}$$

according to Eqs. (3.35) and (3.39). As per Eqs. (2.24), the torsion compliance of a rectangular cross-section hinge segment i with constant out-of-plane width w and variable in-plane thickness $t(x)$ is expressed as:

$$C_{\theta_{xi} - m_{xi}}^{(i)} = \begin{cases} \dfrac{E}{4G} \cdot C_{\theta_{yi} - m_{yi}}^{(i)} = \dfrac{1+\mu}{2} \cdot C_{\theta_{yi} - m_{yi}}^{(i)}, & w \ll t(x) \\[3mm] \dfrac{E}{4G} \cdot C_{\theta_{zi} - m_{zi}}^{(i)} = \dfrac{1+\mu}{2} \cdot C_{\theta_{zi} - m_{zi}}^{(i)}, & w \gg t(x) \end{cases} \tag{3.59}$$

for a homogeneous material with $G = E/[2(1 + \mu)]$. Combining Eqs. (3.57)–(3.59), and also considering Eqs. (3.53) and (3.56), yields:

$$C_{\theta_x - m_x} = \begin{cases} \dfrac{E}{4G} \cdot C_{\theta_y - m_y} = \dfrac{1+\mu}{2} \cdot C_{\theta_y - m_y} = \dfrac{1+\mu}{l} \cdot C_{\theta_y - f_z}, & w \ll t(x) \\[3mm] \dfrac{E}{4G} \cdot C_{\theta_z - m_z} = \dfrac{1+\mu}{2} \cdot C_{\theta_z - m_z} = -\dfrac{1+\mu}{l} \cdot C_{u_y - m_z}, & w \gg t(x) \end{cases}. \tag{3.60}$$

The relationships of Eqs. (3.57) and (3.60) indicate that five full-hinge compliances need to be calculated independently out of the total eight in-plane and out-of-plane compliances defining the elastic properties of a flexible hinge with rectangular cross-section and transverse symmetry. However, when neither $t(x) \ll w$ nor $t(x) \gg w$, the torsional compliance $C_{\theta_x - m_x}$ has to be also calculated separately and independently.

Circular cross-section:

For circular cross-sectional hinges, the following compliances are identical:

$$C_{\theta_z - m_z} = C_{\theta_y - m_y}, C_{\theta_y - f_z} = -C_{u_y - m_z}, C_{u_z - f_z} = C_{u_y - f_y}. \tag{3.61}$$

In addition, as seen in Eqs. (3.57) and (3.60), the following compliance connections can be formed:

$$\begin{cases} C_{\theta_y - m_y} = \dfrac{2}{l} \cdot C_{\theta_y - f_z} = -\dfrac{2}{l} \cdot C_{u_y - m_z}; \\[3mm] C_{\theta_x - m_x} = 2 \cdot C_{\theta_x - m_x}^{(h)} = 2 \sum_{i=1}^{n} C_{\theta_{xi} - m_{xi}}^{(i)} = (1 + \mu) \sum_{i=1}^{n} C_{\theta_{yi} - m_{yi}}^{(i)} = (1 + \mu) \cdot C_{\theta_y - m_y} \end{cases},$$

$$(3.62)$$

where the second Eq. (3.62) also used Eq. (2.19). The compliance connections expressed in Eqs. (3.61) and (3.62) demonstrate that from the eight compliances defining the spatial elastic properties of a flexible hinge with circular cross-section and transverse symmetry, only three compliances are independent.

When using the full displacement vector $[u] = [u_x \ u_y \ \theta_z \ \theta_x \ \theta_y \ u_z]^T$ at the end O of the generic, transversely symmetric hinge of variable circular cross-section, the following compliance matrix, which combines the in-plane and out-of-plane compliance matrices of Eqs. (3.53) and (3.56), together with the compliance identities and connections of Eqs. (3.61) and (3.62), is obtained:

$$[C_O]_{6\times6} = \begin{bmatrix} [C_{O,ip}] & [0]_{3\times3} \\ [0]_{3\times3} & [C_{O,op}] \end{bmatrix} \quad \text{with}$$

$$[C_{O,ip}] = \begin{bmatrix} C_{u_x - f_x} & 0 & 0 \\ 0 & C_{u_y - f_y} & C_{u_y - m_z} \\ 0 & C_{u_y - m_z} & -\dfrac{2}{l} \cdot C_{u_y - m_z} \end{bmatrix}; \qquad (3.63)$$

$$[C_{O,op}] = \begin{bmatrix} -\dfrac{2(1+\mu)}{l} \cdot C_{u_y - m_z} & 0 & 0 \\ 0 & -\dfrac{2}{l} \cdot C_{u_y - m_z} & -C_{u_y - m_z} \\ 0 & -C_{u_y - m_z} & C_{u_z - f_z} \end{bmatrix}$$

Example 3.3

The half-part of a transversely symmetric flexible hinge is formed of two serially connected, axially symmetric segments, as shown in the skeleton representation of Figure 3.9. Considering that all native compliances of the two segments are known, calculate the in- and out-of-plane compliance matrices of the resulting flexible hinge with respect to the central reference frame and the end-point reference frame.

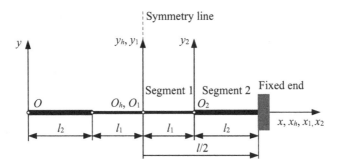

FIGURE 3.9 Skeleton representation of transversely symmetric flexible hinge with half-hinge formed of two serially connected segments.

Solution:

Equations (3.41) and (3.43) provide the in-plane and out-of-plane compliance matrices (with respect to the central frame $O_h x_h y_h z_h$) of a hinge portion that is formed of two segments connected in series. This portion is the half-hinge formed of segments 1 and 2 to the right of the symmetry line in Figure 3.9. Combining Eqs. (3.41) and (3.51) yields the in-plane compliance matrix of the full hinge with respect to the central reference frame:

$$[C_{O_h,ip}] = 2 \cdot \begin{bmatrix} C_{u_x-f_x}^{(1)} + C_{u_x-f_x}^{(2)} & 0 & 0 \\ 0 & \begin{pmatrix} C_{u_y-f_y}^{(1)} + C_{u_y-f_y}^{(2)} - 2l_1 \cdot \\ C_{u_y-m_z}^{(2)} + l_1^2 \cdot C_{\theta_z-m_z}^{(2)} \end{pmatrix} & 0 \\ 0 & 0 & C_{\theta_z-m_z}^{(1)} + C_{\theta_z-m_z}^{(2)} \end{bmatrix}. \quad (3.64)$$

Equations (3.53) and (3.64) yield the in-plane compliance matrix with respect to the end-point reference frame located at the end point O:

$$[C_{O,ip}] = 2 \cdot \begin{bmatrix} C_{u_x-f_x}^{(1)} + C_{u_x-f_x}^{(2)} & 0 & 0 \\ 0 & \begin{pmatrix} C_{u_y-f_y}^{(1)} + C_{u_y-f_y}^{(2)} - 2l_1 \cdot \\ C_{u_y-m_z}^{(2)} + l_1^2 \cdot C_{\theta_z-m_z}^{(2)} \\ + \dfrac{(l_1+l_2)^2}{2} \cdot \\ \left(C_{\theta_z-m_z}^{(1)} + C_{\theta_z-m_z}^{(2)} \right) \end{pmatrix} & \begin{pmatrix} -2(l_1+l_2) \cdot \\ \left(C_{\theta_z-m_z}^{(1)} + C_{\theta_z-m_z}^{(2)} \right) \end{pmatrix} \\ 0 & \begin{pmatrix} -2(l_1+l_2) \cdot \\ \left(C_{\theta_z-m_z}^{(1)} + C_{\theta_z-m_z}^{(2)} \right) \end{pmatrix} & C_{\theta_z-m_z}^{(1)} + C_{\theta_z-m_z}^{(2)} \end{bmatrix}.$$

$$(3.65)$$

Equations (3.39) and (3.54) are combined to produce the out-of-plane compliance matrix referenced to the midpoint O_h:

$$\left[C_{O_h,op}\right] = 2 \cdot \begin{bmatrix} C^{(1)}_{\theta_x-m_x} + C^{(2)}_{\theta_x-m_x} & 0 & 0 \\ 0 & C^{(1)}_{\theta_y-m_y} + C^{(2)}_{\theta_y-m_y} & 0 \\ 0 & 0 & \begin{pmatrix} C^{(1)}_{u_z-f_z} + C^{(2)}_{u_z-f_z} + 2l_1 \cdot \\ C^{(2)}_{\theta_y-f_z} + l_1^2 \cdot C^{(2)}_{\theta_y-m_y} \end{pmatrix} \end{bmatrix}.$$

$$(3.66)$$

Equation (3.66) is now used in conjunction with Eq. (3.56) in order to find the following out-of-plane compliance matrix formulated in the $Oxyz$ reference frame:

$$\left[C_{O,op}\right]$$

$$= 2 \cdot \begin{bmatrix} \begin{pmatrix} C^{(1)}_{\theta_x-m_x} \\ +C^{(2)}_{\theta_x-m_x} \end{pmatrix} & 0 & 0 \\ 0 & C^{(1)}_{\theta_y-m_y} + C^{(2)}_{\theta_y-m_y} & 2(l_1+l_2) \cdot \left(C^{(1)}_{\theta_y-m_y} + C^{(2)}_{\theta_y-m_y}\right) \\ 0 & \begin{pmatrix} 2(l_1+l_2) \cdot \\ \left(C^{(1)}_{\theta_y-m_y} + C^{(2)}_{\theta_y-m_y}\right) \end{pmatrix} & \begin{pmatrix} C^{(1)}_{u_z-f_z} + C^{(2)}_{u_z-f_z} \\ +2l_1 \cdot C^{(2)}_{\theta_y-f_z} + l_1^2 \cdot \\ C^{(2)}_{\theta_y-m_y} + \dfrac{(l_1+l_2)^2}{2} \cdot \\ \left(C^{(1)}_{\theta_y-m_y} + C^{(2)}_{\theta_y-m_y}\right) \end{pmatrix} \end{bmatrix}.$$

$$(3.67)$$

3.2.2.2 Hinge Configurations: Finite Element Confirmation of Analytical Compliances

Several flexible-hinge designs with transverse symmetry and rectangular or circular cross-sections are studied in this section. The analytical model compliance predictions are compared with finite element analysis (FEA) simulation data for several configurations, and the two methods' results generate small relative errors.

In the FEA models, one end is free (the one comprising the origin O of the reference frame $Oxyz$ in Figure 3.10a), while the other end A is fixed.

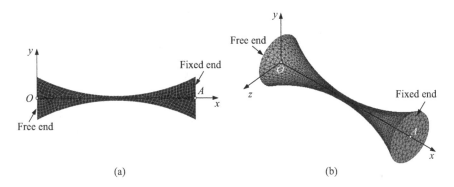

FIGURE 3.10 Finite element models: (a) front view of rectangular cross-section flexible hinge; (b) isometric view of circular cross-section flexible hinge.

For the majority of finite element tests, point loads were applied at O and the corresponding displacements were determined in order to calculate the FEA compliances. The axial compliance was determined by applying a pressure on the free-end surface in order to avoid local effects resulting from a point axial load.

For hinges with rectangular cross-section of constant out-of-plane width w, as shown in Figure 3.10a, two-dimensional, four-node shell elements with six degrees of freedom per node were used to model the various designs. The three-dimensional flexible hinges of circular cross-section – see Figure 3.10b – were meshed by means of ten-node tetrahedron elements with three degrees of freedom per node.

Equations (3.53) and (3.56) were used to calculate the analytic compliances of the full hinges in terms of the half-hinge compliances for various designs. Numerical data for all rectangular cross-section hinge configurations is included in this section, while compliance values of circular cross-section hinges are only provided for circular flexible hinges.

3.2.2.2.1 Circular Flexible Hinge

Figure 3.11a shows the front view of a $(0, \alpha_e)$ circular hinge, whereas Figure 3.11b illustrates a right circular flexible hinge of the $(0, \pi/2)$ type; both configurations identify their planar geometry corresponding to rectangular cross-sections. For designs with circular cross-sections, the diameter d should be used instead of the thickness t. These designs have double symmetry, and as a consequence, their compliances are found from the compliances defining half a segment, as shown in Figure 2.12. For a $(0, \alpha_e)$ design, the half-hinge compliances are found by using $\alpha_0 = 0$ in Eqs. (2.44) and (2.45), which express the in- and out-of-plane compliances of the half-hinge. The compliances of a right half-hinge are provided in Eqs. (A2.3) for rectangular cross-sections and in Eqs. (A2.16) for circular

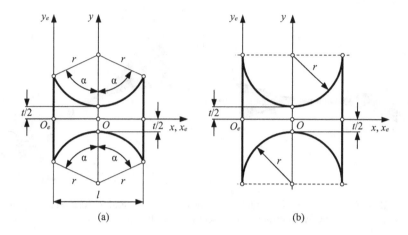

FIGURE 3.11 Front view with geometric parameters of: (a) $(0, \alpha_e)$ circular flexible hinge; (b) $(0, \pi/2)$ right circular flexible hinge.

TABLE 3.1
Circular Hinges of Rectangular Cross-Section with Geometric Parameters

Loads/Displacements	Design #	w (m)	t (m)	r (m)	α (deg)
In-plane	1	0.006	0.001	0.03	30
	2	0.006	0.0005	0.03	30
	3	0.006	0.001	0.03	45
	4	0.006	0.001	0.015	30
	5	0.006	0.001	0.03	90
	6	0.01	0.0015	0.03	90
Out-of-plane	7	0.0002	0.001	0.03	30
	8	0.0002	0.0015	0.03	30
	9	0.0002	0.0015	0.03	90
	10	0.0002	0.0015	0.04	90

cross-sections. The full right design $(0, \pi/2)$ has its compliances provided in the Appendix at the end of this chapter for both the configuration with rectangular cross-section (of constant width w and minimum thickness t) and the variant with circular cross-section (of minimum diameter $d = t$).

Rectangular cross-section:

Table 3.1 lists ten different designs (four of which are right hinges where $\alpha = 90°$) with their parameters, whereas Tables 3.2 and 3.3 display the analytical model and finite element simulation compliance results together with the relative differences between the two models predictions.

It should be noted that designs 1 through 6 have a relatively large out-of-plane width w as these designs are meant for in-plane bending applications. For them,

TABLE 3.2

In-Plane Compliances of Circular Hinges with Rectangular Cross-Section

Design	Results	$C_{u_x-f_x}$ (N⁻¹·m)	$C_{u_y-f_y}$ (N⁻¹·m)	$C_{u_y-m_z}$ (N⁻¹)	$C_{u_z-f_z}$ (N⁻¹·m)	$C_{\theta_y-f_z}$ (N⁻¹)	$C_{\theta_x-m_x}$ (N⁻¹·m⁻¹)
1	A	1.106×10^{-8}	1.506×10^{-5}	9.627×10^{-4}	9.649×10^{-7}	5.53×10^{-5}	4.17×10^{-2}
	FEA	1.14×10^{-8}	1.525×10^{-5}	9.745×10^{-4}	9.777×10^{-7}	5.63×10^{-5}	4.12×10^{-2}
	% diff	3.07	1.46	1.23	1.33	1.81	1.2
2	A	1.69×10^{-8}	8.369×10^{-5}	5.463×10^{-3}	1.481×10^{-6}	8.452×10^{-5}	0.237
	FEA	1.715×10^{-8}	8.374×10^{-5}	5.496×10^{-3}	1.512×10^{-6}	8.583×10^{-5}	0.228
	% diff	1.48	0.6	0.6	2.1	1.55	3.8
3	A	1.187×10^{-8}	2.952×10^{-5}	1.363×10^{-3}	2.04×10^{-6}	8.391×10^{-5}	4.18×10^{-2}
	FEA	1.222×10^{-8}	2.992×10^{-5}	1.38×10^{-3}	2.117×10^{-6}	8.643×10^{-5}	4.11×10^{-2}
	% diff	2.95	1.36	1.25	3.77	3.00	1.67
4	A	7.006×10^{-9}	2.740×10^{-6}	3.375×10^{-3}	1.798×10^{-7}	1.752×10^{-5}	2.92×10^{-2}
	FEA	7.127×10^{-9}	2.797×10^{-6}	3.448×10^{-3}	1.858×10^{-7}	1.78×10^{-5}	2.80×10^{-2}
	% diff	1.73	2.08	2.16	3.34	1.60	4.11
5	A	1.237×10^{-8}	5.844×10^{-5}	1.928×10^{-3}	4.076×10^{-6}	1.237×10^{-4}	4.2×10^{-2}
	FEA	1.297×10^{-8}	5.58×10^{-5}	1.85×10^{-3}	4.3×10^{-6}	1.29×10^{-4}	3.97×10^{-2}
	% diff	4.91	4.73	4.22	5.50	4.28	5.48
6	A	5.866×10^{-9}	1.28×10^{-5}	4.19×10^{-4}	7.18×10^{-7}	2.112×10^{-5}	9.1×10^{-3}
	FEA	5.939×10^{-9}	1.34×10^{-5}	4.38×10^{-4}	7.51×10^{-7}	2.18×10^{-5}	8.8×10^{-3}
	% diff	1.24	4.7	4.53	4.60	3.32	3.3

TABLE 3.3

Out-of-Plane Compliances of Circular Hinges with Rectangular Cross-Section

Design	Results	$C_{u_x-f_x}$ (N⁻¹·m)	$C_{u_y-f_y}$ (N⁻¹·m)	$C_{u_y-m_z}$ (N⁻¹)	$C_{u_z-f_z}$ (N⁻¹·m)	$C_{\theta_x-f_z}$ (N⁻¹)	$C_{\theta_x-m_x}$ (N⁻¹·m⁻¹)
7	A	3.318×10^{-7}	4.519×10^{-4}	2.888×10^{-2}	2.604×10^{-2}	1.493	70.41
	FEA	3.377×10^{-7}	4.567×10^{-4}	2.919×10^{-2}	2.624×10^{-2}	1.503	68.81
	% diff	1.78	1.06	1.07	0.77	0.67	2.27
8	A	2.552×10^{-7}	1.665×10^{-4}	1.044×10^{-2}	2.042×10^{-2}	1.148	52.719
	FEA	2.599×10^{-7}	1.693×10^{-4}	1.061×10^{-2}	2.057×10^{-2}	1.155	50.927
	% diff	1.84	1.68	1.63	0.73	0.61	3.4
9	A	2.933×10^{-7}	6.389×10^{-4}	2.094×10^{-2}	8.741×10^{-2}	2.640	60.22
	FEA	3.1×10^{-7}	6.409×10^{-4}	2.099×10^{-2}	9.073×10^{-2}	2.645	58.74
	% diff	5.69	0.31	0.24	3.80	0.19	2.46
10	A	3.468×10^{-7}	1.31×10^{-3}	3.23×10^{-3}	0.1818	4.162	71.34
	FEA	3.5×10^{-7}	1.38×10^{-3}	4.00×10^{-3}	0.1883	4.193	72.16
	% diff	0.92	5.34	2.38	3.58	0.70	1.15

the independent compliances that directly contribute to the active deformation are $C_{u_y-f_y}$ and $C_{u_y-m_z}$. The other independent compliances $C_{u_x-f_x}$, $C_{u_z-f_z}$ and $C_{\theta_y-f_z}$ are helping to evaluate the parasitic motions. On the contrary, the designs 7, 8, 9 and 10 have the dimension w smaller than the minimum thickness t. They are meant to work actively through out-of-plane deformation, and for them, the compliances $C_{u_z-f_z}$ and $C_{\theta_y-f_z}$ are the active ones – the other compliances are parasitic.

In Tables 3.2 and 3.3, "A" stands for analytic results and "% diff" denotes the relative percentage difference between analytic and FEA results.

It is also worth noting that for designs 1, 2, 3, 4, 5 and 6 (which are all having a relatively large out-of-plane parameter w), the variable thickness $t(x) < w$ for an angular interval $[0, \alpha_1]$ and then $t(x) > w$ for the remaining interval $[\alpha_1, \alpha]$. As per Eqs. (2.20)–(2.22) of Chapter 2, the torsional compliance has to be calculated accordingly with the proper $I_t(x)$.

Circular cross-section

Similar calculations were made for the circular cross-sectional flexible-hinge designs listed in Table 3.4. The analytical and finite element results exhibit, again, little divergence for the three independent compliances, as illustrated in Table 3.5.

TABLE 3.4

Circular Hinges of Circular Cross-Section with Geometric Parameters

Design #	d (m)	r (m)	α (deg)
1	0.001	0.03	30
2	0.0015	0.03	30
3	0.001	0.02	30
4	0.001	0.03	45
5	0.001	0.03	90
6	0.0015	0.03	90
7	0.0015	0.02	90

TABLE 3.5

Compliances of Circular Hinges with Circular Cross-Section

Design	Results	$C_{u_x-f_x}$ (N^{-1}·m)	$C_{u_y-f_y}$ (N^{-1}·m)	$C_{u_y-m_z}$ (N^{-1})
1	A	5.348×10^{-8}	1.263×10^{-4}	8.194×10^{-3}
	FEA	5.395×10^{-8}	1.27×10^{-4}	8.2×10^{-3}
	% diff	0.86	0.58	0.07
2	A	2.87×10^{-8}	3.091×10^{-5}	1.979×10^{-3}
	FEA	2.895×10^{-8}	3.115×10^{-5}	2.0×10^{-3}
	% diff	0.89	0.76	0.9

(Continued)

TABLE 3.5 (*Continued*)
Compliances of Circular Hinges with Circular Cross-Section

Design	Results	$C_{u_x-f_x}$ (N⁻¹·m)	$C_{u_y-f_y}$ (N⁻¹·m)	$C_{u_y-m_z}$ (N⁻¹)
3	A	4.305×10^{-8}	4.637×10^{-5}	4.453×10^{-3}
	FEA	4.343×10^{-8}	4.672×10^{-5}	4.486×10^{-3}
	% diff	0.89	0.77	1.25
4	A	5.399×10^{-8}	2.492×10^{-4}	1.159×10^{-2}
	FEA	5.464×10^{-8}	2.505×10^{-4}	1.16×10^{-2}
	% diff	1.22	0.52	0.51
5	A	5.413×10^{-8}	4.951×10^{-4}	1.639×10^{-2}
	FEA	5.488×10^{-8}	4.976×10^{-4}	1.647×10^{-2}
	% diff	1.39	0.51	0.51
6	A	2.93×10^{-8}	1.2×10^{-4}	3.96×10^{-3}
	FEA	2.983×10^{-8}	1.21×10^{-4}	3.99×10^{-3}
	% diff	1.80	0.77	0.76
7	A	4.395×10^{-8}	1.801×10^{-4}	8.911×10^{-3}
	FEA	4.474×10^{-8}	1.815×10^{-4}	8.979×10^{-3}
	% diff	1.79	0.78	0.76

3.2.2.2.2 Elliptical Flexible Hinge

The front view of a $(0, \alpha_e)$ elliptical flexible hinge of rectangular cross-section with its defining geometry (semi-axes lengths a and b, minimum thickness t and half-segment angle α) is shown in Figure 3.12a, whereas its right counterpart design is sketched in Figure 3.12b. Because of its transverse symmetry, this configuration's compliances are algebraically calculated from the direct compliances that define a half-elliptical segment, like the one pictured in Figure 2.13 – see

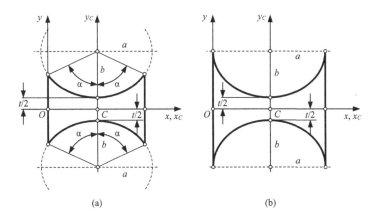

(a)   (b)

FIGURE 3.12 Front view with geometric parameters of: (a) $(0, \alpha_e)$ elliptical flexible hinge; (b) $(0, \pi/2)$ right elliptical flexible hinge.

Eqs. (2.53) through (2.56), where $\varphi_0 = 0$. The Appendix at the end of this chapter includes the closed-form compliance equations of rectangular and circular cross-sectional right elliptical flexible hinges. They were obtained using the compliances of a right elliptic half-segment given in Eqs. (A2.5) and (A2.6).

Table 3.6 displays the geometric parameter values of several elliptic-hinge designs whose compliances were calculated both analytically and by FEA – these compliance values are included in Tables 3.7 and 3.8. Note that only the active compliances are included in Tables 3.7 and 3.8, not the parasitic ones – this means that for designs with large w, the active compliances are $C_{u_y-f_y}$ and $C_{u_y-m_z}$, whereas for thin designs (with small w), the compliances of interest are $C_{u_z-f_z}$ and $C_{\theta_y-f_z}$.

TABLE 3.6

Elliptical Hinges of Rectangular Cross-Section with Geometric Parameters

Loads/Displacements	Design #	w (m)	t (m)	a (m)	b (m)	α (deg)
In-plane	1	0.006	0.001	0.03	0.02	30
	2	0.008	0.001	0.03	0.02	30
	3	0.006	0.001	0.03	0.02	45
	4	0.006	0.001	0.04	0.02	30
	5	0.006	0.001	0.02	0.03	30
	6	0.008	0.001	0.03	0.02	90
	7	0.008	0.0008	0.007	0.00325	90
	8	0.008	0.0012	0.008	0.00325	90
Out-of-plane	9	0.0002	0.0015	0.03	0.02	30
	10	0.0002	0.001	0.03	0.02	90
	11	0.0002	0.001	0.02	0.03	30

TABLE 3.7

In-Plane Compliances of Elliptical Hinges with Rectangular Cross-Sections

Design	Results	$C_{u_y-f_y}$ (N^{-1}·m)	$C_{u_y-m_z}$ (N^{-1})
1	A	1.868×10^{-5}	1.175×10^{-3}
	FEA	1.883×10^{-5}	1.184×10^{-3}
	% diff	0.77	0.77
2	A	1.401×10^{-5}	8.809×10^{-4}
	FEA	1.412×10^{-5}	8.878×10^{-4}
	% diff	0.78	0.77
3	A	3.645×10^{-5}	1.665×10^{-3}
	FEA	3.677×10^{-5}	1.679×10^{-3}
	% diff	0.88	0.86

(Continued)

TABLE 3.7 (*Continued*)
In-Plane Compliances of Elliptical Hinges with Rectangular Cross-Sections

Design	Results	$C_{u_y-f_y}$ (N^{-1}·m)	$C_{u_y-m_z}$ (N^{-1})
4	A	4.424×10^{-5}	2.088×10^{-3}
	FEA	4.443×10^{-5}	2.097×10^{-3}
	% diff	0.43	0.44
5	A	4.476×10^{-4}	4.279×10^{-4}
	FEA	4.596×10^{-4}	4.392×10^{-4}
	% diff	2.61	2.59
6	A	5.387×10^{-5}	1.767×10^{-3}
	FEA	5.416×10^{-5}	1.776×10^{-3}
	% diff	0.54	0.52
7	A	3.022×10^{-6}	4.022×10^{-4}
	FEA	2.96×10^{-6}	3.94×10^{-4}
	% diff	2.05	2.04
8	A	1.102×10^{-6}	1.45×10^{-4}
	FEA	1.12×10^{-6}	1.46×10^{-4}
	% diff	1.63	0.69

TABLE 3.8
Out-of-Plane Compliances of Elliptical Hinges with Rectangular Cross-Section

Design	Results	$C_{u_z-f_z}$ (N^{-1}·m)	$C_{\theta_y-f_z}$ (N^{-1})
9	A	2.372×10^{-2}	1.308
	FEA	2.374×10^{-2}	1.31
	% diff	0.11	0.15
10	A	0.131	3.96
	FEA	0.134	4.007
	% diff	1.85	1.19
11	A	7.717×10^{-3}	0.664
	FEA	7.87×10^{-3}	0.674
	% diff	1.94	1.54

3.2.2.2.3 *Right Circularly Corner-Filleted Flexible Hinge*

The front view of a right circularly corner-filleted flexible hinge of rectangular cross-section together with its geometry is shown in Figure 3.13. The compliances of this design were calculated considering that the half-portion of the hinge

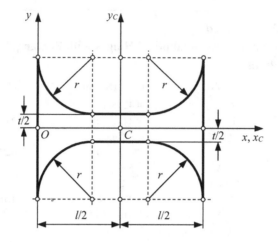

FIGURE 3.13 Front view with geometric parameters of a right circularly corner-filleted flexible hinge.

pictured in Figure 3.13 is formed of one segment (segment 1) of constant thickness t, which is serially connected to segment 2 – a $(0, \pi/2)$ circular segment. The in-plane and out-of-plane compliances of the half-hinge are evaluated by means of Eqs. (3.41) and (3.43). The compliances of segment 1 are provided in Eqs. (2.26) through (2.30), while those of segment 2 are given in Eqs. (A2.3) and (A2.4) of Chapter 2.

The analytical model predictions for several hinge configurations were validated with finite element simulation by testing the in-plane and out-of-plane compliances of the designs whose parameters are listed in Table 3.9. The corresponding compliance results are given in Tables 3.10 and 3.11.

TABLE 3.9

Right Circularly Corner-Filleted Hinges of Rectangular Cross-Section with Geometric Parameters

Loads/Displacements	Design #	l (m)	t (m)	r (m)	w (m)
In-plane	1	0.01	0.001	0.025	0.006
	2	0.014	0.001	0.025	0.006
	3	0.01	0.002	0.025	0.006
	4	0.01	0.001	0.04	0.006
	5	0.01	0.001	0.025	0.004
Out-of-plane	6	0.01	0.001	0.025	0.006
	7	0.014	0.001	0.025	0.006
	8	0.01	0.002	0.025	0.006
	9	0.01	0.001	0.04	0.006
	10	0.01	0.001	0.025	0.004

TABLE 3.10

In-Plane Compliances of Right Circularly Corner-Filleted Hinges with Rectangular Cross-Section

Design	Results	$C_{u_y-f_y}$ (N^{-1}·m)	$C_{u_y-m_z}$ (N^{-1})
1	A	189.46×10^{-8}	322.24×10^{-6}
	FEA	196.74×10^{-8}	332.59×10^{-6}
	% diff	3.70	3.11
2	A	607.26×10^{-8}	717.80×10^{-6}
	FEA	622.82×10^{-8}	732.25×10^{-6}
	% error	2.50	1.97
3	A	273.20×10^{-9}	438.76×10^{-7}
	FEA	289.77×10^{-9}	462.28×10^{-7}
	% error	5.72	5.09
4	A	111.86×10^{-8}	203.83×10^{-6}
	FEA	111.75×10^{-8}	213.26×10^{-6}
	% error	0.10	4.42
5	A	284.19×10^{-8}	483.36×10^{-6}
	FEA	295.21×10^{-9}	498.97×10^{-6}
	% error	3.73	3.13

TABLE 3.11

Out-of-Plane Compliances of Right Circularly Corner-Filleted Hinges with Rectangular Cross-Section

Design	Results	$C_{\theta_y-f_z}$ (N^{-1})	$C_{u_z-f_z}$ (N^{-1}·m)
6	A	147.26×10^{-3}	911.88×10^{-6}
	FEA	152.32×10^{-3}	959.51×10^{-6}
	% diff	3.32	4.96
7	A	308.56×10^{-3}	271.82×10^{-5}
	FEA	305.66×10^{-3}	981.13×10^{-5}
	% error	0.94	3.31
8	A	78.82×10^{-3}	499.47×10^{-6}
	FEA	81.60×10^{-3}	526.37×10^{-6}
	% error	3.41	5.11
9	A	112.49×10^{-3}	678.41×10^{-6}
	FEA	119.92×10^{-3}	728.41×10^{-6}
	% error	4.73	6.86
10	A	536.64×10^{-4}	332.49×10^{-6}
	FEA	556.23×10^{-4}	350.89×10^{-6}
	% error	3.52	5.25

3.2.2.2.4 Right Elliptically Corner-Filleted Flexible Hinge

The front view of a right elliptically corner-filleted flexible hinge of rectangular cross-section with its defining geometric parameters is pictured in Figure 3.14. The analytical compliances were calculated using the procedure outlined for the right circularly corner-filleted hinges. Segment 2 of the half-hinge of Figure 3.14 is a $(0, \pi/2)$ elliptical portion whose compliances are provided in Eqs. (A2.5) and (A2.6).

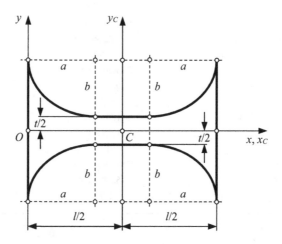

FIGURE 3.14 Front view with geometric parameters of a right elliptically corner-filleted flexible hinge.

Table 3.12 lists the geometric parameters of the rectangular cross-section right elliptically filleted notch hinges that were analyzed, while Tables 3.13 and 3.14 show the analytical and FEA compliances. A length $l = 0.01$ m was used for all designs.

TABLE 3.12

Right Elliptically Corner-Filleted Hinges of Rectangular Cross-Section with Geometric Parameters

Loads/Displacements	Design #	t (m)	w (m)	a (m)	b (m)
In-plane	1	0.001	0.006	0.0025	0.001
	2	0.001	0.006	0.001	0.0025
	3	0.0015	0.008	0.0025	0.001
Out-of-plane	4	0.0002	0.025	0.0025	0.001
	5	0.0002	0.025	0.001	0.0025
	6	0.0002	0.025	0.0025	0.001

TABLE 3.13

In-Plane Compliances of Right Elliptically Corner-Filleted Hinges with Rectangular Cross-Section

Design	Results	$C_{u_y-f_y}$ (N^{-1}·m)	$C_{u_y-m_z}$ (N^{-1})
1	A	2.31×10^{-6}	3.79×10^{-4}
	FEA	2.42×10^{-6}	3.91×10^{-4}
	% diff	4.25	3.06
2	A	2.75×10^{-6}	4.35×10^{-4}
	FEA	2.89×10^{-6}	4.51×10^{-4}
	% error	4.80	3.57
3	A	5.58×10^{-7}	8.87×10^{-5}
	FEA	5.72×10^{-7}	9.06×10^{-5}
	% error	2.55	2.08

TABLE 3.14

Out-of-Plane Compliances of Right Elliptically Corner-Filleted Hinges with Rectangular Cross-Section

Design	Results	$C_{\theta_y-f_z}$ (N^{-1})	$C_{u_z-f_z}$ (N^{-1}·m)
4	A	3.30×10^{-1}	2.10×10^{-3}
	FEA	3.35×10^{-1}	2.15×10^{-3}
	% diff	1.62	2.50
5	A	3.45×10^{-1}	2.23×10^{-3}
	FEA	3.57×10^{-1}	2.34×10^{-3}
	% error	3.35	4.98
6	A	2.27×10^{-1}	1.46×10^{-3}
	FEA	2.31×10^{-1}	1.51×10^{-3}
	% error	1.73	2.70

3.2.2.3 Hinge Configurations: Internal Comparison

This section studies the influence of geometric parameters on the bending compliance of several flexible hinges of either single- or multiple-profile configurations. The approach also enables comparing the deformation capability of the designs under consideration; it allows, as well, to perform shape and dimensional optimization in a straightforward manner.

3.2.2.3.1 *(0, α_e) Single-Profile Hinges*

(0, α_e) symmetric flexible hinges that are generated by single curves, such as circular, elliptic, hyperbolic, parabolic and inverse parabolic, are studied here; these configurations have either rectangular or circular cross-sections.

Rectangular cross-section:

A generic $(0, \alpha_e)$ hinge that is formed by a single curve has its planar profile defined by l, t and c, as shown in Figure 3.15 for a quarter hinge with axial and transverse symmetry. The influence of the geometric parameters l, t and c on the compliance $C_{u_y - f_y}$ is studied here; the analysis also serves at comparing the performance of various hinge configurations. The geometric parameters of a planar curve (a circle, ellipse, etc.) can be expressed in terms of the generic parameters l, c and t, as shown next.

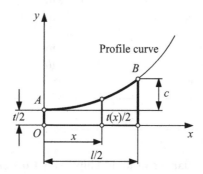

FIGURE 3.15 Planar geometry of the quarter segment of a flexible hinge with axial and transverse symmetry.

For a circular hinge, with $\alpha = \alpha_e$, the following relationships express α, r and $t(x)$ in terms of l, t and c:

$$\alpha = 2\tan^{-1}\left(\frac{2c}{l}\right); \quad r = \frac{c}{1 - \cos\alpha}; \quad t(x) = t + 2r(1 - \cos\varphi). \quad (3.68)$$

where φ ranges in the $[0, \alpha]$ interval.

The following connections apply to a standard elliptic hinge:

$$a = \frac{l}{2\sin\alpha}; \quad b = \frac{c}{1 - \cos\alpha}; \quad t(x) = t + 2b(1 - \cos\varphi). \quad (3.69)$$

The variable thicknesses of hyperbolic, parabolic and inverse parabolic hinges are expressed as:

$$\text{Hyperbolic}: \quad t(x) = \frac{\sqrt{16c(c+t)x^2 + l^2 t^2}}{l}$$

$$\text{Parabolic}: \quad t(x) = \frac{8c}{l^2} \cdot x^2 + t \qquad . \qquad (3.70)$$

$$\text{Inverse parabolic}: \quad t(x) = \frac{l^2 t(t + 2c)}{l^2(t + 2c) - 8cx^2}$$

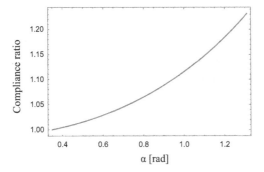

FIGURE 3.16   Plot of compliance ratio $C_{u_y-f_y} / C_{u_y-f_y,\min}$ in terms of the ellipse angle α.

Brief discussion on the elliptical hinge:

As seen in Eqs. (3.68) and (3.70), the geometrical parameters that are needed to calculate the compliances of circular, hyperbolic, parabolic and inverse parabolic hinge configurations are uniquely defined in terms of l, t and c. However, for elliptical flexible hinges, the (first) two connections of Eqs. (3.69) comprise three parameters l, c and α – this indicates that one of the three parameters can be arbitrarily chosen and the other two can be expressed in terms of the arbitrary selection. For instance, if α is selected, the ellipse semi-axes are calculated based on the first two Eqs. (3.69). It can be visualized in Figure 3.15 that for specified values of l and c, there is an infinite number of ellipses passing through the end points A and B. To better substantiate this observation, Figure 3.16 plots the following compliance ratio: $C_{u_y-f_y} / C_{u_y-f_y,\min}$ for an elliptic hinge, where $C_{u_y-f_y,\min}$ represents the smallest compliance corresponding to the minimum value of the angle α in the [20°; 75°] interval, which is $\alpha = 20°$. As the figure illustrates, the compliance $C_{u_y-f_y}$ increases as the angle α increases although the values of l, c and t were kept constant.

The following ranges were considered: $l \rightarrow$ [0.01; 0.05] m, $t \rightarrow$ [0.0005; 0.0015] m and $c \rightarrow$ [0.001; 0.01] m, in order to study the variation of the compliance $C_{u_y-f_y}$ for the flexible hinges defined in Eqs. (3.68)–(3.70). For all these hinge configurations, the compliance increased with an increase in l, a decrease in t and a decrease in c in a monotonic manner. It was also noticed that the hyperbolic design generates the smallest compliance of all the configurations analyzed for the same values of l, c and t. As a consequence, the five hinge configurations were compared based on the ratio of the circular, elliptic, parabolic and inverse parabolic hinge compliances $C_{u_y-f_y}$ to the same compliance of the hyperbolic hinge. Table 3.15 displays the extreme values of these ratios. The subscript "*eh*" means elliptic-to-hyperbolic, and similarly, "*ch*", "*ph*" and "*ih*" relate the circular, parabolic and inverse parabolic hinges to the hyperbolic hinge geometry. An angle $\alpha = 70°$ was used for the elliptic designs. The following constant values were also utilized in conjunction with the parameter ranges: $l = 0.05$ m, $t = 0.0005$ m and $c = 0.002$ m.

As Table 3.15 shows, the most compliant configuration of the four designs that are compared to the hyperbolic version is the inverse parabolic one, for which

TABLE 3.15

Compliance Ratio Comparing Flexible Hinges of Rectangular Cross-Section with Elliptic, Circular, Parabolic and Inverse Parabolic Profiles to the Hyperbolic Profile

	l (m)		t (m)		c (m)	
	$l_{min} = 0.01$	$l_{max} = 0.05$	$t_{min} = 0.0005$	$t_{max} = 0.0015$	$c_{min} = 0.001$	$c_{max} = 0.01$
r_{eh}	2.3	2.3	2.30	1.55	1.77	4.69
r_{ch}	2.04	1.91	1.91	1.32	1.49	4.14
r_{ph}	1.9	1.9	1.90	1.32	1.48	3.84
r_{ih}	4.79	4.79	4.79	2.08	2.75	21.09

the compliance relative to the hyperbolic-hinge compliance is the largest. It can also be seen that, except for the circular design, the compliance ratios of all other hinge configurations are constant when l varies.

Shape/dimension optimization:

The trends highlighted by the data in Table 3.15 enable to perform a relatively simple optimization given the monotonic variation of the compliances and their ratios. The largest values of the $C_{u_y-f_y}$ compliance are obtained when l is maximum and t and c are minimum for all analyzed hinge configurations. Conversely, the smallest compliance values result for l being minimum and t and c being maximum. The maximum compliance of all configurations is produced by the inverse parabolic profile, while the minimum compliance corresponds to the hyperbolic design. The following compliance ratio values related to the extreme values were obtained from the parameter ranges under consideration:

$$\frac{C_{u_y-f_y,max}^{(h)}}{C_{u_y-f_y,min}^{(h)}} = 10,098 \quad \frac{C_{u_y-f_y,max}^{(i)}}{C_{u_y-f_y,min}^{(i)}} = 3,699 \quad \frac{C_{u_y-f_y,max}^{(i)}}{C_{u_y-f_y,min}^{(h)}} = 27,813. \quad (3.71)$$

All these ratios are very large, and they point out that relatively small dimensional changes can lead to substantial variations in the bending compliance of any given hinge configuration (Eq. (3.71) only looks at the hyperbolic – superscript "h" – and the inverse parabolic – superscript "i" – designs, but similar figures are obtained for the other hinges considered here). The last ratio of Eq. (3.71) additionally indicates that choosing a particular shape of a notch hinge over another shape could produce substantial changes in the bending compliance, particularly if dimensions are also altered correspondingly.

Circular cross-section:

A similar study is performed for circular cross-sectional hinges. Note that Eqs. (3.68)–(3.70) are still valid – the only difference is that a variable diameter $d(x)$ should be used instead of the thickness $t(x)$. For elliptical designs, the angle α can be arbitrarily selected to determine the values of a and b from Eqs. (3.69). It is noted again that the compliance $C_{u_y-f_y}$ increases with the angle α in the case of elliptical hinges.

TABLE 3.16

Compliance Ratio Comparing Flexible Hinges of Circular Cross-Section with Circular, Elliptic, Hyperbolic, Parabolic and Inverse Parabolic Profiles

	l (m)		*d* (m)		*c* (m)	
	$l_{min} = 0.01$	$l_{max} = 0.05$	$d_{min} = 0.0005$	$d_{max} = 0.0015$	$c_{min} = 0.001$	$c_{max} = 0.01$
r_{eh}	2.44	2.44	2.44	1.65	1.88	4.97
r_{ch}	2.15	2.01	2.01	1.38	1.56	4.38
r_{ph}	2.00	2.00	2.00	1.38	1.56	4.07
r_{ih}	5.36	5.36	5.36	2.30	3.07	23.46

A parameter analysis and compliance comparison similar to the one used for rectangular cross-sectional hinges have been performed for circular cross-section hinges, and the results are included in Table 3.16. An angle $\alpha = 70°$ was used for the elliptical hinge. The following constant values were utilized: $l = 0.05$ m, $d = 0.0005$ m and $c = 0.002$ m.

The inverse parabolic design is the most compliant, whereas the least compliant is the hyperbolic configuration.

Shape/dimension optimization:

The monotonic variation of the compliances and their ratios, as displayed in Table 3.16, facilitates optimization in a straightforward manner. Similar to the rectangular cross-sectional hinges, the largest values of the $C_{u_y-f_y}$ compliance are obtained when l is maximum and d and c are minimum for all analyzed configurations. The maximum compliance of all configurations is produced by the inverse parabolic profile, while the minimum compliance corresponds to the hyperbolic profile. The following maximum compliance ratio values were obtained for the parameter ranges under consideration:

$$\frac{C_{u_y-f_y,max}^{(h)}}{C_{u_y-f_y,min}^{(h)}} = 30,239 \quad \frac{C_{u_y-f_y,max}^{(i)}}{C_{u_y-f_y,min}^{(i)}} = 11,075 \quad \frac{C_{u_y-f_y,max}^{(i)}}{C_{u_y-f_y,min}^{(h)}} = 92,844. \quad (3.72)$$

The ratios of Eq. (3.72) indicate that relatively small dimensional changes, as well as different hinge shapes, can result in very large variations in the bending compliance.

3.2.2.3.2 Right Corner-Filleted Flexible Hinges

Two classes of right corner-filleted flexible hinges are studied in this section: one comprises configurations defined by single curves, whereas the other class is formed of hinges whose geometric profiles are defined by two (or more) planar curves. All designs have axial and transverse symmetry.

Single-profile hinges:

The standard elliptic hinge is compared to the configurations defined by rational quadratic Bézier curves (ellipse, parabola and hyperbola). Configurations with rectangular and circular cross-sections are discussed separately.

Rectangular cross-section: The quarter model has, again, an envelope of dimensions $l/2$ and $t/2 + c$, as illustrated in Figure 3.15; therefore, the geometric parameters that define the planar geometry of these designs are l, t and c.

For an elliptic hinge, the semi-axes lengths and the variable thickness are expressed as:

$$a = \frac{l}{2}; \quad b = c; \quad t(x) = t + 2b \cdot (1 - \cos\varphi). \tag{3.73}$$

with φ ranging in the $[0, \pi/2]$ interval.

A generic point on a rational quadratic Bézier hinge has the coordinates $x(p)$, $y(p)$ and a thickness $t(p)$, which are expressed based on Eq. (2.82), where $l/2$ is used instead of l, namely:

$$\begin{cases} x(p) = \dfrac{w_1 \cdot l \cdot (1-p) \cdot p + (l/2) \cdot p^2}{(1-p)^2 + 2w_1 \cdot (1-p) \cdot p + p^2}; \\[4mm] y(p) = \dfrac{(t/2) \cdot (1-p)^2 + w_1 \cdot t \cdot (1-p) \cdot p + (c + t/2) \cdot p^2}{(1-p)^2 + 2w_1 \cdot (1-p) \cdot p + p^2}; \quad t(p) = 2y(p) \end{cases} \tag{3.74}$$

The parameter p varies in the $[0, 1]$ interval. As discussed in Chapter 2, the parameter w_1 assumes the following values: $w_1 < 1$ for elliptic curves, $w_1 = 1$ for parabolic profiles and $w_1 > 1$ in the case of hyperbolic shapes.

The compliance $C_{u_y - f_y}$ has been used, again, to form ratios in order to assess the deformation performance of these hinge configurations in terms of the four parameters: l, t, c and w_1. For each parameter, a variation range was used while keeping the other three parameters constant – the constant values were $l = 0.02\,\text{m}$, $t = 0.001\,\text{m}$, $c = 0.004\,\text{m}$ and $w_1 = 0.1$ for ellipses, and $w_1 = 10$ for hyperbolic curves. All compliances increase with an increase in l and w_1 and with a decrease in t and c. It was also noticed that the least compliant configuration is the Bézier elliptic hinge; as a consequence, compliance ratios based on $C_{u_y - f_y}$ were formed having the compliance of the Bézier elliptic hinge in the ratio denominator, while the other three hinge compliances were in the numerator. Table 3.17 shows these compliance ratios.

In Table 3.17, the subscripts "h", "p", "e" and "es" stand for hyperbolic, parabolic, elliptic (all Bézier curves) and standard elliptic, respectively. The most compliant configuration, as given in Table 3.17, is the Bézier hyperbolic hinge, followed by the Bézier parabolic, the standard elliptic and the Bézier elliptic design. It can also be seen that the hinge length variation does not affect the compliance ratio; however, the compliance ratio increases with an increase in c and with a decrease in t.

Circular cross-section: A similar analysis was performed for the right corner-filleted flexible hinges studied in the previous section by considering a circular

TABLE 3.17
Compliance Ratio Comparing Rectangular Cross-Sectional Right Corner-Filleted Hinges of Standard Elliptic and Bézier Profiles

	l (m)		t (m)		c (m)	
	$l_{min} = 0.01$	$l_{max} = 0.05$	$t_{min} = 0.0005$	$t_{max} = 0.0015$	$c_{min} = 0.001$	$c_{max} = 0.008$
r_{he}	11.09	11.09	17.39	8.32	4.02	17.39
r_{pe}	3.75	3.75	4.63	3.26	2.23	4.63
r_{ese}	2.97	2.97	3.56	2.63	1.92	3.56

TABLE 3.18
Compliance Ratio Comparing Circular Cross-Sectional Right Corner-Filleted Hinges of Standard Elliptic and Bézier Profiles

	l (m)		d (m)		c (m)	
	$l_{min} = 0.01$	$l_{max} = 0.05$	$d_{min} = 0.0005$	$d_{max} = 0.0015$	$c_{min} = 0.001$	$c_{max} = 0.008$
r_{he}	14.57	14.57	22.07	11.08	5.33	22.07
r_{pe}	4.32	4.32	5.20	3.80	2.62	5.20
r_{ese}	3.35	3.35	3.94	3.00	2.20	3.94

cross-section this time. All considerations and parameter values/ranges used in the case of the rectangular cross-sectional hinges were also applied here; the only differences are that the diameter d needs to be used instead of the thickness t, and the adequate properties of a circular cross-section need to be utilized. The results and trends in the compliance ratio that are indicated in Table 3.18 are similar to those of Table 3.17.

As Table 3.18 points out, and as the case was with the rectangular cross-sectional configurations, the least compliant design is the Bézier elliptic hinge, followed by the standard elliptic, the Bézier parabolic and the Bézier hyperbolic configurations. The hinge length l variation does not influence the compliance ratios, while increasing c and decreasing d produce an increase of the compliance ratio.

Multiple-profile hinges:

There are numerous possible right corner-filleted hinge designs that are defined by multiple curves – see Figure 3.17, for instance. The configuration of Figure 3.17a, whose quarter segment is formed of a parabolic curve and a circular curve, has been analyzed in Lobontiu et al. [20]; the interested reader may want to study the (new) design of Figure 3.17b. Discussed are here two common right corner-filleted configurations, namely, the circular design of Figure 3.13 and the elliptical configuration of Figure 3.14.

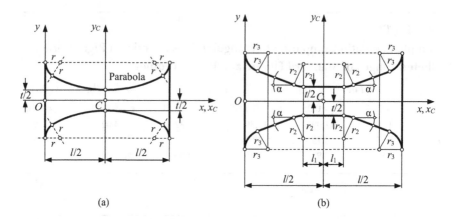

FIGURE 3.17 Planar geometry of right corner-filleted flexible hinges with a profile determined by (a) parabolic-circular curves; (b) straight-circular curves.

The planar geometry of a right circularly corner-filleted flexible hinge is defined by its length l, constant thickness t and fillet radius r. Because $r = c$, this configuration is fully described by l, t and c. A right elliptically corner-filleted hinge is characterized by its ellipse semi-axes lengths a and b in addition to l and t. As seen in Figure 3.14, $b = c$, and therefore, the planar profile of this design is fully defined by l, t, c and a. When the two configurations are compared in terms of their elastic behavior, the elliptic design has one extra parameter, its ellipse length a. As a consequence, the following compliance ratio: $C^c_{u_y - f_y} / C^e_{u_y - f_y}$ (where "e" stands for elliptical and "c" means circular) can be utilized as a relative quantifier of the two hinges' in-plane bending capabilities.

Figure 3.18a is the plot of this compliance ratio with a being the variable (a ranges from $a = c$ to $a = l/2$). The other parameters are constant, namely, $l = 0.05\,\mathrm{m}$, $t = 0.0005\,\mathrm{m}$, $c = 0.005\,\mathrm{m}$, $w = 0.008\,\mathrm{m}$ and $E = 2 \cdot 10^{11}\,\mathrm{N/m^2}$. It can be

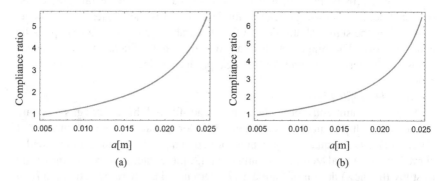

FIGURE 3.18 Plots of compliance ratio $C^c_{u_y - f_y} / C^e_{u_y - f_y}$ in terms of the ellipse semi-axis length a for: (a) rectangular cross-section; (b) circular cross-section.

seen that as a increases, the circular design becomes more bending compliant than the elliptic design. When $a = l/2$ (and the elliptic hinge becomes a right elliptical configuration), the circular design compliance is approximately five times larger than the one of the elliptical configurations. A similar plot is shown in Figure 3.18b for circular cross-sectional designs. The same values of the constant parameters were used (with a diameter d instead of the thickness t), and the conclusions are similar to the ones pertaining to the rectangular cross-sectional configurations. The circular variant can be six times more bending-compliant than its elliptical counterpart.

3.2.2.4 Two-Axis Flexible Hinges

The flexible hinges studied thus far in this chapter had either variable circular cross-section or a rectangular cross-section with one dimension constant and the other variable. The latter design category identified a sensitive axis for hinge bending, which is the primary (desired) deformation of a notch hinge. We discuss here two design groups that enable the hinge to bend around two perpendicular axes. Both configurations have rectangular cross-sections and are defined by two pairs of notches in perpendicular planes. In one design group, the notch pairs are axially collocated and the rectangular cross-section has both dimensions $w(x)$ and $t(x)$ variable. The notch pairs in the other design category are serially connected, and each notch pair has only one cross-sectional dimension variable – either $t(x)$ or $w(x)$.

3.2.2.4.1 Collocated Notches

The hinge of Figure 3.19 is the three-dimensional rendering of a flexible hinge with two pairs of identical and mirrored notches – a similar design is introduced in Lobontiu and Garcia [17]; the hinge of Figure 3.19 is based on the segment studied in Chapter 2 and shown in Figure 2.22 – that segment is half the hinge shown in Figure 3.19, and this final configuration has transverse symmetry in addition to axial symmetry. Therefore, the notches are axially collocated. For this design, two

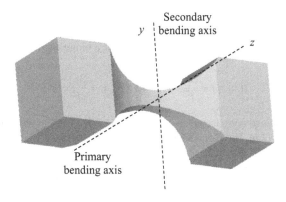

FIGURE 3.19 Three-dimensional view of two-axis flexible hinge with collocated notches.

bending axes can be defined at the hinge symmetry center: one is the primary axis and is associated with the maximum bending capacity, whereas the other axis, the secondary axis, is designed to produce a hinge design that is slightly stiffer in bending. This configuration can be tailored to be sensitive to well-defined bending loads that deform the hinge about the two central axes.

Consider, for instance, that the hinge of Figure 3.19 is formed of two identical halves, each defined by two parabolic notches, as illustrated in Figure 2.23 and discussed in *Example 2.3*. The parameters that define this configuration are the minimum thickness t, the minimum width w, the length l as well as the root dimensions c_t and c_w. It can be assumed that $t < w$ and also $c_t < c_w$, which means that $t(x) < w(x)$. In this situation, the primary bending axis is the z-axis (see Figures 2.23 and 3.19), which is perpendicular to the minimum parameter t, and the secondary bending axis is the y-axis. The compliance ratio $r = C_{u_y-f_y} / C_{u_z-f_z}$ is of interest as it reflects the relationship between the two bending axes.

Example 3.4

Compare the bending performance about the primary axis to that about the secondary axis by graphically analyzing the compliance ratio $r = C_{u_y-f_y} / C_{u_z-f_z}$ for a two-axis flexible hinge with collocated parabolic notches, as in *Example 2.3*. Consider the following parameter ranges: $t \rightarrow$ [0.001; 0.003] m, $w \rightarrow$ [0.003; 0.006] m, $c_t \rightarrow$ [0.001; 0.006] m and $c_w \rightarrow$ [0.006; 0.01] m.

Solution:

The two compliances of this ratio are calculated based on half-hinge compliances as per Eqs. (3.53) and (3.56):

$$
\begin{cases}
C_{u_y-f_y} = 2C_{u_y-f_y}^{(h)} + \dfrac{l^2}{2} \cdot C_{\theta_z-m_z}^{(h)} \\[4mm]
C_{u_z-f_z} = 2C_{u_z-f_z}^{(h)} + \dfrac{l^2}{2} \cdot C_{\theta_y-m_y}^{(h)}
\end{cases}
\tag{3.75}
$$

The variable thickness and width are expressed based on the similar parameters formulated in Eq. (2.98) of *Example 2.3* for a hinge segment of length l:

$$
t(x) = t + \frac{8c_t}{l^2} \cdot x^2; \quad w(x) = w + \frac{8c_w}{l^2} \cdot x^2.
\tag{3.76}
$$

The four half-hinge compliances that intervene in Eqs. (3.75) are determined with the aid of Eqs. (2.6). The plots of Figures 3.20 and 3.21 show the compliance ratio variation in terms of t, w, c_t and c_w. The following constant values were used: $l = 0.015$ m, $t = 0.001$ m, $w = 0.005$ m, $c_t = 0.005$ m and $c_w = 0.006$ m in conjunction with the parameter that was varied. It can be seen that the compliance ratio decreases with an increase in t and c_t; the ratio increases when w and c_w increase.

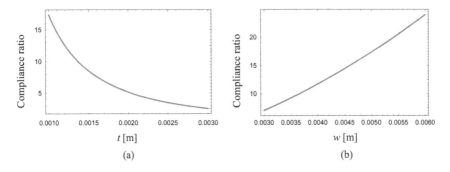

FIGURE 3.20 Plots of compliance ratio in terms of the: (a) minimum thickness t; (b) minimum width w.

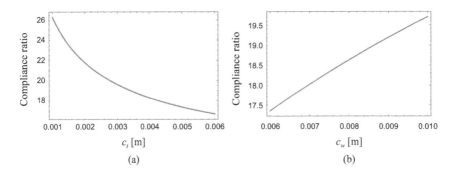

FIGURE 3.21 Plots of compliance ratio in terms of root parameter: (a) c_t; (b) c_w.

3.2.2.4.2 Non-Collocated Notches

When the two notches of the hinge configuration of Figure 3.19 are combined in series, instead of being axially collocated, the flexible hinge of Figure 3.22 results. A similar design with circular notches has been studied in Paros and Weisbord [1]. The geometric details of a generic non-collocated design are shown in Figure 3.23. The hinge is formed of two portions connected in series: each segment has one dimension of its rectangular cross-sectional constant, while the other dimension is variable. It is also assumed that the two segments have the same length l. Assuming, again, that $t < c$ and $c_t < c_w$, the primary bending axis is the one perpendicular on t; the secondary bending, also identified in Figures 3.22 and 3.23, is perpendicular to w.

The in-plane compliance matrix of the full hinge considers bending around the z-axis, which is parallel to the primary sensitive axis, and is calculated in the $Oxyz$ reference frame in Figure 3.23 as:

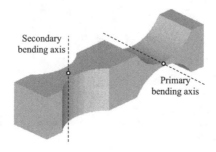

FIGURE 3.22 Three-dimensional view of two-axis flexible hinge with non-collocated notches.

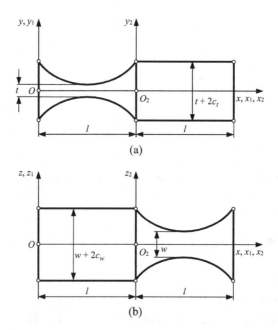

FIGURE 3.23 Geometric parameters of non-collocated flexible hinge: (a) front view; (b) side view.

$$\left[C_{ip}\right]=\left[C_{ip}^{(1)}\right]+\left[T_{ip}^{(2)}\right]^{T}\left[C_{ip}^{(2)}\right]\left[T_{ip}^{(2)}\right] \quad \text{with}$$

$$\left[T_{ip}^{(2)}\right]=\begin{bmatrix} 1 & 0 & 0 \\ 0 & 1 & 0 \\ \Delta y_2 & -\Delta x_2 & 1 \end{bmatrix}=\begin{bmatrix} 1 & 0 & 0 \\ 0 & 1 & 0 \\ 0 & -l & 1 \end{bmatrix}. \tag{3.77}$$

The out-of-plane compliance matrix quantifies bending around the global y-axis, which is parallel to the secondary sensitive axis – this matrix is expressed as:

$$\left[C_{op}\right] = \left[C_{op}^{(1)}\right] + \left[T_{op}^{(2)}\right]^{T} \left[C_{op}^{(2)}\right] \left[T_{op}^{(2)}\right] \quad \text{with}$$

$$\left[T_{op}^{(2)}\right] = \begin{bmatrix} 1 & 0 & -\Delta y_2 \\ 0 & 1 & \Delta x_2 \\ 0 & 0 & 1 \end{bmatrix} = \begin{bmatrix} 1 & 0 & 0 \\ 0 & 1 & l \\ 0 & 0 & 1 \end{bmatrix}. \tag{3.78}$$

Both hinge segments have transverse symmetry, and as a result, these segments' compliances are expressed in terms of their half-segment compliances, as per Eqs. (3.53) and (3.56). As a consequence, the in-plane compliance matrix of Eq. (3.77) becomes:

$$\left[C_{ip}\right] = \begin{bmatrix} C_{u_x-f_x} & 0 & 0 \\ 0 & C_{u_y-f_y} & C_{u_y-m_z} \\ 0 & C_{u_y-m_z} & C_{\theta_z-m_z} \end{bmatrix}, \tag{3.79}$$

with

$$\begin{cases} C_{u_x-f_x} = 2\left(C_{u_x-f_x}^{(1h)} + C_{u_x-f_x}^{(2h)}\right) \\ C_{u_y-f_y} = 2\left(C_{u_y-f_y}^{(1h)} + C_{u_y-f_y}^{(2h)}\right) + \dfrac{l^2}{2} \cdot \left(C_{\theta_z-m_z}^{(1h)} + 9C_{\theta_z-m_z}^{(2h)}\right) \\ C_{u_y-m_z} = -l\left(C_{\theta_z-m_z}^{(1h)} + 3C_{\theta_z-m_z}^{(2h)}\right) \\ C_{\theta_z-m_z} = 2\left(C_{\theta_z-m_z}^{(1h)} + C_{\theta_z-m_z}^{(2h)}\right) \end{cases} \tag{3.80}$$

The out-of-plane compliance matrix of Eq. (3.78) is expressed as:

$$\left[C_{op}\right] = \begin{bmatrix} C_{\theta_x-m_x} & 0 & 0 \\ 0 & C_{\theta_y-m_y} & C_{\theta_y-f_z} \\ 0 & C_{\theta_y-f_z} & C_{u_z-f_z} \end{bmatrix}, \tag{3.81}$$

with

$$\begin{cases} C_{\theta_x-m_x} = 2\left(C_{\theta_x-m_x}^{(1h)} + C_{\theta_x-m_x}^{(2h)}\right) \\ C_{\theta_y-m_y} = 2\left(C_{\theta_y-m_y}^{(1h)} + C_{\theta_y-m_y}^{(2h)}\right) \\ C_{\theta_y-f_z} = l\left(C_{\theta_y-m_y}^{(1h)} + 3C_{\theta_y-m_y}^{(2h)}\right) \\ C_{u_z-f_z} = 2\left(C_{u_z-f_z}^{(1h)} + C_{u_z-f_z}^{(2h)}\right) + \dfrac{l^2}{2} \cdot \left(C_{\theta_y-m_y}^{(1h)} + 9C_{\theta_y-m_y}^{(2h)}\right) \end{cases} \tag{3.82}$$

Example 3.5

Compare the deformation performance in bending around the primary axis of a non-collocated hinge design to its collocated counterpart by using the parabolic profile with the basic geometric parameters of *Example 3.4*.

Solution:

The cross-sectional properties of the two segments' halves in the non-collocated configuration are:

$$A^{(1)} = (w + 2c_w) \cdot t(x); \quad I_z^{(1)} = \frac{(w + 2c_w) \cdot t(x)^3}{12}; \quad I_y^{(1)} = \frac{(w + 2c_w)^3 \cdot t(x)}{12};$$

$$I_t^{(1)} = \frac{(w + 2c_w) \cdot t(x)^3}{3};$$

$$A^{(2)} = (t + 2c_t) \cdot w(x); \quad I_z^{(2)} = \frac{(t + 2c_t)^3 \cdot w(x)}{12}; \quad I_y^{(2)} = \frac{(t + 2c_t) \cdot w(x)^3}{12};$$

$$I_t^{(2)} = \begin{cases} \dfrac{(t + 2c_t) \cdot w(x)^3}{3}, & w(x) \ll t + 2c_t \\[3mm] \dfrac{(t + 2c_t)^3 \cdot w(x)}{3}, & w(x) \gg t + 2c_t \end{cases}. \tag{3.83}$$

The torsional moments of area in Eqs. (3.83) assumed thin sections. However, for the other possible cross-sectional dimensions, Eqs. (2.20) or (2.21) should be used appropriately.

The plots of Figures 3.24 and 3.25 illustrate the variation of the $C_{u_y - f_y}$ ratio (non-collocated compliance to collocated compliance) for the constant values of $l = 0.015$ m, $t = 0.001$ m, $w = 0.005$ m, $c_t = 0.005$ m and $c_w = 0.006$ m – these values were used in conjunction with the parameter that was varied.

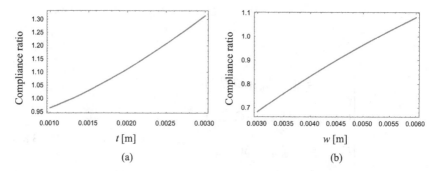

FIGURE 3.24 Plots of compliance ratio in terms of the: (a) minimum thickness t; (b) minimum width w.

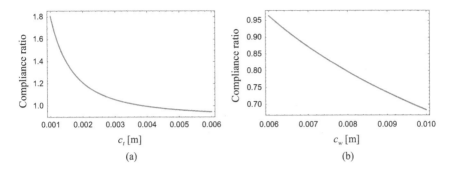

FIGURE 3.25 Plots of compliance ratio in terms of root parameter: (a) c_t; (b) c_w.

As Figure 3.24 indicates, the non-collocated design is slightly less compliant in bending about the primary (z)-axis for values of t and w situated on the lower portion of the selected range – this is shown by the ratio values smaller than 1. However, the compliance ratio increases and assumes values larger than 1 when t and w increase, which indicates that the non-collocated hinge becomes more flexible than the collocated one. The non-collocated design is more compliant than its collocated counterpart when c_t varies, as illustrated in Figure 3.25a, except for a small interval close to the upper limit, where the relationship reverses. The non-collocated variant is less compliant than the collocated hinge for the entire range of c_w, as seen in Figure 3.25b.

3.2.3 FOLDED, SPATIALLY PERIODIC FLEXIBLE HINGES

Folded hinges can be designed to highly increase the deformation capacity of regular, solid flexible hinges by serially connecting several straight-axis segments (of constant or variable cross-section) in a meandering manner, as illustrated in Figure 3.26. The folding segments can be confined by a rectangular envelope (as shown in Figure 3.26a) or by any other envelope curve (as suggested in Figure 3.26b).

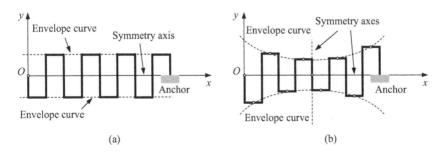

FIGURE 3.26 Skeleton representation of folded flexible hinges with: (a) rectangular envelope; (b) curvilinear envelope.

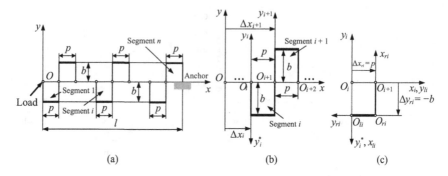

FIGURE 3.27 Rectangular envelope-folded flexible hinge: (a) skeleton representation; (b) geometric details of two neighboring arbitrary hinge segments; (c) basic segment i.

While Chapter 4 studies the class of folded hinges with curvilinear envelope, the analysis here is limited to folded hinges defined by rectangular envelopes. Consider the design of Figure 3.27a showing the skeleton representation of a configuration whose envelope curves are lines parallel to and symmetric with respect to the x axis, which results in a rectangular area of length l and height $2b$ enveloping the full hinge.

Assume that all longer, vertical segments are of constant cross-section and bending-compliant only, whereas the shorter, horizontal segments connecting the longer beams are rigid. This particular design results in units that are geometrically identical; as a consequence, the full hinge has a spatial periodicity along the x axis and its compliances can be evaluated from the compliances of the basic rectangular unit sketched in Figure 3.27c. The in-plane compliance matrix of the full hinge is derived next – they are important to predict the motion capability of the entire hinge. An in-plane load vector $[f] = \begin{bmatrix} f_x & f_y & m_z \end{bmatrix}^T$ applied at the hinge free end O produces the displacement vector $[u] = \begin{bmatrix} u_x & u_y & \theta_z \end{bmatrix}^T$ at the same point. The two vectors are connected by means of a compliance matrix as $[u] = [C_O][f]$.

The compliance matrix of the basic segment i sketched in Figure 3.27c is determined first. It is assumed that this segment is free at the end O_i and fixed at the opposite end O_{i+1}. When axial deformation is neglected, the in-plane compliance matrix of either the left segment O_iO_{li} or the right segment $O_{ri}O_{i+1}$ in their local frames is:

$$\left[C_{O_i}^{(il)} \right] = \left[C_{O_{ri}}^{(ir)} \right] = \begin{bmatrix} 0 & 0 & 0 \\ 0 & \dfrac{b^3}{3EI} & -\dfrac{b^2}{2EI} \\ 0 & -\dfrac{b^2}{2EI} & \dfrac{b}{EI} \end{bmatrix}, \tag{3.84}$$

where E is the material Young's modulus and I is the z-axis area moment of the vertical beams constant cross-section. The in-plane compliance matrix of the basic unit i in the reference frame $O_i x_i y_i z_i$ is calculated by transferring the compliances of the two serially connected beams from their local frames to the $O_i x_i y_i z$ frame and adding them as:

$$\left[C_{O_i}^{(i)} \right] = \left[R^{(il)} \right]^T \left[C_{O_i}^{(il)} \right] \left[R^{(il)} \right] + \left[T_{O_i O_{ri}}^{(ir)} \right]^T \left[R^{(ir)} \right]^T \left[C_{O_i}^{(il)} \right] \left[R^{(ir)} \right] \left[T_{O_i O_{ri}}^{(ir)} \right], \quad (3.85)$$

where the rotation and translation matrices are:

$$\left[R^{(il)} \right] = \begin{bmatrix} \cos\left(-\dfrac{\pi}{2}\right) & \sin\left(-\dfrac{\pi}{2}\right) & 0 \\ -\sin\left(-\dfrac{\pi}{2}\right) & \cos\left(-\dfrac{\pi}{2}\right) & 0 \\ 0 & 0 & 1 \end{bmatrix} = \begin{bmatrix} 0 & -1 & 0 \\ 1 & 0 & 0 \\ 0 & 0 & 1 \end{bmatrix};$$

$$\left[R^{(ir)} \right] = \begin{bmatrix} \cos\left(\dfrac{\pi}{2}\right) & \sin\left(\dfrac{\pi}{2}\right) & 0 \\ -\sin\left(\dfrac{\pi}{2}\right) & \cos\left(\dfrac{\pi}{2}\right) & 0 \\ 0 & 0 & 1 \end{bmatrix} = \begin{bmatrix} 0 & 1 & 0 \\ -1 & 0 & 0 \\ 0 & 0 & 1 \end{bmatrix};$$

$$\left[T_{O_i O_{ri}}^{(ir)} \right] = \begin{bmatrix} 1 & 0 & 0 \\ 0 & 1 & 0 \\ \Delta y_{ri} & -\Delta x_{ri} & 1 \end{bmatrix} = \begin{bmatrix} 1 & 0 & 0 \\ 0 & 1 & 0 \\ -b & -p & 1 \end{bmatrix}. \quad (3.86)$$

The compliance matrix of the same segment can also be expressed in the reference frame: $O_i x_i y_i^* z_i^*$ as:

$$\left[C_{O_i}^{(i*)} \right] = \begin{bmatrix} C_{O_i}^{(i)}(1,1) & -C_{O_i}^{(i)}(1,2) & -C_{O_i}^{(i)}(1,3) \\ -C_{O_i}^{(i)}(1,2) & C_{O_i}^{(i)}(2,2) & C_{O_i}^{(i)}(2,3) \\ -C_{O_i}^{(i)}(1,3) & C_{O_i}^{(i)}(2,3) & C_{O_i p}^{(i)}(3,3) \end{bmatrix} = \left[C_{O_i}^{(i+1)} \right]. \quad (3.87)$$

where the matrix components are taken from $\left[C_{O_i}^{(i)} \right]$ of Eq. (3.85). Note that Eq. (3.87) also expresses the compliance matrix $\left[C_{O_i}^{(i+1)} \right]$ of the segment $i+1$, which is sketched in Figure 3.27b next to the segment i. The minus signs in Eq. (3.87) indicate the components that correspond to an axis that is flipped (either y_i or z_i).

The in-plane compliance matrix at O of the full hinge shown in Figure 3.27a is calculated by adding the translated matrices of Eqs. (3.85) and (3.87) as:

$$[C_O] = \sum_{i=1}^{n} \left(\left[T_{OO_i}^{(i)} \right]^T \left[C_{O_i}^{(i)} \right] \left[T_{OO_i}^{(i)} \right] + \left[T_{OO_i}^{(i+1)} \right]^T \left[C_{O_i}^{(i+1)} \right] \left[T_{OO_i}^{(i+1)} \right] \right). \qquad (3.88)$$

where the translation matrices are:

$$
\left[T_{OO_i}^{(i)} \right] =
\begin{bmatrix}
1 & 0 & 0 \\
0 & 1 & 0 \\
0 & -\Delta x_i & 1
\end{bmatrix}
=
\begin{bmatrix}
1 & 0 & 0 \\
0 & 1 & 0 \\
0 & -(i-1)p & 1
\end{bmatrix};
$$

$$
\left[T_{OO_i}^{(i+1)} \right] =
\begin{bmatrix}
1 & 0 & 0 \\
0 & 1 & 0 \\
0 & -\Delta x_{i+1} & 1
\end{bmatrix}
=
\begin{bmatrix}
1 & 0 & 0 \\
0 & 1 & 0 \\
0 & -i \cdot p & 1
\end{bmatrix}.
$$

$$(3.89)$$

After performing the summation involved in Eq. (3.88), the final expression of the in-plane compliance matrix for the full flexible hinge of Figure 3.27a is:

$$
[C_O] =
\begin{bmatrix}
C_{u_x - f_x} & C_{u_x - f_y} & C_{u_x - m_z} \\
C_{u_x - f_y} & C_{u_y - f_y} & C_{u_y - m_z} \\
C_{u_x - m_z} & C_{u_y - m_z} & C_{\theta_z - m_z}
\end{bmatrix}
$$

$$
= \frac{nb}{EI} \cdot
\begin{bmatrix}
\dfrac{2b^2}{3} & \dfrac{(3n-2)bp}{8} & 0 \\
\dfrac{(3n-2)bp}{8} & \dfrac{(17n^2 + 18n + 16)p^2}{12} & -(n+1)p \\
0 & -(n+1)p & 2
\end{bmatrix}. \qquad (3.90)
$$

Assuming the flexible beams have a constant rectangular cross-section of in-plane thickness t and out-of-plane width w, the area moment is $I = w \cdot t^3 / 12$. It is noteworthy to compare the direct compliance $C_{u_y - f_y}$ of the folded hinge to the similar compliance $C_{u_y - f_y}^s$ of a solid flexure hinge defined by the same dimensions l, b and w, and Young's modulus E. The compliance of the solid hinge is $l^3 / (3EI^s)$, where $I^s = w \cdot 2b^3 / 12$. The compliance ratio is plotted in Figure 3.28 in terms of the number n of identical units. As the figure indicates, the direct compliance decreases

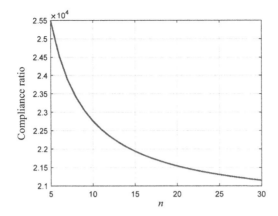

FIGURE 3.28 Plot of compliance ratio $C_{u_y-f_y} / C^s_{u_y-f_y}$ in terms of the number of segments n.

nonlinearly from approximately 25,500 to 21,000 when the number of segments increases from 5 to 60. The following numerical values were used: $b = 0.007$ m, $t = 0.001$ m $p = 0.004$ m, and $l = p \cdot n$.

3.3 PARALLEL-CONNECTION FLEXIBLE HINGES

Traditionally, a flexible hinge is a single member, possibly formed by serially connecting several geometrically dissimilar segments and attached to adjoining rigid links. Expanding upon this concept, two rigid links can also be connected by two or more, identical or dissimilar, flexible segments, as sketched in Figure 3.29. Collectively, the parallel-connection segments can be considered a single hinge from a functional standpoint – see Lobontiu [31], for instance.

Figure 3.30 is the skeleton generic representation of a flexible hinge with n straight-axis segments connected in parallel. Note that each parallel member can be formed by serially combining several individual flexible segments. Also observe that the parallel-connection hinge segments do not have to be geometrically parallel.

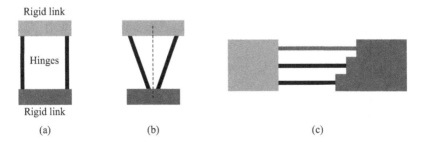

FIGURE 3.29 Parallel flexible hinges with: (a) two geometrically parallel and identical segments; (b) two axially symmetric and identical segments; (c) three (multiple) geometrically parallel segments with dissimilar lengths.

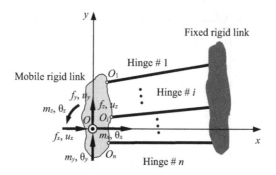

FIGURE 3.30 Parallel-connection flexible hinge identifying loads and displacements at the mobile rigid link.

Consider that one of the rigid links is fixed and the other one is mobile. Under the action of an external load $[f] = [f_x f_y\, m_z\, m_x\, m_y\, f_z]^T$ applied at a point O on the mobile rigid link, the same point experiences the displacement $[u] = [u_x\, u_y\, \theta_z\, \theta_x\, \theta_y\, u_z]^T$ due to the elastic deformations of the n flexible hinges. The deformation and load vectors at O are related as:

$$[u] = [C_O][f]; \quad [C_O] = \left(\sum_{i=1}^{n} \left[C_O^{(i)} \right]^{-1} \right)^{-1}. \qquad (3.91)$$

where $[C_O]$ is the compliance matrix of a single hinge equivalent to the n actual parallel-connection hinges and $\left[C_O^{(i)} \right]$ is the compliance matrix of the generic segment i with respect to the frame centered at O.

The in-plane and the out-of-plane loads/displacements are separated as follows:

$$\begin{cases} \left[u_{ip} \right] = \left[C_{O,ip} \right]\left[f_{ip} \right]; \quad \left[u_{ip} \right] = \begin{bmatrix} u_x & u_y & \theta_z \end{bmatrix}^T \text{ and} \\[2mm] \left[f_{ip} \right] = \begin{bmatrix} f_x & f_y & m_z \end{bmatrix}^T; \\[2mm] \left[u_{op} \right] = \left[C_{O,op} \right]\left[f_{op} \right]; \quad \left[u_{op} \right] = \begin{bmatrix} \theta_x & \theta_y & u_z \end{bmatrix}^T \text{ and} \\[2mm] \left[f_{op} \right] = \begin{bmatrix} m_x & m_y & f_z \end{bmatrix}^T \end{cases} \qquad (3.92)$$

The in- and out-of-plane compliance matrices are calculated based on the corresponding segment compliance matrices as:

$$\left\{ \left[C_{O,ip} \right] = \left(\sum_{i=1}^{n} \left[C_{O,ip}^{(i)} \right]^{-1} \right)^{-1}; \quad \left[C_{O,op} \right] = \left(\sum_{i=1}^{n} \left[C_{O,op}^{(i)} \right]^{-1} \right)^{-1}. \right. \qquad (3.93)$$

We study here three categories of parallel hinges: one category is formed of identical segments that are placed symmetrically with respect to an axis; the second category is formed of identical segments that are geometrically parallel; the third group includes cross-leaf hinges with segments that are offset spatially.

3.3.1 Two-Member Flexible Hinge with Symmetry Axis

A parallel flexible hinge with two identical members that are placed symmetrically with respect to the x-axis is shown in skeleton form in Figure 3.29b and, with more geometric details, in Figure 3.31. Each of the two identical members can be formed of multiple collinear segments that are connected in series. It is assumed that the in- and out-of-plane compliance matrices of the top member are known/defined with respect to the reference frame $O_t x_t y_t z_t$ (where "t" denotes the top segment).

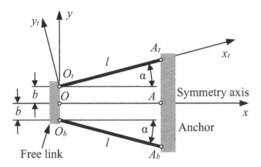

FIGURE 3.31 Skeleton representation of a parallel-connection flexible hinge with two identical and symmetric segments.

3.3.1.1 In-Plane Compliance Matrix

The in-plane compliance matrix of the two identical members in their local frames is assumed to be known; for instance, the matrix of the top member is:

$$\left[C^{(t)}_{O_t,ip} \right] = \begin{bmatrix} C^{(t)}_{u_x-f_x} & 0 & 0 \\ 0 & C^{(t)}_{u_y-f_y} & C^{(t)}_{u_y-m_z} \\ 0 & C^{(t)}_{u_y-m_z} & C^{(t)}_{\theta_z-m_z} \end{bmatrix} \tag{3.94}$$

with respect to the $O_t x_t y_t z_t$ reference frame. This matrix is transferred at O in the global $Oxyz$ reference frame as:

$$\left[C_{O,ip}^{(t)}\right]=\left[T_{ip}^{(t)}\right]^{T}\left[R^{(t)}\right]^{T}\left[C_{O_t,ip}^{(t)}\right]\left[R^{(t)}\right]\left[T_{ip}^{(t)}\right] \quad \text{with}$$

$$\left[T_{ip}^{(t)}\right]=\begin{bmatrix} 1 & 0 & 0 \\ 0 & 1 & 0 \\ \Delta y & -\Delta x & 1 \end{bmatrix}=\begin{bmatrix} 1 & 0 & 0 \\ 0 & 1 & 0 \\ b & 0 & 1 \end{bmatrix} \quad \text{and} \tag{3.95}$$

$$\left[R^{(t)}\right]=\begin{bmatrix} \cos\alpha & \sin\alpha & 0 \\ -\sin\alpha & \cos\alpha & 0 \\ 0 & 0 & 1 \end{bmatrix}.$$

In a similar fashion, the compliance of the bottom member (the superscript "b" means bottom) is transferred to the $Oxyz$ reference frame from its local frame located at O_b as:

$$\left[C_{O,ip}^{(b)}\right]=\left[T_{ip}^{(b)}\right]^{T}\left[R^{(b)}\right]^{T}\left[C_{O_t,ip}^{(t)}\right]\left[R^{(b)}\right]\left[T_{ip}^{(b)}\right] \quad \text{with}$$

$$\left[T_{ip}^{(b)}\right]=\begin{bmatrix} 1 & 0 & 0 \\ 0 & 1 & 0 \\ \Delta y & -\Delta x & 1 \end{bmatrix}=\begin{bmatrix} 1 & 0 & 0 \\ 0 & 1 & 0 \\ -b & 0 & 1 \end{bmatrix} \quad \text{and} \tag{3.96}$$

$$\left[R^{(t)}\right]=\begin{bmatrix} \cos(-\alpha) & \sin(-\alpha) & 0 \\ -\sin(-\alpha) & \cos(-\alpha) & 0 \\ 0 & 0 & 1 \end{bmatrix}.$$

The top and bottom segments are connected in parallel, and therefore, the in-plane compliance of the full, two-member hinge of Figure 3.31 is calculated as:

$$\left[C_{O,ip}\right]=\left(\left[C_{O,ip}^{(t)}\right]^{-1}+\left[C_{O,ip}^{(b)}\right]^{-1}\right)^{-1}=\begin{bmatrix} C_{u_x-f_x} & 0 & 0 \\ 0 & C_{u_y-f_y} & C_{u_y-m_z} \\ 0 & C_{u_y-m_z} & C_{\theta_z-m_z} \end{bmatrix}. \tag{3.97}$$

The elements of the matrix shown in Eq. (3.97) are very large and are not included here.

3.3.1.2 Out-of-Plane Compliance Matrix

A similar approach is used to calculate the out-of-plane compliance matrix of the full hinge with respect to the global frame $Oxyz$. The out-of-plane compliance matrix of the top member is:

$$\left[C_{O_t,op}^{(t)} \right] = \begin{bmatrix} C_{\theta_x-m_x}^{(t)} & 0 & 0 \\ 0 & C_{\theta_y-m_y}^{(t)} & C_{\theta_y-f_z}^{(t)} \\ 0 & C_{\theta_y-f_z}^{(t)} & C_{u_z-f_z}^{(t)} \end{bmatrix}. \tag{3.98}$$

with respect to the $O_t x_t y_t z_t$ reference frame. This matrix is transferred from O_t at O in the global $Oxyz$ reference frame as:

$$\left[C_{O,op}^{(t)} \right] = \left[T_{op}^{(t)} \right]^T \left[R^{(t)} \right]^T \left[C_{O_t,op}^{(t)} \right] \left[R^{(t)} \right] \left[T_{op}^{(t)} \right] \quad \text{with}$$

$$\left[T_{op}^{(t)} \right] = \begin{bmatrix} 1 & 0 & -\Delta y \\ 0 & 1 & \Delta x \\ 0 & 0 & 1 \end{bmatrix} = \begin{bmatrix} 1 & 0 & -b \\ 0 & 1 & 0 \\ 0 & 0 & 1 \end{bmatrix}. \tag{3.99}$$

Similarly, the compliance of the bottom member is transferred to the $Oxyz$ reference frame from its local frame located at O_b as:

$$\left[C_{O,op}^{(b)} \right] = \left[T_{op}^{(b)} \right]^T \left[R^{(b)} \right]^T \left[C_{O_t,op}^{(t)} \right] \left[R^{(b)} \right] \left[T_{op}^{(b)} \right] \quad \text{with}$$

$$\left[T_{op}^{(b)} \right] = \begin{bmatrix} 1 & 0 & -\Delta y \\ 0 & 1 & \Delta x \\ 0 & 0 & 1 \end{bmatrix} = \begin{bmatrix} 1 & 0 & b \\ 0 & 1 & 0 \\ 0 & 0 & 1 \end{bmatrix}. \tag{3.100}$$

The top and bottom segments are connected in parallel, and therefore, the out-of-plane compliance of the two-member flexible hinge of Figure 3.31 is:

$$\left[C_{O,op} \right] = \left(\left[C_{O,op}^{(t)} \right]^{-1} + \left[C_{O,op}^{(b)} \right]^{-1} \right)^{-1} = \begin{bmatrix} C_{\theta_x-m_x} & 0 & 0 \\ 0 & C_{\theta_y-m_y} & C_{\theta_y-f_z} \\ 0 & C_{\theta_y-f_z} & C_{u_z-f_z} \end{bmatrix}. \tag{3.101}$$

The elements of the matrix shown in Eq. (3.103) are too complex and are not included here.

Example 3.6

Consider the two-member parallel hinge of Figure 3.31. The two identical hinge members have a constant rectangular cross-section with the in-plane dimension t and the out-of-plane dimension w. The material is steel with Young's modulus $E = 2 \cdot 10^{11}\,\text{N/m}^2$. Graphically analyze the variation of the compliance ratio $C_{u_y-f_y}^{(t)}/C_{u_y-f_y}$ in terms of α, b, l and t.

Solution:

The following values were used as constants: $\alpha = 20°$, $b = 0.005$ m, $l = 0.03$ m and $t = 0.001$ m. Equations (3.94) through (3.97) were utilized with the basic compliances of a straight-axis element of constant cross-section. The plots of Figures 3.32 and 3.33 show the variation of the compliance ratio in terms of the four geometric parameters. It can be seen that the bending compliance of a single hinge member can be up to 160 times larger than the similar compliance of the two-member flexible hinge for large values of α and l, as well as for small values of b. The single-hinge compliance is around 40 times larger than the two-member flexure for small values of the in-plane thickness t.

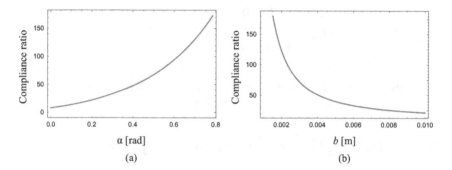

FIGURE 3.32 Plots of the compliance ratio $C_{u_y-f_y}^{(t)}/C_{u_y-f_y}$ in terms of the: (a) angle α; (b) offset distance b.

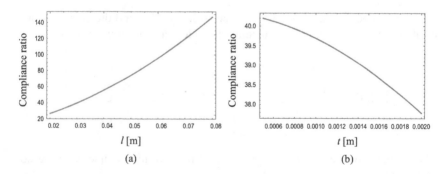

FIGURE 3.33 Plots of the compliance ratio $C_{u_y-f_y}^{(t)}/C_{u_y-f_y}$ in terms of the: (a) length l; (b) thickness t.

3.3.2 Two-Member Flexible Hinge with Identical and Geometrically Parallel Segments

When $\alpha = 0$, the two identical flexure hinges of Figure 3.31 become physically parallel, and the in-plane compliances of Eq. (3.97) are:

$$C_{u_x-f_x} = \frac{C_{u_x-f_x}^{(t)}}{2}; \quad C_{u_y-f_y} = \frac{1}{2}\cdot\left[C_{u_y-f_y}^{(t)} - \frac{b^2\left(C_{u_y-m_z}^{(t)}\right)^2}{C_{u_x-f_x}^{(t)} + b^2 C_{\theta_z-m_z}^{(t)}}\right];$$

$$C_{u_y-m_z} = \frac{1}{2}\cdot\frac{C_{u_x-f_x}^{(t)}\cdot C_{u_y-m_z}^{(t)}}{C_{u_x-f_x}^{(t)} + b^2 C_{\theta_z-m_z}^{(t)}}; \quad C_{\theta_z-m_z} = \frac{1}{2}\cdot\frac{C_{u_x-f_x}^{(t)}\cdot C_{\theta_z-m_z}^{(t)}}{C_{u_x-f_x}^{(t)} + b^2 C_{\theta_z-m_z}^{(t)}}.$$

(3.102)

while the out-of-plane compliances of Eq. (3.101) are:

$$C_{\theta_x-m_x} = \frac{1}{2}\cdot\left\{\frac{C_{\theta_x-m_x}^{(t)}\cdot\left[C_{\theta_y-m_y}^{(t)} C_{u_z-f_z}^{(t)} - \left(C_{\theta_y-f_z}^{(t)}\right)^2\right]}{b^2 C_{\theta_x-m_x}^{(t)} C_{\theta_y-m_y}^{(t)} + C_{\theta_y-m_y}^{(t)} C_{u_z-f_z}^{(t)} - \left(C_{\theta_y-f_z}^{(t)}\right)^2}\right\};$$

$$C_{\theta_y-m_y} = \frac{C_{\theta_y-m_y}^{(t)}}{2}; \quad C_{\theta_y-f_z} = \frac{C_{\theta_y-f_z}^{(t)}}{2}; \quad C_{u_z-f_z} = \frac{C_{u_z-f_z}^{(t)}}{2}.$$

(3.103)

A displacement-amplification device is discussed in Chapter 5 where two identical, geometrically parallel, right circularly corner-filleted flexure hinges are connected in parallel to a rigid lever.

3.3.3 THE CROSSED-LEAF SPRING

The three-member flexible hinge shown in Figure 3.34a is known as a crossed flexure pivot – Wittrick [32], strip bearing – Weinstein [33], or crossed-leaf spring; research related to this design and the modeling of its elastic behavior can also be found in Zhao and Bi [34], Kang and Gweon [35], Pei and Lu [36] or Martin and Robert [37]. The spring consists of three identical thin metal strips, two of which are parallel and perpendicular to the third one. Variants with only two perpendicular strips or more than three are also possible. The length of the strips is l, while the constant cross-section is rectangular with dimensions w and t, as illustrated in Figure 3.34a.

Assume a planar load formed of f_x, f_y and m_z is applied at O in Figure 3.34b along the global frame axes x, y and z. The aim is to evaluate the corresponding displacements at O, which are u_x, u_y and θ_z. There are two identical hinges defined in their local frame $O_2x_2\,y_2z_2$ and a third hinge, rotated at $90°$ from the hinge pair and whose local frame $O_1x_1\,y_1z_1$ coincides with the global frame $Oxyz$. The three hinges are connected in parallel between the two rigid links, one of which is fixed (the anchor), while the other is mobile. The compliance of the three-hinge pivot is calculated as:

$$\left[C_{O,ip}\right] = \left(\left[C_{O,ip}^{(1)}\right]^{-1} + 2\left[C_{O,ip}^{(2)}\right]^{-1}\right)^{-1}.$$

(3.104)

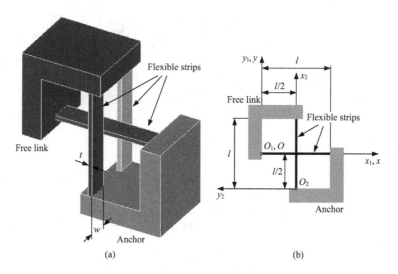

FIGURE 3.34 Crossed-leaf spring: (a) three-dimensional view; (b) front view with reference frames.

with

$$\left[C_{O,ip}^{(1)}\right]=\left[C_{O_1,ip}^{(1)}\right]; \quad \left[C_{O,ip}^{(2)}\right]=\left[T_{OO_2,ip}^{(2)}\right]^{T}\left[R^{(2)}\right]^{T}\left[C_{O_1,ip}^{(1)}\right]\left[R^{(2)}\right]\left[T_{OO_2,ip}^{(2)}\right]. \quad (3.105)$$

Equation (3.105) took into account that the local-frame compliance matrices of hinges 1 and 2 are identical, namely:

$$\left[C_{O_1,ip}^{(1)}\right]=\left[C_{O_2,ip}^{(2)}\right]=\begin{bmatrix} \dfrac{l}{EA} & 0 & 0 \\ 0 & \dfrac{l^3}{3EI} & -\dfrac{l^2}{2EI} \\ 0 & -\dfrac{l^2}{2EI} & \dfrac{l}{EI} \end{bmatrix}=\begin{bmatrix} \dfrac{l}{Ewt} & 0 & 0 \\ 0 & \dfrac{4l^3}{Ewt^3} & -\dfrac{6l^2}{Ewt^3} \\ 0 & -\dfrac{6l^2}{Ewt^3} & \dfrac{12l}{Ewt^3} \end{bmatrix}.$$

$$(3.106)$$

The translation and rotation matrices of Eq. (3.105) are:

$$\left[T_{OO_2,ip}^{(2)}\right]=\begin{bmatrix} 1 & 0 & 0 \\ 0 & 1 & 0 \\ \Delta y_2 & -\Delta x_2 & 1 \end{bmatrix}=\begin{bmatrix} 1 & 0 & 0 \\ 0 & 1 & 0 \\ -l/2 & -l/2 & 1 \end{bmatrix};$$

$$(3.107)$$

$$\left[R^{(2)}\right]=\begin{bmatrix} \cos(\pi/2) & \sin(\pi/2) & 0 \\ -\sin(\pi/2) & \cos(\pi/2) & 0 \\ 0 & 0 & 1 \end{bmatrix}.$$

Substituting Eqs. (3.105)–(3.107) into Eq. (3.104) results in:

$$[C_{O,ip}] = \begin{bmatrix} C_{u_x - f_x} & 0 & 0 \\ 0 & C_{u_y - f_y} & C_{u_y - m_z} \\ 0 & C_{u_y - m_z} & C_{\theta_z - m_z} \end{bmatrix} \text{with} \begin{cases} C_{u_x - f_x} = \dfrac{l^3}{Ewt\left(l^2 + 2t^2\right)}; \\[2mm] C_{u_y - f_y} = \dfrac{2l^3\left(l^2 + t^2\right)}{Ewt^3\left(2l^2 + t^2\right)}; \\[2mm] C_{u_y - m_z} = -\dfrac{2l^2}{Ewt^3}; \quad C_{\theta_z - m_z} = \dfrac{4l}{Ewt^3}. \end{cases}$$

$$(3.108)$$

Compared to Eq. (3.106), Eqs. (3.108) show that the compliances $C_{u_y - m_z}$ and $C_{\theta_z - m_z}$ are three times smaller than the corresponding compliances of a single flexible hinge with geometrical parameters identical to those of the three strips of the crossed-pivot design.

APPENDIX A3: INDEPENDENT CLOSED-FORM COMPLIANCES

A3.1 FLEXIBLE HINGES OF RECTANGULAR CROSS-SECTION WITH CONSTANT OUT-OF-PLANE WIDTH w AND MINIMUM IN-PLANE THICKNESS t

A3.1.1 Right Circular Hinge

$$\begin{cases} C_{u_x - f_x} = \dfrac{1}{Ew} \cdot \left[\dfrac{2(2r+t)}{\sqrt{t \cdot (4r+t)}} \cdot \tan^{-1}\sqrt{1 + \dfrac{4r}{t}} - \dfrac{\pi}{2} \right] \\[4mm] C_{u_y - f_y} = \dfrac{3}{2Ewt \cdot (4r+t)} \cdot \left[\dfrac{\begin{array}{c} 96r^5 + 96r^4 t + 8(4\pi+11)r^3 t^2 + 32(\pi+1)r^2 t^3 \\ + 2(5\pi+2)rt^4 + \pi t^5 \end{array}}{2r+t} \right. \\ \left. \qquad + \dfrac{4(2r+t)^3\left(6r^2 - 4rt - t^2\right)}{\sqrt{t(4r+t)}} \cdot \tan^{-1}\sqrt{1 + \dfrac{4r}{t}} \right] \\[4mm] C_{\theta_z - m_z} = \dfrac{24r}{Ewt^2 \cdot (4r+t)^2} \cdot \left[\dfrac{6r^2 + 4rt + t^2}{2r+t} + \dfrac{6r(2r+t)}{\sqrt{t(4r+t)}} \cdot \tan^{-1}\sqrt{1 + \dfrac{4r}{t}} \right] \end{cases}$$

$$(A3.1)$$

$$\begin{cases} C_{u_z-f_z} = \dfrac{3}{2E \cdot w^3} \cdot \begin{bmatrix} 2(4-\pi)r^2 + 4(1+\pi)rt + \pi t^2 \\[2mm] + \dfrac{4(2r+t)(4r^2 - 4rt - t^2)}{\sqrt{t(4r+t)}} \cdot \tan^{-1}\sqrt{1+\dfrac{4r}{t}} \end{bmatrix} \\[10mm] C_{\theta_y-m_y} = \dfrac{6}{E \cdot w^3} \cdot \begin{bmatrix} \dfrac{4(2r+t)}{\sqrt{t(4r+t)}} \cdot \tan^{-1}\sqrt{1+\dfrac{4r}{t}} - \pi \end{bmatrix} \end{cases} \tag{A3.2}$$

A3.1.2 Right Elliptical Hinge

$$\begin{cases} C_{u_x-f_x} = \dfrac{a}{2Ewb} \cdot \begin{bmatrix} \dfrac{4(2b+t)}{\sqrt{t(4b+t)}} \cdot \tan^{-1}\sqrt{1+\dfrac{4b}{t}} - \pi \end{bmatrix} \\[10mm] C_{u_y-f_y} = \dfrac{3a^3}{2Ewb^3t^2 \cdot (4b+t)^2} \cdot \begin{bmatrix} \dfrac{96b^5 + 96b^4t + 8(11+4\pi)b^3t^2}{2b+t} \\[3mm] + \dfrac{32(1+\pi)b^2t^3 + +2(2+5\pi)bt^4 + \pi t^5}{2b+t} \\[3mm] + \dfrac{(6b^2 - 4bt - t^2) \cdot (2b+t)^3}{\sqrt{t(4b+t)}} \cdot \tan^{-1}\sqrt{1+\dfrac{4b}{t}} \end{bmatrix} \\[18mm] C_{\theta_z-m_z} = \dfrac{24a}{Ewt^2 \cdot (4b+t)^2} \cdot \begin{bmatrix} \dfrac{6b^2 + 4bt + t^2}{2b+t} + \dfrac{6b(2b+t)}{\sqrt{t(4b+t)}} \cdot \tan^{-1}\sqrt{1+\dfrac{4b}{t}} \end{bmatrix} \end{cases}$$
$$\tag{A3.3}$$

$$\begin{cases} C_{u_z-f_z} = \dfrac{3a^3}{2E \cdot w^3 \cdot b^3} \cdot \begin{bmatrix} 2(4-\pi)b^2 + 4(1+\pi)bt + \pi t^2 \\[2mm] + \dfrac{4(2b+t)(4b^2 - 4bt - t^2)}{\sqrt{t(4b+t)}} \cdot \tan^{-1}\left(\sqrt{1+\dfrac{4b}{t}}\right) \end{bmatrix} \\[10mm] C_{\theta_y-m_y} = \dfrac{6a}{E \cdot w^3 \cdot b} \cdot \begin{bmatrix} \dfrac{4(2b+t)}{\sqrt{t(4b+t)}} \cdot \tan^{-1}\sqrt{1+\dfrac{4b}{t}} - \pi \end{bmatrix} \end{cases}$$
$$\tag{A3.4}$$

A3.2 Flexible Hinges of Circular Cross-Section

A3.2.1 Right Circular Hinge

$$
\left\{
\begin{aligned}
C_{u_x-f_x} &= \frac{8r}{\pi Ed(d+4r)}\left[1+\frac{4r}{\sqrt{d(4r+d)}}\cdot\tan^{-1}\sqrt{1+\frac{4r}{d}}\right] \\[2em]
C_{u_y-f_y} &= \frac{256r^3}{3\pi Ed^3(4r+d)^3}\left[\begin{array}{c}\dfrac{2d^4+16d^3r+57d^2r^2+100dr^3+60r^4}{(2r+d)^2} \\[1em] +\dfrac{15r(2r+d)^2}{\sqrt{d(4r+d)}}\cdot\tan^{-1}\sqrt{1+\dfrac{4r}{d}}\end{array}\right] \\[3em]
C_{\theta_y-m_y} &= \frac{128r}{3\pi Ed^3(4r+d)^3}\cdot\left[\begin{array}{c}\dfrac{3d^4+24d^3r+92d^2r^2+176dr^3+120r^4}{(2r+d)^2} \\[1em] +\dfrac{24r(d^2+4dr+5r^2)}{\sqrt{d(4r+d)}}\cdot\tan^{-1}\sqrt{1+\dfrac{4r}{t}}\end{array}\right].
\end{aligned}
\right.
$$

(A3.5)

A3.2.2 Right Elliptical Hinge

$$
\left\{
\begin{aligned}
C_{u_x-f_x} &= \frac{8a}{\pi Ed(4b+d)}\left[1+\frac{4b}{\sqrt{d(4b+d)}}\cdot\tan^{-1}\sqrt{1+\frac{4b}{d}}\right] \\[2em]
C_{u_y-f_y} &= \frac{256a^3}{3\pi E(2b+d)^2\sqrt{d^7(4b+d)^7}}\left[\begin{array}{c}\sqrt{d(4b+d)}\left(60b^4+100b^3d+\right. \\ \left.57b^2d^2+16bd^3+2d^4\right) \\[1em] +15b(2b+d)^4\cdot\tan^{-1}\sqrt{1+\dfrac{4b}{d}}\end{array}\right] \\[3em]
C_{\theta_y-m_y} &= \frac{128a}{3\pi E(2b+d)^2\sqrt{d^7(4b+d)^7}}\cdot\left[\begin{array}{c}\sqrt{d(4b+d)}\left(3d^4+24d^3b+92d^2b^2+\right. \\ \left.176db^3+120b^4\right) \\[1em] +24b(2b+d)^2\left(d^2+4db+5b^2\right) \\[1em] \cdot\tan^{-1}\sqrt{1+\dfrac{4b}{t}}\end{array}\right].
\end{aligned}
\right.
$$

(A3.6)

REFERENCES

1. Paros, J.M. and Weisbord, L., How to design flexure hinges, *Machine Design*, November, 151, 1965.
2. Lobontiu, N. and Paine, J.S.N., Design of circular cross-section corner-filleted flexure hinges for three-dimensional compliant mechanisms, *ASME Journal of Mechanical Design*, 124, 479, 2002.
3. Yong, Y.K. and Lu, T.-F., Comparison of circular flexure hinge design equations and the derivation of empirical stiffness formulations, *2009 IEEE/ASME International Conference on Advanced Intelligent Mechatronics*, Singapore, July 14–17, 510, 2009.
4. Wu, Y. and Zhou, Z., Design calculations for flexure hinges, *Review of Scientific Instruments*, 73 (8), 3101, 2002.
5. Zelenika, S., Munteanu, M.Gh., and De Bona, F., Optimized flexural hinge shapes for microsystems and high-precision applications, *Mechanism and Machine Theory*, 44, 1826, 2009.
6. Schotborg, W.O. et al., Dimensionless design graphs for flexure elements and a comparison between three flexure elements, *Precision Engineering: Journal of the International Societies for Precision Engineering and Nanotechnology*, 29, 41, 2005.
7. Smith, S.T. et al., Elliptical flexure hinges, *Review of Scientific Instruments*, 68 (3), 1474, 1997.
8. Smith, S.T., *Flexures: Elements of Elastic Mechanisms*, Gordon & Breach, Amsterdam, 2000.
9. Lobontiu, N. et al., Parabolic and hyperbolic flexure hinges: flexibility, motion precision and stress characterization based on compliance closed-form equations, *Precision Engineering*, 26, 183, 2002.
10. Lobontiu, N. et al., Design of conic-section flexure hinges based on closed-form compliance equations, *Mechanism and Machine Theory*, 37, 477, 2002.
11. Lobontiu, N., Garcia, E., and Canfield, S., Torsional stiffness of several variable rectangular cross-section flexure hinges for macro-scale and MEMS applications, *Smart Materials and Structures*, 13, 12, 2004.
12. Chen, G., Shao, X., and Huang, X., A new generalized model for elliptical arc flexure hinges, *Review of Scientific Instruments*, 79, 095103-1, 2008.
13. Chen, G. et al., A generalized model for conic flexure hinges, *Review of Scientific Instruments*, 80, 055106-1, 2009.
14. Linβ, S., Erbe, T., and Zentner, L., On polynomial flexure hinges for increased deflection and an approach for simplified manufacturing, *13th World Congress in Mechanism and Machine Science*, Guananjuato, Mexico, June 19–25, 1, 2011.
15. Li, Q., Pan, C., and Xu, X., Closed-form compliance equations for power-function-shaped flexure hinge based on unit-load method, *Precision Engineering*, 37, 135, 2013.
16. Vallance, R.R., Haghighian, B., and Marsh, E.R., A unified geometric model for designing elastic pivots, *Precision Engineering*, 32, 278, 2008.
17. Lobontiu, N. and Garcia, E., Two-axis flexure hinges with axially collocated and symmetric notches, *Computers & Structures*, 81, 1329, 2003.
18. Lobontiu, N. et al., Corner filleted flexure hinges, *ASME Journal of Mechanical Design*, 123, 346, 2001.
19. Meng, Q., Li, Y., and Xu, J., New empirical stiffness equations for corner-filleted flexure hinges, *Mechanical Sciences*, 4, 345, 2013.
20. Lobontiu, N. et al., Planar compliances of symmetric notch flexure hinges: the right circularly corner-filleted parabolic design, *IEEE Transactions on Automation Science and Engineering*, 11 (1), 169, 2014.

21. Lin, R., Zhan, X., and Fatikow, S., Hybrid flexure hinges, *Review of Scientific Instruments*, 84, 085004-1, 2013.
22. Wu, J. et al., Modeling and analysis of conical-shaped notch flexure hinges based on NURBS, *Mechanism and Machine Theory*, 128, 560, 2018.
23. Tian, Y., Shirinzadeh, B., and Zhang, D., Closed-form compliance equations of filleted V-shaped flexure hinges for compliant mechanism design, *Precision Engineering*, 34, 408, 2010.
24. Tseytlin, Y., Note: Rotational compliance and instantaneous center of rotation in segmented and V-shaped notch hinges, *Review of Scientific Instruments*, 83, 026102-1, 2012.
25. Lobontiu, N. et al., A generalized analytical compliance model for transversely symmetric three-segment flexure hinges, *Review of Scientific Instruments*, 82, 105116-1, 2011.
26. Lobontiu, N., Note: Bending compliances of generalized symmetric notch flexure hinges, *Review of Scientific Instruments*, 83, 016107-1, 2012.
27. Lobontiu, N., Symmetry-based compliance model of multisegment notch flexure hinges, *Mechanics Based Design of Structures and Machines*, 40, 185, 2012.
28. De Bona, F. and Munteanu, M.Gh., Optimized flexural hinges for compliant mechanisms, *Analog Integrated Circuits and Signal Processing*, 44, 163, 2005.
29. Liu, C.-F. and Shih, C.-J., Multiobjective design optimization of flexure hinges for enhancing the performance of micro-compliant mechanisms, *Journal of Chinese Institute of Engineers*, 28 (6), 999, 2005.
30. Li, B. and Gou, J., Performance optimization of flexure hinges using genetic algorithms, *Applied Mechanics of Materials*, 121–126, 4543, 2012.
31. Lobontiu, N., Modeling and design of planar parallel-connection flexible hinges for in- and out-of-plane mechanism applications, *Precision Engineering*, 42, 113, 2015.
32. Wittrick, W.H., The theory of symmetrical crossed flexure pivots, *Australian Journal of Science Research*, 1 (2), 121, 1948.
33. Weinstein, W.D., Flexure-pivot bearings, *Machine Design*, 37 (16), 136, 1965.
34. Zhao, H. and Bi, S., Accuracy characteristics of the generalized cross-spring pivot, *Mechanism and Machine Theory*, 45, 1434, 2010.
35. Kang, D. and Gweon, D., Analysis and design of a cartwheel-type flexure hinge, *Precision Engineering*, 37, 33, 2013.
36. Pei, X. and Lu, J., ADLIF: a new large-displacement beam-based flexure joint, *Mechanical Sciences*, 183, 2011.
37. Martin, J. and Robert, M., Novel flexible pivot with large angular range and small center shift to be integrated into a bio-inspired robotic hand, *Journal of Intelligent Material Systems and Structures*, 22, 1431, 2011.

4 Compliances of Curvilinear-Axis Flexible Hinges

Flexible hinges with curved/circular axis are employed in applications where coupling between axial and bending deformation, as well as between bending and torsion, are necessary. Analytical modeling and design of circular-axis hinges for in-plane mechanism applications with either small or large deformations have been studied by Hsiao and Lin [1], Lee and Ziegert [2], Shisheng et al. [3], Callegari et al. [4], Howell [5] or Jagirda [6], for instance. Similar analysis/design methods have also been developed to characterize the out-of-plane (diaphragm) behavior of circular-axis elastic members in a wide array of applications, including opto-mechanics, precision positioning or robotics – see Trylinski [7], Sydenham [8], Vukobratovich et al. [9], Kyusogin and Sagawa [10], Hermann [11], Mohamad et al. [12], Duong and Kazerounian [13], Shielpiekandula and Youcef-Toumi [14], Yong and Reza-Mohemani [15], Awtar [16], Awtar and Slocum [17], Trease et al. [18], Ishaquddin et al. [19] and Kress et al. [20]. Finite element methods as detailed in Ishaquddin et al. [19], Kress et al. [20], also see Chapter 8 – and compliance matrix-based approaches – as detailed in Lobontiu and Cullin [21], Lobontiu et al. [22], Lobontiu [23–26], Lobontiu et al. [27], Lobontiu [28] – have been utilized to model curvilinear-axis flexible hinges and mechanisms incorporating them.

This chapter formulates the in-plane and out-of-plane compliance matrices of planar flexible hinges that are fully formed of curvilinear-axis segments or combine curvilinear-axis segments with straight-axis segments. By means of translation, addition, rotation and mirroring, the local-frame compliance matrices of the individual segments, like those studied in Chapter 2, are mixed either in series or in parallel to produce the compliance matrices of the final hinge configurations. Flexible-hinge designs, which are symmetric or antisymmetric with respect to an axis, are studied in this chapter by means of compliance matrices of simpler hinge portions. Folded hinges are also analyzed; they comprise curvilinear-axis segments in arrangements that are inscribed in geometric envelopes of straight/curvilinear profile or may be of radial design.

4.1 COMPLIANCE MATRIX TRANSFORMATIONS

In order to formulate the in-plane and out-of-plane compliance matrices of flexible hinges that incorporate curvilinear-axis segments, the local-frame segment compliance matrices need to be transferred to the global frame; this is mainly achieved by matrix translation and rotation, but it can also include mirroring.

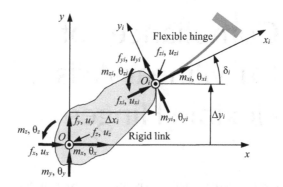

FIGURE 4.1   Rigid link connected to planar curvilinear-axis flexible hinge in global and local reference frames under the action of load displaced from its free end.

Assume that a load formed of f_x, f_y, m_z, m_x, m_y and f_z is applied at a point O that is rigidly connected to the free end O_i of a generic curvilinear-axis flexible hinge, which is fixed at the opposite end, as illustrated in Figure 4.1. The aim is to evaluate the displacements u_x, u_y, θ_z, θ_x, θ_y and u_z in the global reference frame $Oxyz$ in terms of the load at O and to express the flexible-hinge compliance matrices with respect to O based on the hinge native compliance matrices that are formulated in the $O_i x_i y_i z_i$ local frame. The load/displacement vectors are formally separated into in-plane and out-of-plane components.

4.1.1   In-Plane Compliances

The in-plane loads f_x, f_y and m_z statically transfer to O_i from their original location O as:

$$\begin{cases} f_{xi} = f_x \cdot \cos\delta_i + f_y \cdot \sin\delta_i \\ f_{yi} = -f_x \cdot \sin\delta_i + f_y \cdot \cos\delta_i \, . \\ m_{zi} = f_x \cdot \Delta y_i - f_y \cdot \Delta x_i + m_z \end{cases} \tag{4.1}$$

Equation (4.1) can be written in matrix form as:

$$\left[f_{O_i,ip}^{(i)} \right] = \left[R^{(i)} \right] \left[T_{OO_i,ip}^{(i)} \right] \left[f_{O,ip}^{(i)} \right]$$

$$\text{with:} \begin{cases} \left[f_{O_i,ip}^{(i)} \right] = \begin{bmatrix} f_{xi} & f_{yi} & m_{zi} \end{bmatrix}^T; \quad \left[f_{O,ip}^{(i)} \right] = \begin{bmatrix} f_x & f_y & m_z \end{bmatrix}^T \\ \\ \left[R^{(i)} \right] = \begin{bmatrix} \cos\delta_i & \sin\delta_i & 0 \\ -\sin\delta_i & \cos\delta_i & 0 \\ 0 & 0 & 1 \end{bmatrix}; \quad \left[T_{OO_i,ip}^{(i)} \right] = \begin{bmatrix} 1 & 0 & 0 \\ 0 & 1 & 0 \\ \Delta y_i & -\Delta x_i & 1 \end{bmatrix}. \end{cases}$$

$$\tag{4.2}$$

where $[R^{(i)}]$ is the rotation matrix and $\left[T_{OO_i,ip}^{(i)} \right]$ is the in-plane translation matrix.

The global-frame displacements u_x, u_y and θ_z are expressed in terms of the local-frame displacements u_{xi}, u_{yi} and θ_{zi} as follows:

$$\begin{cases} u_x = u_{xi} \cdot \cos\delta_i - u_{yi} \cdot \sin\delta_i + \Delta y_i \cdot \theta_{zi} \\ u_y = u_{xi} \cdot \sin\delta_i + u_{yi} \cdot \cos\delta_i - \Delta x_i \cdot \theta_{zi} \\ \theta_z = \theta_{xi} \end{cases} \tag{4.3}$$

Equations (4.3) are rewritten in the matrix form as:

$$\left[u_{O,ip}^{(i)} \right] = \left[T_{OO_i,ip}^{(i)} \right]^T \left[R^{(i)} \right]^T \left[u_{O_i,ip}^{(i)} \right] \text{ with } \begin{cases} \left[u_{O_i,ip}^{(i)} \right] = \left[\begin{array}{ccc} u_{xi} & u_{yi} & \theta_{zi} \end{array} \right]^T \\ \left[u_{O,ip}^{(i)} \right] = \left[\begin{array}{ccc} u_x & u_y & \theta_z \end{array} \right]^T \end{cases} \tag{4.4}$$

The local-frame loads and displacements are related by means of the hinge in-plane compliance matrix by the known relationship:

$$\left[u_{O_i,ip}^{(i)} \right] = \left[C_{OO_i,ip}^{(i)} \right] \left[f_{O_i,ip}^{(i)} \right]. \tag{4.5}$$

Combining Eqs. (4.2), (4.4) and (4.5) yields:

$$\left[u_{O,ip}^{(i)} \right] = \left[C_{O,ip}^{(i)} \right] \left[f_{O,ip}^{(i)} \right] \text{ with } \left[C_{O,ip}^{(i)} \right] = \left[T_{OO_i,ip}^{(i)} \right]^T \left[R^{(i)} \right]^T \left[C_{O_i,ip}^{(i)} \right] \left[R^{(i)} \right] \left[T_{OO_i,ip}^{(i)} \right]. \tag{4.6}$$

where $\left[C_{O,ip}^{(i)} \right]$ is the global-frame, in-plane compliance matrix of the generic curvilinear-axis flexible hinge.

4.1.2 OUT-OF-PLANE COMPLIANCES

A similar procedure is applied to determine the relationship between global-frame out-of-plane loads and displacements. The out-of-plane loads m_x, m_y and f_z are statically reduced (moved) to O_i from their original location O as:

$$\begin{cases} m_{xi} = m_x \cdot \cos\delta_i + m_y \cdot \sin\delta_i + f_z \cdot \left(\Delta x_i \cdot \sin\delta_i - \Delta y_i \cdot \cos\delta_i \right) \\ m_{yi} = -m_x \cdot \sin\delta_i + m_y \cdot \cos\delta_i + f_z \cdot \left(\Delta x_i \cdot \cos\delta_i + \Delta y_i \cdot \sin\delta_i \right) \\ f_{zi} = f_x \end{cases} \tag{4.7}$$

which are collected in the matrix form:

$$\left[f_{O_i,op}^{(i)} \right] = \left[R^{(i)} \right] \left[T_{OO_i,op}^{(i)} \right] \left[f_{O,op}^{(i)} \right]$$

$$\left[f_{O_i,op}^{(i)} \right] = \left[\begin{array}{ccc} m_{xi} & m_{yi} & f_{zi} \end{array} \right]^T ; \quad \left[f_{O,op}^{(i)} \right] = \left[\begin{array}{ccc} m_x & m_y & f_z \end{array} \right]^T$$

with :

$$\left[T_{OO_i,op}^{(i)} \right] = \left[\begin{array}{ccc} 1 & 0 & -\Delta y_i \\ 0 & 1 & \Delta x_i \\ 0 & 0 & 1 \end{array} \right]$$

$$(4.8)$$

The rotation matrix $[R^{(i)}]$ is calculated in Eq. (4.2) and $\left[T_{OO_i,op}^{(i)} \right]$ is the out-of-plane translation matrix. The displacements θ_x, θ_y and u_z are expressed in terms of the displacements θ_{xi}, θ_{yi} and u_{zi} as:

$$\begin{cases} \theta_x = \theta_{xi} \cdot \cos\delta_i - \theta_{yi} \cdot \sin\delta_i \\ \theta_y = \theta_{xi} \cdot \sin\delta_i + \theta_{yi} \cdot \cos\delta_i \\ u_z = \theta_{xi} \cdot (\Delta x_i \cdot \sin\delta_i - \Delta y_i \cdot \cos\delta_i) + \theta_{yi} \cdot (\Delta x_i \cdot \cos\delta_i + \Delta y_i \cdot \sin\delta_i) + u_{zi} \end{cases}$$

$$. \quad (4.9)$$

Equations (4.9) can also be written in matrix form as:

$$\left[u_{O,op}^{(i)} \right] = \left[T_{OO_i,op}^{(i)} \right]^T \left[R^{(i)} \right]^T \left[u_{O_i,op}^{(i)} \right] \quad \text{with} \quad \begin{cases} \left[u_{O_i,op}^{(i)} \right] = \left[\begin{array}{ccc} \theta_{xi} & \theta_{yi} & u_{zi} \end{array} \right]^T \\ \left[u_{O,ip}^{(i)} \right] = \left[\begin{array}{ccc} u_x & u_y & \theta_z \end{array} \right]^T \end{cases}$$

$$. \quad (4.10)$$

By definition, the out-of-plane compliance matrix connects the corresponding loads and displacements in the local frame as:

$$\left[u_{O_i,op}^{(i)} \right] = \left[C_{O_i,op}^{(i)} \right] \left[f_{O_i,op}^{(i)} \right].$$

$$(4.11)$$

Equations (4.8), (4.10) and (4.11) are combined into:

$$\left[u_{O,op}^{(i)} \right] = \left[C_{O,op}^{(i)} \right] \left[f_{O,op}^{(i)} \right] \text{ with } \left[C_{O,op}^{(i)} \right] = \left[T_{OO_i,op}^{(i)} \right]^T \left[R^{(i)} \right]^T \left[C_{O_i,op}^{(i)} \right] \left[R^{(i)} \right] \left[T_{OO_i,op}^{(i)} \right].$$

$$(4.12)$$

where $\left[C_{O,op}^{(i)} \right]$ is the global-frame, out-of-plane compliance matrix of the generic curvilinear-axis flexure hinge.

It is important to note that the transformation equations derived in this section for a curvilinear-axis hinge are also valid for a straight-axis hinge. The only difference is that the in-plane and out-of-plane, local-frame compliance matrices of a straight-axis hinge need to be used instead of those of a curvilinear-axis hinge.

4.2 SERIES-CONNECTION FLEXIBLE HINGES

This section formulates compliance matrices for flexible hinges that result from serially connecting several basic curvilinear-axis segments; straight-axis segments can also be incorporated into these hinges. Hinge configurations that have a symmetry axis, an antisymmetry axis or are, simply, without symmetry are studied here. Discussed are also folded hinges defined by envelope curves that are either straight-axis or curvilinear-axis, as well as folded flexible hinges of radial configuration. Like in Chapter 3, this chapter utilizes the *compliance matrix approach* based on which the individual flexible segments/hinges composing a complex series hinge/chain configuration are considered free-fixed members – this assumption enables to operate with the segment compliance matrices derived in Chapter 2.

4.2.1 COMPLIANCE MATRICES OF FLEXIBLE HINGES WITHOUT SYMMETRY

The generic planar flexible hinge of Figure 4.2 has one end free and the other end fixed. The hinge is formed by serially connecting n different segments whose in-plane and out-of-plane compliance matrices are formulated in their local frames; the positions of all individual segments are also known with respect to the global reference frame $Oxyz$, which is positioned at the free end of the hinge. The serially connected flexible segments composing the hinge have curvilinear centroidal axes, but some of the segments can be of straight longitudinal axis. This section derives the in-plane and out-of-plane, global-frame compliance matrices of the full, series-connection flexible hinge.

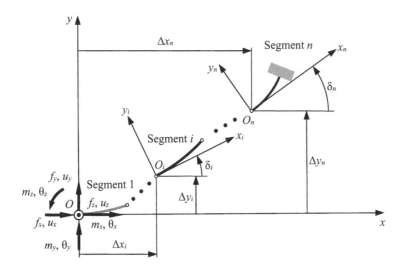

FIGURE 4.2 Fixed-free planar multiple-segment serial flexible hinge.

4.2.1.1 In-Plane Compliance Matrix

In the global frame, the in-plane displacements and loads at the free end O (which is also the global-frame origin) are connected by means of a compliance matrix (to be determined) as:

$$\left[u_{O,ip}\right]=\left[C_{O,ip}\right]\left[f_{O,ip}\right] \quad \text{with} \quad \begin{cases} \left[u_{O,ip}\right]=\begin{bmatrix} u_x & u_y & \theta_z \end{bmatrix}^T \\ \left[f_{O,ip}\right]=\begin{bmatrix} f_x & f_y & m_z \end{bmatrix}^T \end{cases}. \quad (4.13)$$

The displacement $[u_{O,ip}]$ is calculated by summing all corresponding individual displacements pertaining to the n serially connected segments based on Eq. (4.6) as:

$$\left[u_{O,ip}\right]=\sum_{i=1}^{n}\left[u_{O,ip}^{(i)}\right]=\sum_{i=1}^{n}\left(\left[C_{O,ip}^{(i)}\right]\cdot\left[f_{O,ip}\right]\right)$$

$$=\left(\sum_{i=1}^{n}\left(\left[T_{OO_i,ip}^{(i)}\right]^T\left[R^{(i)}\right]^T\left[C_{O_i,ip}^{(i)}\right]\left[R^{(i)}\right]\left[T_{OO_i,ip}^{(i)}\right]\right)\right)\cdot\left[f_{O,ip}\right]. \quad (4.14)$$

Comparing Eqs. (4.13) and (4.14) shows that the in-plane compliance matrix of the full serial hinge of Figure 4.2 is:

$$\left[C_{O,ip}\right]=\sum_{i=1}^{n}\left(\left[T_{OO_i,ip}^{(i)}\right]^T\left[R^{(i)}\right]^T\left[C_{O_i,ip}^{(i)}\right]\left[R^{(i)}\right]\left[T_{OO_i,ip}^{(i)}\right]\right). \quad (4.15)$$

4.2.1.2 Out-of-Plane Compliance Matrix

The out-of-plane compliance matrix of the series-connection flexible hinge connects the out-of-plane displacements and loads that are applied at O in the form:

$$\left[u_{O,op}\right]=\left[C_{O,op}\right]\left[f_{O,op}\right] \quad \text{with} \quad \begin{cases} \left[u_{O,op}\right]=\begin{bmatrix} \theta_x & \theta_y & u_z \end{bmatrix}^T \\ \left[f_{O,ip}\right]=\begin{bmatrix} m_x & m_y & f_z \end{bmatrix}^T \end{cases}. \quad (4.16)$$

Because of the series connection, the displacement $[u_{O,op}]$ is determined with the aid of Eqs. (4.12) by adding all individual global-frame, out-of-plane displacements of the hinge segments, namely:

$$\left[u_{O,op}\right]=\sum_{i=1}^{n}\left[u_{O,op}^{(i)}\right]=\sum_{i=1}^{n}\left(\left[C_{O,op}^{(i)}\right]\cdot\left[f_{O,op}\right]\right)$$

$$=\left(\sum_{i=1}^{n}\left(\left[T_{OO_i,op}^{(i)}\right]^T\left[R^{(i)}\right]^T\left[C_{O_i,op}^{(i)}\right]\left[R^{(i)}\right]\left[T_{OO_i,op}^{(i)}\right]\right)\right)\cdot\left[f_{O,op}\right]. \quad (4.17)$$

The full-hinge out-of-plane compliance matrix is identified by comparing Eqs. (4.16) and (4.17) as:

$$[C_{O,op}] = \sum_{i=1}^{n} \left(\left[T_{OO_i,op}^{(i)} \right]^T \left[R^{(i)} \right]^T \left[C_{O_i,op}^{(i)} \right] \left[R^{(i)} \right] \left[T_{OO_i,op}^{(i)} \right] \right). \qquad (4.18)$$

Example 4.1

The planar flexible hinge of Figure 4.3 is formed by serially connecting two quarter-circle segments of radii R_1 and R_2. The hinge has a constant rectangular cross-section with an in-plane thickness t and an out-of-plane width w. Graphically study the variation of the compliance ratio $C_{u_x-f_x}/C_{u_y-f_y}$, in terms of the two radii for $t = 0.001$ m, $w = 0.008$ m and $E = 2 \cdot 10^{11}$ N/m^2. The two compliances are evaluated in the global reference frame $Oxyz$.

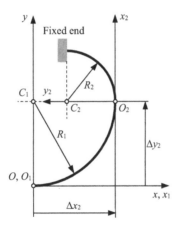

FIGURE 4.3 Skeleton representation of planar flexible hinge formed of two quarter-circle segments.

Solution:

The in-plane compliance matrix of this hinge is:

$$[C_{O,ip}] = \left[C_{O,ip}^{(1)} \right] + \left[T_{OO_2,ip}^{(2)} \right]^T \left[R^{(2)} \right]^T \left[C_{O_2,ip}^{(2)} \right] \left[R^{(2)} \right] \left[T_{OO_2,ip}^{(2)} \right]$$

$$= \begin{bmatrix} C_{u_x-f_x} & C_{u_x-f_y} & C_{u_x-m_z} \\ C_{u_x-f_y} & C_{u_y-f_y} & C_{u_y-m_z} \\ C_{u_x-m_z} & C_{u_y-m_z} & C_{\theta_z-m_z} \end{bmatrix}. \qquad (4.19)$$

where $\left[C_{O_1,ip}^{(1)}\right] \equiv \left[C_{O,ip}^{(1)}\right], \left[C_{O_2,ip}^{(2)}\right]$ are the local-frame, in-plane compliance matrices of the two serially connected segments. They are provided in Eqs. (A2.22) through (A2.26). The translation and rotation matrices that realize the transfer of segment 2 compliance matrix from its local frame $O_2x_2y_2z_2$ to the global frame $Oxyz$ are:

$$\left[T_{OO_2.ip}^{(2)}\right] = \begin{bmatrix} 1 & 0 & 0 \\ 0 & 1 & 0 \\ \Delta y_2 & -\Delta x_2 & 1 \end{bmatrix} = \begin{bmatrix} 1 & 0 & 0 \\ 0 & 1 & 0 \\ R_1 & -R_1 & 1 \end{bmatrix};$$

$$\left[R^{(2)}\right] = \begin{bmatrix} \cos(\pi/2) & \sin(\pi/2) & 0 \\ -\sin(\pi/2) & \cos(\pi/2) & 0 \\ 0 & 0 & 1 \end{bmatrix} = \begin{bmatrix} 0 & 1 & 0 \\ -1 & 0 & 0 \\ 0 & 0 & 1 \end{bmatrix}.$$

(4.20)

The compliance ratio of interest $C_{u_x-f_x}/C_{u_y-f_y}$ is plotted in Figure 4.4 in terms of R_1 and R_2.

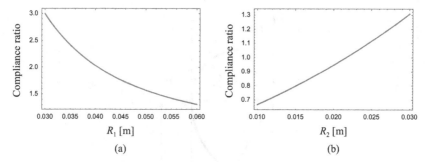

FIGURE 4.4   Plots of compliance ratio $C_{u_x-f_x}/C_{u_y-f_y}$ in terms of the: (a) radius R_1; (b) radius R_2.

The plot of Figure 4.4a was drawn for $R_2 = 0.03\,\text{m}$, while the plot of Figure 4.4b is graphed for $R_1 = 0.06\,\text{m}$. It can be seen in Figure 4.4a that the compliance $C_{u_x-f_x}$ is always larger (up to three times larger when R_1 is small) than $C_{u_y-f_y}$ for the numerical range of R_1. However, for values of R_1 larger than the particular ones used in the plot of Figure 4.4a, the compliance ratio becomes smaller than 1 – that situation is not captured in Figure 4.4a. The compliance $C_{u_x-f_x}$ is smaller than $C_{u_y-f_y}$ for small values of R_2, but $C_{u_x-f_x}$ becomes larger than $C_{u_y-f_y}$ for larger values of R_2, as illustrated in Figure 4.4b.

4.2.2   Compliance Matrices of Flexible Hinges with Symmetry Axis

This section analyzes the class of multiple-segment, curvilinear-axis flexible hinges that are symmetric with respect to an axis. The in-plane and out-of-plane compliance matrices of a generic hinge are derived first based on the corresponding matrices that define half the hinge. The resulting compliance matrices are subsequently used to study several particular symmetric hinge configurations.

4.2.2.1 Generic Hinge and Compliance Matrices

Curvilinear-axis flexible hinges are frequently symmetric with respect to an axis in order to facilitate similar motion capabilities when the load direction alternates and/or to use geometric symmetry in conjunction with load symmetry to generate motion along predesigned directions.

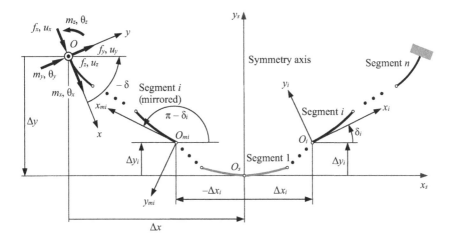

FIGURE 4.5 Skeleton representation of a generic planar, symmetric, multiple-segment flexible hinge.

A generic planar flexible hinge with multiple serially connected curvilinear-axis segments, which is symmetric with respect to an axis, is represented in Figure 4.5. Any individual hinge segment i ($i = 1$ to n) in the right half-plane $O_s x_s y_s z_s$ has a counterpart that is placed symmetrically (or mirrored) with respect to the y_s axis in the left half-plane. The in-plane and out-of-plane compliance matrices could be calculated using the algorithm presented in Section 4.2.1 by summing all segment compliance matrices after transforming them (through rotation and translation) in terms of the global reference frame $Oxyz$ that is placed at the free end of the hinge with the load being applied at the same point. However, it is possible to first calculate the hinge compliance matrices with respect to the $O_s x_s y_s z_s$ reference frame.

The in-plane compliance matrix of the full hinge with respect to the $O_s x_s y_s z_s$ frame is calculated by means of Eqs. (4.15) and by pairing each hinge segment i on the right-hand semi-plane to its mirrored counterpart that is located in the left-hand semi-plane, namely:

$$\left[C_{O_s,ip} \right] = \sum_{i=1}^{n} \left(\begin{aligned} & \left[T_{O_s O_{mi},ip}^{(m,i)} \right]^T \left[R^{(m,i)} \right]^T \left[C_{O_{mi},ip}^{(m,i)} \right] \left[R^{(m,i)} \right] \left[T_{O_s O_{mi},ip}^{(m,i)} \right] \\ & + \left[T_{O_s O_i,ip}^{(i)} \right]^T \left[R^{(i)} \right]^T \left[C_{O_i,ip}^{(i)} \right] \left[R^{(i)} \right] \left[T_{O_s O_i,ip}^{(i)} \right] \end{aligned} \right). \tag{4.21}$$

where the subscript "s" denotes the symmetry reference frame and the "m, i" superscript indicates the segment that is the mirrored counterpart of segment i. Note that the mirrored segment i uses the local axis identified by y_{mi} in Figure 4.5 in order to enable rotation of its frame by $\pi - \delta_i$ from the x_s axis. The compliance matrices of segments i and mi are:

$$\left[C_{O_i,ip}^{(i)} \right] = \begin{bmatrix} C_{O_i,u_{xi}-f_{xi}}^{(i)} & C_{O_i,u_{xi}-f_{yi}}^{(i)} & C_{O_i,u_{xi}-m_{zi}}^{(i)} \\ C_{O_i,u_{xi}-f_{yi}}^{(i)} & C_{O_i,u_{yi}-f_{yi}}^{(i)} & C_{O_i,u_{yi}-m_{zi}}^{(i)} \\ C_{O_i,u_{xi}-m_{zi}}^{(i)} & C_{O_i,u_{yi}-m_{zi}}^{(i)} & C_{O_i,\theta_{zi}-m_{zi}}^{(i)} \end{bmatrix};$$

$$\left[C_{O_{mi},ip}^{(m,i)} \right] = \begin{bmatrix} C_{O_i,u_{xi}-f_{xi}}^{(i)} & -C_{O_i,u_{xi}-f_{yi}}^{(i)} & -C_{O_i,u_{xi}-m_{zi}}^{(i)} \\ -C_{O_i,u_{xi}-f_{yi}}^{(i)} & C_{O_i,u_{yi}-f_{yi}}^{(i)} & C_{O_i,u_{yi}-m_{zi}}^{(i)} \\ -C_{O_i,u_{xi}-m_{zi}}^{(i)} & C_{O_i,u_{yi}-m_{zi}}^{(i)} & C_{O_i,\theta_{zi}-m_{zi}}^{(i)} \end{bmatrix}.$$

(4.22)

As explained in Section 2.2.1, the compliances corresponding to the pairs x_{mi}-y_{mi} and x_{mi}-z_{mi} in $\left[C_{O_{mi},ip}^{(m,i)} \right]$ are the similar compliances of $\left[C_{O_i,ip}^{(i)} \right]$ with a minus sign in front of them. This is because the mirrored segment i has its local axes y_{mi} and z_{mi} in directions opposite to the local axes y_i and z_i of the original segment i. The rotation and translation matrices of Eq. (4.21) are:

$$\left[R^{(i)} \right] = \begin{bmatrix} \cos\delta_i & \sin\delta_i & 0 \\ -\sin\delta_i & \cos\delta_i & 0 \\ 0 & 0 & 1 \end{bmatrix};$$

$$\left[R^{(mi)} \right] = \begin{bmatrix} \cos(\pi-\delta_i) & \sin(\pi-\delta_i) & 0 \\ -\sin(\pi-\delta_i) & \cos(\pi-\delta_i) & 0 \\ 0 & 0 & 1 \end{bmatrix};$$

(4.23)

$$\left[T_{O_sO_i,ip}^{(i)} \right] = \begin{bmatrix} 1 & 0 & 0 \\ 0 & 1 & 0 \\ \Delta y_i & -\Delta x_i & 1 \end{bmatrix}; \quad \left[T_{O_sO_{mi},ip}^{(m,i)} \right] = \begin{bmatrix} 1 & 0 & 0 \\ 0 & 1 & 0 \\ \Delta y_i & \Delta x_i & 1 \end{bmatrix}.$$

Making the calculations of Eqs. (4.21)–(4.23) results in:

$$
\left[C_{O_s,ip}\right] = 2 \cdot
\begin{bmatrix}
\displaystyle\sum_{i=1}^{n} C_{O_s,u_x-f_x}^{(i)} & 0 & \displaystyle\sum_{i=1}^{n} C_{O_s,u_x-m_z}^{(i)} \\[2ex]
0 & \displaystyle\sum_{i=1}^{n} C_{O_s,u_y-f_y}^{(i)} & 0 \\[2ex]
\displaystyle\sum_{i=1}^{n} C_{O_s,u_x-m_z}^{(i)} & 0 & \displaystyle\sum_{i=1}^{n} C_{O_s,\theta_z-m_z}^{(i)}
\end{bmatrix}.
\tag{4.24}
$$

where $C_{O_s,u_x-f_x}^{(i)}$, $C_{O_s,u_x-m_z}^{(i)}$, $C_{O_s,u_y-f_y}^{(i)}$ and $C_{O_s,\theta_z-m_z}^{(i)}$ are the symbolic compliances of the generic segment i evaluated in the reference frame centered at O_s. The full-hinge in-plane compliance matrix of Eq. (4.24) has zero terms that denote no interaction (no coupling) between y_s (the symmetry axis) and either x_s or z_s. The compliance matrix of Eq. (4.24) can now be transferred to the global frame of Figure 4.5 by using the equation:

$$
\left[C_{O,ip}\right] = \left[T_{OO_s,ip}\right]^T [R]^T \left[C_{O_s,ip}\right][R]\left[T_{OO_s,ip}\right].
\tag{4.25}
$$

The rotation matrix $[R]$ is calculated as in Eq. (4.2) with δ shown in Figure 4.5 and used instead of δ_i; the translation matrix is also determined by means of Eq. (4.2) with the offsets Δx and Δy indicated in Figure 4.5 and used instead of Δx_i and Δy_i.

A similar procedure is applied to evaluate the out-of-plane compliance matrix. In the central reference frame, the full-hinge compliance matrix is calculated as

$$
\left[C_{O_s,op}\right] = \sum_{i=1}^{n}
\begin{pmatrix}
\left[T_{O_sO_{mi},op}^{(m,i)}\right]^T \left[R^{(m,i)}\right]^T \left[C_{O_{mi},op}^{(m,i)}\right]\left[R^{(m,i)}\right]\left[T_{O_sO_{mi},op}^{(m,i)}\right] \\[2ex]
+ \left[T_{O_sO_i,op}^{(i)}\right]^T \left[R^{(i)}\right]^T \left[C_{O_i,op}^{(i)}\right]\left[R^{(i)}\right]\left[T_{O_sO_i,op}^{(i)}\right]
\end{pmatrix}.
\tag{4.26}
$$

where the out-of-plane compliance matrices of segments i and mi are:

$$
\left[C_{O_i,op}^{(i)} \right] =
\begin{bmatrix}
C_{O_i,\theta_{xi}-m_{xi}}^{(i)} & C_{O_i,\theta_{xi}-m_{yi}}^{(i)} & C_{O_i,\theta_{xi}-f_{zi}}^{(i)} \\
C_{O_i,\theta_{xi}-m_{yi}}^{(i)} & C_{O_i,\theta_{yi}-m_{yi}}^{(i)} & C_{O_i,\theta_{yi}-f_{zi}}^{(i)} \\
C_{O_i,\theta_{xi}-f_{zi}}^{(i)} & C_{O_i,\theta_{yi}-f_{zi}}^{(i)} & C_{O_i,u_{zi}-f_{zi}}^{(i)}
\end{bmatrix};
$$

$$
\left[C_{O_{mi},op}^{(m,i)} \right] =
\begin{bmatrix}
C_{O_i,\theta_{xi}-m_{xi}}^{(i)} & -C_{O_i,\theta_{xi}-m_{yi}}^{(i)} & -C_{O_i,\theta_{xi}-f_{zi}}^{(i)} \\
-C_{O_i,\theta_{xi}-m_{yi}}^{(i)} & C_{O_i,\theta_{yi}-m_{yi}}^{(i)} & C_{O_i,\theta_{yi}-f_{zi}}^{(i)} \\
-C_{O_i,\theta_{xi}-f_{zi}}^{(i)} & C_{O_i,\theta_{yi}-f_{zi}}^{(i)} & C_{O_i,u_{zi}-f_{zi}}^{(i)}
\end{bmatrix}. \quad (4.27)
$$

The translation matrices of Eq. (4.26) are:

$$
\left[T_{O_sO_i,op}^{(i)} \right] =
\begin{bmatrix}
1 & 0 & -\Delta y_i \\
0 & 1 & \Delta x_i \\
0 & 0 & 1
\end{bmatrix};
\quad
\left[T_{O_sO_{mi},ip}^{(m,i)} \right] =
\begin{bmatrix}
1 & 0 & -\Delta y_i \\
0 & 1 & -\Delta x_i \\
0 & 0 & 1
\end{bmatrix}. \quad (4.28)
$$

Substituting the matrices expressed in the left-hand side of Eqs. (4.27) and (4.28) into Eq. (4.26) results in:

$$
\left[C_{O_s,op} \right] = 2 \cdot
\begin{bmatrix}
\sum_{i=1}^{n} C_{O_s,\theta_x-m_x}^{(i)} & 0 & \sum_{i=1}^{n} C_{O_s,\theta_x-f_z}^{(i)} \\
0 & \sum_{i=1}^{n} C_{O_s,\theta_y-m_y}^{(i)} & 0 \\
\sum_{i=1}^{n} C_{O_s,\theta_x-f_z}^{(i)} & 0 & \sum_{i=1}^{n} C_{O_s,u_z-f_z}^{(i)}
\end{bmatrix}.
$$

$$(4.29)$$

where $C_{O_s,\theta_x-m_x}^{(i)}$, $C_{O_s,\theta_x-f_z}^{(i)}$, $C_{O_s,\theta_y-m_y}^{(i)}$ and $C_{O_s,u_z-f_z}^{(i)}$ are out-of-plane symbolic compliances of the generic segment i that are calculated with respect to the central system $O_s x_s y_s z_s$. Like in the planar-deformation case, the full-hinge out-of-plane compliance matrix has zero components that identify decoupling between the symmetry axis y_s and either x_s or z_s. The compliance matrix of Eq. (4.29) is transformed into the following global-frame matrix:

$$
\left[C_{O,op} \right] = \left[T_{OO_s,op} \right]^T [R]^T \left[C_{O_s,op} \right] [R] \left[T_{OO_s,op} \right]. \quad (4.30)
$$

where $\left[T_{OO_s,op} \right]$ is calculated with Eq. (4.8) by using the offsets shown in Figure 4.5.

4.2.2.2 Flexible-Hinge Configurations

A few design examples of curvilinear-axis hinges having a symmetry axis are discussed in this section. Some of the designs have a single centroidal axis (for instance, a circle), whereas a few other designs are formed by serially connecting hinge segments of different centroidal axes, both curvilinear and straight.

4.2.2.2.1 Circular-Axis, Right Flexible Hinge

Figure 4.6a shows the planar geometry of a circular-axis, right-circular flexible hinge, which is the counterpart of the straight-axis, right-circular flexible hinge studied in Chapter 3 and illustrated in Figure 3.11b. This configuration, which is defined by a single, circular median curve, is obtained from the half-hinge of Figure 2.28, which identifies all the geometric parameters. The design has a rectangular cross-section of constant out-of-plane width w. In-depth details on the in-plane compliances of this hinge design are provided in Lobontiu and Cullin [21]. Finite element simulation and experimental measurements (a photograph of the experimental setup is pictured in Figure 4.6b) did acceptably match the analytical model compliance predictions. It should be noted that the circular axial line of this hinge is actually different from the centroidal curve; because the cross-sectional properties were evaluated with respect to the axial line, the relatively small deviations of the analytical results from the finite element and experimental predictions (as reported in Lobontiu and Cullin [21]) are explicable.

It is of interest to compare the y direction and the x direction in-plane compliances, which can be achieved by means of the ratio $C_{u_y-f_y}/C_{u_x-f_x}$, for instance. Note that the x and y directions denote the respective global-frame axes identified in Figures 4.5 and 4.6. The basic in-plane compliances of the half-hinge are calculated

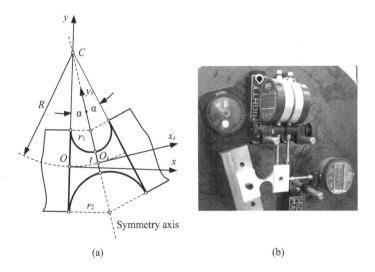

(a)   (b)

FIGURE 4.6 (a) Front-view geometry of symmetric, circular-axis, right flexible hinge; (b) experimental setup.

by means of Eqs. (2.121), (2.124), (2.145), (2.146) and (2.147). Equation (4.22) are subsequently applied to evaluate the compliances $C_{u_x-f_x}$ and $C_{u_y-f_y}$. The offsets in the in-plane translation matrix of Eqs. (4.2) and (4.25) – where the subscript i is dropped because the half is formed of a single segment – are $\Delta x = R \cdot \sin\alpha$ and $\Delta y = R \cdot (1 - \cos\alpha)$; the rotation matrix of the same Eqs. (4.2) and (4.25) uses $\delta = \alpha$. The following constant values: $t = 0.001$ m, $w = 0.006$ m, $E = 2 \cdot 10^{11}$ N/m², $R = 0.08$ m and $\alpha = 30°$, were used to produce the plots of Figure 4.7, which show the compliance ratio variation in terms of α and R, respectively. The angle α variation range was [10°; 45°], while R varied from 0.03 to 0.08 m. It is noticed that the y-direction compliance is larger than the x-direction one. For small values of the angle α, $C_{u_y-f_y}$ is substantially larger than $C_{u_x-f_x}$, as seen in Figure 4.7a; however, the compliance ratio is less sensitive to the radius variation, as illustrated in Figure 4.7b.

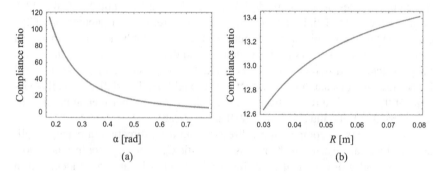

(a)   (b)

FIGURE 4.7 Plots of compliance ratio $C_{u_y-f_y}/C_{u_x-f_x}$ in terms of the (a) angle α; (b) radius R.

4.2.2.2.2 Circular-Axis, Right Circularly Corner-Filleted Flexible Hinge

The symmetric flexible hinge of Figure 4.8a is another example that uses a single circular curve as its median axis. The half-part, which is represented in Figure 4.8b, is formed of a constant-thickness portion extending over a center angle α_1, and a variable-thickness segment of center angle α_2; this segment is the one of Figure 2.28, also used as half the hinge of Figure 4.6a. The in-plane compliances of this configuration have been studied in Lobontiu et al. [22], while its out-of-plane elastic response was studied in Lobontiu [23].

Based on Eq. (4.24), the in-plane compliance matrix of the full hinge with respect to the middle reference frame $O_s x_s y_s z_s$ is:

$$\left[C_{O_s,ip} \right] = \left[C_{O_s,ip}^{(1)} \right] + \left[T_{O_s O_2,ip}^{(2)} \right]^T \left[R^{(2)} \right]^T \left[C_{O_2,ip}^{(2)} \right] \left[R^{(2)} \right] \left[T_{O_s O_2,ip}^{(2)} \right]. \quad (4.31)$$

where $\left[C_{O_s,ip}^{(1)} \right]$ and $\left[C_{O_2,ip}^{(2)} \right]$ are the local-frame, in-plane compliance matrices of the two segments sketched in Figure 4.8b. The in-plane compliances of $\left[C_{O_s,ip}^{(1)} \right]$

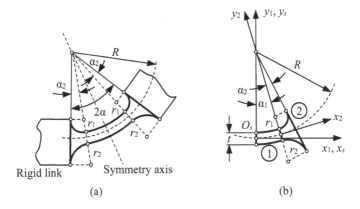

FIGURE 4.8 Geometry of symmetric, circular-axis, right circularly corner-filleted flexible hinge: (a) full hinge; (b) half-hinge.

are provided in Eqs. (A2.22) through (A2.26), whereas the compliances of $\left[C_{O_2,ip}^{(2)} \right]$ are calculated based on the definition of Eqs. (2.121) through (2.125) with the geometric parameters and variable thickness given in Eqs. (2.145)–(2.147). The translation matrix of Eq. (4.31) is:

$$\left[T_{O_sO_2,ip}^{(2)} \right] = \begin{bmatrix} 1 & 0 & 0 \\ 0 & 1 & 0 \\ -R \cdot (1 - \cos\alpha_1) & -R \cdot \sin\alpha_1 & 1 \end{bmatrix}. \tag{4.32}$$

The full-hinge, global-frame, in-plane compliance matrix is calculated as in Eq. (4.25) with:

$$\left[T_{OO_s,ip} \right] = \begin{bmatrix} 1 & 0 & 0 \\ 0 & 1 & 0 \\ \Delta y & -\Delta x & 1 \end{bmatrix}$$

$$= \begin{bmatrix} 1 & 0 & 0 \\ 0 & 1 & 0 \\ R\left[1 - \cos(\alpha_1 + \alpha_2)\right] & -R\sin(\alpha_1 + \alpha_2) & 1 \end{bmatrix}$$

$$[R] = \begin{bmatrix} \cos(\alpha_1 + \alpha_2) & \sin(\alpha_1 + \alpha_2) & 0 \\ -\sin(\alpha_1 + \alpha_2) & \cos(\alpha_1 + \alpha_2) & 0 \\ 0 & 0 & 1 \end{bmatrix}. \tag{4.33}$$

Similarly and using Eq. (4.29), the out-of-plane compliance matrix of the full hinge with respect to the middle reference frame $O_s x_s y_s z_s$ is:

$$\left[C_{O_s,op}\right]=\left[C^{(1)}_{O_s,op}\right]+\left[T^{(2)}_{O_s O_2,op}\right]^T\left[R^{(2)}\right]^T\left[C^{(2)}_{O_2,op}\right]\left[R^{(2)}\right]\left[T^{(2)}_{O_s O_2,op}\right]. \quad (4.34)$$

The translation matrix of Eq. (4.34) is calculated as:

$$\left[T^{(2)}_{O_s O_2,op}\right]=\begin{bmatrix} 1 & 0 & R\cdot(1-\cos\alpha_1) \\ 0 & 1 & R\cdot\sin\alpha_1 \\ 0 & 0 & 1 \end{bmatrix}. \quad (4.35)$$

The full-hinge out-of-plane compliance matrix is calculated in the global frame $Oxyz$ as in Eq. (4.30) with:

$$\left[T_{OO_s,op}\right]=\begin{bmatrix} 1 & 0 & -\Delta y \\ 0 & 1 & \Delta x \\ 0 & 0 & 1 \end{bmatrix}=\begin{bmatrix} 1 & 0 & -R[1-\cos(\alpha_1+\alpha_2)] \\ 0 & 1 & R\sin(\alpha_1+\alpha_2) \\ 0 & 0 & 1 \end{bmatrix}. \quad (4.36)$$

Very good match is reported in Lobontiu et al. [22] and Lobontiu [23] between the analytical model, experimental test and finite element compliance values.

The in-plane compliance $C_{\theta_z - m_z}$ of this design is compared to the same compliance of a straight-axis, right circularly corner-filleted hinge – see also Lobontiu et al. [22] for a similar comparison. It is assumed that the two designs have the same length $l = 2R\cdot\alpha$ (with $\alpha = \alpha_1 + \alpha_2$) and that the fillet radius of the straight-axis design is $r = r_1$ (see Figure 4.8). The compliance ratio $C^s_{\theta_z - m_z}/C_{\theta_z - m_z}$ (where the superscript "s" stands for the straight-axis design) was graphically analyzed in terms of the angle α_2 – the resulting plot is shown in Figure 4.9a, as well as in terms of the circular-axis radius R – the corresponding plot is graphed in Figure 4.9b. Both simulations considered hinges with a constant out-of-plane width w and a minimum in-plane thickness t. The following constant parameter values were used in the plot of Figure 4.9a: $w = 0.006\,\mathrm{m}$, $t = 0.001\,\mathrm{m}$, $R = 0.08\,\mathrm{m}$, $\alpha = 30°$ and $E = 2\cdot10^{11}\,\mathrm{N/m^2}$. For the plot of Figure 4.9b, the radius $r = r_1 = 0.002\,\mathrm{m}$ was used. The angle α_2 depends on the radii R and r_1, as per Eq. (2.145), namely: $\alpha_2 = \sin^{-1}[r_1/(R - r_1 - t/2)]$. As Figure 4.9a shows it, the compliance ratio $C^s_{\theta_z - m_z}/C_{\theta_z - m_z}$ increases nonlinearly with the angular opening of the filleted portion of the circular-axis design, which shows that the straight-axis design's compliance becomes larger than the one of the circular-axis configuration as α_2 increases. The same compliance ratio is almost insensitive to the variation of R, as illustrated in Figure 4.9b.

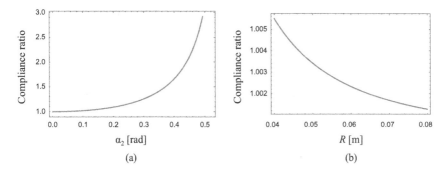

FIGURE 4.9 Plots of compliance ratio $C^s_{\theta_z - m_z} / C_{\theta_z - m_z}$ in terms of the (a) angle α_2; (b) radius R.

The out-of-plane compliant behavior of the design depicted in Figure 4.8 is compared to the in-plane response of the same design, as well as to the out-of-plane elastic performance of a similar, constant cross-sectional hinge – details are offered in Lobontiu [23].

4.2.2.2.3 Design with Circular-Axis Segment of Constant Cross-Section and Straight-Axis, Right Circularly Corner-Filleted Segment

The design of Figure 4.10a, whose out-of-plane compliances are derived and studied in Lobontiu [24], has its half-part formed of a constant cross-section, circular-axis segment extending over a center angle α and a straight-axis, right circular segment – these two segments are shown in Figure 4.10b and c; they both have the same constant out-of-plane width w.

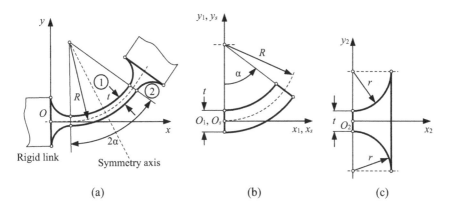

FIGURE 4.10 Symmetric right circularly corner-filleted flexible hinge: (a) full configuration; (b) circular-axis, constant-thickness segment 1; (c) straight-axis right circular segment 2 (rotated).

Equations (4.29) and (4.34) are used to calculate the out-of-plane compliance matrix of the full hinge with respect to the reference frame centered at the midpoint O_s (see Figure 4.10b). The translation and rotation matrices enabling the transfer from O_s to O (and allowing the calculation of $\left[C_{O,op}\right]$ per Eq. (4.30)) are:

$$\left[T_{OO_s,op}\right] = \begin{bmatrix} 1 & 0 & -\Delta y \\ 0 & 1 & \Delta x \\ 0 & 0 & 1 \end{bmatrix} = \begin{bmatrix} 1 & 0 & -R(1-\cos\alpha) \\ 0 & 1 & R\sin\alpha + r \\ 0 & 0 & 1 \end{bmatrix}. \quad (4.37)$$

$$[R] = \begin{bmatrix} \cos\alpha & \sin\alpha & 0 \\ -\sin\alpha & \cos\alpha & 0 \\ 0 & 0 & 1 \end{bmatrix}. \quad (4.38)$$

As discussed in Lobontiu [24], the analytical compliance results match the finite element simulation for several designs. The influence of the parameters R, α, r and w on the out-of-plane deformation capabilities of the flexible hinge of Figure 4.10a is studied here by graphing the compliance ratio $C_{\theta_y-m_y}/C_{\theta_x-m_x}$ as illustrated in Figures 4.11 and 4.12. It can be seen that $C_{\theta_y-m_y} > C_{\theta_x-m_x}$ for all values of the four geometric parameters. The following constant values were used: $R = 0.03$ m, $r = 0.005$ m, $t = 0.004$ m, $w = 0.001$ m, $E = 2 \cdot 10^{11}$ N/m^2, $\mu = 0.3$.

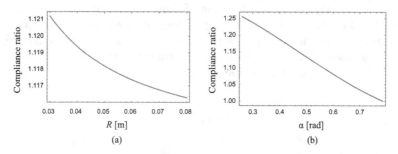

FIGURE 4.11 Plots of compliance ratio $C_{\theta_y-m_y}/C_{\theta_x-m_x}$ in terms of the: (a) radius R; (b) center angle α.

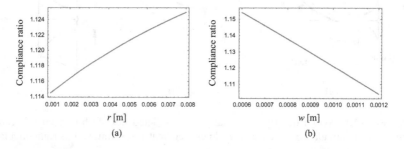

FIGURE 4.12 Plots of compliance ratio $C_{\theta_y-m_y}/C_{\theta_x-m_x}$ in terms of the: (a) fillet radius r; (b) out-of-plane width w.

4.2.2.2.4 Right Elliptically Filleted Flexible Hinge with Parabolic-Axis and Straight-Axis Segments

A design similar to the previous one and shown in Figure 4.10a is discussed in Lobontiu 2015 [25] and is depicted in Figure 4.13a. The half-part of this symmetric flexible hinge consists of a constant-thickness, parabolic-axis segment, which is serially connected to a straight-axis right elliptically corner-filleted segment – these segments are shown in Figure 4.13b and c. The full hinge has a rectangular cross-section with a constant out-of-plane width w.

The parabolic segment 1 of Figure 4.13b has its centroidal axis defined by the equation $y_1 = a_p \cdot x_1^2$ in its local frame $O_1 x_1 y_1 z_1$. The segment has an x-axis length l_{x1} and a constant in-plane thickness t. The planar profile of the straight-axis segment 2 of Figure 4.13c is defined by the ellipse semi-axes lengths a and b, and the minimum thickness t.

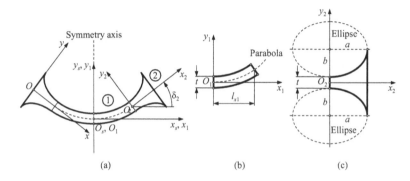

(a)   (b)   (c)

FIGURE 4.13 Symmetric right circularly corner-filleted flexible hinge: (a) full configuration; (b) parabolic-axis, constant-thickness segment 1; (c) straight-axis right elliptically corner-filleted segment 2 (rotated).

The in-plane compliance matrix of the full hinge with respect to the central reference frame $O_s x_s y_s z_s$ is calculated based on Eqs. (4.24) and (4.31). The translation and rotation matrices enabling the transfer from O_s to O (and calculation of $[C_{O,ip}]$ as in Eq. (4.30)) are:

$$[T_{OO_s,ip}] = \begin{bmatrix} 1 & 0 & 0 \\ 0 & 1 & 0 \\ \Delta y & -\Delta x & 1 \end{bmatrix} = \begin{bmatrix} 1 & 0 & 0 \\ 0 & 1 & 0 \\ a_p(l_{x1})^2 & -(a+l_{x1}) & 1 \end{bmatrix}. \quad (4.39)$$

$$[R] = \begin{bmatrix} \cos\delta_2 & \sin\delta_2 & 0 \\ -\sin\delta_2 & \cos\delta_2 & 0 \\ 0 & 0 & 1 \end{bmatrix} \quad (4.40)$$

with $\delta_2 = \tan^{-1}(2a_p \cdot l_{x1})$.

The in-plane response of the flexible hinge shown in Figure 4.13a is evaluated by plotting the compliance ratio $C_{u_y-f_y}/C_{u_x-f_x}$ in terms of relevant planar geometric parameters. Figures 4.14–4.16 show the compliance ratio variation with respect to a, b, a_p, l_{x1} and t for the following constant values of the parameters: $w = 0.01$ m, $a = 0.015$ m, $b = 0.005$ m, $a_p = 20$ m^{-1}, $l_{x1} = 0.01$ m and $t = 0.001$ m; a value $E = 2 \cdot 10^{11}$ N/m^2 was also used for these plots.

It can be seen that $C_{u_y-f_y} > C_{u_x-f_x}$ for all five parameters over their variation ranges.

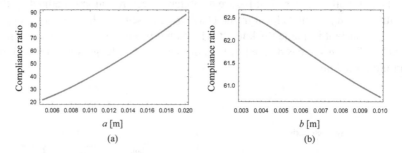

FIGURE 4.14 Plots of compliance ratio $C_{u_y-f_y}/C_{u_x-f_x}$ in terms of the elliptical semi-axis length: (a) a; (b) b.

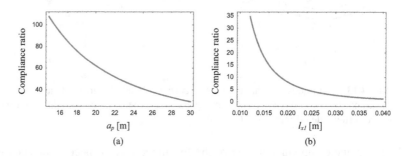

FIGURE 4.15 Plots of compliance ratio $C_{u_y-f_y}/C_{u_x-f_x}$ in terms of the: (a) parabolic parameter a_p; (b) length l_{x1}.

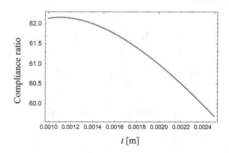

FIGURE 4.16 Plot of compliance ratio $C_{u_y-f_y}/C_{u_x-f_x}$ in terms of the in-plane thickness t.

4.2.3 COMPLIANCE MATRICES OF FLEXIBLE HINGES WITH ANTISYMMETRY AXIS

Multiple-segment flexible hinges may have an antisymmetry axis, case where the full-hinge compliance matrices can be, again, evaluated based on the compliance matrices of half the hinge. This section derives a generic compliance model for antisymmetric flexible hinges and discusses a particular design from this category.

4.2.3.1 Generic Hinge and Compliance Matrices

A generic flexible hinge with antisymmetry axis is sketched in Figure 4.17.

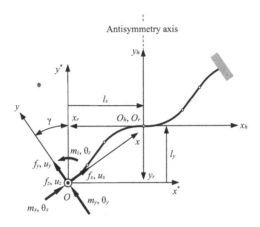

FIGURE 4.17 Skeleton representation of generic flexible hinge formed of two segments that are antisymmetric with respect to an axis.

The hinge can be obtained physically from one half, like the one shown in the reference frame $O_h x_h y_h z_h$ to the right of the y_h axis. Its antisymmetric counterpart is obtained by rotating the original half-part 180° clockwise around point O_h and the full flexure hinge is formed by serially connecting the two parts. Assuming that the compliance matrices of the original half flexible segment are determined in its reference frame $O_h x_h y_h z_h$, the aim is to calculate the compliances of the full hinge in the global reference frame $Oxyz$.

The full-hinge, in-plane compliance matrix is first determined in the reference frame $Ox^* y^* z^*$, which has its y^* axis parallel to the antisymmetry axis y_h, namely:

$$\left[C_{O,ip}^* \right] = \left[T_{OO_h,ip}^{(h)} \right]^T \left(\left[C_{O_h,ip}^{(h)} \right] + \left[R^{(r)} \right]^T \left[C_{O_r,ip}^{(h)} \right] \left[R^{(r)} \right] \right) \left[T_{OO_h,ip}^{(h)} \right]$$

$$\text{with} \quad \left[C_{O_h,ip}^{(h)} \right] = \left[C_{O_r,ip}^{(h)} \right].$$

(4.41)

The first term in the parenthesis of the right-hand side, $\left[C_{O_h,ip}^{(h)} \right]$, is the compliance matrix of the right half-hinge in Figure 4.17 – this is translated from its local frame $O_h x_h y_h z_h$ to the parallel axis, global frame $Ox^* y^* z^*$ (the superscript "h"

stands for half). Similarly, the second term in the parenthesis in the right-hand side of Eq. (4.41) – with the superscript "r" meaning rotated – represents the compliance matrix of the 180°-rotated left half-hinge; this compliance matrix is transferred through translation and rotation from its local reference frame $O_r x_r y_r z_r$ to the same $O x^* y^* z^*$ frame. Note that the two halves are identical when represented in their local frames $O_h x_h y_h z_h$ and $O_r x_r y_r z_r$, and as a consequence, the same matrix $\left[C_{O_h,ip}^{(h)} \right]$ defining the half-hinge in its local frame is used in both terms. It should also be remarked that the translation matrices of the two segments are identical since the two local frames are collocated. The displacements needed to evaluate the translation matrix of Eq. (4.41) are: $\Delta x = l_x$ and $\Delta y = l_y$. The translation and rotation matrices of Eq. (4.41) are:

$$
\left[T_{OO_h,ip}^{(h)} \right] = \begin{bmatrix} 1 & 0 & 0 \\ 0 & 1 & 0 \\ l_y & -l_x & 1 \end{bmatrix}; \quad \left[R^{(r)} \right] = \begin{bmatrix} \cos\pi & \sin\pi & 0 \\ -\sin\pi & \cos\pi & 0 \\ 0 & 0 & 1 \end{bmatrix}
$$

$$
= \begin{bmatrix} -1 & 0 & 0 \\ 0 & -1 & 0 \\ 0 & 0 & 1 \end{bmatrix}. \tag{4.42}
$$

The full-hinge, global-frame, in-plane compliance matrix is obtained from the matrix of Eq. (4.41) by means of a rotation transformation:

$$
\left[C_{O,ip} \right] = [R]^T \left[C_{O,ip}^* \right][R] = \begin{bmatrix} C_{u_x-f_x} & C_{u_x-f_y} & C_{u_x-m_z} \\ C_{u_x-f_y} & C_{u_y-f_y} & C_{u_y-m_z} \\ C_{u_x-m_z} & C_{u_y-m_z} & C_{\theta_z-m_z} \end{bmatrix}
$$

$$
\text{with} \quad [R] = \begin{bmatrix} \cos(-\gamma) & \sin(-\gamma) & 0 \\ -\sin(-\gamma) & \cos(-\gamma) & 0 \\ 0 & 0 & 1 \end{bmatrix} = \begin{bmatrix} \cos(\gamma) & -\sin(\gamma) & 0 \\ \sin(\gamma) & \cos(\gamma) & 0 \\ 0 & 0 & 1 \end{bmatrix}.
$$

$$
\tag{4.43}
$$

The rotation matrix $[R]$ of Eq. (4.43) is calculated using Eq. (4.2) with $\delta = -\gamma$ (see Figure 4.17).

A similar approach is used to determine the out-of-plane compliances of the generic, antisymmetric hinge of Figure 4.17. The full-hinge, out-of-plane compliance matrix is calculated in the reference frame $O x^* y^* z^*$ as:

$$\left[C^*_{O,op}\right]=\left[T^{(h)}_{OO_h,op}\right]^T\left(\left[C^{(h)}_{O_h,op}\right]+\left[R^{(r)}\right]^T\left[C^{(r)}_{O_r,op}\right]\left[R^{(r)}\right]\right)\left[T^{(h)}_{OO_h,op}\right]$$

$$\text{with } \left[C^{(h)}_{O_h,ip}\right]=\left[C^{(r)}_{O_r,ip}\right] \text{ and } \left[T^{(h)}_{OO_h,op}\right]=\begin{bmatrix} 1 & 0 & -l_y \\ 0 & 1 & l_x \\ 0 & 0 & 1 \end{bmatrix}. \tag{4.44}$$

The full-hinge, global-frame, out-of-plane compliance matrix is:

$$\left[C_{O,op}\right]=[R]^T\left[C^*_{O,op}\right][R]=\begin{bmatrix} C_{\theta_x-m_x} & C_{\theta_x-m_y} & C_{\theta_x-f_z} \\ C_{\theta_x-m_y} & C_{\theta_y-m_y} & C_{\theta_y-f_z} \\ C_{\theta_x-f_z} & C_{\theta_y-f_z} & C_{u_z-f_z} \end{bmatrix}. \tag{4.45}$$

4.2.3.2 Flexible-Hinge Configuration

Figure 4.18 illustrates a flexible serpentine hinge with antisymmetry axis whose in-plane compliance matrix was formulated and analyzed in Lobontiu [26]. The basic half-hinge is formed of two constant cross-section straight-axis segments – one constant cross-section circular-axis segment and one variable cross-section root circular-axis segment, which has a filleted configuration and which is used to connect the hinge to a rigid link.

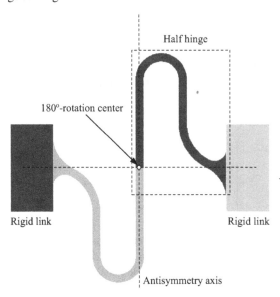

FIGURE 4.18 Planar geometry of serpentine flexible hinge with antisymmetry axis.

The in-plane and out-of-plane compliance matrices are derived here, and the planar elastic response of the serpentine hinge is characterized in terms of defining geometric parameters. The hinge is shown in skeleton form in Figure 4.19a, while the basic half-hinge is depicted in Figure 4.19b together with its component segments.

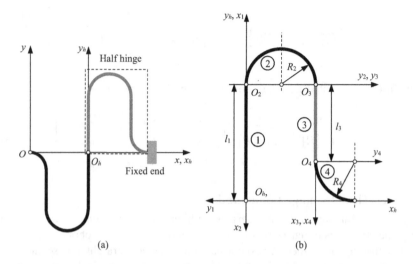

FIGURE 4.19 Skeleton representation of constant cross-sectional serpentine flexible hinge: (a) full-hinge configuration; (b) basic half-hinge with individual segments and geometry.

Assuming the in-plane compliance matrix of the basic half-hinge is available, the full-hinge compliance matrix with respect to the global frame $Oxyz$ is calculated based on Eqs. (4.41)–(4.43) by taking into account that no (mathematical) matrix rotation is needed at O since the reference frames $O_h x_h y_h z_h$ and $Oxyz$ are parallel; as a consequence, the global-frame, in-plane compliance matrix is:

$$\left[C_{O,ip}\right]=\left[T_{OO_h,ip}^{(h)}\right]^T\left(\left[C_{O_h,ip}^{(h)}\right]+\left[R^{(r)}\right]^T\left[C_{O_h,ip}^{(h)}\right]\left[R^{(r)}\right]\right)\left[T_{OO_h,ip}^{(h)}\right]$$

$$=2\begin{bmatrix} C_{u_x-f_x}^{(h)} & C_{u_x-f_y}^{(h)} & 0 \\ C_{u_x-f_y}^{(h)} & C_{u_y-f_y}^{(h)}+l_x^2\cdot C_{\theta_z-m_z}^{(h)} & -l_x\cdot C_{\theta_z-m_z}^{(h)} \\ 0 & -l_x\cdot C_{\theta_z-m_z}^{(h)} & C_{\theta_z-m_z}^{(h)} \end{bmatrix}. \qquad (4.46)$$

The components of the Eq. (4.46) matrix characterize the half-antisymmetric flexible hinge of Figure 4.19b. Equation (4.46) considered that $l_y = 0$ in the in-plane translation matrix of Eq. (4.42), while $l_x = 2R_2 + R_4$, as seen in Figure 4.19b. The in-plane compliance matrix of the half-hinge pictured in Figure 4.19b is expressed in terms of the four segments' in-plane compliance matrices as:

$$\left[C_{O_h,ip}^{(h)}\right]=\left[R^{(1)}\right]^T\left[C_{O_1,ip}^{(1)}\right]\left[R^{(1)}\right]+\sum_{i=2}^{4}\left[T_{O_hO_i,ip}^{(i)}\right]^T\left[R^{(i)}\right]^T\left[C_{O_i,ip}^{(i)}\right]\left[R^{(i)}\right]\left[T_{O_hO_i,ip}^{(i)}\right].$$

$$(4.47)$$

where the rotation matrices are:

$$\left[R^{(1)} \right] = \begin{bmatrix} \cos(\pi/2) & \sin(\pi/2) & 0 \\ -\sin(\pi/2) & \cos(\pi/2) & 0 \\ 0 & 0 & 1 \end{bmatrix};$$

$$\left[R^{(2)} \right] = \left[R^{(3)} \right] = \left[R^{(4)} \right] = \begin{bmatrix} \cos(-\pi/2) & \sin(-\pi/2) & 0 \\ -\sin(-\pi/2) & \cos(-\pi/2) & 0 \\ 0 & 0 & 1 \end{bmatrix}. \quad (4.48)$$

and the translation matrices are:

$$\left[T_{O_h O_i,ip}^{(i)} \right] = \begin{bmatrix} 1 & 0 & 0 \\ 0 & 1 & 0 \\ \Delta y_i & -\Delta x_i & 1 \end{bmatrix} \quad \text{with} \quad \begin{cases} \Delta x_2 = 0, \Delta y_2 = l_1 \\ \Delta x_3 = 2R_2, \Delta y_3 = l_1 \\ \Delta x_4 = 2R_2, \Delta y_4 = l_1 - l_3 \end{cases}. \quad (4.49)$$

The in-plane compliance matrices of the straight-axis segments 1 and 3 are provided in Eqs. (2.26). The in-plane compliance matrix and components of the circular-axis segment 2 are calculated by means of Eqs. (A2.22) through (A2.26) and Eq. (2.107). Lastly, the in-plane compliance matrix of segment 4 is formed using Eqs. (2.102) through (2.106) together with Eqs. (2.145)–(2.147).

A similar approach enables to calculate of the out-of-plane compliance matrix of the full hinge in terms of the $Oxyz$ reference frame with the aid of Eqs. (4.44) and (4.45) as:

$$\left[C_{O,op} \right] = \left[T_{OO_h,op}^{(h)} \right]^T \left(\left[C_{O_h,op}^{(h)} \right] + \left[R^{(r)} \right]^T \left[C_{O_h,op}^{(h)} \right] \left[R^{(r)} \right] \right) \left[T_{OO_h,op}^{(h)} \right]$$

$$= 2 \begin{bmatrix} C_{\theta_x-m_x}^{(h)} & C_{\theta_x-m_y}^{(h)} & l_x C_{\theta_x-m_y}^{(h)} \\ C_{\theta_x-m_y}^{(h)} & C_{\theta_y-m_y}^{(h)} & l_x C_{\theta_y-m_y}^{(h)} \\ l_x C_{\theta_x-m_y}^{(h)} & l_x C_{\theta_y-m_y}^{(h)} & C_{u_z-f_z}^{(h)} + l_x^2 C_{\theta_y-m_y}^{(h)} \end{bmatrix}. \quad (4.50)$$

where, again, the individual compliances of Eq. (4.50) represent the half-hinge compliances calculated in the $O_h x_h y_h z_h$ reference frame. The out-of-plane compliance matrix of the half-hinge shown in Figure 4.19b is calculated by adding the transformed matrices of segments 1 through 4, namely:

$$\left[C_{O_h,op}^{(h)} \right] = \left[R^{(1)} \right]^T \left[C_{O_1,op}^{(1)} \right] \left[R^{(1)} \right] + \sum_{i=2}^{4} \left[T_{O_h O_i,op}^{(i)} \right]^T \left[R^{(i)} \right]^T \left[C_{O_i,op}^{(i)} \right] \left[R^{(i)} \right] \left[T_{O_h O_i,op}^{(i)} \right].$$

$$(4.51)$$

The translation matrices of Eq. (4.51) are calculated as:

$$\left[T^{(i)}_{O_sO_i,ip}\right] = \begin{bmatrix} 1 & 0 & -\Delta y_i \\ 0 & 1 & \Delta x_i \\ 0 & 0 & 1 \end{bmatrix}. \tag{4.52}$$

with the offsets provided in Eq. (4.49). The out-of-plane compliance matrices of the two straight-axis segments 1 and 3 are given in Eqs. (2.27). The out-of-plane compliance matrix of segment 2 is assembled by means of Eqs. (A2.27) through (A2.32) and Eq. (2.117), while the similar matrix of segment 4 is evaluated using (2.109) through (2.116) in combination with Eqs. (2.145)–(2.147).

Note that for the circular-axis segment 2, the local frame $O_2x_2\,y_2z_2$ has its regular x axis swapped from the regular position, which is opposing x_2 of Figure 4.19b; as a result, both the in-plane and out-of-plane compliance matrices of segment 2 use minus signs in their regular x-y and x-z compliances, as also captured in Eqs. (2.107) and (2.117), namely:

$$\left[C^{(2)}_{O_2,ip}\right] = \begin{bmatrix} C^{(2)}_{O_2,u_x-f_x} & -C^{(2)}_{O_2,u_x-f_y} & -C^{(2)}_{O_2,u_x-m_z} \\ -C^{(2)}_{O_2,u_x-f_y} & C^{(2)}_{O_2,u_y-f_y} & C^{(2)}_{O_2,u_y-m_z} \\ -C^{(2)}_{O_2,u_x-m_z} & C^{(2)}_{O_2,u_y-m_z} & C^{(2)}_{O_2,\theta_z-m_z} \end{bmatrix};$$

$$\left[C^{(2)}_{O_2,op}\right] = \begin{bmatrix} C^{(2)}_{O_2,\theta_x-m_x} & -C^{(2)}_{O_2,\theta_x-m_y} & -C^{(2)}_{O_2,\theta_x-f_z} \\ -C^{(2)}_{O_2,\theta_x-m_y} & C^{(2)}_{O_2,\theta_y-m_y} & C^{(2)}_{O_2,\theta_y-f_z} \\ -C^{(2)}_{O_2,\theta_x-f_z} & C^{(2)}_{O_2,\theta_y-f_z} & C^{(2)}_{O_2,u_z-f_z} \end{bmatrix}. \tag{4.53}$$

Example 4.2

Identify the geometric parameters of the flexible hinge shown in Figure 4.19 that result in the maximum compliance $C_{u_x-f_x}$ and the minimum compliance $C_{u_y-f_y}$ when considering the following parameter ranges: $l_1 \rightarrow [0.015; 0.03]$ m, $R_2 \rightarrow [0.003; 0.015]$ m and $R_4 \rightarrow [0.002; 0.01]$ m. Assume that the constant cross-section is rectangular with the out-of-plane width $w = 0.01$ m and in-plane (minimum) thickness $t = 0.002$ m. The material Young's modulus is $E = 2.1 \cdot 10^{11}$ N/m².

Solution:

Maximizing $C_{u_x-f_x}$ while minimizing $C_{u_y-f_y}$ is akin to maximizing the compliance ratio $C_{u_x-f_x}/C_{u_y-f_y}$. To help identify the parameters l_1, R_2 and R_4 that achieve this objective, the compliance ratio is plotted in terms of these three parameters in Figures 4.20 and 4.21. Note that $l_3 = l_1 - R_4$.

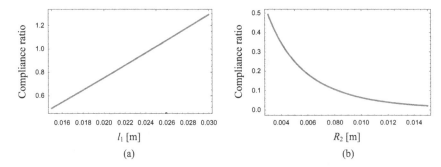

FIGURE 4.20 Plots of compliance ratio $C_{u_x-f_x}/C_{u_y-f_y}$ in terms of the: (a) length l_1; (b) radius R_2.

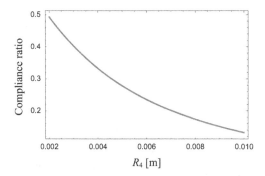

FIGURE 4.21 Plot of compliance ratio $C_{u_x-f_x}/C_{u_y-f_y}$ in terms of the radius R_4.

As shown in Figures 4.20 and 4.21, the compliance ratio varies monotonically with the three geometric parameters; the ratio is maximum when l_1 reaches the upper limit of its interval, which is $l_1 = 0.03$ m, R_2 is at the lower limit, $R_2 = 0.003$ m, and R_4 assumes its lowest value, $R_4 = 0.002$ m; for these parameter values, the $C_{u_x-f_x}/C_{u_y-f_y}$ ratio is equal to 1.294.

4.2.4 Folded Flexible Hinges

Section 3.2.3 introduces the folded hinges of rectangular envelopes that are formed of straight-axis flexible segments connected by rigid connectors – see Figure 3.27. Folded hinges of either rectangular or curvilinear geometric envelope can also be designed by including curvilinear-axis flexible segments in combination with other straight-axis segments, as illustrated in Figure 4.22. This section discusses the generic configurations of Figure 4.22 – more details can be found in Lobontiu et al. [27].

In-plane compliance matrices are formulated for the two families of folded hinges shown in Figure 4.22.

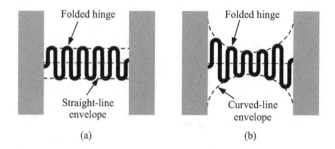

FIGURE 4.22   Straight-axis folded flexible hinges with axial symmetry and: (a) straight-line (rectangular) envelope; (b) curvilinear-line envelope.

4.2.4.1   Straight-Line (Rectangular) Envelope

4.2.4.1.1   Generic Configuration

A folded hinge that is inscribed in a rectangular area of dimensions l and $2y_e$ is sketched in Figure 4.23a. The full hinge is the series combination of the half-portion situated to the right of the midpoint O_h and the other half, located to the

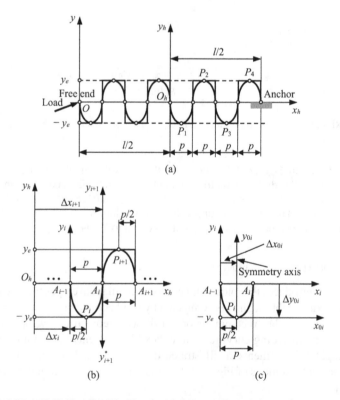

FIGURE 4.23   (a) Folded flexible hinge with rectangular envelope; (b) geometry of two neighboring generic unit segments; (c) geometry of segment i.

left of O_h; the left half results by an $180°$ rotation of the right half around O_h, and, therefore, the full hinge is antisymmetric with respect to the y_h axis. Several equidistant points $P_1, P_2, ..., P_n$ are positioned on the two envelope curves y_e and $-y_e$ on the right hinge portion. The axial pitch distance between two consecutive points is p. A generic point P_i (such as the one of Figure 4.23b) has the following coordinates in the $O_h x_h y_h z_h$ reference frame:

$$x_i = (i-1)p + \frac{p}{2} = \frac{2i-1}{2} \cdot p; \quad y_i = (-1)^i y_e. \tag{4.54}$$

with $i = 1, 2, ..., n$. Consider the generic hinge segment i of Figure 4.23c, which is inscribed in a rectangle of dimensions p and y_e. This segment is symmetric with respect to the $P_i y_{0i}$ axis and is formed of the curvilinear segments $A_{i-1}P_i$ and $P_i A_i$.

Because the two segments of Figure 4.23c are connected in series, the compliance matrix of the segment $A_{i-1}A_i$ with respect to a frame centered at P_i is expressed as:

$$\left[C_{P_i}^{(i)} \right] = \left[C_{P_i}^{(i,r)} \right] + \left[C_{P_i}^{(i,l)} \right] =
\begin{bmatrix}
C_{uxi-fxi}^{(i,r)} & C_{uxi-fyi}^{(i,r)} & C_{uxi-mzi}^{(i,r)} \\
C_{uxi-fyi}^{(i,r)} & C_{uyi-fyi}^{(i,r)} & C_{uyi-mzi}^{(i,r)} \\
C_{uxi-mzi}^{(i,r)} & C_{uyi-mzi}^{(i,r)} & C_{\theta zi-mzi}^{(i,r)}
\end{bmatrix}$$

$$+
\begin{bmatrix}
C_{uxi-fxi}^{(i,r)} & -C_{uxi-fyi}^{(i,r)} & C_{uxi-mzi}^{(i,r)} \\
-C_{uxi-fyi}^{(i,r)} & C_{uyi-fyi}^{(i,r)} & -C_{uyi-mzi}^{(i,r)} \\
C_{uxi-mzi}^{(i,r)} & -C_{uyi-mzi}^{(i,r)} & C_{\theta zi-mzi}^{(i,r)}
\end{bmatrix}
= 2
\begin{bmatrix}
C_{uxi-fxi}^{(i,r)} & 0 & C_{uxi-mzi}^{(i,r)} \\
0 & C_{uyi-fyi}^{(i,r)} & 0 \\
C_{uxi-mzi}^{(i,r)} & 0 & C_{\theta zi-mzi}^{(i,r)}
\end{bmatrix}.$$

$$\tag{4.55}$$

where "r" and "l" mean right and left, respectively. The compliance matrix of the left segment is obtained from the one of the right segment with the appropriate sign changes accounting for the mirroring about the $P_i y_{0i}$ symmetry axis. The compliance matrix of segment i is transferred to the frame $A_{i-1} x_i y_i z_i$ of Figure 4.23c as:

$$\left[C_{A_{i-1}}^{(i)} \right] = \left[T_{A_{i-1}P_i}^{(i*)} \right]^T \left[C_{P_i}^{(i)} \right] \left[T_{A_{i-1}P_i}^{(i*)} \right] =
\begin{bmatrix}
C_{A_{i-1},ux-fx}^{(i)} & C_{A_{i-1},ux-fy}^{(i)} & C_{A_{i-1},ux-mz}^{(i)} \\
C_{A_{i-1},ux-fy}^{(i)} & C_{A_{i-1},uy-fy}^{(i)} & C_{A_{i-1},uy-mz}^{(i)} \\
C_{A_{i-1},ux-mz}^{(i)} & C_{A_{i-1},uy-mz}^{(i)} & C_{A_{i-1},\theta z-mz}^{(i)}
\end{bmatrix}$$

with $\left[T_{A_{i-1}P_i}^{(i*)} \right] =
\begin{bmatrix}
1 & 0 & 0 \\
0 & 1 & 0 \\
\Delta y_{O_i} & -\Delta x_{O_i} & 1
\end{bmatrix}
=
\begin{bmatrix}
1 & 0 & 0 \\
0 & 1 & 0 \\
-y_e & -p/2 & 1
\end{bmatrix}.$

$$\tag{4.56}$$

The next segment $i+1$, which is structurally identical to the segment i except from being mirrored with respect to the x axis, has the following in-plane compliance matrix:

$$\left[C_{A_i}^{(i)} \right] = \begin{bmatrix} C_{A_{i-1},u_x-f_x}^{(i)} & -C_{A_{i-1},u_x-f_y}^{(i)} & -C_{A_{i-1},u_x-m_z}^{(i)} \\ -C_{A_{i-1},u_x-f_y}^{(i)} & C_{A_{i-1},u_y-f_y}^{(i)} & C_{A_{i-1},u_y-m_z}^{(i)} \\ -C_{A_{i-1},u_x-m_z}^{(i)} & C_{A_{i-1},u_y-m_z}^{(i)} & C_{A_{i-1},\theta_z-m_z}^{(i)} \end{bmatrix}. \qquad (4.57)$$

The compliance matrix of the right half of the full hinge – the one to the right of O_h in Figure 4.23a – is calculated by adding the compliance matrices of all n individual segments as:

$$\left[C_{O_h}^{(h)} \right] = \sum_{i=1}^{n} \left[T_{O_h A_{i-1}}^{(i)} \right]^T \left[C_{A_{i-1}}^{(i)} \right] \left[T_{O_h A_{i-1}}^{(i)} \right] = \begin{bmatrix} C_{O_h,u_x-f_x}^{(h)} & C_{O_h,u_x-f_y}^{(h)} & C_{O_h,u_x-m_z}^{(h)} \\ C_{O_h,u_x-f_y}^{(h)} & C_{O_h,u_y-f_y}^{(h)} & C_{O_h,u_y-m_z}^{(h)} \\ C_{O_h,u_x-m_z}^{(h)} & C_{O_h,u_y-m_z}^{(h)} & C_{O_h,\theta_z-m_z}^{(h)} \end{bmatrix}$$

with $\left[T_{O_h A_{i-1}}^{(i)} \right] = \begin{bmatrix} 1 & 0 & 0 \\ 0 & 1 & 0 \\ \Delta y_i & -\Delta x_i & 1 \end{bmatrix} = \begin{bmatrix} 1 & 0 & 0 \\ 0 & 1 & 0 \\ 0 & -(i-1)\cdot p & 1 \end{bmatrix}.$

$$(4.58)$$

The compliance matrix of the left half of the hinge is obtained by a 180° rotation of the right half compliance matrix in order to express it in the $O_h x_h y_h z_h$ reference frame. The matrices of the two halves are then added, and the resulting matrix is translated at O – the global-frame origin where the load is applied and where the displacements need to be calculated; this final matrix is:

$$[C_O] = [T_{OO_h}]^T \left(\left[C_{O_h}^{(h)} \right] + [R]^T \left[C_{O_h}^{(h)} \right] [R] \right) [T_{OO_h}] = \begin{bmatrix} C_{O,u_x-f_x} & C_{O,u_x-f_y} & C_{O,u_x-m_z} \\ C_{O,u_x-f_y} & C_{O,u_y-f_y} & C_{O,u_y-m_z} \\ C_{O,u_x-m_z} & C_{O,u_y-m_z} & C_{O,\theta_z-m_z} \end{bmatrix}$$

with $[T_{OO_h}] = \begin{bmatrix} 1 & 0 & 0 \\ 0 & 1 & 0 \\ 0 & -\Delta x & 1 \end{bmatrix} = \begin{bmatrix} 1 & 0 & 0 \\ 0 & 1 & 0 \\ 0 & -l/2 & 1 \end{bmatrix};$

$$[R] = \begin{bmatrix} \cos\pi & \sin\pi & 0 \\ -\sin\pi & \cos\pi & 0 \\ 0 & 0 & 1 \end{bmatrix}.$$

$$(4.59)$$

The explicit compliances of $[C_o]$ expressed in Eq. (4.59) are provided in Lobontiu et al. [27].

4.2.4.1.2 Design with Basic Segment Formed of Elliptical and Straight-Line Portions

Consider the design shown in Figure 4.24a, which is formed of any number of identical units. The right portion of a basic segment is sketched in Figure 4.24b, and it comprises two members: one-quarter of an ellipse (the semi-axes lengths are a and b) and a straight-line segment of length h.

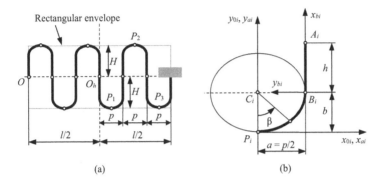

(a)   (b)

FIGURE 4.24 (a) Rectangular-envelope folded flexible hinge; (b) geometry of basic half-segment formed of a quarter ellipse and a straight line.

The in-plane compliance matrix of the half-segment with respect to the reference frame centered at P_i is obtained by adding the transformed compliance matrices of the two serially connected segments:

$$\left[C_{P_i} \right] = \left[C_{P_i}^{(i,e)} \right] + \left[T_{P_i B_i}^{(i,s)} \right]^T \left[R^{(i,s)} \right]^T \left[C_{B_i}^{(i,s)} \right] \left[R^{(i,s)} \right] \left[T_{P_i B_i}^{(i,s)} \right] \text{ with}$$

$$\left[T_{P_i B_i}^{(i,s)} \right] = \begin{bmatrix} 1 & 0 & 0 \\ 0 & 1 & 0 \\ b & -p/2 & 1 \end{bmatrix}; \left[R^{(i,s)} \right] = \begin{bmatrix} \cos(\pi/2) & \sin(\pi/2) & 0 \\ -\sin(\pi/2) & \cos(\pi/2) & 0 \\ 0 & 0 & 1 \end{bmatrix}.$$

(4.60)

where "e" stands for the first (elliptical) portion and "s" for the second (straight line) segment. The in-plane compliances of the elliptical-axis segment are provided in Eqs. (2.133) through (2.137), while the straight-line segment compliances are given in Eqs. (2.26).

Equations (4.55) through (4.60) are subsequently used to express the in-plane compliance matrix of a rectangular envelope-folded hinge with the half-basic segment of Figure 4.24b. All segments composing the folded hinge have the same constant rectangular cross-section defined by in-plane thickness t and out-of-plane width w. The variation of the compliances $C_{u_x - f_x}$ and $C_{u_y - f_y}$ are studied in terms of the parameters t, h and a, which are identified in Figure 4.24b; this can be achieved by analyzing the compliance ratio $C_{u_y-f_y}/C_{u_x-f_x}$. Figures 4.25 and 4.26 plot the variation of the compliance ratio in terms of these parameters. For each plot, the constant values of the other parameters were: $n = 6$, $a = 0.008\,$m, $b = a/2$, $t = 0.001\,$m, $w = 0.008\,$m and $h = 0.015\,$m. A Young's modulus $E = 2{\cdot}10^{11}\,$N/m^2 was also used. As the two figures indicate, the y axis compliance is larger than the x-axis compliance for all three parameters over their selected ranges.

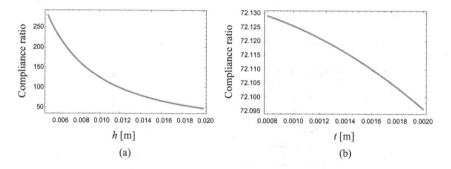

FIGURE 4.25 Plots of compliance ratio $C_{u_y-f_y}/C_{u_x-f_x}$ in terms of the (a) parameter h; (b) in-plane thickness t.

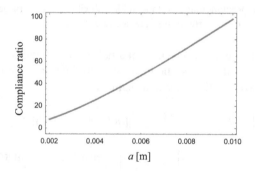

FIGURE 4.26 Plot of compliance ratio $C_{u_y-f_y}/C_{u_x-f_x}$ in terms of the ellipse semi-axis length a.

4.2.4.2 Curvilinear-Line Envelope

4.2.4.2.1 Generic Configuration

The two symmetric curves $y_e(x)$ and $-y_e(x)$ of Figure 4.27a are envelopes of the skeleton folded hinge sketched in the same figure. Two adjacent segments, i and $i + 1$, are shown in Figure 4.27b. The coordinates x_i and y_i of a generic point P_i (like the one shown in Figure 4.27b) are calculated with Eq. (4.54).

The compliance matrix of the generic segment i with respect to the frame $A_{i-1}x_iy_iz_i$ of Figure 4.23c is calculated by means of Eqs. (4.55)–(4.57) with $y_e(x_i)$ instead of y_e. The compliance matrix of the right half of the hinge – see

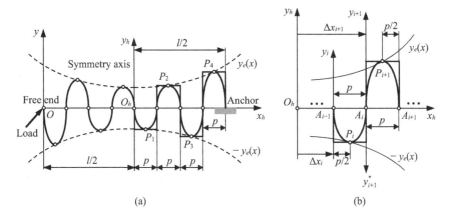

FIGURE 4.27 (a) Folded flexible hinge with symmetric, curvilinear envelopes; (b) geometry of two neighboring generic unit segments.

Figure 4.27a – is determined by adding the compliance matrices of all the individual symmetric segments (after translations from their local frames to the common origin O_h) as per Eq. (4.58). The compliance matrix of the full flexure hinge of Figure 4.27a is calculated in the reference frame $Oxyz$ with Eq. (4.59).

4.2.4.2.2 Design with Circular-Line Envelope and Basic Unit Formed of Elliptical and Straight-Line Segments

As an illustration of the compliance calculation procedure for a generic, curvilinear-line-envelope folded hinge, the particular design of Figure 4.28 is briefly discussed here – more details are offered in Lobontiu et al. [27].

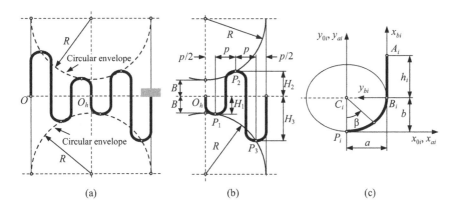

FIGURE 4.28 Folded flexible hinge with circular envelope: (a) skeleton hinge representation; (b) right half of hinge; (c) basic half-segment formed of a quarter ellipse and a straight line.

The two envelope curves are circular segments of radius R disposed symmetrically with respect to the axial direction. The basic symmetric segment, the right half of which is shown in Figure 4.28c, is formed of one elliptic quarter with semi-axes lengths a and b, and a straight line of length h_i. As Figures 4.28b and c indicate, the pitch distance is $p = 2a$ and the envelope radius is $R = l/2 = n \cdot p = 2n \cdot a$ for a half-portion formed of n scaled units. The dimensions $H_1, H_2, \ldots, H_n$, as well as $h_1, h_2, \ldots, h_n$ of Figure 4.28b and c, are:

$$H_i = R + B - \sqrt{R^2 - (2i - 1)^2 a^2}; \quad h_i = H_i - b. \qquad (4.61)$$

with $i = 1, 2, \ldots, n$. The midpoints P_i of the elliptical portions are located on the circular envelopes. The compliance matrix of the half-segment adds the compliance matrices of the quarter ellipse and the straight line (after rotating + translating the native matrix of this portion), as in Eqs. (4.60).

Because $R = n \cdot p = 2n \cdot a$, the envelope radius R is related to the number of segments n, the long semi-axis length a and the pitch distance p. Figure 4.29 plots the variation of the rotary compliance ratio $C_{\theta_z - m_z} / C_{\theta_z - m_z}^{(1)}$ in terms of R for $B = 0.002\,\text{m}$, $n = 3$ $a = 0.01\,\text{m}$, $b = a/2$, $t = 0.001\,\text{m}$, $w = 0.008\,\text{m}$ and $E = 2 \cdot 10^{11}\,\text{N/m}^2$. $C_{\theta_z - m_z}^{(1)}$ is the compliance of the first unit located to the right of O_h in Figure 4.28b. The compliance decreases nonlinearly with increasing R, as shown in Figure 4.29.

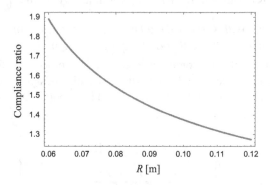

FIGURE 4.29   Plot of compliance ratio $C_{\theta_z - m_z} / C_{\theta_z - m_z}^{(1)}$ in terms of the envelope radius R.

4.2.4.3   Radial Folded Hinge with Rigid Connectors

The planar flexible hinge of Figure 4.30 is inscribed in a circular sector of angle α and is formed of several ($i = 1, 2, \ldots, n$) circular-axis flexible segments interconnected by radial shorter and identical segments, which are assumed rigid – all these segments are connected in series. The rigid segments have a length g, and therefore, the circular flexible portions are equidistant; the radii of two consecutive circular segments are connected as: $R_{i+1} = R_i + g$ or $R_{i+1} = R_1 + (i - 1)g$. Assuming the load vector is applied at the hinge end O, while the opposite end is anchored, the in-plane and out-of-plane compliance matrices of this flexible hinge are formulated in this section.

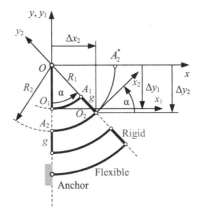

FIGURE 4.30 Planar geometry of folded hinge formed of flexible, concentric, circular-axis segments and rigid radial connectors.

4.2.4.3.1 In-Plane Compliance Matrix

The in-plane compliance matrix of a circular-axis segment i of center angle α is expressed in its local frame $O_i x_i y_i z_i$ (not shown in Figure 4.30) as:

$$\left[C_{O_i,ip}^{(i)} \right] = \begin{bmatrix} C_{O_i,u_x-f_x}^{(i)} & C_{O_i,u_x-f_y}^{(i)} & C_{O_i,u_x-m_z}^{(i)} \\ C_{O_i,u_x-f_y}^{(i)} & C_{O_i,u_y-f_y}^{(i)} & C_{O_i,u_y-m_z}^{(i)} \\ C_{O_i,u_x-m_z}^{(i)} & C_{O_i,u_y-m_z}^{(i)} & C_{O_i,\theta_z-m_z}^{(i)} \end{bmatrix}. \tag{4.62}$$

where the matrix components are given explicitly in Eqs. (A2.22) through (A2.26). Consider segment 2 ($O_2 A_2$) in Figure 4.30 plays the role of the generic segment i. The odd-number circular-axis segments 1, 3, ..., $2i - 1$, ... have their y_i axes collinear with the global y axis. For these segments, the in-plane compliance matrices are expressed in the global frame $Oxyz$ by means of a translation as:

$$\left[C_{O,ip}^{(i)} \right] = \left[T_{OO_i,ip}^{(i)} \right]^T \left[C_{O_i,ip}^{(i)} \right] \left[T_{OO_i,ip}^{(i)} \right] \quad \text{with} \quad \left[T_{OO_i,ip}^{(i)} \right] = \begin{bmatrix} 1 & 0 & 0 \\ 0 & 1 & 0 \\ \Delta y_i & -\Delta x_i & 1 \end{bmatrix}$$

$$= \begin{bmatrix} 1 & 0 & 0 \\ 0 & 1 & 0 \\ -R_i & 0 & 1 \end{bmatrix}. \tag{4.63}$$

As illustrated in Figure 4.30, the even-number circular-axis segments 2, 4, ..., $2i$, ... have their local frames both offset and rotated with respect to the global frame; as a consequence, their in-plane global-frame compliance matrices are formulated based on the corresponding local-frame matrices as:

$$
\left[C^{(i^*)}_{O,ip} \right] = \begin{bmatrix} C^{(i)}_{O_i,u_x-f_x} & -C^{(i)}_{O_i,u_x-f_y} & C^{(i)}_{O_i,u_x-m_z} \\ -C^{(i)}_{O_i,u_x-f_y} & C^{(i)}_{O_i,u_y-f_y} & -C^{(i)}_{O_i,u_y-m_z} \\ C^{(i)}_{O_i,u_x-m_z} & -C^{(i)}_{O_i,u_y-m_z} & C^{(i)}_{O_i,\theta_z-m_z} \end{bmatrix}
\tag{4.64}
$$

$$
\left[C^{(i)}_{O,ip} \right] = \left[T^{(i)}_{OO_i,ip} \right]^T \left[R^{(i)} \right]^T \left[C^{(i^*)}_{O_i,ip} \right] \left[R^{(i)} \right] \left[T^{(i)}_{OO_i,ip} \right]
$$

with

$$
\left[T^{(i)}_{OO_i,ip} \right] = \begin{bmatrix} 1 & 0 & 0 \\ 0 & 1 & 0 \\ \Delta y_i & -\Delta x_i & 1 \end{bmatrix} = \begin{bmatrix} 1 & 0 & 0 \\ 0 & 1 & 0 \\ -R_i\cos\alpha & -R_i\sin\alpha & 1 \end{bmatrix};
$$

$$
\left[R^{(i)} \right] = \begin{bmatrix} \cos\alpha & \sin\alpha & 0 \\ -\sin\alpha & \cos\alpha & 0 \\ 0 & 0 & 1 \end{bmatrix}.
\tag{4.65}
$$

Note that the global-frame matrix $\left[C^{(i)}_{O,ip} \right]$ is calculated from the local-frame compliance matrix $\left[C^{(i^*)}_{O,ip} \right]$ through translation and rotation. Because the even-number segments are mirrored with respect to their local y axes – see segment O_2A_2 in Figure 4.30, $\left[C^{(i^*)}_{O,ip} \right]$ results from $\left[C^{(i)}_{O_i,ip} \right]$ of Eq. (4.62) by using minus signs in front of the components that combine the x-y and y-z axes.

The in-plane compliance matrix of the full hinge is the sum of all compliance matrices of Eqs. (4.63) and (4.64) since the flexible components are linked in series:

$$
\left[C_{O,ip} \right] = \sum_{i=1}^{n} \left[C^{(i)}_{O,ip} \right].
\tag{4.66}
$$

For constant rectangular cross-sectional circular-axis segments with in-plane width w and out-of-plane thickness t, the compliances of the full-hinge compliance matrix depend on the parameters n, R_1, g, α, w and t. All compliances increase when n, R_1, g and α increase, and w and t decrease. The following compliance ratio is analyzed:

$$
r_{u_y} = \frac{C_{u_y-f_y}}{C^0_{u_y-f_y}} \quad \text{with} \quad C^0_{u_y-f_y} = \lim_{\substack{\alpha\to\alpha_0 \\ w\to w_0}} C_{u_y-f_y}.
\tag{4.67}
$$

where the compliance with the superscript "0" represents the minimum value. For α ranging in the [15°; 120°] interval, the minimum compliance occurs at $\alpha_0 = 15°$, whereas for w in the [0.001; 0.003] m range, the minimum compliance corresponds to $w_0 = 0.003$ m. The other constant parameters were: $n = 6$, $R_1 = 0.03$ m, $g = 0.015$ m, $t = 0.001$ m, $\alpha = 50°$ and $w = 0.005$ m. The compliance ratio of Eq. (4.67) is plotted in Figure 4.31a in terms of α and in Figure 4.31b as a function of the cross-sectional width w – the ratio varies nonlinearly with both α and w.

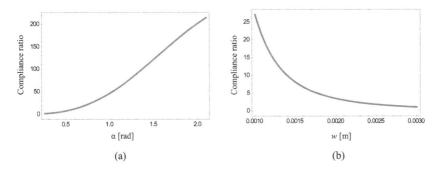

FIGURE 4.31 Plots of compliance ratio $C_{u_y-f_y}/C^0_{u_y-f_y}$ in terms of the (a) center angle α; (b) segment width w.

4.2.4.3.2 Out-of-Plane Compliance Matrix

The local-frame, out-of-plane compliance matrix of a segment i is:

$$\left[C^{(i)}_{O_i,op} \right] = \begin{bmatrix} C^{(i)}_{O_i,\theta_x-m_x} & C^{(i)}_{O_i,\theta_x-m_y} & C^{(i)}_{O_i,\theta_x-f_z} \\ C^{(i)}_{O_i,\theta_x-m_y} & C^{(i)}_{O_i,\theta_y-m_y} & C^{(i)}_{O_i,\theta_y-f_z} \\ C^{(i)}_{O_i,\theta_x-f_z} & C^{(i)}_{O_i,\theta_y-f_z} & C^{(i)}_{O_i,u_z-f_z} \end{bmatrix}. \tag{4.68}$$

and its components are provided in Eqs. (A2.27) through (A2.32). Using a reasoning similar to the one applied to the in-plane compliance matrix, the global-frame compliance matrix of odd-number circular segments is:

$$\left[C^{(i)}_{O,op} \right] = \left[T^{(i)}_{OO_i,op} \right]^T \left[C^{(i)}_{O_i,op} \right] \left[T^{(i)}_{OO_i,op} \right] \quad \text{with} \quad \left[T^{(i)}_{OO_i,op} \right] = \begin{bmatrix} 1 & 0 & -\Delta y_i \\ 0 & 1 & \Delta x_i \\ 0 & 0 & 1 \end{bmatrix}$$

$$= \begin{bmatrix} 1 & 0 & R_i \\ 0 & 1 & 0 \\ 0 & 0 & 1 \end{bmatrix}, \tag{4.69}$$

whereas the compliance matrix of even-number segments is transferred at the global frame using the equations:

$$\left[C_{O_i,op}^{(i*)}\right] = \begin{bmatrix} C_{O_i,\theta_x-m_x}^{(i)} & -C_{O_i,\theta_x-m_y}^{(i)} & C_{O_i,\theta_x-f_z}^{(i)} \\ -C_{O_i,\theta_x-m_y}^{(i)} & C_{O_i,\theta_y-m_y}^{(i)} & -C_{O_i,\theta_y-f_z}^{(i)} \\ C_{O_i,\theta_x-f_z}^{(i)} & -C_{O_i,\theta_y-f_z}^{(i)} & C_{O_i,u_z-f_z}^{(i)} \end{bmatrix} \tag{4.70}$$

$$\left[C_{O,op}^{(i)}\right] = \left[T_{OO_i,op}^{(i)}\right]^T \left[R^{(i)}\right]^T \left[C_{O_i,op}^{(i*)}\right] \left[R^{(i)}\right] \left[T_{OO_i,op}^{(i)}\right],$$

with

$$\left[T_{OO_i,ip}^{(i)}\right] = \begin{bmatrix} 1 & 0 & -\Delta y_i \\ 0 & 1 & \Delta x_i \\ 0 & 0 & 1 \end{bmatrix} = \begin{bmatrix} 1 & 0 & R_i\cos\alpha \\ 0 & 1 & R_i\sin\alpha \\ 0 & 0 & 1 \end{bmatrix}. \tag{4.71}$$

Equations (4.64) and (4.70) were obtained using the same reasoning. The full-hinge compliance $[C_{O,op}]$ is expressed as in Eq. (4.66) by using "op" instead of "ip" in the subscript.

The out-of-plane compliances of the full-hinge matrix depend upon the parameters n, R_1, g, α, w and t. All compliances increase when n, R_1, g and α increase, and w and t decrease. Figure 4.32a and b illustrates the variation in the following compliance ratio:

$$r_{u_z} = \frac{C_{u_z-f_z}}{C_{u_z-f_z}^0} \quad \text{with} \quad C_{u_z-f_z}^0 = \lim_{\substack{\alpha\to\alpha_0 \\ t\to t_0}} C_{u_z-f_z}. \tag{4.72}$$

where the compliance with the superscript "0" represents the minimum value. For α ranging in the [15°; 120°] interval, the minimum compliance occurs at $\alpha_0 = 15°$, whereas for t spanning the [0.0005; 0.002] m range, the minimum compliance corresponds to $t_0 = 0.002$ m. The other constant parameters were: $n = 6$, $R_1 = 0.03$ m, $g = 0.015$ m, $t = 0.001$ m, $\alpha = 50°$ and $w = 0.005$ m. The compliance ratio of Eq. (4.72) varies nonlinearly with both α and t.

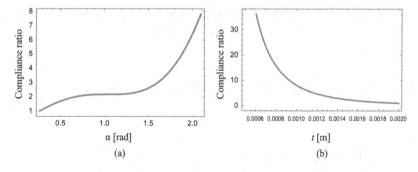

FIGURE 4.32 Plots of compliance ratio $C_{u_z-f_z}/C_{u_z-f_z}^0$ in terms of the: (a) center angle α; (b) thickness t.

TABLE 4.1
Variation of Representative Compliances (in m/N · 10^{-4}) with the Number of Circular Segments n

n	2	3	4	5	6
$C_{u_y-f_y}$	0.108	0.305	0.690	1.356	2.413
$C_{u_z-f_z}$	4.963	14.038	31.762	62.389	111.024

The variation in terms of n (the number of circular segments) of the two representative compliances that have been studied here is presented in Table 4.1. It can be seen that a three times increase in the number of circular segments increases the two compliances more than five times. All other constant parameters have the values utilized for the plots of Figures 4.31 and 4.32.

4.3 PARALLEL-CONNECTION FLEXIBLE HINGES

A few examples of hinges comprising two or more curved-axis flexible segments connected in parallel are sketched in Figure 4.33.

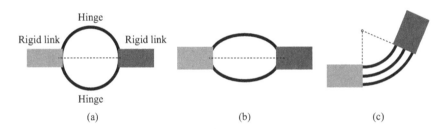

FIGURE 4.33 Parallel-connection flexible hinges: (a) axially symmetric configuration with circular-axis segments; (b) axially symmetric design with curvilinear-axis segments; (c) configuration with multiple, concentric circular-axis segments.

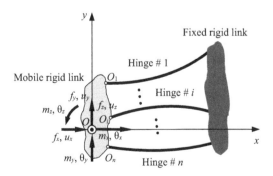

FIGURE 4.34 Parallel connection of planar curvilinear-axis flexible-hinge segments with loads and displacements of mobile rigid link.

Figure 4.34 is the generic representation of a planar flexible hinge with multiple curvilinear-axis segments connected in parallel. Note that each parallel member can be formed by serially connecting several individual segments.

The n planar, curvilinear-axis flexible-hinge segments (shown in skeleton form in Figure 4.34) connect two adjacent rigid links, one of which is assumed fixed and the other one movable. As demonstrated in Chapter 3, the in-plane and out-of-plane loads and displacements corresponding to the rigid mobile link are expressed as in Eq. (3.92), which is rewritten here:

$$
\begin{cases}
[u_{O,ip}] = [C_{O,ip}][f_{ip}] \text{ with } [u_{O,ip}] = \begin{bmatrix} u_x & u_y & \theta_z \end{bmatrix}^T \text{ and} \\
[f_{ip}] = \begin{bmatrix} f_x & f_y & m_z \end{bmatrix}^T ; \\
[u_{O,op}] = [C_{O,op}][f_{op}] \text{ with } [u_{O,op}] = \begin{bmatrix} \theta_x & \theta_y & u_z \end{bmatrix}^T \text{ and} \\
[f_{op}] = \begin{bmatrix} m_x & m_y & f_z \end{bmatrix}^T
\end{cases}
\tag{4.73}
$$

The compliance matrices of Eq. (4.73) are the ones given in Eq. (3.93):

$$
[C_{O,ip}] = \left(\sum_{i=1}^{n} [C_{O,ip}^{(i)}]^{-1} \right)^{-1} ; \quad
[C_{O,op}] = \left(\sum_{i=1}^{n} [C_{O,op}^{(i)}]^{-1} \right)^{-1} .
\tag{4.74}
$$

Studied next is the parallel-connection hinge of Figure 4.35a – a similar presentation can be found in Lobontiu (2015) [28]. This generic configuration has two symmetry axes and therefore, a symmetry center C. Because of the double symmetry, the compliance matrices of the full hinge in a global frame centered at O (which is considered the point where the external load is applied) can be

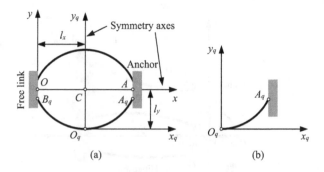

(a)   (b)

FIGURE 4.35 Skeleton representation of: (a) two-member parallel-connection flexible hinge with symmetry center; (b) lower-right quarter hinge segment.

expressed in terms of the compliance matrices of one-quarter of the full hinge. Figure 4.35b sketches the lower-right quarter segment in its local frame centered at O_q. This segment, which stretches between O_q and A_q, can be a single segment or can be formed of multiple segments connected in series; it is assumed that the in- and out-of-plane compliance matrices of the quarter segment are known/defined with respect to the reference frame $O_q x_q y_q z_q$.

4.3.1 IN-PLANE COMPLIANCE MATRIX

The lower member of the two-member parallel hinge of Figure 4.35a is formed by serially connecting the right lower quarter comprised between points O_q and A_q to the identical quarter stretching between B_q and O_q (this left quarter results by mirroring the right quarter with respect to the symmetry axis y_q). Due to mirroring, the compliance matrix of the left lower quarter with respect to the $O_q x_q y_q z_q$ frame results from the one of the right quarter segments (illustrated in Figure 4.35b) by changing the signs of the cross-terms involving the axes pairs x_q-y_q and y_q-z_q. The two segments are connected in series, and therefore, their compliance matrices with respect to the frame centered at O_q are added, resulting in:

$$\left[C_{O_q,ip}^{(h,l)}\right]=\left[C_{O_q,ip}^{(q,r)}\right]+\left[C_{O,ip}^{(q,l)}\right]=\begin{bmatrix} C_{u_x-f_x}^{(q,r)} & C_{u_x-f_y}^{(q,r)} & C_{u_x-m_z}^{(q,r)} \\ C_{u_x-f_y}^{(q,r)} & C_{u_y-f_y}^{(q,r)} & C_{u_y-m_z}^{(q,r)} \\ C_{u_x-m_z}^{(q,r)} & C_{u_y-m_z}^{(q,r)} & C_{\theta_z-m_z}^{(q,r)} \end{bmatrix}$$

$$+\begin{bmatrix} C_{u_x-f_x}^{(q,r)} & -C_{u_x-f_y}^{(q,r)} & C_{u_x-m_z}^{(q,r)} \\ -C_{u_x-f_y}^{(q,r)} & C_{u_y-f_y}^{(q,r)} & -C_{u_y-m_z}^{(q,r)} \\ C_{u_x-m_z}^{(q,r)} & -C_{u_y-m_z}^{(q,r)} & C_{\theta_z-m_z}^{(q,r)} \end{bmatrix}=2\begin{bmatrix} C_{u_x-f_x}^{(q,r)} & 0 & C_{u_x-m_z}^{(q,r)} \\ 0 & C_{u_y-f_y}^{(q,r)} & 0 \\ C_{u_x-m_z}^{(q,r)} & 0 & C_{\theta_z-m_z}^{(q,r)} \end{bmatrix}.$$

$$(4.75)$$

where the subscript "h, l" stands for left half, and "q, r" and "q, l" signify right quarter and left quarter, respectively. The compliance matrix of the lower half-segment is translated to the global frame $Oxyz$ of Figure 4.35a. This latter compliance matrix is:

$$\left[C_{O,ip}^{(h,l)}\right]=\left[T_{OO_q,ip}\right]^T\left[C_{O_q,ip}^{(h,l)}\right]\left[T_{OO_q,ip}\right] \text{ with } \left[T_{OO_q,ip}\right]=\begin{bmatrix} 1 & 0 & 0 \\ 0 & 1 & 0 \\ \Delta y & -\Delta x & 1 \end{bmatrix}$$

$$=\begin{bmatrix} 1 & 0 & 0 \\ 0 & 1 & 0 \\ -l_y & -l_x & 1 \end{bmatrix}.$$

$$(4.76)$$

Because the upper half-segment of the two-member parallel hinge is the lower half-segment mirrored with respect to the symmetry x axis, the in-plane compliance matrix of the upper half-segment is obtained from the one of the lower half segment as:

$$
\left[C_{O,ip}^{(h,u)} \right] = \begin{bmatrix}
C_{O,ip}^{(h,l)}(1,1) & -C_{O,ip}^{(h,l)}(1,2) & -C_{O,ip}^{(h,l)}(1,3) \\
-C_{O,ip}^{(h,l)}(1,2) & C_{O,ip}^{(h,l)}(2,2) & C_{O,ip}^{(h,l)}(2,3) \\
-C_{O,ip}^{(h,l)}(1,3) & C_{O,ip}^{(h,l)}(2,3) & C_{O,ip}^{(h,l)}(3,3)
\end{bmatrix}. \tag{4.77}
$$

where "h, u" signifies the upper half-portion. The upper and lower half-segments are connected in parallel, and therefore, the in-plane compliance matrix of the full, two-member hinge of Figure 4.35a is calculated as:

$$
\left[C_{O,ip} \right] = \left(\left[C_{O,ip}^{(h,u)} \right]^{-1} + \left[C_{O,ip}^{(h,l)} \right]^{-1} \right)^{-1} = \begin{bmatrix}
C_{u_x - f_x} & 0 & 0 \\
0 & C_{u_y - f_y} & C_{u_y - m_z} \\
0 & C_{u_y - m_z} & C_{\theta_z - m_z}
\end{bmatrix}. \tag{4.78}
$$

with

$$
\begin{cases}
C_{u_x - f_x} = C_{u_x - f_x}^{(q,r)} - \dfrac{\left(C_{u_x - m_z}^{(q,r)} \right)^2}{C_{\theta_z - m_z}^{(q,r)}}; \\[4mm]
C_{u_y - f_y} = C_{u_y - f_y}^{(q,r)} + \dfrac{\left[C_{u_x - f_x}^{(q,r)} \cdot C_{\theta_z - m_z}^{(q,r)} - \left(C_{u_x - m_z}^{(q,r)} \right)^2 \right] \cdot (\Delta x)^2}{C_{u_x - f_x}^{(q,r)} + \left(2 C_{u_x - m_z}^{(q,r)} + C_{\theta_z - m_z}^{(q,r)} \cdot \Delta y \right) \cdot \Delta y} = C_{u_y - f_y}^{(q,r)} \\[4mm]
\qquad + \dfrac{\left[C_{u_x - f_x}^{(q,r)} \cdot C_{\theta_z - m_z}^{(q,r)} - \left(C_{u_x - m_z}^{(q,r)} \right)^2 \right] \cdot (l_x)^2}{C_{u_x - f_x}^{(q,r)} + \left(C_{\theta_z - m_z}^{(q,r)} \cdot l_y - 2 C_{u_x - m_z}^{(q,r)} \right) \cdot l_y}; \\[4mm]
C_{u_y - m_z} = \dfrac{\left[\left(C_{u_x - m_z}^{(q,r)} \right)^2 - C_{u_x - f_x}^{(q,r)} \cdot C_{\theta_z - m_z}^{(q,r)} \right] \cdot \Delta x}{C_{u_x - f_x}^{(q,r)} + \left(2 C_{u_x - m_z}^{(q,r)} + C_{\theta_z - m_z}^{(q,r)} \cdot \Delta y \right) \cdot \Delta y} = \dfrac{\left[\left(C_{u_x - m_z}^{(q,r)} \right)^2 - C_{u_x - f_x}^{(q,r)} \cdot C_{\theta_z - m_z}^{(q,r)} \right] \cdot l_x}{C_{u_x - f_x}^{(q,r)} + \left(C_{\theta_z - m_z}^{(q,r)} \cdot l_y - 2 C_{u_x - m_z}^{(q,r)} \right) \cdot l_y}; \\[4mm]
C_{\theta_z - m_z} = \dfrac{C_{u_x - f_x}^{(q,r)} \cdot C_{\theta_z - m_z}^{(q,r)} - \left(C_{u_x - m_z}^{(q,r)} \right)^2}{C_{u_x - f_x}^{(q,r)} + \left(2 C_{u_x - m_z}^{(q,r)} + C_{\theta_z - m_z}^{(q,r)} \cdot \Delta y \right) \cdot \Delta y} = \dfrac{C_{u_x - f_x}^{(q,r)} \cdot C_{\theta_z - m_z}^{(q,r)} - \left(C_{u_x - m_z}^{(q,r)} \right)^2}{C_{u_x - f_x}^{(q,r)} + \left(C_{\theta_z - m_z}^{(q,r)} \cdot l_y - 2 C_{u_x - m_z}^{(q,r)} \right) \cdot l_y}
\end{cases}
$$

$$\tag{4.79}$$

4.3.2 OUT-OF-PLANE COMPLIANCE MATRIX

A similar reasoning and procedure are applied to calculate the out-of-plane compliance matrix of the full hinge with respect to the global frame $Oxyz$. The compliance matrix of the lower member with respect to the frame centered at O_q

is calculated by adding the compliance matrices of the right and left quarter segments forming the lower half, namely:

$$\left[C_{O_q,op}^{(h,l)}\right]=\left[C_{O_q,op}^{(q,r)}\right]+\left[C_{O,op}^{(q,l)}\right]=\begin{bmatrix} C_{\theta_x-m_x}^{(q,r)} & C_{\theta_x-m_y}^{(q,r)} & C_{\theta_x-f_z}^{(q,r)} \\ C_{\theta_x-m_y}^{(q,r)} & C_{\theta_y-m_y}^{(q,r)} & C_{\theta_y-f_z}^{(q,r)} \\ C_{\theta_x-f_z}^{(q,r)} & C_{\theta_y-f_z}^{(q,r)} & C_{u_z-f_z}^{(q,r)} \end{bmatrix}$$

$$+\begin{bmatrix} C_{\theta_x-m_x}^{(q,r)} & -C_{\theta_x-m_y}^{(q,r)} & C_{\theta_x-f_z}^{(q,r)} \\ -C_{\theta_x-m_y}^{(q,r)} & C_{\theta_y-m_y}^{(q,r)} & -C_{\theta_y-f_z}^{(q,r)} \\ C_{\theta_x-f_z}^{(q,r)} & -C_{\theta_y-f_z}^{(q,r)} & C_{u_z-f_z}^{(q,r)} \end{bmatrix}=2\begin{bmatrix} C_{\theta_x-m_x}^{(q,r)} & 0 & C_{\theta_x-f_z}^{(q,r)} \\ 0 & C_{\theta_y-m_y}^{(q,r)} & 0 \\ C_{\theta_x-f_z}^{(q,r)} & 0 & C_{u_z-f_z}^{(q,r)} \end{bmatrix}.$$

$$(4.80)$$

The compliance matrix of the lower half-segment is translated to the global frame Oxy, which yields:

$$\left[C_{O,op}^{(h,l)}\right]=\left[T_{OO_q,op}\right]^T\left[C_{O_q,op}^{(h,l)}\right]\left[T_{OO_q,op}\right] \quad \text{with} \quad \left[T_{OO_q,op}\right]=\begin{bmatrix} 1 & 0 & -\Delta y \\ 0 & 1 & \Delta x \\ 0 & 0 & 1 \end{bmatrix}$$

$$=\begin{bmatrix} 1 & 0 & l_y \\ 0 & 1 & l_x \\ 0 & 0 & 1 \end{bmatrix}.$$

$$(4.81)$$

The out-of-plane compliance matrix of the upper half-segment is obtained from the one of the lower half-segments as:

$$\left[C_{O,op}^{(h,u)}\right]=\begin{bmatrix} C_{O,op}^{(h,l)}(1,1) & -C_{O,op}^{(h,l)}(1,2) & -C_{O,op}^{(h,l)}(1,3) \\ -C_{O,op}^{(h,l)}(1,2) & C_{O,op}^{(h,l)}(2,2) & C_{O,op}^{(h,l)}(2,3) \\ -C_{O,op}^{(h,l)}(1,3) & C_{O,op}^{(h,l)}(2,3) & C_{O,op}^{(h,l)}(3,3) \end{bmatrix}. \qquad (4.82)$$

The out-of-plane compliance matrix of the full hinge is calculated by combining the upper- and lower halves compliances in parallel as:

$$\left[C_{O,op}\right]=\left(\left[C_{O,op}^{(h,u)}\right]^{-1}+\left[C_{O,op}^{(h,l)}\right]^{-1}\right)^{-1}=\begin{bmatrix} C_{\theta_x-m_x} & 0 & 0 \\ 0 & C_{\theta_y-m_y} & C_{\theta_y-f_z} \\ 0 & C_{\theta_y-f_z} & C_{u_z-f_z} \end{bmatrix}. \qquad (4.83)$$

with

$$
\begin{cases}
C_{\theta_x - m_x} = \dfrac{C_{\theta_x - m_x}^{(q,r)} \cdot C_{u_z - f_z}^{(q,r)} - \left(C_{\theta_x - f_z}^{(q,r)}\right)^2}{C_{u_z - f_z}^{(q,r)} + \left(C_{\theta_x - m_x}^{(q,r)} \cdot \Delta y - 2C_{\theta_x - f_z}^{(q,r)}\right) \cdot \Delta y} = \dfrac{C_{\theta_x - m_x}^{(q,r)} \cdot C_{u_z - f_z}^{(q,r)} - \left(C_{\theta_x - f_z}^{(q,r)}\right)^2}{C_{u_z - f_z}^{(q,r)} + \left(2C_{\theta_x - f_z}^{(q,r)} - C_{\theta_x - m_x}^{(q,r)} \cdot l_y\right) \cdot l_y}; \\[4mm]
C_{\theta_y - m_y} = C_{\theta_y - m_y}^{(q,r)}; \\[3mm]
C_{\theta_y - f_z} = C_{\theta_y - m_y}^{(q,r)} \cdot \Delta x = C_{\theta_y - m_y}^{(q,r)} \cdot l_x; \\[3mm]
C_{u_z - f_z} = C_{u_z - f_z}^{(q,r)} + C_{\theta_y - m_y}^{(q,r)} \cdot (\Delta x)^2 - \dfrac{\left(C_{\theta_x - f_z}^{(q,r)}\right)^2}{C_{\theta_x - m_x}^{(q,r)}} = C_{u_z - f_z}^{(q,r)} + C_{\theta_y - m_y}^{(q,r)} \cdot (l_x)^2 - \dfrac{\left(C_{\theta_x - f_z}^{(q,r)}\right)^2}{C_{\theta_x - m_x}^{(q,r)}}
\end{cases}
$$

$$(4.84)$$

Example 4.3

Consider the two-member parallel hinge of Figure 4.36a, which is formed of two circular-axis half-segments. The hinge cross-section is constant rectangular with the in-plane thickness t and the out-of-plane width w. The material is steel with Young's modulus $E = 2 \cdot 10^{11} \, \text{N/m}^2$ and $G = 7.7 \cdot 10^{10} \, \text{N/m}^2$. Graphically analyze the variation of the compliance ratio $C_{u_y - f_y} / C_{u_z - f_z}$ in terms of all geometric parameters. Use the following constant parameters: $R = 0.02 \, \text{m}$, $\alpha = 60°$, $h = 0.005 \, \text{m}$, $t = 0.001 \, \text{m}$ and $w = 0.008 \, \text{m}$.

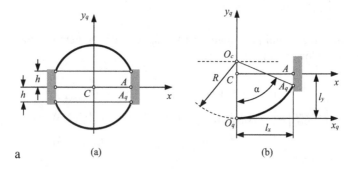

FIGURE 4.36 Skeleton representation of: (a) two-member parallel-connection flexible hinge with symmetry center and circular-axis segments; (b) lower-right quarter hinge segment with geometry.

Solution:

The device of Figure 4.36a was analyzed by finite element analysis in Lee and Ziegert [2], among others. Comparing the dimensions of Figure 4.36a and b – the latter one shows the geometric parameters defining the lower-right quarter – it follows that the two offset distances of Figure 4.35a are:

$$l_x = R \cdot \sin\alpha; \quad l_y = h + R \cdot (1 - \cos\alpha). \tag{4.85}$$

Using Eqs. (4.75) through (4.85) in conjunction with Eqs. (A2.22) through (A2.32), which define the compliances of a constant cross-sectional circular-axis segment of angle α, the plots of Figures 4.37–4.39 are obtained.

Note that the compliance ratio is proportional to the in-plane compliance and inversely proportional to the out-of-plane compliance. Larger values of this ratio indicate good in-plane deformation capabilities (due to $C_{u_y-f_y}$) coupled with relatively small sensitivity to out-of-plane parasitic loading (which is represented by $C_{u_z-f_z}$). While the ratio can reach a maximum value of 2.4 for the selected geometric parameters' ranges, it can be seen that the ratio increases with an increase in R and w, as well as with a decrease in h, α and t.

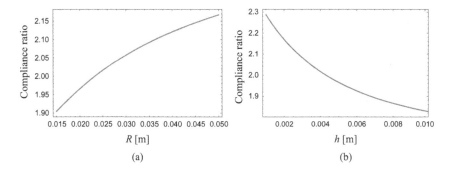

FIGURE 4.37 Plots of compliance ratio $C_{u_y-f_y}/C_{u_z-f_z}$ in terms of the: (a) radius R; (b) offset distance h.

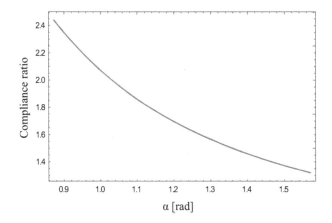

FIGURE 4.38 Plot of compliance ratio $C_{u_y-f_y}/C_{u_z-f_z}$ in terms of the quarter center angle α.

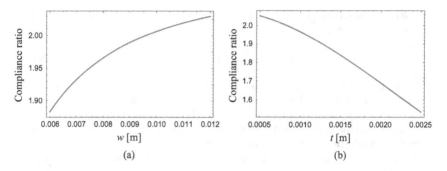

FIGURE 4.39 Plots of compliance ratio $C_{u_y - f_y}/C_{u_z - f_z}$ in terms of the: (a) out-of-plane width w; (b) in-plane thickness t.

REFERENCES

1. Hsiao, F.Z. and Lin, T.W., Analysis of a novel flexure hinge with three degrees of freedom, *Review of Scientific Instruments*, 72 (2), 1565, 2001.
2. Lee, V. and Ziegert, J., Hybrid bi-directional flexure joints, *Proceedings of ASPE 2007 Annual Meeting*, Dallas, TX, October 14–19, 2007.
3. Shisheng, B., Shanshan, Z., and Xiaofeng, Z., Dimensionless design graphs for three types of annulus-shaped flexure hinges, *Precision Engineering*, 34, 659, 2010.
4. Callegari, M., Cammarata, A., and Gabrielli, A., Analysis and design of a spherical mechanism with flexure hinges, *ASME Journal of Mechanical Design*, 131 (5), 051003, 2009.
5. Howell, L., *Compliant Mechanisms*, John Wiley and Sons, Hoboken, NJ, 2001.
6. Jagirda, S., *Kinematics of curved flexible beam*, M.Sc. Thesis, University of South Florida, Tampa, FL, 2006.
7. Trylinski, V., *Fine Mechanisms and Precision Instruments*, Pergamon Press, Oxford, 1971.
8. Sydenham, P.H., Elastic design of fine mechanism in instruments, *Journal of Physics – Scientific Instruments*, 17, 922, 1984.
9. Vukobratovich, D. et al., Slit diaphragm flexures for optomechanics, *Proceedings of SPIE: Optomechanical and Precision Instrument Design*, 2542, 2, 1995.
10. Kyusogin, A. and Sagawa, D., Development of linear and rotary movement mechanism by using flexible strips, *Bulletin of Japan Society of Precision Engineering*, 22 (4), 309, 1988.
11. Hermann, G., A low cost submicron measuring probe, *Proceedings of the 5th IEEE International Symposium on Applied Computational Intelligence and Informatics*, Timisoara, Romania, 2009.
12. Mohamad, N., Iovenitti, P., and Vinay, T., Modeling and optimisation of a spring supported diaphragm capacitive MEMS microphone, *Engineering*, 2, 762, 2010.
13. Duong, L. and Kazerounian, K., Design improvement of the mechanical coupling diaphragms for aerospace applications, *Mechanics Based Design of Structures and Machines*, 35, 467, 2007.
14. Shielpiekandula, V. and Youcef-Toumi, K., Characterization of dynamic behavior of flexure-based mechanisms for precision angular alignment, *American Control Conference*, Seattle, WA, June 11–13, 3005, 2008.

15. Yong, Y.K. and Reza-Mohemani, S.O., A z-scanner design for high-speed scanning probe microscopy, *Proceedings of IEEE International Conference on Robotics and Automation*, 4780, Anchorage, AK, 2010.
16. Awtar, S., *Synthesis and analysis of parallel kinematic XY flexure mechanisms*, Sc.D. Thesis, Massachusetts Institute of Technology, Cambridge, MA, 2004.
17. Awtar, S. and Slocum, A.H., Design of flexure stages based on symmetric diaphragm flexure, *Proceedings of ASPE 2005 Annual Meeting*, Norfolk, VA, October 9–14, 2005.
18. Trease, B.P., Moon, Y.-M., and Kota, S., Design of large-displacement compliant joints, *ASME Journal of Mechanical Design*, 127, 788, 2005.
19. Ishaquddin, M., Raveendranath, P., and Reddy, J.N., Coupled polynomial field approach for elimination of flexure and torsion locking phenomena in the Timoshenko and Euler-Bernoulli curved beam elements, *Finite Elements in Analysis and Design*, 65, 17, 2013.
20. Kress, G., Sauter, M., and Ermanni, P., Complex-shaped finite element, *Finite Elements in Analysis and Design*, 43, 112, 2006.
21. Lobontiu, N. and Cullin, M., In-plane elastic response of two-segment circular-axis symmetric notch flexure hinges: the right circular design, *Precision Engineering*, 37, 542, 2013.
22. Lobontiu, N. et al., Planar compliances of thin circular-axis notch flexure hinges with midpoint radial symmetry, *Mechanics Based Design of Structures and Machines*, 41 (2), 202, 2013.
23. Lobontiu, N., Out-of-plane (diaphragm) compliances of circular-axis notch flexible hinges with midpoint radial symmetry, *Mechanics Based Design of Structures and Machines*, 42 (1), 517, 2014.
24. Lobontiu, N., Compliance-based modeling and design of straight-axis/circular-axis flexible hinges with small out-of-plane deformations, *Mechanism and Machine Theory*, 80 (C), 166, 2014.
25. Lobontiu, N., Planar flexible hinges with curvilinear-axis segments for mechanisms of in-plane and out-of-plane operation, *ASME Journal of Mechanical Design*, 137, 012302–1, 2015.
26. Lobontiu, N., In-plane compliances of planar flexure hinges with serially-connected straight-axis and circular-axis segments, *ASME Journal of Mechanical Design*, 136, 122301–1, 2014.
27. Lobontiu, N., Wight-Crask, J., and Kawagley, C., Straight-axis folded flexure hinges: in-plane elastic response, *Precision Engineering: Journal of the International Societies for Precision Engineering and Nanotechnology*, 57C, 54, 2019.
28. Lobontiu, N., Modeling and design of planar parallel-connection flexible hinges for in- and out-of-plane mechanism applications, *Precision Engineering*, 42, 113, 2015.

5 Quasi-Static Response of Serial Flexible-Hinge Mechanisms

This chapter studies the quasi-static response of serial mechanisms that comprise flexible hinges and rigid links by using the compliance matrix approach. Collinear and planar (two-dimensional) mechanisms are analyzed and designed using mathematical models that incorporate the in-plane and/or out-of-plane compliance matrices of straight- and curvilinear-axis flexible hinges, such as those presented in Chapters 2–4. The quasi-static model assumes that the external load is applied gradually and slowly from zero to its nominal value, which allows the exclusion of any dynamic effects in evaluating the final (deformed) state of the mechanism. Various boundary, loading and symmetry conditions are considered. Serial spatial (three-dimensional) mechanisms are also qualified in terms of their quasi-static behavior. Applications include symmetric devices, displacement-amplification and precision-positioning flexible-hinge mechanisms.

Modeling procedures that are based on small-deformation compliance/stiffness, as well as illustrative applications of serial hinge mechanisms, are provided by Smith [1], Xu [2], Henin et al. [3], Zhang and Zhu [4] or Zentner and Linβ [5]. Serial, flexible-hinge chains can be used as standalone (open-loop) mechanisms that use levers to achieve amplified input displacements – as discussed by Boronkay and Mei [6], Williams and Turcic [7], Jouaneh and Yang [8] and Lobontiu [9]. However, serial flexible chains are also incorporated into more complex, parallel or serial/parallel mechanisms, of which the most common applications are precision displacement-amplification devices – see Du et al. [10], Kim et al. [11], Quyang et al. [12], Choi et al. [13], Lobontiu and Garcia [14,15], Ma et al. [16], Mottard and St-Amant [17], Scire and Teague [18], Her and Chang [19], Park and Yang [20], Mukhopadyay et al. [21], Zettl et al. [22], Furukawa and Mizuno [23], Kim et al. [24], Chang and Du [25], Tian et al. [26], Choi et al. [27,28], Tian et al. [29], Wu et al. [30], Pham and Chen [31] and Ryu et al. [32]. Flexible-hinge precision-positioning stages were studied by Yong and Lu [33], Ryu et al. [34], Hua and Xianmin [35], Yue et al. [36] and Wang et al. [37], among many others. As reported in the literature, the quasi-static modeling of serial flexible mechanisms undergoing relatively small deformations/motions has mainly been performed by the finite element method, by lumped-parameter procedures (such as the loop-closure method) and by the compliance matrix approach, which is the method utilized in this chapter – see also Lobontiu [9,38].

5.1 PLANAR (2D) MECHANISMS

Mechanisms whose structure is planar (two-dimensional, 2D) are modeled in this section based on a compliance matrix approach. Load/displacement models are derived for simpler serial chains formed of one flexible hinge and one rigid link, as well as for more complex chains comprising multiple links.

5.1.1 SERIAL CHAIN WITH ONE FLEXIBLE HINGE AND ONE RIGID LINK

This section derives and discusses the quasi-static model of the basic serial chain sketched in Figure 5.1, which is formed of a flexible hinge and a mobile rigid link. The model requires translation and/or rotation transformation of the hinge in-plane and out-of-plane compliance matrices. Both properly constrained (determinate) and overconstrained (indeterminate) configurations are analyzed here.

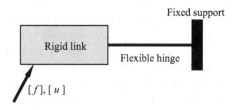

FIGURE 5.1 Skeleton representation of flexible hinge connected to a movable rigid link and a fixed support.

This planar configuration is the simplest possible, and the resulting mathematical model is fundamental to the analysis and design of more complex flexible-hinge mechanisms. The first part of the section establishes vector relationships allowing to transfer loads and displacements between two arbitrary points on a rigid link. The second part derives an algorithm that transforms the local-frame compliance matrices of a generic flexible hinge in order to evaluate resulting displacements and loads at various points of interest on the mobile rigid link that is adjacent to the hinge.

5.1.1.1 Rigid-Link Load and Displacement Transformation

Chapters 3 and 4 introduce the in-plane and out-of-plane transformations of compliance matrices in relation to flexible hinges that are formed by combining basic hinge segments in series and/or in parallel. This section, which is similar to the corresponding sections in Chapters 3 and 4 covering the compliance matrix transformation, describes the matrix transformations pertaining to planar compliant mechanisms that combine rigid links and flexible hinges.

5.1.1.1.1 In-Plane Transformations

Consider the rigid link of Figure 5.2a undergoes a general (planar) motion in the plane $Oxyz$. Also, consider that the planar load consisting of f_{x1}, f_{y1} and m_{z1} acts at point A of the rigid link and generates displacements u_{x1}, u_{y1} and θ_{z1} at the

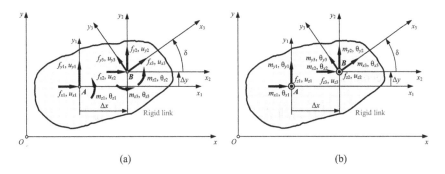

(a) (b)

FIGURE 5.2 Load and displacement transfer in a planar rigid link undergoing: (a) in-plane motion; (b) out-of-plane motion.

same point. The objective is to express the transferred loads and displacements at a different point B in terms of a reference frame that is rotated with respect to the one attached to point A.

Loads:

As per Eq. (3.2), the original load at A is first translated at B as:

$$\left[f_{2,ip} \right] = \left[T_{AB,ip} \right]\left[f_{1,ip} \right] \quad \text{or} \quad \left[f_{1,ip} \right] = \left[T_{AB,ip} \right]^{-1}\left[f_{2,ip} \right] \quad \text{with}$$

$$\left[f_{1,ip} \right] = \left[\begin{array}{ccc} f_{x1} & f_{y1} & m_{z1} \end{array} \right]^{T}; \quad \left[f_{2,ip} \right] = \left[\begin{array}{ccc} f_{x2} & f_{y2} & m_{z2} \end{array} \right]^{T}; \tag{5.1}$$

$$\left[T_{AB,ip} \right] = \left[\begin{array}{ccc} 1 & 0 & 0 \\ 0 & 1 & 0 \\ \Delta y & -\Delta x & 1 \end{array} \right],$$

where $[T_{AB,ip}]$ is the in-plane translation matrix realizing the load offset from A to B.

The load vector of Eq. (5.1) is now projected on the rotated reference frame $Bx_3y_3z_3$ as:

$$\left[f_{3,ip} \right] = [R]\left[f_{2,ip} \right] \quad \text{or} \quad \left[f_{2,ip} \right] = [R]^{-1}\left[f_{3,ip} \right] = [R]^{T}\left[f_{3,ip} \right] \quad \text{with}$$

$$\left[f_{3,ip} \right] = \left[\begin{array}{ccc} f_{x3} & f_{y3} & m_{z3} \end{array} \right]^{T}; \quad [R] = \left[\begin{array}{ccc} \cos\delta & \sin\delta & 0 \\ -\sin\delta & \cos\delta & 0 \\ 0 & 0 & 1 \end{array} \right], \tag{5.2}$$

where $[R]$ is the rotation matrix and is defined in Eq. (4.2). Combining Eqs. (5.1) and (5.2) yields:

$$\left[f_{3,ip} \right] = [R]\left[T_{AB,ip} \right]\left[f_{1,ip} \right] \quad \text{or}$$

$$\left[f_{1,ip} \right] = \left[T_{AB,ip} \right]^{-1}[R]^{-1}\left[f_{3,ip} \right] = \left[T_{AB,ip} \right]^{-1}[R]^{T}\left[f_{3,ip} \right]. \tag{5.3}$$

Defining a transformation matrix $[Tr_{AB,ip}]$ as:

$$[Tr_{AB,ip}] = [R][T_{AB,ip}]. \tag{5.4}$$

Eq. (5.3) becomes:

$$[f_{3,ip}] = [Tr_{AB,ip}][f_{1,ip}] \quad \text{or} \quad [f_{1,ip}] = [Tr_{AB,ip}]^{-1}[f_{3,ip}]. \tag{5.5}$$

Displacements:

The displacements at B in the $Bx_3y_3z_3$ reference frame, which are u_{x3}, u_{y3} and θ_{z3}, are now expressed in terms of the original displacements at A: u_{x1}, u_{y1} and θ_{z1}. The displacements at B that are recorded with respect to the reference frame $Bx_2y_2z_2$ (which is a translated frame with respect to $Ax_1y_1z_1$) are connected to the displacements at A as in Eq. (3.6), namely:

$$[u_{1,ip}] = [T_{AB,ip}]^T [u_{2,ip}] \quad \text{or} \quad [u_{2,ip}] = ([T_{AB,ip}]^T)^{-1}[u_{1,ip}] \quad \text{with}$$

$$[u_{1,ip}] = \begin{bmatrix} u_{x1} & u_{y1} & \theta_{z1} \end{bmatrix}^T; \quad [u_{2,ip}] = \begin{bmatrix} u_{x2} & u_{y2} & \theta_{z2} \end{bmatrix}^T. \tag{5.6}$$

The displacements in the $Bx_2y_2z_2$ and $Bx_3y_3z_3$ rotated frames are related as in Eq. (4.4) – with an identity translation matrix:

$$[u_{3,ip}] = [R][u_{2,ip}] \quad \text{or} \quad [u_{2,ip}] = [R]^{-1}[u_{3,ip}] = [R]^T[u_{3,ip}] \quad \text{with}$$

$$[u_{3,ip}] = \begin{bmatrix} u_{x3} & u_{y3} & \theta_{z3} \end{bmatrix}^T. \tag{5.7}$$

Combining Eqs. (5.6) and (5.7) results in the following displacement vector connections:

$$[u_{3,ip}] = [R]([T_{AB,ip}]^T)^{-1}[u_{1,ip}] \quad \text{or}$$

$$[u_{1,ip}] = [T_{AB,ip}]^T[R]^T[u_{3,ip}] = [Tr_{AB,ip}]^T[u_{3,ip}]. \tag{5.8}$$

5.1.1.1.2 Out-of-Plane Transformations

Consider now Figure 5.2b, which depicts the same rigid link being acted upon by a load vector at A that contains the three out-of-plane components m_{x1}, m_{y1} and f_{z1} in the frame $Ax_1y_1z_1$; this load produces, correspondingly, the displacements θ_{x1}, θ_{y1} and u_{z1} at the same point A on the rigid link. Load and displacement relationships are sought connecting the source at A to the transfer point B where the reference is the rotated frame $Bx_3y_3z_3$.

Loads

The out-of-plane load at A is first translated to B using the procedure discussed when deriving Eq. (3.9):

$$[f_{2,op}] = [T_{AB,op}][f_{1,op}] \quad \text{or} \quad [f_{1,op}] = [T_{AB,op}]^{-1}[f_{2,op}] \quad \text{with}$$

$$[f_{1,op}] = \begin{bmatrix} m_{x1} & m_{y1} & f_{z1} \end{bmatrix}^T; \quad [f_{2,op}] = \begin{bmatrix} m_{x2} & m_{y2} & f_{z2} \end{bmatrix}^T;$$

$$[T_{AB,op}] = \begin{bmatrix} 1 & 0 & -\Delta y \\ 0 & 1 & \Delta x \\ 0 & 0 & 1 \end{bmatrix}, \tag{5.9}$$

where $[T_{AB,op}]$ is the out-of-plane translation matrix realizing the translation from A to B. Comparing this matrix to the in-plane translation matrix of Eq. (5.1), it can be seen that:

$$[T_{AB,op}]^{-1} = [T_{AB,ip}]^T. \tag{5.10}$$

The loads corresponding to the $Bx_2y_2z_2$ frame that were obtained in Eq. (5.9) are projected to the rotated reference frame $Bx_3y_3z_3$ axes as in Eq. (4.8) – with an identity translation matrix, namely:

$$[f_{3,op}] = [R][f_{2,op}] \quad \text{or} \quad [f_{2,op}] = [R]^{-1}[f_{3,op}] = [R]^T[f_{3,op}] \quad \text{with}$$

$$[f_{3,op}] = \begin{bmatrix} m_{x3} & m_{y3} & f_{z3} \end{bmatrix}^T, \tag{5.11}$$

where the rotation matrix $[R]$ is defined in Eq. (5.2). Combination of Eqs. (5.9) and (5.11) results in:

$$[f_{3,op}] = [R][T_{AB,op}][f_{1,op}] \quad \text{or}$$

$$[f_{1,op}] = [T_{AB,op}]^{-1}[R]^{-1}[f_{3,op}] = [T_{AB,op}]^{-1}[R]^T[f_{3,op}]. \tag{5.12}$$

The out-of-plane transformation matrix $[Tr_{AB,op}]$ is defined as:

$$[Tr_{AB,op}] = [R][T_{AB,op}], \tag{5.13}$$

which changes Eq. (5.12) into:

$$[f_{3,op}] = [Tr_{AB,op}][f_{1,op}] \quad \text{or} \quad [f_{1,op}] = [T_{AB,ip}]^T[R]^T[f_{3,op}] = [Tr_{AB,ip}][f_{3,op}]. \tag{5.14}$$

Displacements:

Similar to the in-plane displacements transfer, the original displacements at A are translated to B by means of the equations that follow the derivation of Eqs. (3.13), namely:

$$\left[u_{1,op}\right]=\left[T_{AB,op}\right]^{T}\left[u_{2,op}\right] \quad \text{or} \quad \left[u_{2,op}\right]=\left(\left[T_{AB,op}\right]^{T}\right)^{-1}\left[u_{1,op}\right] \quad \text{with}$$

$$\left[u_{1,op}\right]=\left[\begin{array}{ccc} \theta_{x1} & \theta_{y1} & u_{z1} \end{array}\right]^{T}; \quad \left[u_{2,op}\right]=\left[\begin{array}{ccc} \theta_{x2} & \theta_{y2} & u_{z2} \end{array}\right]^{T}. \tag{5.15}$$

As per the definition Eq. (4.10) and using an identity translation matrix, the following relationships connect the displacements in the rotated frames $Bx_2y_2z_2$ and $Bx_3y_3z_3$:

$$\left[u_{3,op}\right]=\left[R\right]\left[u_{2,op}\right] \quad \text{or} \quad \left[u_{2,op}\right]=\left[R\right]^{-1}\left[u_{3,op}\right]=\left[R\right]^{T}\left[u_{3,op}\right] \quad \text{with}$$

$$\left[u_{3,op}\right]=\left[\begin{array}{ccc} \theta_{x3} & \theta_{y3} & u_{z3} \end{array}\right]^{T}. \tag{5.16}$$

Equations (5.15) and (5.16) are combined as:

$$\left[u_{3,op}\right]=\left[R\right]\left(\left[T_{AB,op}\right]^{T}\right)^{-1}\left[u_{1,op}\right]=\left[R\right]\left[T_{AB,ip}\right]\left[u_{1,op}\right]=\left[Tr_{AB,ip}\right]\left[u_{1,op}\right]$$

or $\qquad\qquad\qquad\qquad\qquad\qquad\qquad\qquad\qquad\qquad\qquad\qquad$ (5.17)

$$\left[u_{1,op}\right]=\left[T_{AB,op}\right]^{T}\left[R\right]^{T}\left[u_{3,op}\right]=\left[Tr_{AB,op}\right]^{T}\left[u_{3,ip}\right].$$

Some connections between loads and displacements can be formulated using in-plane and out-of-plane transformation matrices. For instance, the second Eq. (5.3) can be combined with Eq. (5.10) and results in:

$$\left[f_{1,ip}\right]=\left[T_{AB,ip}\right]^{-1}\left[R\right]^{-1}\left[f_{3,ip}\right]=\left[T_{AB,op}\right]^{T}\left[R\right]^{T}\left[f_{3,ip}\right]=\left[Tr_{AB,op}\right]^{T}\left[f_{3,ip}\right], \tag{5.18}$$

whereas the first Eq. (5.8) is reformulated as:

$$\left[u_{3,ip}\right]=\left[R\right]\left(\left[T_{AB,ip}\right]^{T}\right)^{-1}\left[u_{1,ip}\right]=\left[R\right]\left[T_{AB,op}\right]\left[u_{1,ip}\right]=\left[Tr_{AB,op}\right]\left[u_{1,ip}\right]. \tag{5.19}$$

with the aid of the same Eq. (5.10).

5.1.1.2 Fixed-Free Chain

The serial chain of Figure 5.3 is composed of a rigid link and a fixed-end flexible hinge. The in-plane and out-of-plane loads acting on the rigid link at F deform the flexible hinge and, as a consequence, the rigid link displaces. The displacements at another point P of the rigid link need to be determined.

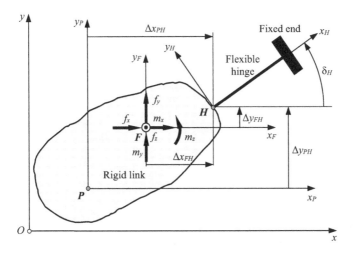

FIGURE 5.3 Schematic representation of planar serial chain with one rigid link, one flexible hinge and one-point external loading.

5.1.1.2.1 In-Plane Transformations

The in-plane load $[f_{F,ip}] = [f_x f_y m_z]^T$ is applied at F on the rigid link, which rigidly connects to the flexible hinge at H. The subsequent hinge deformation produces a displacement $[u_{P,ip}] = [u_x u_y \theta_z]^T$ at an arbitrary point P on the rigid link. As per Eqs. (5.3) and (5.4), the original load at F is transferred to the root H of the hinge by means of a transformation matrix as:

$$\left[f_{H,ip} \right] = \left[Tr_{FH,ip} \right]\left[f_{F,ip} \right] = [R]\left[T_{FH,ip} \right]\left[f_{F,ip} \right], \tag{5.20}$$

where the rotation and translation matrices realizing the transfer from F to H are:

$$[R] = \begin{bmatrix} \cos\delta_H & \sin\delta_H & 0 \\ -\sin\delta_H & \cos\delta_H & 0 \\ 0 & 0 & 1 \end{bmatrix}; \left[T_{FH,ip} \right] = \begin{bmatrix} 1 & 0 & 0 \\ 0 & 1 & 0 \\ \Delta y_{FH} & -\Delta x_{FH} & 1 \end{bmatrix}. \tag{5.21}$$

The relationship between the flexible hinge's displacements/deformations and loads in the native (local) reference frame $O_H x_H y_H z_H$ is:

$$\left[u_{H,ip} \right] = \left[C_{ip} \right]\left[f_{H,ip} \right] = \left[C_{ip} \right]\left[Tr_{FH,ip} \right]\left[f_{F,ip} \right], \tag{5.22}$$

where $[C_{ip}]$ is the hinge in-plane compliance matrix. Equation (5.22) was obtained by also using Eq. (5.20). The resulting displacement at H (which is shared by both the hinge and the adjacent rigid link) is transferred at P based on Eq. (5.8) as:

$$\left[u_{P,ip}\right]=\left[Tr_{PH,ip}\right]^{T}\left[u_{H,ip}\right]=\left[Tr_{PH,ip}\right]^{T}\left[C_{ip}\right]\left[Tr_{FH,ip}\right]\left[f_{F,ip}\right]$$

$$=\left[T_{PH,ip}\right]^{T}\left[R\right]^{T}\left[C_{ip}\right]\left[R\right]\left[T_{FH,ip}\right]\left[f_{F,ip}\right], \qquad (5.23)$$

where Eq. (5.22) was also utilized and where the new translation matrix is:

$$\left[T_{PH,ip}\right]=\begin{bmatrix} 1 & 0 & 0 \\ 0 & 1 & 0 \\ \Delta y_{PH} & -\Delta x_{PH} & 1 \end{bmatrix}. \qquad (5.24)$$

Equation (5.23) is written in short form as:

$$\left[u_{P,ip}\right]=\left[C_{PFH,ip}\right]\left[f_{F,ip}\right] \quad \text{with} \quad \left[C_{PFH,ip}\right]=\left[Tr_{PH,ip}\right]^{T}\left[C_{ip}\right]\left[Tr_{FH,ip}\right]. \qquad (5.25)$$

The three-letter subscript in the compliance matrix $[C_{PFH,ip}]$ indicates the three relevant points, namely, P where the displacements are calculated, F where the external load is applied and H, which is the moving end of flexible hinge (also shared by the rigid link).

When the displacements are to be calculated at the load application point F (which means when P and F are identical), Eq. (5.25) is written in a two-letter subscript notation:

$$\left[u_{F,ip}\right]=\left[C_{FH,ip}\right]\left[f_{F,ip}\right] \quad \text{with}$$

$$\left[C_{FH,ip}\right]=\left[Tr_{FH,ip}\right]^{T}\left[C_{ip}\right]\left[Tr_{FH,ip}\right]=\left[T_{FH,ip}\right]^{T}\left[R\right]^{T}\left[C_{ip}\right]\left[R\right]\left[T_{FH,ip}\right]. \qquad (5.26)$$

5.1.1.2.2   Out-of-Plane Transformations

Consider now that the out-of-plane load $[f_{F,op}] = [m_x \; m_y \; f_z]^T$ is applied at point F on the rigid link of Figure 5.3. Due to the hinge deformation, the resulting displacement $[u_{P,op}] = [\theta_x \; \theta_y \; u_z]^T$ at the generic point P on the rigid link needs to be determined. The load at F is transferred to H as per Eqs. (5.12) and (5.14):

$$\left[f_{H,op}\right]=\left[Tr_{FH,op}\right]\left[f_{F,op}\right]=\left[R\right]\left[T_{FH,op}\right]\left[f_{F,op}\right], \qquad (5.27)$$

where the new translation matrix is:

$$\left[T_{FH,op}\right]=\begin{bmatrix} 1 & 0 & -\Delta y_{FH} \\ 0 & 1 & \Delta x_{FH} \\ 0 & 0 & 1 \end{bmatrix}, \qquad (5.28)$$

The out-of-plane displacements and loads at H are related by means of the flexible hinge's compliance matrix $[C_{op}]$ defined in the native reference frame $O_H x_H y_H z_H$, namely:

$$\left[u_{H,op} \right] = \left[C_{op} \right]\left[f_{H,op} \right] = \left[C_{op} \right]\left[Tr_{FH,op} \right]\left[f_{H,op} \right], \tag{5.29}$$

where Eq. (5.27) was used. The displacement at H is transferred at P based on Eq. (5.17) as:

$$\left[u_{P,op} \right] = \left[Tr_{PH,op} \right]^{T}\left[u_{H,op} \right] = \left[Tr_{PH,op} \right]^{T}\left[C_{op} \right]\left[Tr_{FH,op} \right]\left[f_{F,op} \right]$$
$$= \left[T_{PH,op} \right]^{T}[R]^{T}\left[C_{op} \right][R]\left[T_{FH,op} \right]\left[f_{F,op} \right]. \tag{5.30}$$

Equation (5.29) was also used to obtain Eq. (5.30), where the new translation matrix is:

$$\left[T_{PH,op} \right] = \begin{bmatrix} 1 & 0 & -\Delta y_{PH} \\ 0 & 1 & \Delta x_{PH} \\ 0 & 0 & 1 \end{bmatrix}. \tag{5.31}$$

Equation (5.30) is written in short form as:

$$\left[u_{P,op} \right] = \left[C_{PFH,op} \right]\left[f_{F,op} \right] \quad \text{with} \quad \left[C_{PFH,op} \right] = \left[Tr_{PH,op} \right]^{T}\left[C_{op} \right]\left[Tr_{FH,op} \right]. \tag{5.32}$$

Similar to the in-plane load/deformation, when points P and F coincide, Eq. (5.32) is written in the form:

$$\left[u_{F,op} \right] = \left[C_{FH,op} \right]\left[f_{F,op} \right] \quad \text{with}$$
$$\left[C_{FH,op} \right] = \left[Tr_{FH,op} \right]^{T}\left[C_{op} \right]\left[Tr_{FH,op} \right] = \left[T_{FH,op} \right]^{T}[R]^{T}\left[C_{op} \right][R]\left[T_{FH,op} \right]. \tag{5.33}$$

5.1.1.3 Overconstrained Chains

The fixed-free serial chain studied in Section 5.1.1.2 is statically determinate (properly constrained) as it introduces three reaction loads at the fixed end (in either in-plane or out-of-plane loading), which can be determined from the static equilibrium equations. An overconstrained chain has redundant boundary conditions, which result in more unknown reactions than the number of static equilibrium equations – the system is therefore statically indeterminate. Using compatibility conditions (consisting of zero displacements at support points) alongside the static equilibrium equations, the overconstrained system is equivalently transformed into a statically determinate one. Two examples are analyzed next to illustrate the procedure.

Example 5.1

A unidirectional, non-amplified actuator has its symmetric elastic frame designed as shown in Figure 5.4a. It consists of a middle rigid link where the load is applied along the symmetry axis and two identical flexible hinges of right elliptically corner-filleted configuration – see Figure 3.14. Determine the stiffness k_y along the actuation direction knowing the hinge geometric parameters $a = 0.01$ m, $b = 0.005$ m, $l = 0.03$ m, $t = 0.001$ m, $w = 0.008$ m, the length parameter $l_1 = 0.01$ m, as well as the hinge material Young's modulus $E = 2.1 \cdot 10^{11}$ N/m².

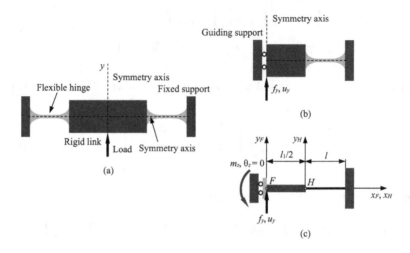

FIGURE 5.4 Symmetric mechanism with two identical right elliptically corner-filleted flexible hinges, one connecting rigid link and symmetric loading: (a) schematic of full mechanism; (b) schematic of half-mechanism; (c) skeleton representation of half-mechanism.

Solution:

Due to both geometry and load symmetry, the half-mechanism of Figure 5.4b can be analyzed instead of the original device of Figure 5.4a. The skeleton representation of the simpler half-mechanism is sketched in Figure 5.4c. This design falls under the particular case of the general configuration analyzed in this section with the point of load application and the point of displacements being collocated, $F \equiv P$. In addition, as Figure 5.4c points out, the hinge inclination angle is $\delta_H = 0$, which renders the rotation matrix $[R_{FH}]$ into a unity matrix. As a consequence, Eq. (5.26) simplifies to:

$$[u_{F,ip}] = [C_{FH,ip}][f_{F,ip}] \quad \text{with} \quad \begin{cases} [C_{FH,ip}] = [T_{FH,ip}]^T [C_{ip}][T_{FH,ip}] \\ [C_{ip}] = \begin{bmatrix} C_{u_x-f_x} & 0 & 0 \\ 0 & C_{u_y-f_y} & C_{u_y-m_z} \\ 0 & C_{u_y-m_z} & C_{\theta_z-m_z} \end{bmatrix} \end{cases}. \quad (5.34)$$

The translation matrix is calculated as per Eq. (5.21) with $\Delta x_{FH} = l_1/2$ and $\Delta y_{FH} = 0$. The transformed compliance of Eq. (5.34) becomes:

$$
\left[C_{FH,ip} \right] =
\begin{bmatrix}
C_{u_x-f_x} & 0 & 0 \\
0 & C_{u_y-f_y} - l_1 \cdot C_{u_y-m_z} + \left(\dfrac{l_1}{2}\right)^2 \cdot C_{\theta_z-m_z} & C_{u_y-m_z} - \dfrac{l_1}{2} \cdot C_{\theta_z-m_z} \\
0 & C_{u_y-m_z} - \dfrac{l_1}{2} \cdot C_{\theta_z-m_z} & C_{\theta_z-m_z}
\end{bmatrix}.
$$

$$(5.35)$$

The matrix of Eq. (5.35) connects the displacement vector at F, $[u_{F,ip}]$, to the load applied at the same point, $[f_{F,ip}]$, in the explicit form:

$$
\begin{bmatrix} 0 \\ u_y \\ 0 \end{bmatrix} =
\begin{bmatrix}
C_{u_x-f_x} & 0 & 0 \\
0 & \left(\begin{aligned} &C_{u_y-f_y} - l_1 \cdot C_{u_y-m_z} \\ &+ \left(\dfrac{l_1}{2}\right)^2 \cdot C_{\theta_z-m_z} \end{aligned} \right) & C_{u_y-m_z} - \dfrac{l_1}{2} \cdot C_{\theta_z-m_z} \\
0 & C_{u_y-m_z} - \dfrac{l_1}{2} \cdot C_{\theta_z-m_z} & C_{\theta_z-m_z}
\end{bmatrix}
\begin{bmatrix} f_x \\ f_y \\ m_z \end{bmatrix}.
\qquad (5.36)
$$

Equation (5.36) considered the zero displacement along the x axis, $u_x = 0$ and the zero rotation along the z axis $\theta_z = 0$ at point P, which is guided – see Figure 5.4c. The first algebraic equation of Eq. (5.36) shows that $f_x = 0$. The remaining system consists of two algebraic equations. The last equation allows expressing the reaction moment m_z in terms of the actuation force f_y as:

$$
m_z = \left(\frac{l_1}{2} - \frac{C_{u_y-m_z}}{C_{\theta_z-m_z}} \right) \cdot f_y.
\qquad (5.37)
$$

The moment m_z of Eq. (5.37) is substituted in the remaining nontrivial Eq. (5.36) to yield:

$$
u_y = \left(C_{u_y-f_y} - \frac{\left(C_{u_y-m_z} \right)^2}{C_{\theta_z-m_z}} \right) \cdot f_y.
\qquad (5.38)
$$

Because the original mechanism of Figure 5.4a comprises two identical portions with the flexible hinge of one-half connected in parallel to the hinge of the other (identical and mirrored) half, the stiffness of the mechanism is twice the stiffness of half the mechanism. Therefore, the mechanism stiffness is:

$$
k_y = 2 \cdot \frac{f_y}{u_y} = \frac{2 C_{\theta_z-m_z}}{C_{u_y-f_y} \cdot C_{\theta_z-m_z} - \left(C_{u_y-m_z} \right)^2}.
\qquad (5.39)
$$

For the particular design of a straight-axis flexible hinge with axial and transverse symmetry, like the one of this example, the full-hinge compliances of Eq. (5.39) are calculated by the method presented in Section 3.2.2.2.4. With the given numerical values, the stiffness of Eq. (5.39) is $k_y = 10{,}268{,}549$ N/m.

Example 5.2

A stage designed for out-of-plane motion is formed of a central rigid platform and three identical circular-axis flexible hinges (of median radius R) that are placed equidistantly in a radial manner, as illustrated in Figure 5.5a – this particular design was studied in Lobontiu [38]. Determine the out-of-plane translation (piston-motion) stiffness k_z of this device along the z axis for a generic circular-axis hinge. Consider the particular design of flexible hinges with a constant rectangular cross-section – see Figure 2.25. Known are $R = 0.04$ m, in-plane hinge thickness $t = 0.008$ m, out-of-plane hinge width $w = 0.001$ m, $\alpha = 60°$, Young's modulus $E = 2 \cdot 10^{11}$ N/m² and Poisson's ratio $\mu = 0.3$.

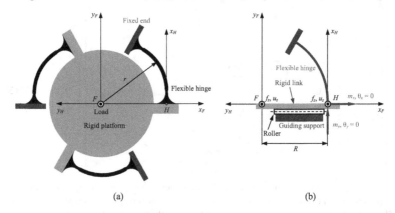

(a)   (b)

FIGURE 5.5   Planar mechanism with rigid platform and three identical circular-axis flexible hinges undergoing z-axis (piston) motion under the action of a central force: (a) schematic representation of the mechanism; (b) skeleton representation of one hinge and its adjacent rigid link with boundary conditions.

Solution:

Due to geometry and load symmetry, the portion corresponding to one flexible hinge – see Figure 5.5b – is analyzed instead of the full mechanism of Figure 5.5a with its three identical flexible hinges. Because of the stage z-axis translation produced by a central force perpendicular to the device plane, the end point H of the hinge (which also belongs to the rigid link separated from the original platform) undergoes the same translation motion as the platform, which means that the only nonzero out-of-plane displacement at H is u_z. Preventing the rotations around the local x_H and y_H axes are the reaction moments m_x and m_y, which need to be determined.

As shown in Figure 5.5b, the point of load application F and the point of displacements evaluation P are collocated. It is also seen that $\delta_H = \pi/2$, which yields the following rotation matrix $[R_{FH}]$:

$$[R] = \begin{bmatrix} 0 & 1 & 0 \\ -1 & 0 & 0 \\ 0 & 0 & 1 \end{bmatrix}. \tag{5.40}$$

The out-of-plane translation matrix is calculated according to the formulation of Eq. (5.28) with $\Delta x_{FH} = R$ and $\Delta y_{FH} = 0$. The native out-of-plane compliance matrix of a generic circular-axis flexure hinge is:

$$\left[C_{op} \right] = \begin{bmatrix} C_{\theta_x - m_x} & C_{\theta_x - m_y} & C_{\theta_x - f_z} \\ C_{\theta_x - m_y} & C_{\theta_y - m_y} & C_{\theta_y - f_z} \\ C_{\theta_x - f_z} & C_{\theta_y - f_z} & C_{u_z - f_z} \end{bmatrix}. \tag{5.41}$$

This matrix is used in conjunction with the rotation matrix of Eq. (5.40) and the translation matrix defined in Eq. (5.28) with the particular offsets of this example to generate the following transformed compliance matrix as per Eq. (5.33):

$$\left[C_{FH,op} \right] = \begin{bmatrix} C_{\theta_y - m_y} & -C_{\theta_x - m_y} & -\left(C_{\theta_y - f_z} + R \cdot C_{\theta_x - m_y} \right) \\ -C_{\theta_x - m_y} & C_{\theta_x - m_x} & C_{\theta_x - f_z} + R \cdot C_{\theta_x - m_x} \\ -\left(C_{\theta_y - f_z} + R \cdot C_{\theta_x - m_y} \right) & C_{\theta_x - f_z} + R \cdot C_{\theta_x - m_x} & \begin{matrix} \left(C_{u_z - f_z} + R \right. \\ \left. \cdot \left(2C_{\theta_x - f_z} + R \cdot C_{\theta_x - m_x} \right) \right) \end{matrix} \end{bmatrix}. \tag{5.42}$$

The displacements at F are connected to the loads at the same point by means of the transformed compliance matrix of Eq. (5.42) as:

$$\begin{bmatrix} 0 \\ 0 \\ u_z \end{bmatrix}$$

$$= \begin{bmatrix} C_{\theta_y - m_y} & -C_{\theta_x - m_y} & -\left(C_{\theta_y - f_z} + R \cdot C_{\theta_x - m_y} \right) \\ -C_{\theta_x - m_y} & C_{\theta_x - m_x} & C_{\theta_x - f_z} + R \cdot C_{\theta_x - m_x} \\ -\left(C_{\theta_y - f_z} + R \cdot C_{\theta_x - m_y} \right) & C_{\theta_x - f_z} + R \cdot C_{\theta_x - m_x} & C_{u_z - f_z} + R \cdot \left(2C_{\theta_x - f_z} + R \cdot C_{\theta_x - m_x} \right) \end{bmatrix}$$

$$\cdot \begin{bmatrix} m_x \\ m_y \\ f_z \end{bmatrix}. \tag{5.43}$$

Equation (5.43) considered that the rotations at F about the x and y directions are zero, namely: $\theta_x = 0$ and $\theta_y = 0$. The first two algebraic equations of the vector Eq. (5.43) provide the reaction moments m_x and m_y in terms of the external load f_z as:

$$
\begin{cases}
m_x = \dfrac{C_{\theta_x - f_z} \cdot C_{\theta_x - m_y} - C_{\theta_y - f_z} \cdot C_{\theta_x - m_x}}{\left(C_{\theta_x - m_y}\right)^2 - C_{\theta_x - m_x} \cdot C_{\theta_y - m_y}} \cdot f_z; \\[4mm]
m_y = \left[\dfrac{C_{\theta_x - f_z} \cdot C_{\theta_y - m_y} - C_{\theta_y - f_z} \cdot C_{\theta_x - m_y}}{\left(C_{\theta_x - m_y}\right)^2 - C_{\theta_x - m_x} \cdot C_{\theta_y - m_y}} - R\right] \cdot f_z
\end{cases}
\tag{5.44}
$$

Substituting these expressions into the third individual Eq. (5.43) allows expressing the ratio f_z/u_z, which is the stiffness corresponding to one flexible hinge. Because there are three identical hinges coupled in parallel in the mechanism of Figure 5.5a, the total stiffness is three times as large, namely:

$$
k_z = 3 \cdot \frac{f_z}{u_z}
$$

$$
= \frac{3\left[\left(C_{\theta_x - m_y}\right)^2 - C_{\theta_x - m_x} \cdot C_{\theta_y - m_y}\right]}{\left[\left(C_{\theta_y - f_z}\right)^2 - C_{\theta_y - m_y} \cdot C_{u_z - f_z}\right] \cdot C_{\theta_x - m_x} + \left(C_{\theta_x - f_z}\right)^2 \cdot C_{\theta_y - m_y} + \left(C_{\theta_x - m_y}\right)^2 \cdot C_{u_z - f_z} - 2C_{\theta_x - m_y} \cdot C_{\theta_x - f_x} \cdot C_{\theta_y - f_z}}.
$$

$$
\tag{5.45}
$$

The compliance Eqs. (A2.27) through (A2.32) together with the numerical values of this example result in $k_z = 393{,}070\,\text{N/m}$.

5.1.2 MULTIPLE-LINK, FLEXIBLE-HINGE SERIAL CHAINS

This section studies serial mechanisms that are formed of multiple flexible hinges and rigid links. Both properly constrained (determinate) and overconstrained (indeterminate) configurations are considered.

A brief analysis is presented first of the common external supports that are utilized in flexible-hinge mechanisms together with the associated boundary conditions and reaction loads. Compliant mechanisms are commonly designed as monolithic (from one piece) with free, guided or fixed boundary conditions at various points supporting/constraining the mechanisms. Depending on whether these support points are used in conjunction with external in-plane or out-of-plane loading, the degrees of freedom that are zero and the corresponding unknown reaction loads are different. Figure 5.6 depicts the fixed and guided support

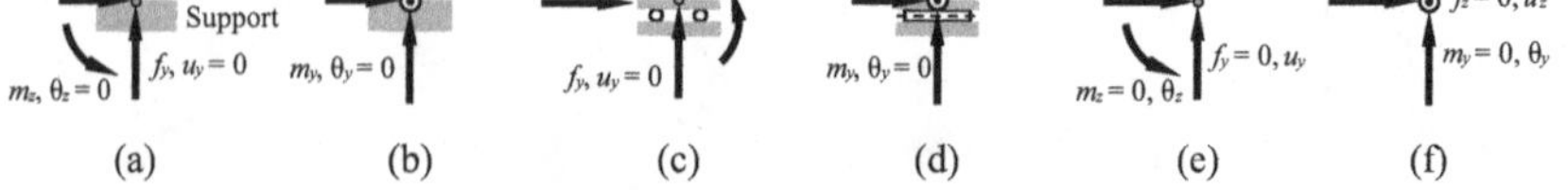

FIGURE 5.6 Constraints and reactions/displacements corresponding to: (a) in-plane fixed support; (b) out-of-plane fixed support; (c) in-plane guided support; (d) out-of-plane guided support; (e) in-plane free point; (f) out-of-plane free point.

points, as well as the free points, for both in-plane and out-of-plane motions by indicating the zero displacements and the corresponding reaction loads.

In order to transfer motion, compliant mechanisms rely on connecting rigid links (that confer rigidity and stability to the mechanism) to flexible links (hinges) that, through their elastic deformation, enable the adjacent rigid links to undergo relative displacements. Such connections of multiple rigid links with flexible hinges result oftentimes in serial chains, like the lever sketched in Figure 5.7a, which is subjected to in-plane external forcing (actuation and payload) and undergoes in-plane displacements.

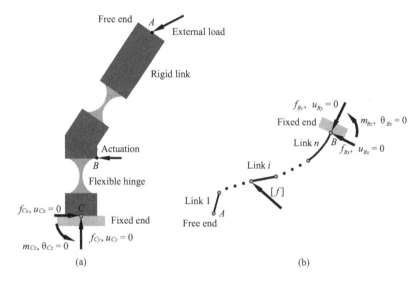

FIGURE 5.7 Fixed-free serial chain with planar actuation, external loading, reactions and boundary conditions: (a) lever arm; (b) skeleton representation of a generic serial chain.

The chain of Figure 5.7a has its end A free (unconstrained) and its opposite end C fixed (fully constrained). Due to the in-plane forcing, the fixed support at C induces three (unknown) reactions, f_{Cx}, f_{Cy} and m_{Cz} – see Figure 5.6a; as a consequence, the system is statically determinate (or properly constrained) because the three reactions can be calculated using the three available static equilibrium conditions/equations. Figure 5.7b shows the skeleton representation of a similar fixed-free, n-link statically determinate planar flexible mechanism under generic in-plane loading. For fixed-free mechanisms, it is not necessary to determine the reaction loads (unless this is required) in order to evaluate all relevant displacements/loads.

Any combination of a free end (as shown in Figure 5.6e and f) with the opposite end having boundary conditions other than fixed results in an insufficiently constrained (movable) mechanism. Conversely, a combination of two ends with boundary conditions that are either fixed or guided results in overconstrained systems that induce more than three unknown reactions in either in-plane or

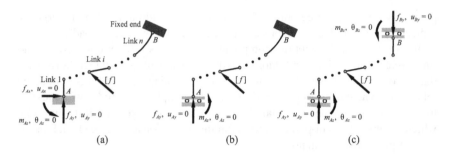

FIGURE 5.8 Skeleton representations of planar overconstrained serial flexible chains under in-plane loading with: (a) two fixed ends; (b) one guided end and one fixed end; (c) two guided ends.

out-of-plane motion; such systems are statically indeterminate. Figure 5.8 shows the skeleton representations of a few overconstrained mechanisms with in-plane load and deformation, but they can be similarly overconstrained when acted upon by out-of-plane loads.

Unlike the statically determinate (properly constrained) structure of a serial chain with one end free and the other end fixed, any of the configurations shown in Figure 5.8 require the determination of (some) reaction loads at one of the two ends in order to enable the calculation of other displacement/load components of interest. For the chain of Figure 5.8a, the reactions at either of the two fixed ends are needed first. For instance, the three reactions at end A: f_{Ax}, f_{Ay} and m_{Az}, are determined by enforcing the three zero boundary conditions at the same point, namely: $u_{Ax} = 0$, $u_{Ay} = 0$ and $\theta_{Az} = 0$. The chain of Figure 5.8b has two unknown reactions at A: f_{Ay} and m_{Az}, which can be found by solving the additional zero-displacement equations: $u_{Ay} = 0$ and $\theta_{Az} = 0$. Eventually, the guided-guided over-constrained serial chain of Figure 5.8c generates four unknown reactions. One needs to conveniently combine the zero-displacement boundary conditions at points $A\left(u_{Ay} = 0, \theta_{Az} = 0\right)$ and $B\left(u_{By} = 0, \theta_{Bz} = 0\right)$ with the three static equilibrium equations in order to obtain four equations allowing to solve for the four unknown reactions f_{Ay}, m_{Az}, f_{By} and m_{Bz}.

5.1.2.1 Fixed-Free Chains

Load-displacement relationships and corresponding compliance matrices are formulated here for planar serial chains formed of rigid links and flexible hinges based on the basic pair of a rigid link and flexible-hinge algorithm introduced in a previous section. The serial chains studied here are free at one end and fixed at the opposite end, and therefore are statically determinate. In order to simplify the mathematical model, it is considered that all segments making up a serial chain are flexible and possess compliance matrices. This assumption encompasses both flexible links (hinges), which are inherently deformable, and rigid links for which zero compliance matrices need to be used. One subcategory of problems considers that there is a single load acting on the chain, whereas another subcategory studies the situation with multiple loads applied at various points on the serial chain.

5.1.2.1.1   Single Load

The single load can be applied either at the free end of the serial chain or at any other point located on the chain between the free and the fixed ends.

Load applied at the free end:

The load $[f_F]$, which is formed of an in-plane component $[f_{F,ip}]$ and an out-of-plane component $[f_{F,op}]$, acts at the free end of a serial chain formed of n flexible links, as illustrated in Figure 5.9. The aim is to calculate the in-plane and out-of-plane displacements (with appropriate compliance matrices) at the free end, as well as at an arbitrary internal point.

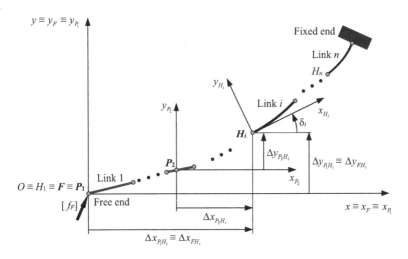

FIGURE 5.9   Fixed-free serial chain with multiple flexible hinges and a single load applied at the free end.

<u>In-plane displacements and compliance matrices</u>:   The displacement at P_1 (which coincides with F and the chain free end) is calculated by adding all individual displacements produced by the single load $[f_{F,ip}]$ acting at F in conjunction with every flexure hinge that is located between the free end and the fixed end of the chain. The total in-plane deformation is therefore calculated as:

$$\left[u_{P_1,ip} \right] = \left[u_{F,ip} \right] = \left(\sum_{i=1}^{n} \left[C_{FH_i,ip} \right] \right) \left[f_{F,ip} \right]. \tag{5.46}$$

Equation (5.46) states that the compliance matrix of a series connection of flexible hinges is the sum of individual-segment compliance matrices, namely:

$$\left[C_{P_1,ip} \right] = \left[C_{F,ip} \right] = \left(\sum_{i=1}^{n} \left[C_{FH_i,ip} \right] \right), \tag{5.47}$$

which is also established in Chapters 3 and 4 in the context of multiple-segment flexible hinges. Using Eq. (5.26), the individual compliance matrices of Eq. (5.46) are expressed as:

$$\left[C_{FH_i,ip} \right] = \left[Tr_{P_1H_i,ip} \right]^T \left[C_{i,ip} \right] \left[Tr_{FH_i,ip} \right] = \left[Tr_{FH_i,ip} \right]^T \left[C_{i,ip} \right] \left[Tr_{FH_i,ip} \right]$$

$$= \left[T_{FH_i,ip} \right]^T \left[R_{FH_i} \right]^T \left[C_{i,ip} \right] \left[R_{FH_i} \right] \left[T_{FH_i,ip} \right], \tag{5.48}$$

where $[C_{i,ip}]$ is the local-frame, in-plane compliance matrix of the generic flexible hinge i. The translation and the rotation matrices of Eq. (5.48) are:

$$\left[T_{FH_i,ip} \right] = \begin{bmatrix} 1 & 0 & 0 \\ 0 & 1 & 0 \\ \Delta y_{FH_i} & -\Delta x_{FH_i} & 1 \end{bmatrix}; \quad \left[R_{FH_i} \right] = \begin{bmatrix} \cos\delta_i & \sin\delta_i & 0 \\ -\sin\delta_i & \cos\delta_i & 0 \\ 0 & 0 & 1 \end{bmatrix},$$

$$\tag{5.49}$$

with the offsets and rotation angle identified in Figure 5.9.

Consider now that the in-plane displacements need to be calculated at a generic point P_2 that is located between the free and fixed end points of the serial chain. The nonbinding assumption can be made that the next flexible hinge (link) past P_2 is link p_2 (not identified in Figure 5.9). In this case, the in-plane load acting at the free end of the chain needs to be combined with all the flexible links that are situated past P_2, and which are links p_2, $p_2 + 1$, ..., n. Using the principle of superposition, the displacement at P_2 adds all partial deformations incurred by these links under the action of the free-end load, namely:

$$\left[u_{P_2,ip} \right] = \left(\sum_{i=p_2}^{n} \left[C_{P_2FH_i,ip} \right] \right) \left[f_{F,ip} \right] = \left[C_{P_2,ip} \right] \left[f_{F,ip} \right]. \tag{5.50}$$

Equation (5.50) indicates that a single compliance matrix can be used instead of all the compliance matrices describing the flexible hinges p_2, $p_2 + 1$, ..., n. The individual compliance matrices of Eq. (5.50) are:

$$\left[C_{P_2FH_i,ip} \right] = \left[Tr_{P_2H_i,ip} \right]^T \left[C_{i,ip} \right] \left[Tr_{FH_i,ip} \right] = \left[T_{P_2H_i,ip} \right]^T \left[R_{P_2H_i} \right]^T \left[C_{i,ip} \right] \left[R_{FH_i} \right] \left[T_{FH_i,ip} \right]$$

$$= \left[T_{P_2H_i,ip} \right]^T \left[R_{FH_i} \right]^T \left[C_{i,ip} \right] \left[R_{FH_i} \right] \left[T_{FH_i,ip} \right], \tag{5.51}$$

where $\left[R_{P_2H_i} \right] = \left[R_{FH_i} \right]$ because the reference frames at $P_1 \equiv F$ and P_2 have their corresponding axes parallel. The new translation matrix of Eq. (5.51) is:

$$\left[T_{P_2H_i,ip} \right] = \begin{bmatrix} 1 & 0 & 0 \\ 0 & 1 & 0 \\ \Delta y_{P_2H_i} & -\Delta x_{P_2H_i} & 1 \end{bmatrix}. \tag{5.52}$$

<u>Out-of-plane displacements and compliance matrices:</u> A reasoning similar to the one used for the in-plane derivation enables to express the total out-of-plane displacement at P_1 as the sum of individual displacements produced by the out-of-plane load at F over every flexible hinge situated between the free and the fixed end points:

$$\left[u_{P_1,op} \right] = \left[u_{F,op} \right] = \left(\sum_{i=1}^{n} \left[C_{FH_i,op} \right] \right) \left[f_{F,op} \right] = \left[C_{F,op} \right] \left[f_{F,op} \right] = \left[C_{P_1,op} \right] \left[f_{F,op} \right],$$

(5.53)

which indicates that the compliance matrix is the sum of individual-segment compliance matrices for a series chain. The individual compliance matrices of Eq. (5.53) are calculated based on Eq. (5.33) as:

$$\left[C_{FH_i,op} \right] = \left[Tr_{P_1H_i,op} \right]^T \left[C_{i,op} \right] \left[Tr_{FH_i,op} \right] = \left[Tr_{FH_i,op} \right]^T \left[C_{i,op} \right] \left[Tr_{FH_i,op} \right]$$

$$= \left[T_{FH_i,op} \right]^T \left[R_{FH_i} \right]^T \left[C_{i,op} \right] \left[R_{FH_i} \right] \left[T_{FH_i,op} \right],$$

(5.54)

where the translation matrix is:

$$\left[T_{FH_i,op} \right] = \begin{bmatrix} 1 & 0 & -\Delta y_{FH_i} \\ 0 & 1 & \Delta x_{FH_i} \\ 0 & 0 & 1 \end{bmatrix}.$$

(5.55)

The displacement at the internal point P_2 adds all partial deformations of the links situated between P_2 and the fixed end under the action of the free-end load, namely:

$$\left[u_{P_2,op} \right] = \left(\sum_{i=p_2}^{n} \left[C_{P_2FH_i,op} \right] \right) \left[f_{F,op} \right] = \left[C_{P_2,op} \right] \left[f_{F,op} \right],$$

(5.56)

which shows that one compliance matrix substitutes all the compliance matrices pertaining to the flexible hinges p_2, $p_2 + 1$, ..., n. The individual compliance matrices of Eq. (5.56) are:

$$\left[C_{P_2FH_i,op} \right] = \left[Tr_{P_2H_i,op} \right]^T \left[C_{i,op} \right] \left[Tr_{FH_i,op} \right]$$

$$= \left[T_{P_2H_i,op} \right]^T \left[R_{P_2H_i} \right]^T \left[C_{i,op} \right] \left[R_{FH_i} \right] \left[T_{FH_i,op} \right]$$

$$= \left[T_{P_2H_i,op} \right]^T \left[R_{FH_i} \right]^T \left[C_{i,op} \right] \left[R_{FH_i} \right] \left[T_{FH_i,op} \right].$$

(5.57)

The new translation matrix of Eq. (5.57) is:

$$\left[T_{P_2 H_i,op}\right]=\begin{bmatrix} 1 & 0 & -\Delta y_{P_2 H_i} \\ 0 & 1 & \Delta x_{P_2 H_i} \\ 0 & 0 & 1 \end{bmatrix}. \qquad (5.58)$$

Example 5.3

Evaluate the in-plane and out-of-plane compliance matrices with respect to the free end A of the serial mechanism sketched in Figure 5.10, which consists of two identical straight-axis flexible hinges and three rigid links. The hinges and the connecting rigid link are collinear.

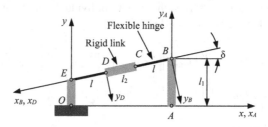

FIGURE 5.10 Planar serial mechanism with two identical and collinear straight-axis flexible hinges and three rigid links.

Solution:

The in-plane compliance matrix of the serial chain is calculated with respect to the free end point A as:

$$\left[C_{A,ip}\right]=\left[Tr_{AB,ip}\right]^{T}\left[C_{ip}\right]\left[Tr_{AB,ip}\right]+\left[Tr_{AD,ip}\right]^{T}\left[C_{ip}\right]\left[Tr_{AD,ip}\right], \qquad (5.59)$$

where $[C_{ip}]$ is the local-frame, in-plane compliance matrix of the hinge. The two transformation matrices of Eq. (5.59) realize the load and displacement transfer between points A and B, and A and D, respectively; they are calculated as:

$$\left[Tr_{AB,ip}\right]=\left[R_{AB}\right]\left[T_{AB,ip}\right]; \quad \left[Tr_{AD,ip}\right]=\left[R_{AD}\right]\left[T_{AD,ip}\right]=\left[R_{AB}\right]\left[T_{AD,ip}\right], \quad (5.60)$$

where

$$\left[T_{AB,ip}\right]=\begin{bmatrix} 1 & 0 & 0 \\ 0 & 1 & 0 \\ \Delta y_{AB} & -\Delta x_{AB} & 1 \end{bmatrix}; \quad \left[T_{AD,ip}\right]=\begin{bmatrix} 1 & 0 & 0 \\ 0 & 1 & 0 \\ \Delta y_{AD} & -\Delta x_{AD} & 1 \end{bmatrix};$$

$$\left[R_{AB}\right]=\left[R_{AD}\right]=\begin{bmatrix} \cos(\pi+\delta) & \sin(\pi+\delta) & 0 \\ -\sin(\pi+\delta) & \cos(\pi+\delta) & 0 \\ 0 & 0 & 1 \end{bmatrix} \qquad (5.61)$$

with $\Delta x_{AB} = 0$, $\Delta y_{AB} = l_1$, $\Delta x_{AD} = (l + l_2) \cdot \cos\delta$, $\Delta y_{AD} = l_1 - (l + l_2) \cdot \sin\delta$. Combining Eqs. (5.59)–(5.61) yields:

$$\left[C_{A,ip} \right] = \begin{bmatrix} C_{A,u_x-f_x} & C_{A,u_x-f_y} & C_{A,u_x-m_z} \\ C_{A,u_x-f_y} & C_{A,u_y-f_y} & C_{A,u_y-m_z} \\ C_{A,u_x-m_z} & C_{A,u_y-m_z} & C_{A,\theta_z-m_z} \end{bmatrix};$$

(5.62)

$$\left[C_{ip} \right] = \begin{bmatrix} C_{u_x-f_x} & 0 & 0 \\ 0 & C_{u_y-f_y} & C_{u_y-m_z} \\ 0 & C_{u_y-m_z} & C_{\theta_z-m_z} \end{bmatrix}.$$

The individual compliances of $[C_{A,ip}]$ are too complex and are not included here.

A similar procedure is followed to determine the out-of-plane compliance matrix of the device sketched in Figure 5.10 with respect to A, which is:

$$\left[C_{A,op} \right] = \left[Tr_{AB,op} \right]^T \left[C_{op} \right] \left[Tr_{AB,op} \right] + \left[Tr_{AD,op} \right]^T \left[C_{op} \right] \left[Tr_{AD,op} \right], \quad (5.63)$$

where $[C_{op}]$ is the out-of-plane compliance matrix of the flexible hinge in its local frame. The out-of-plane transformation matrices of Eq. (5.63) are calculated as:

$$\left[Tr_{AB,op} \right] = \left[R_{AB} \right] \left[T_{AB,op} \right]; \quad \left[Tr_{AD,op} \right] = \left[R_{AB} \right] \left[T_{AD,op} \right], \quad (5.64)$$

with the rotation matrix given in Eq. (5.61) and the translation matrices being:

$$\left[T_{AB,op} \right] = \begin{bmatrix} 1 & 0 & -\Delta y_{AB} \\ 0 & 1 & \Delta x_{AB} \\ 0 & 0 & 1 \end{bmatrix}; \quad \left[T_{AD,op} \right] = \begin{bmatrix} 1 & 0 & -\Delta y_{AD} \\ 0 & 1 & \Delta x_{AD} \\ 0 & 0 & 1 \end{bmatrix}. (5.65)$$

Combining Eq. (5.63) and (5.65) results in:

$$\left[C_{A,op} \right] = \begin{bmatrix} C_{A,\theta_x-m_x} & C_{A,\theta_x-m_y} & C_{A,\theta_x-f_z} \\ C_{A,\theta_x-m_y} & C_{A,\theta_y-m_y} & C_{A,\theta_y-f_z} \\ C_{A,\theta_x-f_z} & C_{A,\theta_y-f_y} & C_{A,\theta_y-f_z} \end{bmatrix}; \left[C_{op} \right] = \begin{bmatrix} C_{\theta_x-m_x} & 0 & 0 \\ 0 & C_{\theta_y-m_y} & C_{\theta_y-f_z} \\ 0 & C_{\theta_y-f_z} & C_{u_z-f_z} \end{bmatrix},$$

(5.66)

but the components of $[C_{A,op}]$ are quite involved and not given here.

Load applied at an internal point:

The load $[f_F]$ is now applied at a generic point F situated between the free and fixed end points, as illustrated in Figure 5.11. Displacements need to be determined at a point P_1 situated before F, at F, as well as at another point P_2 past F on the sequence that starts from the free end of the serial chain. In order to evaluate both in-plane and out-of-plane displacements, the load is correspondingly partitioned into the similar components $[f_{F,ip}]$ and $[f_{F,op}]$.

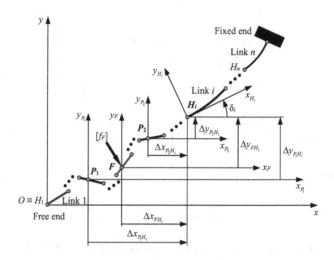

FIGURE 5.11 Fixed-free serial chain with multiple flexible hinges and a single load applied at an internal point.

In-plane displacements and compliance matrices: The displacement at P_1 in Figure 5.11 adds all deformations produced by the in-plane load $[f_{F,ip}]$ applied at F in conjunction with the compliance matrices of all the flexible hinges that are situated between point F and the fixed end of the serial chain. Assuming that the first flexible hinge past the loaded link is hinge number f, it follows that the contributions from hinges $f, f+1, \ldots, n$ have to be included in the displacement vector at P_1, namely:

$$\left[u_{P_1,ip}\right]=\left(\sum_{i=f}^{n}\left[C_{P_1FH_i,ip}\right]\right)\left[f_{F,ip}\right]=\left[C_{P_1,ip}\right]\left[f_{F,ip}\right]. \tag{5.67}$$

The individual compliance matrices of Eq. (5.67) are:

$$\left[C_{P_1FH_i,ip}\right]=\left[Tr_{P_1H_i,ip}\right]^{T}\left[C_{i,ip}\right]\left[Tr_{FH_i,ip}\right]=\left[T_{P_1H_i,ip}\right]^{T}\left[R_{P_1H_i}\right]^{T}\left[C_{i,ip}\right]\left[R_{FH_i}\right]\left[T_{FH_i,ip}\right]$$

$$=\left[T_{P_1H_i,ip}\right]^{T}\left[R_{FH_i}\right]^{T}\left[C_{i,ip}\right]\left[R_{FH_i}\right]\left[T_{FH_i,ip}\right]. \tag{5.68}$$

The new translation matrix of Eq. (5.68) is:

$$\left[T_{P_1H_i,ip}\right]=\begin{bmatrix} 1 & 0 & 0 \\ 0 & 1 & 0 \\ \Delta y_{P_1H_i} & -\Delta x_{P_1H_i} & 1 \end{bmatrix}. \tag{5.69}$$

When the displacement point of interest is F (where the load is also applied), the calculation procedure is identical to the one presented in Eqs. (5.46) through (5.49), where F coincided with the free end of the serial chain. The only difference is that, instead of considering all n flexible hinges, only the hinges $f, f+1, \ldots, n$ need to be accounted for, and therefore, the lower limit of 1 in the summations of Eqs. (5.46) and (5.47) is replaced by f.

Also, similar to the case where the load is applied at the free end and the displacement point of interest is located between F and the fixed end point, the current problem (where displacements at P_2 are needed for F not coinciding with the free end point) is solved by means of Eqs. (5.50)–(5.52).

Another problem that can be solved with this approach is to relate the in-plane displacements that are produced at two points on the serial chain, such as P_1 and P_2 in Figure 5.11. Using Eqs. (5.67) enables expressing the displacement at P_1 as:

$$[u_{P_1,ip}] = \left(\sum_{i=f}^{n} [C_{P_1FH_i,ip}] \right) [f_{F,ip}] = [C_{P_1,ip}][f_{F,ip}] \quad \text{or} \tag{5.70}$$

$$[f_{F,ip}] = [C_{P_1,ip}]^{-1}[u_{P_1,ip}] = [K_{P_1,ip}][u_{P_1,ip}],$$

where $[K_{P_1,ip}] = [C_{P_1,ip}]^{-1}$ is the in-plane stiffness matrix corresponding to all flexible segments situated between P_1 and the fixed end. Considering the modality of evaluating the displacement at P_2, it follows that the displacements at the two points of interest are related as:

$$[u_{P_2,ip}] = \left(\sum_{i=p_2}^{n} [C_{P_2FH_i,ip}] \right) [f_{F,ip}] = [C_{P_2,ip}][f_{F,ip}]$$

$$= [C_{P_2,ip}][K_{P_1,ip}][u_{P_1,ip}]. \tag{5.71}$$

Obviously, the following relationship is applicable resulting from inverting the matrices in Eq. (5.71):

$$[u_{P_1,ip}] = ([C_{P_2,ip}][K_{P_1,ip}])^{-1}[u_{P_2,ip}] = [C_{P_1,ip}][K_{P_2,ip}][u_{P_2,ip}], \tag{5.72}$$

where $[K_{P_2,ip}] = [C_{P_2,ip}]^{-1}$ is the in-plane stiffness matrix corresponding to all flexible segments situated between P_2 and the fixed end.

Out-of-plane displacements and compliance matrices: The out-of-plane displacements produced by $[f_{F,op}]$ at P_1 in Figure 5.11 are calculated similarly to the in-plane displacements at the same point, namely:

$$[u_{P_1,op}] = \left(\sum_{i=f}^{n} [C_{P_1FH_i,op}] \right) [f_{F,op}] = [C_{P_1,op}][f_{F,op}], \tag{5.73}$$

which indicates that a single compliance matrix, denoted by $\left[C_{P_1,op} \right]$, can be used instead of all the compliance matrices defining the flexible hinges $f, f+1, \ldots, n$. The individual compliance matrices of Eq. (5.73) are:

$$
\begin{aligned}
\left[C_{P_1FH_i,op} \right] &= \left[Tr_{P_1H_i,op} \right]^T \left[C_{i,op} \right] \left[Tr_{FH_i,op} \right] \\
&= \left[T_{P_1H_i,op} \right]^T \left[R_{P_1H_i} \right]^T \left[C_{i,op} \right] \left[R_{FH_i} \right] \left[T_{FH_i,op} \right] \\
&= \left[T_{P_1H_i,op} \right]^T \left[R_{FH_i} \right]^T \left[C_{i,op} \right] \left[R_{FH_i} \right] \left[T_{FH_i,op} \right].
\end{aligned} \tag{5.74}
$$

The new translation matrix of Eq. (5.74) is:

$$
\left[T_{P_1H_i,op} \right] = \begin{bmatrix} 1 & 0 & -\Delta y_{P_1H_i} \\ 0 & 1 & \Delta x_{P_1H_i} \\ 0 & 0 & 1 \end{bmatrix}. \tag{5.75}
$$

The displacement at F (which is the load application point) is evaluated using a procedure identical to the one of Eqs. (5.53)–(5.55), where F was the free end of the serial chain. The only difference is that one needs to consider the hinges $f, f+1, \ldots, n$ instead of all the hinges; as a consequence, the lower limit of 1 in the summations of Eqs. (5.53) is replaced by f.

Also, similar to the case when the load F is applied at the free end and the displacement point of interest is located between F and the fixed end point, Eqs. (5.56)–(5.58) apply without any change to the current case where displacements produced by F are needed at a point P_2 located past F.

The out-of-plane displacements at the generic point are:

$$
\begin{aligned}
\left[u_{P_1,op} \right] &= \left(\sum_{i=f}^{n} \left[C_{P_1FH_i,op} \right] \right) \left[f_{F,op} \right] = \left[C_{P_1,op} \right] \left[f_{F,op} \right] \quad \text{or} \\
\left[f_{F,op} \right] &= \left[C_{P_1,op} \right]^{-1} \left[u_{P_1,op} \right] = \left[K_{P_1,op} \right] \left[u_{P_1,op} \right],
\end{aligned} \tag{5.76}
$$

where $\left[K_{P_1,op} \right] = \left[C_{P_1,op} \right]^{-1}$ is the out-of-plane stiffness matrix corresponding to the flexible segments past point P_1. The displacements at the two points of interest are related as:

$$
\begin{aligned}
\left[u_{P_2,op} \right] &= \left(\sum_{i=p_2}^{n} \left[C_{P_2FH_i,op} \right] \right) \left[f_{F,op} \right] = \left[C_{P_2,op} \right] \left[f_{F,op} \right] \\
&= \left[C_{P_2,op} \right] \left[K_{P_1,op} \right] \left[u_{P_1,op} \right]
\end{aligned} \tag{5.77}
$$

equation, which can be written reversely as:

$$\left[u_{P_1,op}\right]=\left(\left[C_{P_2,op}\right]\left[K_{P_1,op}\right]\right)^{-1}\left[u_{P_2,op}\right]=\left[C_{P_1,op}\right]\left[K_{P_2,op}\right]\left[u_{P_2,op}\right], \quad (5.78)$$

where $\left[K_{P_2,op}\right]=\left[C_{P_2,op}\right]^{-1}$ is the out-of-plane stiffness matrix corresponding to the flexible segments past point P_2.

Conclusion:

The two cases that have been discussed in this section highlight the following generalized algorithm: to calculate the displacements at a generic point P on a fixed-free serial chain comprising n flexible hinges and acted upon by a single load, the load applied at the generic point F needs to be combined with the compliance matrices of all flexible hinges situated between whichever of the points P and F is closer to the chain's fixed end and the fixed end. The displacement at P is determined by adding up all these displacements after transferring them to the point of interest P.

Example 5.4

A force f_{By} acts at point B on the planar mechanism sketched in Figure 5.12, which is made of two perpendicular links: the rigid lever AC and the flexible hinge CD. Calculate the displacements u_{Ay} and u_{By} at points A and B, as well as the displacement amplification u_{Ay}/u_{By}. Utilize a right circularly corner-filleted flexible hinge – see Figure 3.13. The hinge has a constant out-of-plane width w. The parameters have the following numerical values: $l = 0.02\,\text{m}$, $r = 0.002\,\text{m}$, $t = 0.001\,\text{m}$, $w = 0.009\,\text{m}$, $l_A = 0.08\,\text{m}$, $l_B = 0.01\,\text{m}$, $E = 2\cdot10^{11}\,\text{N/m}^2$ and $f_{By} = 50\,\text{N}$.

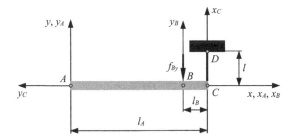

FIGURE 5.12 Planar serial mechanism with one rigid link and one straight-axis flexible hinge.

Solution:

The in-plane displacement vectors at A and B are expressed as:

$$\left[u_{A,ip}\right]=\left[C_{ABC,ip}\right]\left[f_B\right]; \quad \left[u_{B,ip}\right]=\left[C_{BC,ip}\right]\left[f_B\right]. \quad (5.79)$$

The compliance matrix $[C_{ABC,ip}]$ is calculated as:

$$\left[C_{ABC,ip}\right]=\left[Tr_{AC,ip}\right]^{T}\left[C_{ip}\right]\left[Tr_{BC,ip}\right]=\left[T_{AC,ip}\right]^{T}\left[R_{AC}\right]^{T}\left[C_{ip}\right]\left[R_{BC}\right]\left[T_{BC,ip}\right]$$

$$=\begin{bmatrix} C_{ABC,u_x-f_x} & C_{ABC,u_x-f_y} & C_{ABC,u_x-m_z} \\ C_{ABC,u_x-f_y} & C_{ABC,u_y-f_y} & C_{ABC,u_y-m_z} \\ C_{ABC,u_x-m_z} & C_{ABC,u_y-m_z} & C_{ABC,\theta_z-m_z} \end{bmatrix}, \tag{5.80}$$

where

$$\left[T_{AC,ip}\right]=\begin{bmatrix} 1 & 0 & 0 \\ 0 & 1 & 0 \\ \Delta y_{AC} & -\Delta x_{AC} & 1 \end{bmatrix}=\begin{bmatrix} 1 & 0 & 0 \\ 0 & 1 & 0 \\ 0 & -l_A & 1 \end{bmatrix};$$

$$\left[R_{AC}\right]=\left[R_{BC}\right]=\begin{bmatrix} \cos(\pi/2) & \sin(\pi/2) & 0 \\ -\sin(\pi/2) & \cos(\pi/2) & 0 \\ 0 & 0 & 1 \end{bmatrix}=\begin{bmatrix} 0 & 1 & 0 \\ -1 & 0 & 0 \\ 0 & 0 & 1 \end{bmatrix}; \tag{5.81}$$

$$\left[T_{BC,ip}\right]=\begin{bmatrix} 1 & 0 & 0 \\ 0 & 1 & 0 \\ \Delta y_{BC} & -\Delta x_{BC} & 1 \end{bmatrix}=\begin{bmatrix} 1 & 0 & 0 \\ 0 & 1 & 0 \\ 0 & -l_B & 1 \end{bmatrix}.$$

Similarly, the compliance matrix $[C_{BC,ip}]$ is calculated as:

$$\left[C_{BC,ip}\right]=\left[Tr_{BC,ip}\right]^{T}\left[C_{ip}\right]\left[Tr_{BC,ip}\right]=\left[T_{BC,ip}\right]^{T}\left[R_{BC}\right]^{T}\left[C_{ip}\right]\left[R_{BC}\right]\left[T_{BC,ip}\right]$$

$$=\begin{bmatrix} C_{BC,u_x-f_x} & C_{BC,u_x-f_y} & C_{BC,u_x-m_z} \\ C_{BC,u_x-f_y} & C_{BC,u_y-f_y} & C_{BC,u_y-m_z} \\ C_{BC,u_x-m_z} & C_{ABC,u_y-m_z} & C_{BC,\theta_z-m_z} \end{bmatrix}. \tag{5.82}$$

with the rotation and translation matrices of Eq. (5.81). The in-plane compliance matrix of the right circularly corner-filleted flexible hinge is evaluated based on the calculation procedure outlined in Section 3.2.2.2.3. Considering that the load vector at B is $[f_B] = [0 - f_{By}\ 0]^T$, and the displacement vectors at A and B are $[u_A] = [u_{Ax}\ u_{Ay}\ \theta_{Az}]^T$, $[u_B] = [u_{Bx}\ u_{By}\ \theta_{Bz}]^T$, the two displacements of interest, as well as their ratio, are determined as:

$$u_{Ay} = -C_{ABC,u_y-f_y} \cdot f_{By}; \quad u_{By} = -C_{BC,u_y-f_y} \cdot f_{By}; \quad u_{Ay}/u_{By} = \frac{C_{ABC,u_y-f_y}}{C_{BC,u_y-f_y}}. \tag{5.83}$$

The following numerical values are obtained: $u_{Ay} = -0.037$ m, $u_{By} = -0.000586$ m and $u_{Ay}/u_{By} = 63.94$.

5.1.2.1.2 Multiple Loads

Several loads $\left[f_{F_j} \right]$, $j = 1$, 2, ..., m act on a fixed-free serial chain, as illustrated in Figure 5.13. The displacements at a generic point will be generated by superimposing the actions of each individual load over the flexible hinges that are situated either past the displacement point or the load application point, whichever is closer to the fixed end. While the displacement point of interest can occupy various positions along the chain, the analysis is limited here to the point P_1, which is situated in front of (before) all m external loads in a sequence that starts from the free end.

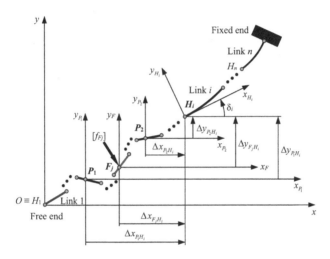

FIGURE 5.13 Fixed-free serial chain with multiple flexible hinges and rigid links, and multiple loads.

In-plane displacements and compliance matrices:

The total in-plane displacement at P_1 adds the displacements produced by all the loads through all the flexible hinges that are located between a load application point and the chain's fixed end, namely:

$$\left[u_{P_1,ip} \right] = \left(\sum_{i=f_1}^{n} \left[C_{P_1 F_1 H_i,ip} \right] \right) \left[f_{F_1,ip} \right] + \left(\sum_{i=f_2}^{n} \left[C_{P_1 F_2 H_i,ip} \right] \right) \left[f_{F_2,ip} \right] + \cdots$$

$$+ \left(\sum_{i=f_m}^{n} \left[C_{P_1 F_m H_i,ip} \right] \right) \left[f_{F_m,ip} \right]$$

$$= \sum_{j=1}^{m} \left(\left(\sum_{i=f_j}^{n} \left[C_{P_1 F_j H_i,ip} \right] \right) \left[f_{F_j,ip} \right] \right) = \sum_{j=1}^{m} \left(\left[C_{P_1 F_j,ip} \right] \left[f_{F_j,ip} \right] \right). \quad (5.84)$$

Equation (5.84) assumed the flexible hinges f_j, $f_j + 1$, ..., n are placed past the location of the load $\left[f_{F_j,ip} \right]$, which is applied at point F_j. The individual compliance matrices of Eq. (5.84) are:

$$\left[C_{P_1 F_j H_i,ip} \right] = \left[Tr_{P_1 H_i,ip} \right]^T \left[C_{i,ip} \right] \left[Tr_{F_j H_i,ip} \right]$$

$$= \left[T_{P_1 H_i,ip} \right]^T \left[R_{P_1 H_i} \right]^T \left[C_{i,ip} \right] \left[R_{F_j H_i} \right] \left[T_{F_j H_i,ip} \right]$$

$$= \left[T_{P_1 H_i,ip} \right]^T \left[R_{F_j H_i} \right]^T \left[C_{i,ip} \right] \left[R_{F_j H_i} \right] \left[T_{F_j H_i,ip} \right]. \tag{5.85}$$

$\left[C_{i,ip} \right]$ is the local-frame, in-plane compliance matrix of hinge i. The new translation matrix of Eq. (5.85) is:

$$\left[T_{F_j H_i,ip} \right] = \begin{bmatrix} 1 & 0 & 0 \\ 0 & 1 & 0 \\ \Delta y_{F_j H_i} & -\Delta x_{F_j H_i} & 1 \end{bmatrix}, \tag{5.86}$$

while $\left[T_{P_1 H_i,ip} \right]$ is provided in Eq. (5.69).

Out-of-plane displacements and compliance matrices:

A similar procedure is applied for the out-of-plane case when the displacements at P_1 are calculated as:

$$\left[u_{P_1,op} \right] = \left(\sum_{i=f_1}^{n} \left[C_{P_1 F_1 H_i,op} \right] \right) \left[f_{F_1,op} \right] + \left(\sum_{i=f_2}^{n} \left[C_{P_1 F_2 H_i,op} \right] \right) \left[f_{F_2,op} \right] + \cdots$$

$$+ \left(\sum_{i=f_m}^{n} \left[C_{P_1 F_m H_i,op} \right] \right) \left[f_{F_m,op} \right]$$

$$= \sum_{j=1}^{m} \left(\left(\sum_{i=f_j}^{n} \left[C_{P_1 F_j H_i,op} \right] \right) \left[f_{F_j,op} \right] \right) = \sum_{j=1}^{m} \left(\left[C_{P_1 F_j,op} \right] \left[f_{F_j,op} \right] \right). \tag{5.87}$$

The individual compliance matrices of Eq. (5.87) are:

$$\left[C_{P_1 F_j H_i,op} \right] = \left[Tr_{P_1 H_i,op} \right]^T \left[C_{i,op} \right] \left[Tr_{F_j H_i,op} \right]$$

$$= \left[T_{P_1 H_i,op} \right]^T \left[R_{P_1 H_i} \right]^T \left[C_{i,op} \right] \left[R_{F_j H_i} \right] \left[T_{F_j H_i,op} \right]$$

$$= \left[T_{P_1 H_i,op} \right]^T \left[R_{F_j H_i} \right]^T \left[C_{i,op} \right] \left[R_{F_j H_i} \right] \left[T_{F_j H_i,op} \right]. \tag{5.88}$$

$\left[C_{i,op}\right]$ is the local-frame, out-of-plane compliance matrix of hinge i. The new translation matrix of Eq. (5.88) is:

$$\left[T_{F_jH_i,op}\right]=\begin{bmatrix} 1 & 0 & -\Delta y_{F_jH_i} \\ 0 & 1 & \Delta x_{F_jH_i} \\ 0 & 0 & 1 \end{bmatrix}. \tag{5.89}$$

Example 5.5

i. Calculate the free-end, in-plane displacements of the lever mechanism illustrated in Figure 5.14, which is acted upon by an external force $f_{Ax} = 20\,N$ and an actuation force $f_D = 300\,N$. Consider that the identical flexible hinges are of constant circular cross-section and that known are the hinge diameter $d = 0.002\,m$, the hinge length $l = 0.01\,m$, $l_{AB} = 0.08\,m$, $l_{CD} = 0.04\,m$, $\delta = 45°$, $E = 0.8 \cdot 10^{11}\,N/m^2$ and $\mu = 0.33$;

ii. Evaluate the minimum value of the actuation force f_D that produces a horizontal displacement $u_{Ax} = 0$ at A for f_{Ax} provided at (i). Also, calculate the vertical displacement u_{Ay} and the rotation θ_{Az} for this load condition;

iii. Evaluate the maximum value of a force f_{Az} perpendicular to the lever's plane applied at A that results in an out-of-plane deflection $u_{Az} = 0.0005\,m$. Also, calculate the rotations at A around the x and y axes, θ_{Ax} and θ_{Ay} pertaining to this load condition. Consider that all the in-plane loads are removed.

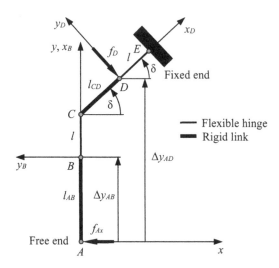

FIGURE 5.14 Fixed-free planar serial chain with two flexible hinges, two rigid links and two external loads.

Solution:

i. The in-plane displacement at A adds the contributions of f_{Ax} over the two hinges and of f_D over the hinge at the fixed root of the chain, namely:

$$[u_{A,ip}] = \left([Tr_{AB,ip}]^T [C_{ip}][Tr_{AB,ip}] + [Tr_{AD,ip}]^T [C_{ip}][Tr_{AD,ip}] \right)[f_{A,ip}]$$

$$+ [Tr_{AD,ip}]^T [C_{ip}][f_{D,ip}], \tag{5.90}$$

where $[C_{ip}]$ is the in-plane compliance matrix of the two identical flexible hinges. The first term in the parenthesis multiplying the vector $[f_{A,ip}]$ indicates that $[f_{A,ip}]$ is transferred from A to B by means of the transformation matrix $[Tr_{AB,ip}]$, and then, the resulting deformation at B due to the hinge BC is relocated at A by means of $[Tr_{AB,ip}]^T$. Similar notations are used in the second term of the parenthesis to transfer the same load from A to D and then to transfer the resulting deformation produced by the hinge DE back to A. The load $[f_{D,ip}]$ already resides at D so it needs no relocation; only the deformation produced by it through the hinge DE is transferred at A by means of $[Tr_{AD,ip}]^T$. The transformation matrices in Eq. (5.90) are:

$$[Tr_{AB,ip}] = [R_{AB}][T_{AB,ip}]; \quad [Tr_{AD,ip}] = [R_{AD}][T_{AD,ip}], \tag{5.91}$$

with

$$[R_{AB}] = \begin{bmatrix} \cos(\pi/2) & \sin(\pi/2) & 0 \\ -\sin(\pi/2) & \cos(\pi/2) & 0 \\ 0 & 0 & 1 \end{bmatrix}; \quad [R_{AD}] = \begin{bmatrix} \cos(\delta) & \sin(\delta) & 0 \\ -\sin(\delta) & \cos(\delta) & 0 \\ 0 & 0 & 1 \end{bmatrix}$$

$$[T_{AB,ip}] = \begin{bmatrix} 1 & 0 & 0 \\ 0 & 1 & 0 \\ \Delta y_{AB} & -\Delta x_{AB} & 1 \end{bmatrix}; \quad [T_{AD,ip}] = \begin{bmatrix} 1 & 0 & 0 \\ 0 & 1 & 0 \\ \Delta y_{AD} & -\Delta x_{AD} & 1 \end{bmatrix}. \tag{5.92}$$

The offsets of Eq. (5.92) are: $\Delta x_{AB} = 0$, $\Delta y_{AB} = l_{AB}$, $\Delta x_{AD} = l_{CD}\cdot\cos\delta$, $\Delta y_{AD} = l_{AB} + l + l_{CD}\cdot\sin\delta$. The force vectors of Eq. (5.90) are:

$$[f_{A,ip}] = \begin{bmatrix} -f_{Ax} & 0 & 0 \end{bmatrix}^T; \quad [f_{D,ip}] = \begin{bmatrix} f_D\cdot\sin\delta & -f_D\cdot\cos\delta & 0 \end{bmatrix}^T. \tag{5.93}$$

Using the compliances of a constant circular cross-sectional flexure hinge of Eq. (2.33), as well as the numerical values provided here, the following results are obtained: $u_{Ax} = -0.0495\,\text{m}$, $u_{Ay} = 0.0068\,\text{m}$ and $\theta_{Az} = -0.489\,\text{rad}$.

ii. The value of the force f_D that statically balances the effect of the external force f_{Ax} is found by setting the x-displacement component of the vector $[u_{A,ip}]$ to zero. As a consequence, the following results from Eq. (5.90):

$$\left[f_{D,ip}\right]=\left(\left[Tr_{AD,ip}\right]^T\left[C_{ip}\right]\right)^{-1}$$
$$\times\left(\left[u_{A,ip}\right]-\left(\left[Tr_{AB,ip}\right]^T\left[C_{ip}\right]\left[Tr_{AB,ip}\right]+\left[Tr_{AD,ip}\right]^T\left[C_{ip}\right]\left[Tr_{AD,ip}\right]\right)\left[f_{A,ip}\right]\right).$$

(5.94)

Combining Eqs. (5.93) and (5.94) results in three algebraic equations that are solved for $f_D = 1{,}015.09\,\text{N}$, $u_{Ay} = -0.0065\,\text{m}$ and $\theta_{Az} = -0.087\,\text{rad}$.

iii. The only force vector acting on the lever is the out-of-plane one at A defined as:

$$\left[f_{A,op}\right]=\begin{bmatrix} 0 & 0 & f_{Az} \end{bmatrix}^T.$$

(5.95)

and which produces the following out-of-plane displacement at A:

$$\left[u_{A,op}\right]=\left[C_{A,op}\right]\left[f_{A,op}\right]$$

with

(5.96)

$$\left[C_{A,op}\right]=\left(\left[Tr_{AB,op}\right]^T\left[C_{op}\right]\left[Tr_{AB,op}\right]+\left[Tr_{AD,op}\right]^T\left[C_{op}\right]\left[Tr_{AD,op}\right]\right),$$

where

$$\left[Tr_{AB,op}\right]=\left[R_{AB}\right]\left[T_{AB,op}\right]; \quad \left[Tr_{AD,op}\right]=\left[R_{AD}\right]\left[T_{AD,op}\right].$$

(5.97)

The out-of-plane translation matrices are:

$$\left[T_{AB,op}\right]=\begin{bmatrix} 1 & 0 & -\Delta y_{AB} \\ 0 & 1 & \Delta x_{AB} \\ 0 & 0 & 1 \end{bmatrix}; \left[T_{AD,op}\right]=\begin{bmatrix} 1 & 0 & -\Delta y_{AD} \\ 0 & 1 & \Delta x_{AD} \\ 0 & 0 & 1 \end{bmatrix},$$

(5.98)

and the out-of-plane displacement vector at A is:

$$\left[u_{A,op}\right]=\begin{bmatrix} \theta_{Ax} & \theta_{Ay} & u_{Az} \end{bmatrix}^T.$$

(5.99)

Combining now Eqs. (5.95) through (5.99) yields:

$$\begin{bmatrix} \theta_{Ax} \\ \theta_{Ay} \\ u_{Az} \end{bmatrix}=\begin{bmatrix} C_{A,\theta_x-m_x} & C_{A,\theta_x-m_y} & C_{A,\theta_x-f_z} \\ C_{A,\theta_x-m_y} & C_{A,\theta_y-m_y} & C_{A,\theta_y-f_z} \\ C_{A,\theta_x-f_z} & C_{A,\theta_y-f_z} & C_{A,u_z-f_z} \end{bmatrix}\begin{bmatrix} 0 \\ 0 \\ f_{Az} \end{bmatrix}.$$

(5.100)

The last algebraic equation in the matrix Eq. (5.100) allows solving for f_{Az} as:

$$f_{Az} = \frac{u_{Az}}{C_{A,u_z-f_z}}.$$

(5.101)

The other two algebraic equation Eq. (5.100) are combined with Eq. (5.101) to yield the out-of-plane rotations at A as:

$$\begin{cases} \theta_{Ax} = C_{A,\theta_x-f_z} \cdot f_{Az} = \dfrac{C_{A,\theta_x-f_z}}{C_{A,u_z-f_z}} \cdot u_{Az} \\[4mm] \theta_{Ay} = C_{A,\theta_y-f_z} \cdot f_{Az} = \dfrac{C_{A,\theta_y-f_z}}{C_{A,u_z-f_z}} \cdot u_{Az} \end{cases}.$$

(5.102)

With the numerical parameters and the $[C_{op}]$ matrix defined by its components of Eqs. (2.33), the results are $f_{Az} = 0.128\,\text{N}$, $\theta_{Ax} = -0.0045\,\text{rad}$ and $\theta_{Ay} = 0.00035\,\text{rad}$.

Symmetric chains:
Planar flexible-hinge mechanisms are frequently designed with geometry and load symmetry in order to generate displacements along specified directions, since the symmetry results in boundary conditions preventing/favoring displacements along certain axes. In such mechanisms, as discussed later in this chapter, a symmetric mechanism is formed by serially connecting two identical chains that are mirrored with respect to a line, which becomes the symmetry axis. Consider, for instance, the planar serial mechanisms of Figure 5.15; they are composed of chain 1 (comprising several flexible and rigid segments, but represented as a single chain) and chain 2, which is the mirror of chain 1 with respect to an axis passing through the joining point A – as a result, this line becomes a symmetry axis for the overall portion formed of the two chains connected in series.

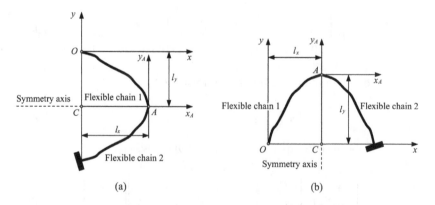

(a)   (b)

FIGURE 5.15 Two identical and mirrored flexible chains, which are serially connected and display symmetry with respect to: (a) a horizontal axis; (b) a vertical axis.

Symmetry with respect to the x axis:

Assume that the in-plane compliance matrix of the flexible chain 1 in Figure 5.15a is known with respect to the reference frame $Ax_Ay_Az_A$ (whose axis Ax_A is the symmetry axis) as:

$$\left[C_{A,ip}^{(1)}\right] = \begin{bmatrix} C_{u_x-f_x}^{(1)} & C_{u_x-f_y}^{(1)} & C_{u_x-m_z}^{(1)} \\ C_{u_x-f_y}^{(1)} & C_{u_y-f_y}^{(1)} & C_{u_y-m_z}^{(1)} \\ C_{u_x-m_z}^{(1)} & C_{u_y-m_z}^{(1)} & C_{\theta_z-m_z}^{(1)} \end{bmatrix}. \tag{5.103}$$

Because chain 2 is chain 1 mirrored with respect to the Ax_A axis, its in-plane compliance matrix in the same reference frame centered at A results from the matrix of chain 1 given in Eq. (5.103) by changing the signs of the compliances that connect the axes x to y and x to z, namely:

$$\left[C_{A,ip}^{(2)}\right] = \begin{bmatrix} C_{u_x-f_x}^{(1)} & -C_{u_x-f_y}^{(1)} & -C_{u_x-m_z}^{(1)} \\ -C_{u_x-f_y}^{(1)} & C_{u_y-f_y}^{(1)} & C_{u_y-m_z}^{(1)} \\ -C_{u_x-m_z}^{(1)} & C_{u_y-m_z}^{(1)} & C_{\theta_z-m_z}^{(1)} \end{bmatrix}. \tag{5.104}$$

The compliance matrix of the entire segment resulting from serially combining the two chains with respect to the reference frame centered at A is the sum of the two compliances expressed in Eqs. (5.103) and (5.104):

$$\left[C_{A,ip}\right] = \left[C_{A,ip}^{(1)}\right] + \left[C_{A,ip}^{(2)}\right] = 2 \cdot \begin{bmatrix} C_{u_x-f_x}^{(1)} & 0 & 0 \\ 0 & C_{u_y-f_y}^{(1)} & C_{u_y-m_z}^{(1)} \\ 0 & C_{u_y-m_z}^{(1)} & C_{\theta_z-m_z}^{(1)} \end{bmatrix}. \tag{5.105}$$

Oftentimes, the compliance matrix of the full segment needs to be evaluated with respect to a translated reference frame that is located at one of the chain's ends; for instance, the compliance matrix is calculated with reference to the $Oxyz$ frame as:

$$\left[C_{O,ip}\right] = \left[T_{OA,ip}\right]^T \left[C_{A,ip}\right] \left[T_{OA,ip}\right] \quad \text{with} \quad \left[T_{OA,ip}\right] = \begin{bmatrix} 1 & 0 & 0 \\ 0 & 1 & 0 \\ \Delta y_{OA} & -\Delta x_{OA} & 1 \end{bmatrix}, \tag{5.106}$$

with $\Delta x_{OA} = l_x$ and $\Delta y_{OA} = -l_y$. Combining Eqs. (5.105) and (5.106) results in:

$$\left[C_{O,ip} \right]$$

$$= 2 \cdot \begin{bmatrix} C^{(1)}_{u_x-f_x} + l_y^2 \cdot C^{(1)}_{\theta_z-m_z} & l_y \cdot \left(l_x \cdot C^{(1)}_{\theta_z-m_z} - C^{(1)}_{u_y-m_z} \right) & -l_y \cdot C_{\theta_z-m_z} \\ l_y \cdot \left(l_x \cdot C^{(1)}_{\theta_z-m_z} - C^{(1)}_{u_y-m_z} \right) & \begin{pmatrix} C^{(1)}_{u_y-f_y} + l_x \cdot \\ \left(l_x \cdot C^{(1)}_{\theta_z-m_z} - 2C^{(1)}_{u_y-m_z} \right) \end{pmatrix} & C^{(1)}_{u_y-m_z} - l_x \cdot C^{(1)}_{\theta_z-m_z} \\ -l_y \cdot C_{\theta_z-m_z} & C^{(1)}_{u_y-m_z} - l_x \cdot C^{(1)}_{\theta_z-m_z} & C^{(1)}_{\theta_z-m_z} \end{bmatrix} \cdot$$

$$(5.107)$$

A similar approach is used to calculate the out-of-plane compliance of the serial chain depicted in Figure 5.15a. The compliances of chains 1 and 2 are expressed in the reference frame centered at A as:

$$\left[C^{(1)}_{A,op} \right] = \begin{bmatrix} C^{(1)}_{\theta_x-m_x} & C^{(1)}_{\theta_x-m_y} & C^{(1)}_{\theta_x-f_z} \\ C^{(1)}_{\theta_x-m_y} & C^{(1)}_{\theta_y-m_y} & C^{(1)}_{\theta_y-f_z} \\ C^{(1)}_{\theta_x-f_z} & C^{(1)}_{\theta_y-f_z} & C^{(1)}_{u_z-f_z} \end{bmatrix} ;$$

$$(5.108)$$

$$\left[C^{(2)}_{A,op} \right] = \begin{bmatrix} C^{(1)}_{\theta_x-m_x} & -C^{(1)}_{\theta_x-m_y} & -C^{(1)}_{\theta_x-f_z} \\ -C^{(1)}_{\theta_x-m_y} & C^{(1)}_{\theta_y-m_y} & C^{(1)}_{\theta_y-f_z} \\ -C^{(1)}_{\theta_x-f_z} & C^{(1)}_{\theta_y-f_z} & C^{(1)}_{u_z-f_z} \end{bmatrix} \cdot$$

The out-of-plane compliance matrix of the entire segment with respect to the reference frame centered at A is:

$$\left[C_{A,op} \right] = \left[C^{(1)}_{A,op} \right] + \left[C^{(2)}_{A,op} \right] = 2 \cdot \begin{bmatrix} C^{(1)}_{\theta_x-m_x} & 0 & 0 \\ 0 & C^{(1)}_{\theta_y-m_y} & C^{(1)}_{\theta_y-f_z} \\ 0 & C^{(1)}_{\theta_y-f_z} & C^{(1)}_{u_z-f_z} \end{bmatrix} . \quad (5.109)$$

The compliance matrix of Eq. (5.109) is translated at the end point O as:

$$\left[C_{O,op} \right] = \left[T_{OA,op} \right]^T \left[C_{A,op} \right] \left[T_{OA,op} \right] \quad \text{with} \quad \left[T_{OA,op} \right] = \begin{bmatrix} 1 & 0 & -\Delta y_{OA} \\ 0 & 1 & \Delta x_{OA} \\ 0 & 0 & 1 \end{bmatrix} .$$

$$(5.110)$$

Substituting Eq. (5.109) into Eq. (5.110) results in:

$$
\left[C_{O,op} \right] = 2 \cdot
\begin{bmatrix}
C_{\theta_x-m_x}^{(1)} & 0 & l_y \cdot C_{\theta_x-m_x}^{(1)} \\[2mm]
0 & C_{\theta_y-m_y}^{(1)} & C_{\theta_y-f_z}^{(1)} + l_x \cdot C_{\theta_y-m_y}^{(1)} \\[2mm]
l_y \cdot C_{\theta_x-m_x}^{(1)} & C_{\theta_y-f_z}^{(1)} + l_x \cdot C_{\theta_y-m_y}^{(1)} & \begin{aligned} &\left(C_{u_z-f_z}^{(1)} + l_x \cdot \right. \\ &\left(l_x \cdot C_{\theta_y-m_y}^{(1)} + 2 \cdot C_{\theta_y-f_z}^{(1)} \right) \\ &\left. + l_y^2 \cdot C_{\theta_x-m_x}^{(1)} \right) \end{aligned}
\end{bmatrix}.
$$

(5.111)

Symmetry with respect to the y axis:

The generic flexible segment of Figure 5.15b is symmetric with respect to the y_A axis. The in-plane and out-of-plane compliance matrices of the mirrored segment 2 are expressed in terms of those of the (basic) segment 1 (both matrices are formulated in the $Ax_Ay_Az_A$ frame):

$$
\left[C_{A,ip}^{(2)} \right] =
\begin{bmatrix}
C_{u_x-f_x}^{(1)} & -C_{u_x-f_y}^{(1)} & C_{u_x-m_z}^{(1)} \\[2mm]
-C_{u_x-f_y}^{(1)} & C_{u_y-f_y}^{(1)} & -C_{u_y-m_z}^{(1)} \\[2mm]
C_{u_x-m_z}^{(1)} & -C_{u_y-m_z}^{(1)} & C_{\theta_z-m_z}^{(1)}
\end{bmatrix};
$$

$$
\left[C_{A,op}^{(2)} \right] =
\begin{bmatrix}
C_{\theta_x-m_x}^{(1)} & -C_{\theta_x-m_y}^{(1)} & C_{\theta_x-f_z}^{(1)} \\[2mm]
-C_{\theta_x-m_y}^{(1)} & C_{\theta_y-m_y}^{(1)} & -C_{\theta_y-f_z}^{(1)} \\[2mm]
C_{\theta_x-f_z}^{(1)} & -C_{\theta_y-f_z}^{(1)} & C_{u_z-f_z}^{(1)}
\end{bmatrix}.
$$

(5.112)

by using minus signs for the xy and yz compliances. After adding the two segment compliance matrices and translating the resulting matrices from A to O, the full-mechanism compliance matrices are expressed in the end-point reference frame $Oxyz$ as:

$$
\left[C_{O,ip} \right]
$$

$$
= 2 \cdot
\begin{bmatrix}
\begin{aligned} &\left(C_{u_x-f_x}^{(1)} + l_y \cdot \right. \\ &\left. \left(l_y \cdot C_{\theta_z-m_z}^{(1)} + 2C_{u_x-m_z}^{(1)} \right) \right) \end{aligned} & \begin{aligned} &\left(-l_x \cdot \right. \\ &\left. \left(l_y \cdot C_{\theta_z-m_z}^{(1)} + C_{u_x-m_z}^{(1)} \right) \right) \end{aligned} & l_y \cdot C_{\theta_z-m_z}^{(1)} + C_{u_x-m_z}^{(1)} \\[4mm]
-l_x \cdot \left(l_y \cdot C_{\theta_z-m_z}^{(1)} + C_{u_x-m_z}^{(1)} \right) & C_{u_y-f_y}^{(1)} + l_x^2 \cdot C_{\theta_z-m_z}^{(1)} & -l_x \cdot C_{\theta_z-m_z}^{(1)} \\[4mm]
l_y \cdot C_{\theta_z-m_z}^{(1)} + C_{u_x-m_z}^{(1)} & -l_x \cdot C_{\theta_z-m_z}^{(1)} & C_{\theta_z-m_z}^{(1)}
\end{bmatrix}.
$$

(5.113)

$$\left[C_{O,op}\right]$$

$$= 2 \cdot \begin{bmatrix} C_{\theta_x - m_x}^{(1)} & 0 & C_{\theta_x - f_z}^{(1)} - l_y \cdot C_{\theta_x - m_x}^{(1)} \\ 0 & C_{\theta_y - m_y}^{(1)} & l_x \cdot C_{\theta_y - m_y}^{(1)} \\ C_{\theta_x - f_z}^{(1)} - l_y \cdot C_{\theta_x - m_x}^{(1)} & l_x \cdot C_{\theta_y - m_y}^{(1)} & \begin{pmatrix} C_{u_z - f_z}^{(1)} + l_x^2 \cdot C_{\theta_y - m_y}^{(1)} \\ + l_y \cdot \left(l_y \cdot C_{\theta_x - m_x}^{(1)} - 2 \cdot C_{\theta_x - f_z}^{(1)} \right) \end{pmatrix} \end{bmatrix}.$$

$$(5.114)$$

Note that $\Delta x_{OA} = l_x$ and $\Delta y_{OA} = l_y$ in the translation matrices defined in Eqs. (5.106) and (5.110).

Example 5.6

Evaluate the in-plane and out-of-plane compliance matrices of the fixed-free serial segment of Figure 5.16a with respect to the end-point reference frame $Oxyz$. The mechanism is formed of two serially connected identical flexible chains that are mirrored with respect to a horizontal axis; the upper segment, which is pictured in Figure 5.16b, comprises two identical straight-axis flexible hinges and three rigid links.

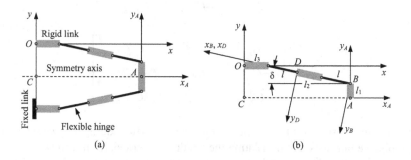

(a) (b)

FIGURE 5.16 Serial chain with axial symmetry: (a) full configuration; (b) basic (half)-mechanism comprising two identical and collinear straight-axis flexible hinges with defining geometry.

Solution:

The in-plane compliance matrix of the serial half-segment AO of Figure 5.16b recorded with respect to the reference frame $Ax_A y_A z_A$ is calculated as:

$$\left[C_{A,ip}\right] = \left[Tr_{AB,ip}\right]^T \left[C_{ip}\right]\left[Tr_{AB,ip}\right] + \left[Tr_{AD,ip}\right]^T \left[C_{ip}\right]\left[Tr_{AD,ip}\right], \quad (5.115)$$

with

$$\left[Tr_{AB,ip}\right] = \left[R_{AB}\right]\left[T_{AB,ip}\right]; \quad \left[Tr_{AD,ip}\right] = \left[R_{AD}\right]\left[T_{AD,ip}\right]. \quad (5.116)$$

$[C_{ip}]$ is the in-plane compliance matrix of any of the two hinges in its local frame. The translation matrices $[T_{AB,ip}]$ and $[T_{AD,ip}]$, and the rotation matrices $[R_{AB}]$ and $[R_{AD}]$ realizing the transfer of the two hinge compliance matrices from B to A and D to A are:

$$
[T_{AB,ip}] = \begin{bmatrix} 1 & 0 & 0 \\ 0 & 1 & 0 \\ \Delta y_{AB} & -\Delta x_{AB} & 1 \end{bmatrix}; \quad [T_{AD,ip}] = \begin{bmatrix} 1 & 0 & 0 \\ 0 & 1 & 0 \\ \Delta y_{AD} & -\Delta x_{AD} & 1 \end{bmatrix};
$$

$$
[R_{AB}] = [R_{AD}] = \begin{bmatrix} \cos(\pi - \delta) & \sin(\pi - \delta) & 0 \\ -\sin(\pi - \delta) & \cos(\pi - \delta) & 0 \\ 0 & 0 & 1 \end{bmatrix}. \tag{5.117}
$$

with $\Delta x_{AB} = 0$, $\Delta y_{AB} = l_1$, $\Delta x_{AD} = -(l + l_2)\cdot\cos\delta$ and $\Delta y_{AD} = l_1 + (l + l_2)\cdot\sin\delta$. Using the in-plane compliance matrix of a generic straight-axis flexure hinge and combining it with Eqs. (5.115)–(5.117) produce the in-plane compliance matrix of the full chain of Figure 5.16a with respect to the reference frame centered at A as given in Eq. (5.105) and with the following components:

$$
\begin{cases}
C^{(1)}_{u_x-f_x} = 2\cos^2\delta \cdot C_{u_x-f_x} + 2\sin^2\delta \cdot C_{u_y-f_y} - 2\left[2l_1 + (l + l_2)\cdot\sin\delta\right]\cdot\sin\delta \cdot C_{u_y-m_z} \\
\qquad + \left[2l_1^2 + 2l_1\cdot(l + l_2)\cdot\sin\delta + (l + l_2)^2\cdot\sin^2\delta\right]\cdot C_{\theta_z-m_z}; \\
C^{(1)}_{u_y-f_y} = 2\sin^2\delta \cdot C_{u_x-f_x} + 2\cos^2\delta \cdot C_{u_y-f_y} - 2(l + l_2)\cdot\cos^2\delta \cdot C_{u_y-m_z} \\
\qquad + (l + l_2)^2\cdot\cos^2\delta \cdot C_{\theta_z-m_z}; \\
C^{(1)}_{u_y-m_z} - 2\cdot\cos\delta \cdot C_{u_y-m_z} + (l + l_2)\cdot\cos\delta \cdot C_{\theta_z-m_z}; \\
C^{(1)}_{\theta_z-m_z} = 2\cdot C_{\theta_z-m_z}.
\end{cases} \tag{5.118}
$$

These compliances can now be substituted in Eq. (5.107) to obtain the components of the in-plane compliance matrix of the full chain with respect to the end point O.

The out-of-plane compliance matrix of the serial half-chain AO of Figure 5.16b with respect to the reference frame $Ax_A y_A z_A$ is calculated as:

$$
[C_{A,op}] = [Tr_{AB,op}]^T[C_{op}][Tr_{AB,op}] + [Tr_{AD,op}]^T[C_{op}][Tr_{AD,op}], \tag{5.119}
$$

where $[C_{op}]$ is the hinge out-of-plane compliance matrix. The transformation matrices of Eq. (5.119) are:

$$
[Tr_{AB,op}] = [R_{AB}][T_{AB,op}]; \quad [Tr_{AD,op}] = [R_{AD,ip}][T_{AD,op}]. \tag{5.120}
$$

The rotation matrices of Eq. (5.120) are evaluated in Eq. (5.117), whereas the out-of-plane translation matrices are:

$$
\left[T_{AB,op}\right] = \begin{bmatrix} 1 & 0 & -\Delta y_{AB} \\ 0 & 1 & \Delta x_{AB} \\ 0 & 0 & 1 \end{bmatrix}; \quad \left[T_{AD,op}\right] = \begin{bmatrix} 1 & 0 & -\Delta y_{AD} \\ 0 & 1 & \Delta x_{AD} \\ 0 & 0 & 1 \end{bmatrix}. \quad (5.121)
$$

Equations (5.119)–(5.121) are combined to obtain the following components of the out-of-plane compliance matrix of Eq. (5.109):

$$
\begin{cases}
C^{(1)}_{\theta_x - m_x} = \left[1 + \cos(2\delta)\right] \cdot C_{\theta_x - m_x} + \left[1 - \cos(2\delta)\right] \cdot C_{\theta_y - m_y}; \\
C^{(1)}_{\theta_y - m_y} = \left[1 - \cos(2\delta)\right] \cdot C_{\theta_x - m_x} + \left[1 + \cos(2\delta)\right] \cdot C_{\theta_y - m_y}; \\
C^{(1)}_{\theta_y - f_z} = l_1 \cdot \sin(2\delta) \cdot C_{\theta_x - m_x} - (l + l_2 + 2l_1 \cdot \sin\delta) \cdot \cos\delta \cdot C_{\theta_y - m_y} - 2\cos\delta \cdot C_{\theta_y - f_z}; \\
C^{(1)}_{u_z - f_z} = l_1^2 \cdot \left[1 + \cos(2\delta)\right] C_{\theta_x - m_x} \\
\qquad\qquad + \left[(l + l_2)^2 + (1 - \cos(2\delta)) \cdot l_1^2 + 2l_1 \cdot (l + l_2) \cdot \sin\delta \right] \cdot C_{\theta_y - m_y} \\
\qquad\qquad + 2(l + l_2 + 2l_1 \cdot \sin\delta) \cdot C_{\theta_y - f_z} + 2 \cdot C_{u_z - f_z}.
\end{cases}
$$

$$(5.122)$$

They can be utilized in Eq. (5.114) to express the out-of-plane compliance matrix in the *Oxyz* reference frame.

5.1.2.2 Overconstrained Chains

Planar serial flexible-hinge mechanisms may be overconstrained when the external supports result in a number of reactions that are in excess of the number of static equilibrium equations. In those situations, it is necessary to formulate additional load/displacement equations accounting for zero displacements in order to solve for the extra reaction loads. With the additional reactions determined, these chains are equivalently transformed into statically determinate (properly constrained) serial chains acted upon by multiple loads (the external loads and the reaction loads), and can be modeled by the procedure presented in the previous sections. For the sake of simplicity, it is considered that the supplemental constraints are applied at the free end, while the opposite end remains fixed.

Several external loads $\left[f_{Fj}\right]$, $j = 1, 2, \ldots, m$ act on the generic planar serial chain sketched in Figure 5.17. Without reducing the generality of the discussion, assume that the point of interest, P (where displacements need to be evaluated), is in front of all the m applied loads. Also assume that the previously free end point O is constrained, and therefore, there is a reaction load $[f_O]$ at that point, which needs to be determined before being able to calculate the displacements at P. For each unknown reaction component, there is a zero displacement at point O

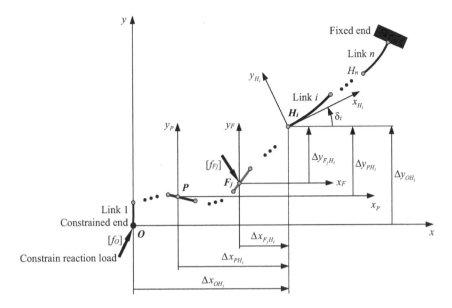

FIGURE 5.17 Serial chain with multiple flexible hinges and rigid links, multiple loads, with one end constrained and the other end fixed.

and these zero-displacement equations are utilized to evaluate the corresponding reaction components at O in terms of the externally applied multiple loads. The displacements at P are subsequently calculated by superimposing the deformations of each individual external load over the flexible hinges that are situated past their application points to the deformations resulting from applying the reaction load at O on the flexible hinges that are situated between (and past) point P and the fixed end of the chain.

The in-plane and out-of-plane quasi-static response of the mechanism shown in Figure 5.17 is analyzed next considering that point O is either fixed or guided.

5.1.2.2.1 In-Plane Model

The loads and zero boundary conditions pertaining to a fixed end O are depicted in Figure 5.18a, while the reaction loads and zero boundary conditions for a guided end at O are shown in Figure 5.18b when the serial chain of Figure 5.17 is acted upon by in-plane loads only.

Fixed end at O:

The displacements at O result by superimposing the displacement of the reaction load vector $\left[f_{O,ip} \right]$ at O to that produced by all the loads through all the flexible hinges that are located between the load application point and the chain's fixed end, namely:

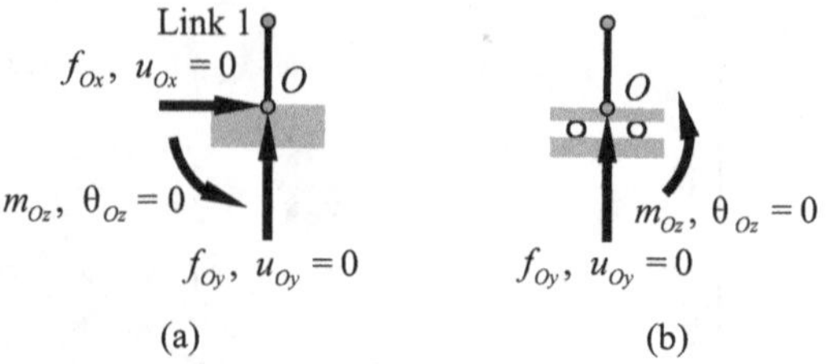

FIGURE 5.18 Constrained end O of serial chain acted upon by in-plane loads: (a) fixed support; (b) guided support.

$$\left[u_{O,ip} \right] = \left(\sum_{i=1}^{n} \left[C_{OH_i,ip} \right] \right) \left[f_{O,ip} \right] + \sum_{j=1}^{m} \left(\left(\sum_{i=f_j}^{n} \left[C_{OF_jH_i,ip} \right] \right) \left[f_{F_j,ip} \right] \right)$$

$$= \left[C_{O,ip} \right] \left[f_{O,ip} \right] + \sum_{j=1}^{m} \left(\left[C_{OF_j,ip} \right] \left[f_{F_j,ip} \right] \right), \tag{5.123}$$

The individual compliance matrices of Eq. (5.123) pertaining to the known external loads are:

$$\left[C_{OF_jH_i,ip} \right] = \left[Tr_{OH_i,ip} \right]^T \left[C_{i,ip} \right] \left[Tr_{F_jH_i,ip} \right] = \left[T_{OH_i,ip} \right]^T \left[R_{OH_i} \right]^T \left[C_{i,ip} \right] \left[R_{F_jH_i} \right] \left[T_{F_jH_i,ip} \right]$$

$$= \left[T_{OH_i,ip} \right]^T \left[R_{F_jH_i} \right]^T \left[C_{i,ip} \right] \left[R_{F_jH_i} \right] \left[T_{F_jH_i,ip} \right]. \tag{5.124}$$

where the translation and rotation matrices are defined in Eqs. (5.49) and (5.86). The compliance matrices related to the load at O are:

$$\left[C_{OH_i,ip} \right] = \left[Tr_{OH_i,ip} \right]^T \left[C_{i,ip} \right] \left[Tr_{OH_i,ip} \right] = \left[T_{OH_i,ip} \right]^T \left[R_{OH_i} \right]^T \left[C_{i,ip} \right] \left[R_{OH_i} \right] \left[T_{OH_i,ip} \right]. \tag{5.125}$$

Considering that $\left[u_{O,ip} \right] = \left[\begin{array}{ccc} u_{O_x} = 0 & u_{O_y} = 0 & \theta_{O_z} = 0 \end{array} \right]^T$, Eq. (5.123) allows solving for the reaction load at O as:

$$\left[f_{O,ip} \right] = -\left[C_{O,ip} \right]^{-1} \sum_{j=1}^{m} \left(\left[C_{OF_j,ip} \right] \left[f_{F_j,ip} \right] \right). \tag{5.126}$$

The in-plane displacement vector at P is calculated as:

$$\left[u_{P,ip}\right]=\left(\sum_{i=f_p}^{n}\left[C_{POH_i,ip}\right]\right)\left[f_{O,ip}\right]+\sum_{j=1}^{m}\left(\left(\sum_{i=f_j}^{n}\left[C_{PF_jH_i,ip}\right]\right)\left[f_{F_j,ip}\right]\right)$$

$$=\left[C_{P,ip}\right]\left[f_{O,ip}\right]+\sum_{j=1}^{m}\left(\left[C_{PF_j,ip}\right]\left[f_{F_j,ip}\right]\right)$$

$$=-\left[C_{P,ip}\right]\left[\overset{\bullet}{C}_{O,ip}\right]^{-1}\sum_{j=1}^{m}\left(\left[C_{OF_j,ip}\right]\left[f_{F_j,ip}\right]\right)+\sum_{j=1}^{m}\left(\left[C_{PF_j,ip}\right]\left[f_{F_j,ip}\right]\right). \quad (5.127)$$

The $\left[C_{P,ip}\right]$ matrix considers there are f_P flexible hinges past point P.

Example 5.7

The suspension mechanism shown in skeleton representation in Figure 5.19 is formed of rigid and flexible links that are all of straight axis and collinear. An in-plane load vector $[f_E] = [f_{Ex}\ f_{Ey}\ 0]^T$ acts at the midpoint E of the symmetric mechanism. Evaluate the in-plane displacements at E. Consider the identical flexible hinges have a constant rectangular cross-section with in-plane thickness $t = 0.001$ m, out-of-plane width $w = 0.008$ m, lengths $l_{h1} = 0.02$ m and $l_{h2} = 0.015$ m, and Young's modulus $E = 0.8 \cdot 10^{11}$ N/m². Known are also $l_1 = 0.04$ m, $l_2 = 0.03$ m, $f_{Ex} = 10$ N and $f_{Ey} = -200$ N.

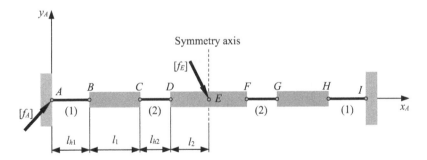

FIGURE 5.19 Fixed-fixed serial flexible chain with collinear links and acted upon by an in-plane load.

Solution:

The in-plane displacements at A are expressed in terms of the reaction load at A and the external load at E in short form as:

$$\left[u_{A,ip}\right]=\left[C_{AA,ip}\right]\left[f_{A,ip}\right]+\left[C_{AE,ip}\right]\left[f_{E,ip}\right], \quad (5.128)$$

where the compliance matrices are:

$$\begin{cases} \left[C_{AA,ip}\right] = \left[C_{AB,ip}\right] + \left[C_{AC,ip}\right] + \left[C_{AF,ip}\right] + \left[C_{AH,ip}\right] \\ \left[C_{AE,ip}\right] = \left[C_{AEF,ip}\right] + \left[C_{AEH,ip}\right] \end{cases} . \qquad (5.129)$$

The compliance matrices in the right-hand side of Eq. (5.129) are determined based on the in-plane compliances of the two different flexible hinges denoted by (1) and (2) in Figure 5.19, namely:

$$\begin{cases} \left[C_{AB,ip}\right] = \left[C_{ip}^{(1)}\right]; \\ \left[C_{AC,ip}\right] = \left[T_{AC,ip}\right]^{T}\left[C_{ip}^{(2)}\right]\left[T_{AC,ip}\right]; \\ \left[C_{AF,ip}\right] = \left[T_{AF,ip}\right]^{T}\left[C_{ip}^{(2)}\right]\left[T_{AF,ip}\right]; \\ \left[C_{AH,ip}\right] = \left[T_{AH,ip}\right]^{T}\left[C_{ip}^{(1)}\right]\left[T_{AH,ip}\right]; \\ \left[C_{AEF,ip}\right] = \left[T_{AF,ip}\right]^{T}\left[C_{ip}^{(2)}\right]\left[T_{EF,ip}\right]; \\ \left[C_{AEH,ip}\right] = \left[T_{AH,ip}\right]^{T}\left[C_{ip}^{(1)}\right]\left[T_{EH,ip}\right]; \end{cases} \qquad (5.130)$$

with the translation matrices being generically defined as:

$$\left[T_{ij,ip}\right] = \begin{bmatrix} 1 & 0 & 0 \\ 0 & 1 & 0 \\ 0 & -\Delta x_{ij} & 1 \end{bmatrix}, \qquad (5.131)$$

where $i = A$ or E and $j = C$, F or H. The x-axis offsets of Eq. (5.131) are: $\Delta x_{AC} = l_{h1} + l_1$, $\quad \Delta x_{AF} = l_{h1} + l_{h2} + l_1 + 2 \cdot l_2$, $\quad \Delta x_{AH} = l_{h1} + 2 \cdot l_{h2} + 2 \cdot l_1 + 2 \cdot l_2$ and $\Delta x_{EF} = l_2$, $\Delta x_{EH} = l_{h2} + l_1 + l_2$. Because the in-plane displacements at A are zero, the reaction load at A is found from Eq. (5.128) as:

$$\left[f_{A,ip}\right] = -\left[C_{AA,ip}\right]^{-1}\left[C_{AE,ip}\right]\left[f_{E,ip}\right]. \qquad (5.132)$$

The in-plane displacement at E is:

$$\left[u_{E,ip}\right] = \left[C_{EA,ip}\right]\left[f_{A,ip}\right] + \left[C_{EE,ip}\right]\left[f_{E,ip}\right], \qquad (5.133)$$

with

$$\begin{cases} \left[C_{EA,ip} \right] = \left[C_{EAF,ip} \right] + \left[C_{EAH,ip} \right] \\ \left[C_{EE,ip} \right] = \left[C_{EF,ip} \right] + \left[C_{EH,ip} \right]. \end{cases} \tag{5.134}$$

The in-plane compliance matrices in the right-hand sides of Eq. (5.134) are:

$$\begin{cases} \left[C_{EAF,ip} \right] = \left[T_{EF,ip} \right]^{T} \left[C_{ip}^{(2)} \right] \left[T_{AF,ip} \right] \\ \left[C_{EAH,ip} \right] = \left[T_{EH,ip} \right]^{T} \left[C_{ip}^{(1)} \right] \left[T_{AH,ip} \right] \\ \left[C_{EF,ip} \right] = \left[T_{EF,ip} \right]^{T} \left[C_{ip}^{(2)} \right] \left[T_{EF,ip} \right] \\ \left[C_{EH,ip} \right] = \left[T_{EH,ip} \right]^{T} \left[C_{ip}^{(1)} \right] \left[T_{EH,ip} \right] \end{cases} \tag{5.135}$$

where the translation matrices are defined generically in Eq. (5.131) with $i = E$ or A and $j = F$ or H. Taking into account that the reaction load at A is expressed in terms of the external load at E in Eq. (5.132), the displacement of Eq. (5.133) becomes:

$$\left[u_{E,ip} \right] = \left[C_{E,ip} \right] \left[f_{E,ip} \right] \quad \text{with} \quad \left[C_{E,ip} \right] = -\left[C_{EA,ip} \right] \left[C_{AA,ip} \right]^{-1} \left[C_{AE,ip} \right] + \left[C_{EE,ip} \right]. \tag{5.136}$$

The flexible-hinge compliances were calculated with the aid of Eqs. (2.26). The numerical results are $u_{Ex} = 3.125 \cdot 10^{-7}$ m, $u_{Ey} = 0.038$ m and $\theta_{Ez} = -1.124$ rad.

Guided end at O:

When the end O is guided instead of fixed, the in-plane displacement vector at O has its x component different from zero due to the unconstrained motion allowed along x by the guide, which means: $\left[u_{O,ip} \right] = \left[\begin{matrix} u_{Ox} \neq 0 & u_{Oy} = 0 & \theta_{Oz} = 0 \end{matrix} \right]^{T}$.

As a consequence, and as indicated in Figure 5.18b, there are only two nonzero reaction loads, f_{Oy} and m_{Oz}. Equation (5.123) enables expressing the reaction loads f_{Oy} and m_{Oz}, as well as the displacement u_{Ox} in terms of the externally applied loads. Following a procedure similar to the one used for the fixed boundary conditions at O, the displacements at P can now be determined by means of equations similar to Eqs. (5.123) through (5.127).

Example 5.8

The suspension mechanism of Figure 5.19 is acted upon by a force $f_E = 200$ N that is aligned with the symmetry axis. Evaluate the resulting displacement at E for the same geometric and material parameters of *Example 5.7*.

Solution:

Because of both geometry and load symmetry, one can study the elastic response of half the mechanism of Figure 5.19, as illustrated in Figure 5.20.

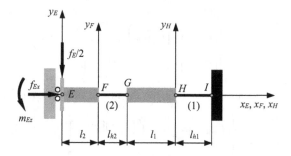

FIGURE 5.20 Fixed-guided serial flexible chain with collinear links and acted upon by a force along the guiding axis.

The horizontal displacement and rotation at the guided end E are zero, which are $u_{Ex} = 0$ and $\theta_{Ez} = 0$. The following relationship connects the loads at E ($f_E/2$ – the external load – and reactions – f_{Ex} and m_{Ez}):

$$
\begin{bmatrix} 0 \\ u_{Ey} \\ 0 \end{bmatrix} = \left[C_{E,ip}^{(h)} \right] \begin{bmatrix} f_{Ex} \\ -f_E/2 \\ m_{Ez} \end{bmatrix} = \begin{bmatrix} C_{E,u_x-f_x}^{(h)} & 0 & 0 \\ 0 & C_{E,u_y-f_y}^{(h)} & C_{E,u_y-m_z}^{(h)} \\ 0 & C_{E,u_y-m_z}^{(h)} & C_{E,\theta_z-m_z}^{(h)} \end{bmatrix} \begin{bmatrix} f_{Ex} \\ -f_E/2 \\ m_{Ez} \end{bmatrix},
$$

(5.137)

where the compliance matrix $\left[C_{E,ip}^{(h)} \right]$ is calculated for the half-segment EI in the reference frame located at E, as shown in Figure 5.20. This matrix combines the compliance matrices of the two flexible hinges, namely:

$$
\left[C_{E,ip}^{(h)} \right] = \left[C_{EF,ip}^{(2)} \right] + \left[C_{EH,ip}^{(1)} \right] = \left[T_{EF,ip}^{(2)} \right]^T \left[C_{ip}^{(2)} \right] \left[T_{EF,ip}^{(2)} \right] + \left[T_{EH,ip}^{(1)} \right]^T \left[C_{ip}^{(1)} \right] \left[T_{EH,ip}^{(1)} \right], (5.138)
$$

with

$$
\left[T_{EF,ip}^{(2)} \right] = \begin{bmatrix} 1 & 0 & 0 \\ 0 & 1 & 0 \\ 0 & -l_2 & 1 \end{bmatrix}; \quad \left[T_{EH,ip}^{(1)} \right] = \begin{bmatrix} 1 & 0 & 0 \\ 0 & 1 & 0 \\ 0 & -(l_1+l_2+l_{h2}) & 1 \end{bmatrix}.
$$

(5.139)

The first algebraic Eq. (5.137) results in the x-reaction at E being zero because: $C_{E,u_x-f_x}^{(h)} \cdot f_{Ex} = 0$. The other two algebraic Eqs. (5.137) allow solving for the other reaction m_{Ez} and for the displacement u_{Ey}:

$$
m_{Ez} = \frac{C_{E,u_y-m_z}^{(h)}}{C_{E,\theta_z-m_z}^{(h)}} \cdot \frac{f_E}{2}; \quad u_{Ey} = \left[\frac{\left(C_{E,u_y-m_z}^{(h)} \right)^2}{C_{E,\theta_z-m_z}^{(h)}} - C_{E,u_y-f_y}^{(h)} \right] \cdot \frac{f_E}{2}.
$$

(5.140)

The displacement of interest is $u_{Ey} = -0.046 \, \text{m}$.

5.1.2.2.2 Out-of-Plane Model

The reaction loads and zero boundary conditions pertaining to a fixed or guided end O are depicted in Figure 5.21a and b for out-of-plane external loads.

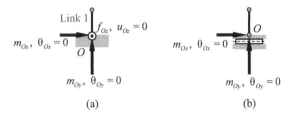

(a)   (b)

FIGURE 5.21 Constrained end O of serial chain acted upon by out-of-plane loads: (a) fixed support; (b) guided support.

Fixed end at O:

The out-of-plane displacements at O are produced by superimposing the deformation effects of the reaction load vector $\left[f_{O,op}\right]$ at O with those of the external loads through all the flexible hinges that are located between the load application point and the chain's fixed end, namely:

$$\left[u_{O,op}\right] = \left(\sum_{i=1}^{n}\left[C_{OH_i,op}\right]\right)\left[f_{O,op}\right] + \sum_{j=1}^{m}\left(\left(\sum_{i=f_j}^{n}\left[C_{OF_jH_i,op}\right]\right)\left[f_{F_j,op}\right]\right)$$

$$= \left[C_{O,op}\right]\left[f_{O,op}\right] + \sum_{j=1}^{m}\left(\left[C_{OF_j,op}\right]\left[f_{F_j,op}\right]\right). \tag{5.141}$$

The compliance matrices that are related to the external loads in Eq. (5.141) are:

$$\left[C_{OF_jH_i,op}\right] = \left[Tr_{OH_i,op}\right]^T\left[C_{i,op}\right]\left[Tr_{F_jH_i,op}\right]$$

$$= \left[T_{OH_i,op}\right]^T\left[R_{OH_i}\right]^T\left[C_{i,op}\right]\left[R_{F_jH_i}\right]\left[T_{F_jH_i,op}\right]$$

$$= \left[T_{OH_i,op}\right]^T\left[R_{F_jH_i}\right]^T\left[C_{i,op}\right]\left[R_{F_jH_i}\right]\left[T_{F_jH_i,op}\right]. \tag{5.142}$$

The rotation and translation matrices of Eq. (5.142) are defined in Eqs. (5.49), (5.55) and (5.89).

The compliance matrices connected to the reaction load at O are:

$$\left[C_{OH_i,op}\right] = \left[Tr_{OH_i,op}\right]^T\left[C_{i,op}\right]\left[Tr_{OH_i,op}\right] = \left[T_{OH_i,op}\right]^T\left[R_{OH_i}\right]^T\left[C_{i,op}\right]\left[R_{OH_i}\right]\left[T_{OH_i,op}\right]. \tag{5.143}$$

Because $\left[u_{O,op}\right]=\begin{bmatrix} \theta_{O_x}=0 & \theta_{O_y}=0 & u_{O_z}=0 \end{bmatrix}^T$. the reaction load is found from Eq. (5.141) as:

$$\left[f_{O,op}\right]=-\left[C_{O,op}\right]^{-1}\sum_{j=1}^{m}\left(\left[C_{OF_j,op}\right]\left[f_{F_j,op}\right]\right). \tag{5.144}$$

The out-of-plane displacements at P are determined as:

$$\left[u_{P,op}\right]=\left(\sum_{i=f_p}^{n}\left[C_{POH_i,op}\right]\right)\left[f_{O,op}\right]+\sum_{j=1}^{m}\left(\left(\sum_{i=f_j}^{n}\left[C_{PF_jH_i,op}\right]\right)\left[f_{F_j,op}\right]\right)$$

$$=\left[C_{P,op}\right]\left[f_{O,op}\right]+\sum_{j=1}^{m}\left(\left[C_{PF_j,op}\right]\left[f_{F_j,op}\right]\right)$$

$$=-\left[C_{P,op}\right]\left[C_{O,op}\right]^{-1}\sum_{j=1}^{m}\left(\left[C_{OF_j,op}\right]\left[f_{F_j,op}\right]\right)+\sum_{j=1}^{m}\left(\left[C_{PF_j,op}\right]\left[f_{F_j,op}\right]\right).$$

$$\tag{5.145}$$

The $\left[C_{P,op}\right]$ matrix considers there are f_p flexible hinges past point P.

Guided end at O:

For a guided end at O, as shown in Figure 5.21b, the z-axis translation is nonzero, while the other rotations are zero, namely: $\left[u_{O,op}\right]=\begin{bmatrix} \theta_{O_x}=0 & \theta_{O_y}=0 & u_{O_z}\neq 0 \end{bmatrix}^T$. Consequently, the nonzero reaction loads that correspond to the associated zero displacements are m_{O_x} and m_{O_y}. The calculation procedure, which is not detailed here, resembles the one performed for the fixed-end chain. The reaction loads m_{O_x} and m_{O_y}, as well as the displacement u_{O_z}, are first expressed from Eq. (5.141) in terms of the externally applied loads, and then, the out-of-plane displacements at P are evaluated by means of equations similar to Eqs. (5.142) through (5.145).

5.1.3 DISPLACEMENT-AMPLIFICATION MECHANISMS REDUCIBLE TO SERIAL FLEXIBLE CHAINS

Displacement-amplification devices with flexible hinges are utilized to augment the usually small input displacement levels, such as those produced by piezoelectric actuators. Some of these mechanisms can have two symmetry axes, but they can also possess a single symmetry axis. Figure 5.22 illustrates two hinge mechanisms, which possess two perpendicular symmetry axes; the actuation is applied along one symmetry axis (the horizontal direction in Figure 5.22), and the output displacement (also aligned with the external load direction) is produced along the

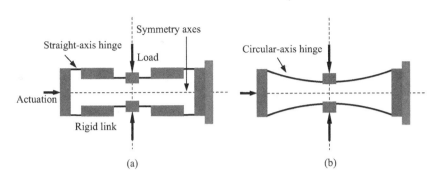

FIGURE 5.22 Displacement-amplification mechanisms with geometry and load symmetry, and: (a) straight-axis flexible hinges; (b) curvilinear-axis flexure hinges.

other symmetry axis (the vertical direction in the same Figure 5.22). As discussed in this section, flexible-hinge mechanisms (such as displacement-amplification devices) with geometry and load symmetry can be modeled by only studying a serial portion of the mechanism. Chapter 6 studies flexible-hinge displacement-amplification mechanisms that are analyzed as parallel mechanisms in the absence of symmetry; the same chapter also mentions publications that are related to this class of mechanisms.

5.1.3.1 Mechanisms with Two Symmetry Axes

A generic displacement-amplification mechanism is the one sketched in Figure 5.23a. The device possesses double symmetry and utilizes four identical chains that are formed of serially connected flexible hinges and rigid segments. The actuation (forces), as well as the external load (forces), act along the two symmetry lines; the assumption here is that the two components of the actuation are identical and opposite, and the same assumption is valid for the two external load components. Due to the relatively small inclination angle of the flexible chains with respect to the horizontal direction, the output displacement (which is along the vertical symmetry axis) is larger than the input displacement occurring along the horizontal symmetry axis, which lends the name of "displacement amplification" to this mechanism category.

Due to both geometry and load symmetry about two perpendicular axes, the quarter mechanism shown with load and boundary conditions in Figure 5.23b can be used instead of the original, full mechanism of Figure 5.23a. This simpler quarter mechanism is actually a serial chain formed of two end rigid links connected to the middle flexible chain. The end boundary conditions are guided, as symbolized in the same Figure 5.23b. Assume that known are the lengths of the two rigid links, l_1 and l_2, the distances between the two end points O and C (which are denoted by Δx_{CO} and Δy_{CO} in Figure 5.23b with $\Delta x_{CO} > \Delta y_{CO}$ in order to realize displacement amplification), as well as the actuation force f_a and the external load force f_l corresponding to the quarter mechanism. This mechanism

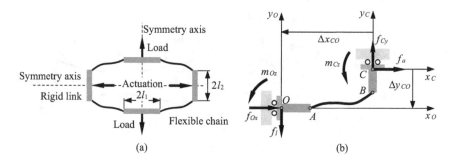

FIGURE 5.23 Double-symmetry displacement-amplification mechanism with four identical flexible chains: (a) skeleton representation; (b) quarter mechanism with boundary conditions.

is characterized by several performance qualifiers in terms of its regular, in-plane operation. One objective is to express the planar *displacement* (*geometric*) *amplification* (short-form notation is *d.a.*), which is defined as the ratio of the *y*-displacement at *O* to the *x*-displacement at *C*, namely:

$$d.a. = \frac{\text{output displacement}}{\text{input displacement}} = \frac{u_{Oy}}{u_{Cx}}. \tag{5.146}$$

It is of interest to also evaluate the minimum level of actuation f_a that can counteract an existing external load or, conversely, to determine the *block load*, which results in zero output displacement for a given actuation load; the block load is determined as:

$$\text{block load} = f_{l,\max} = f_l\big|_{u_{Oy}=0}. \tag{5.147}$$

Another qualifier is the *input stiffness* of the elastic frame in the absence of the external loading ($f_l = 0$). The full mechanism of Figure 5.23a can be designed by stacking two quarter-mechanism portions (like the ones to the right of the vertical symmetry axis in Figure 5.23a), which means combining them in series. The stiffness of the resulting (half) mechanism is half the quarter-mechanism stiffness. The full mechanism is obtained by connecting in parallel the two identical half-portions that are to the right and the left of the vertical symmetry axis. As a result, the stiffness of the full mechanism is twice the stiffness of the half-mechanism (which is half the stiffness of the quarter portion), with the net result that the stiffness of the full mechanism sketched in Figure 5.23a is the same as the stiffness of the quarter mechanism of Figure 5.23b in terms of the input (actuation) load. The input stiffness is therefore defined as:

$$k_{in} = \frac{\text{input force}}{\text{input displacement}} = \frac{f_a}{u_{Cx}}. \tag{5.148}$$

Similarly, one can define the *output stiffness* of this mechanism as being identical to the one of the quarter mechanism when only the external load is taken into consideration ($f_a = 0$) as:

$$k_{out} = \frac{\text{load force}}{\text{output displacement}} = \frac{f_l}{u_{Oy}}. \tag{5.149}$$

Eventually, the *cross-stiffness* can be evaluated by relating either the input force to the output displacement (when considering the external load force is zero) or by means of the output force in relation to the input displacement (under the assumption of zero actuation force), namely:

$$k_c = \frac{\text{input force}}{\text{output displacement}} = \frac{f_a}{u_{Oy}} = \frac{\text{load force}}{\text{input displacement}} = \frac{f_l}{u_{Cx}}. \tag{5.150}$$

It will be demonstrated that, actually, the cross-stiffness is the same when calculated by either of the two variants of Eq. (5.150).

As per Eq. (5.146), the displacement amplification is evaluated by means of the output and input displacements resulting from the action of the actuation load f_a and the external load f_l shown in Figure 5.23b. To evaluate the two unknown displacements, it is necessary to first determine the reaction loads set at the end points O and C. It should be noted that all subsequent displacements and loads are in the plane of the mechanism, and therefore, the subscript "*ip*" is eliminated to simplify notation. The relative displacements at the points O and C are expressed by means of the corresponding loads applied at either O or C and the compliance of the serial chain of Figure 5.23b as:

$$[u_O] - [u_C] = [C_O][f_O]; \quad [u_C] - [u_O] = [C_C][f_C], \tag{5.151}$$

which yields

$$[C_O][f_O] + [C_C][f_C] = [0]. \tag{5.152}$$

The first Eq. (5.151) assumes that point C is fixed, while the second Eq. (5.151) assumes that point O is fixed. The in-plane displacement and load vectors at O and C are:

$$[u_O] = \begin{bmatrix} u_{Ox} = 0 & u_{Oy} & \theta_{Oz} = 0 \end{bmatrix}^T; \quad [u_C] = \begin{bmatrix} u_{Cx} & u_{Cy} = 0 & \theta_{Cz} = 0 \end{bmatrix}^T$$

$$[f_O] = \begin{bmatrix} f_{Ox} & -f_l & m_{Oz} \end{bmatrix}^T; \quad [f_C] = \begin{bmatrix} f_a & f_{Cy} & m_{Cz} \end{bmatrix}^T, \tag{5.153}$$

with f_{Ox}, m_{Oz}, f_{Cy} and m_{Cz}. being the unknown reactions at O and C. At quasi-static equilibrium, the reaction forces result from the known actuation and external load forces, as seen in Figure 5.23b:

$$f_{Ox} = -f_a; \quad f_{Cy} = f_l. \tag{5.154}$$

Equations (5.152)–(5.154) are combined into:

$$[C_O]\begin{bmatrix} -f_a \\ -f_l \\ m_{Oz} \end{bmatrix} + [C_C]\begin{bmatrix} f_a \\ f_l \\ m_{Cz} \end{bmatrix} = \begin{bmatrix} 0 \\ 0 \\ 0 \end{bmatrix}. \tag{5.155}$$

The two in-plane compliance matrices of Eq. (5.155) define the serial chain connecting the points O and C; based on the offsets shown in Figure 5.23b and because the two reference frames associated with these points are not rotated, the two in-plane compliance matrices are related as:

$$[C_C] = [T_{CO}]^T [C_O][T_{CO}] \quad \text{with} \quad [T_{CO}] = \begin{bmatrix} 1 & 0 & 0 \\ 0 & 1 & 0 \\ \Delta y_{CO} & -\Delta x_{CO} & 1 \end{bmatrix}. \tag{5.156}$$

As a consequence, Eq. (5.155) can be reformulated considering Eq. (5.156) by means of only one compliance matrix, namely, $[C_O]$. Because $[C_O]$ is formed by adding the transformed compliance matrices of the serially connected flexible segments making up together the chain OC and because all those transformed matrices are symmetric (it is known that a matrix $[B]^T[C]\,[B]$ with $[C]$ symmetric is also symmetric), it follows that $[C_O]$ is symmetric and can be written as:

$$[C_O] = \begin{bmatrix} C_{O,u_x-f_x} & C_{O,u_x-f_y} & C_{O,u_x-m_z} \\ C_{O,u_x-f_y} & C_{O,u_y-f_y} & C_{O,u_y-m_z} \\ C_{O,u_x-m_z} & C_{O,u_y-m_z} & C_{O,\theta_z-m_z} \end{bmatrix}. \tag{5.157}$$

Combining Eqs. (5.155)–(5.157) yields a system of three algebraic equations with two unknowns, the two reaction moments at O and C. The reactions are solved from two of the resulting equations as:

$$\begin{cases} m_{Oz} = \dfrac{C_{O,u_x-m_z}}{C_{O,\theta_z-m_z}} \cdot f_a + \dfrac{C_{O,u_y-m_z}}{C_{O,\theta_z-m_z}} \cdot f_l \\[4mm] m_{Cz} = -\left(\dfrac{C_{O,u_x-m_z}}{C_{O,\theta_z-m_z}} + \Delta y_{CO}\right) \cdot f_a - \left(\dfrac{C_{O,u_y-m_z}}{C_{O,\theta_z-m_z}} - \Delta x_{CO}\right) \cdot f_l \end{cases}. \tag{5.158}$$

It can be shown that the third algebraic equation that was not utilized to find the unknowns is verified when using the two reaction moments of Eq. (5.158).

The first Eq. (5.151) is written in the form:

$$\begin{bmatrix} -u_{Cx} \\ u_{Oy} \\ 0 \end{bmatrix} = [C_O] \begin{bmatrix} -f_a \\ -f_l \\ m_{Oz} \end{bmatrix}, \tag{5.159}$$

and by means of the moment of Eq. (5.158), the two displacements of Eq. (5.159) are calculated as:

$$\begin{cases} u_{Cx} = a_{11} \cdot f_a + a_{12} \cdot f_l \\ u_{Oy} = -a_{12} \cdot f_a + a_{22} \cdot f_l \end{cases}, \tag{5.160}$$

with

$$\begin{cases} a_{11} = \dfrac{C_{O,u_x-f_x} \cdot C_{O,\theta_z-m_z} - \left(C_{O,u_x-m_z} \right)^2}{C_{O,\theta_z-m_z}}; \\[4mm] a_{12} = \dfrac{C_{O,u_x-f_y} \cdot C_{O,\theta_z-m_z} - C_{O,u_x-m_z} \cdot C_{O,u_y-m_z}}{C_{O,\theta_z-m_z}};. \\[4mm] a_{22} = \dfrac{\left(C_{O,u_y-m_z} \right)^2 - C_{O,u_y-f_y} \cdot C_{O,\theta_z-m_z}}{C_{O,\theta_z-m_z}} \end{cases} \tag{5.161}$$

Equations (5.160) are used now to express the performance qualifiers defined at the beginning of this section. Consider, for instance, that $f_l = 0$. From the two Eqs. (5.160), the displacement amplification defined in Eq. (5.146) becomes:

$$d.a. = \frac{|u_{Oy}|}{|u_{Cx}|} = \left| \frac{a_{12}}{a_{11}} \right|. \tag{5.162}$$

Similarly, the input stiffness and the cross-stiffness are determined from the two Eqs. (5.160) based on the definitions given in Eqs. (5.148) and (5.150) as:

$$k_{in} = \frac{f_a}{|u_{Cx}|} = \frac{1}{|a_{11}|}; \quad k_c = \frac{f_a}{|u_{Oy}|} = \frac{1}{|a_{12}|}. \tag{5.163}$$

The block load formulated in Eq. (5.147) is calculated from the second Eq. (5.160), namely:

$$f_{l,max} = \frac{a_{12}}{a_{22}} \cdot f_a. \tag{5.164}$$

Eventually, when considering that $f_a = 0$, the output stiffness defined in Eq. (5.149) is found from the second Eq. (5.160) as:

$$k_{out} = \frac{f_l}{|u_{Oy}|} = \frac{1}{|a_{22}|}. \tag{5.165}$$

Inspecting the expressions of the displacement amplification, input stiffness, cross-stiffness and output stiffness, the following relationships can be formulated between them:

$$d.a. = \frac{k_{in}}{k_c}; \quad f_{l,max} = \frac{k_{out}}{k_c} \cdot f_a. \tag{5.166}$$

Example 5.9

Evaluate the displacement amplification, the input, output and cross-stiffnesses, and the block load (as a fraction of the actuation load) for the mechanisms whose quarter models are shown in Figure 5.24. The identical straight-axis flexible hinges have a constant rectangular cross-section with an in-plane thickness $t = 0.001$ m and an out-of-plane width $w = 0.007$ m; their length is $l_h = 0.02$ m, and Young's modulus is $E = 0.6 \cdot 10^{11}$ N/m². Known are also $l = 0.08$ m and $\delta = 8°$.

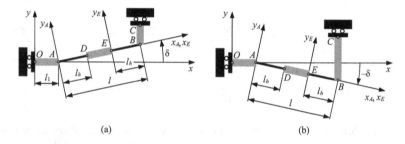

(a)   (b)

FIGURE 5.24 Quarter model of double-symmetry displacement-amplification mechanism with two identical and collinear straight-axis flexible hinges of (a) inward inclination; (b) outward inclination.

Solution:

The quarter models are formed of two straight-axis flexible hinges connected by a rigid segment with the three segments aligned and having a total length of l. Two other rigid segments are placed at the end of this three-segment portion. The compliance matrix of the serial segment AB recorded with respect to the reference frame $Ax_A y_A z_A$ is calculated as:

$$[C_A] = [C] + [T_{AE}]^T [C][T_{AE}], \tag{5.167}$$

where $[C]$ is the local-frame, in-plane compliance of any of the two hinges and $[T_{AE}]$ is the translation matrix, which realizes the transfer of the hinge EB compliance matrix from E to A; the two matrices are:

$$[C] = \begin{bmatrix} C_{u_x-f_x} & 0 & 0 \\ 0 & C_{u_y-f_y} & C_{u_y-m_z} \\ 0 & C_{u_y-m_z} & C_{\theta_z-m_z} \end{bmatrix}; \quad [T_{AE}] = \begin{bmatrix} 1 & 0 & 0 \\ 0 & 1 & 0 \\ 0 & -(l-l_f) & 1 \end{bmatrix}. \tag{5.168}$$

For the mechanism of Figure 5.24a, the compliance matrix of the segment AB is expressed in terms of the reference frame $Oxyz$ as:

$$[C_O] = [T_{OA}]^T [R]^T [C_A][R][T_{OA}]$$

$$\text{with} \quad [T_{OA}] = \begin{bmatrix} 1 & 0 & 0 \\ 0 & 1 & 0 \\ 0 & -l_1 & 1 \end{bmatrix}; \quad [R] = \begin{bmatrix} \cos\delta & \sin\delta & 0 \\ -\sin\delta & \cos\delta & 0 \\ 0 & 0 & 1 \end{bmatrix}. \tag{5.169}$$

Using the general definition of Eqs. (5.161), as well as Eqs. (5.167)–(5.169), the coefficients of interest are determined as:

$$\begin{cases} a_{11} = 2C_{u_x-f_x} \cdot \cos^2\delta + \left[2C_{u_y-f_y} - \dfrac{2\left(C_{u_y-m_z}\right)^2}{C_{\theta_z-m_z}} + \dfrac{1}{2} \cdot C_{\theta_z-m_z} \cdot \left(l-l_f\right)^2 \right] \cdot \sin^2\delta; \\[3mm] a_{12} = \left\{ \dfrac{2\left(C_{u_y-m_z}\right)^2}{C_{\theta_z-m_z}} - \left[2\left(C_{u_y-f_y} - C_{u_x-f_x}\right) + \dfrac{1}{2} \cdot C_{\theta_z-m_z} \cdot \left(l-l_f\right)^2 \right] \right\} \cdot \sin\delta \cdot \cos\delta; \\[3mm] a_{22} = -2C_{u_x-f_x} \cdot \sin^2\delta - \left[2C_{u_y-f_y} - \dfrac{2\left(C_{u_y-m_z}\right)^2}{C_{\theta_z-m_z}} + \dfrac{1}{2} \cdot C_{\theta_z-m_z} \cdot \left(l-l_f\right)^2 \right] \cdot \cos^2\delta \end{cases} \tag{5.170}$$

For the configuration with outward inclination of Figure 5.24b, the calculation and parameters are identical to the ones pertaining to the inward-inclination mechanism of Figure 5.24a except that the inclination angle is $-\delta$ instead of δ. The coefficients a_{11} and a_{22} are the ones of Eqs. (5.170), whereas the new coefficient a_{12} is the one of Eq. (5.170) with a minus (–) sign, namely:

$$a_{12} = \left[2\left(C_{u_y-f_y} - C_{u_x-f_x}\right) + \dfrac{1}{2} \cdot C_{\theta_z-m_z} \cdot \left(l-l_f\right)^2 - \dfrac{2\left(C_{u_y-m_z}\right)^2}{C_{\theta_z-m_z}} \right] \cdot \sin\delta \cdot \cos\delta. \tag{5.171}$$

The following results are obtained for both configurations: $d.a. = 7.083$, $k_{in} = 48{,}183.9\,\text{N/m}$, $k_c = 6{,}803.02\,\text{N/m}$, $k_{out} = 956.02\,\text{N/m}$ and $f_{l,max} = 0.14 \cdot f_a$.

Example 5.10

Solve *Example 5.9* for the mechanisms whose quarter models are displayed in Figure 5.25; each quarter uses a single circular-axis flexible hinge – see Figure 2.25. The hinge has a constant rectangular cross-section defined by in-plane thickness $t = 0.001\,\text{m}$ and out-of-plane width $w = 0.007\,\text{m}$. The median radius is $R = 0.04\,\text{m}$, the opening angle is $\alpha = 20°$, and the Young's modulus is $E = 0.6 \cdot 10^{11}\,\text{N/m}^2$. Ignore the axial (normal) deformations of the flexible hinges.

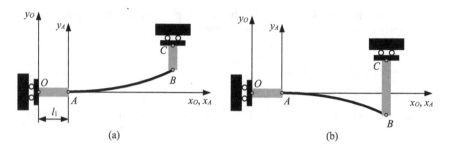

(a)   (b)

FIGURE 5.25 Quarter model of double-symmetry displacement-amplification device with one circular-axis flexible hinge of (a) internal curvature center; (b) external curvature center.

Solution:

Figure 5.25a sketches the quarter model of a mechanism with one circular-axis flexible hinge, whereas Figure 5.25b depicts a similar model with the flexure hinge direction reversed (the hinge is flipped with respect to the Ax_A axis). For the mechanism of Figure 5.25a, the in-plane compliance matrix $[C]$ of the segment AB is actually the compliance of the flexible hinge connecting those two points and is also the compliance matrix $[C_A]$:

$$[C_A]=[C]=\begin{bmatrix} C_{u_x-f_x} & C_{u_x-f_y} & C_{u_x-m_z} \\ C_{u_x-f_y} & C_{u_y-f_y} & C_{u_y-m_z} \\ C_{u_x-m_z} & C_{u_y-m_z} & C_{\theta_z-m_z} \end{bmatrix}, \tag{5.172}$$

where the bending-only compliances are given in Eqs. (A2.22) through (A2.26). Equation (5.169) is used with an identity rotation matrix $[R]$ to express $[C_O]$. As a consequence, the coefficients generically defined in Eq. (5.161) become:

$$a_{11}=C_{u_x-f_x}-\frac{\left(C_{u_x-m_z}\right)^2}{C_{\theta_z-m_z}}; \quad a_{12}=C_{u_x-f_y}-\frac{C_{u_x-m_z}\cdot C_{u_y-m_z}}{C_{\theta_z-m_z}}; \quad a_{22}=\frac{\left(C_{u_y-m_z}\right)^2-C_{u_y-f_y}\cdot C_{\theta_z-m_z}}{C_{\theta_z-m_z}}.$$
$$\tag{5.173}$$

For the outward-inclination configuration of Figure 5.25b, the in-plane compliance matrix of the curved-axis flexure hinge results from Eq. (5.172) considering that the two hinges are mirrored with respect to the Ax_A axis:

$$[C]=\begin{bmatrix} C_{u_x-f_x} & -C_{u_x-f_y} & -C_{u_x-m_z} \\ -C_{u_x-f_y} & C_{u_y-f_y} & C_{u_y-m_z} \\ -C_{u_x-m_z} & C_{u_y-m_z} & C_{\theta_z-m_z} \end{bmatrix}. \tag{5.174}$$

The coefficients a_{11} and a_{22} are the ones of Eq. (5.173), while a_{12} becomes:

$$a_{12}=\frac{C_{u_x-m_z}\cdot C_{u_y-m_z}}{C_{\theta_z-m_z}}-C_{u_x-f_y}, \tag{5.175}$$

which is a_{12} of Eq. (5.173) with a minus sign.

The numerical values of the performance qualifiers of the two mechanisms shown in Figure 5.25 are: $d.a. = 5.312$, $k_{in} = 4{,}832{,}010\,\text{N/m}$, $k_c = 909{,}612\,\text{N/m}$, $k_{out} = 160{,}053\,\text{N/m}$ and $f_{l,max} = 0.176 \cdot f_a$.

5.1.3.2 Mechanisms with One Symmetry Axis

A generic displacement-amplification mechanism possessing one symmetry axis is shown schematically in Figure 5.26a. It is assumed that the two actuation loads are equal, opposite and collinear as they are acting along a direction that is perpendicular to the mechanism symmetry axis. The geometry and load symmetry of this device result in a translatory motion of the rigid link of length $2l_1$ along the symmetry axis. Due to the inclination of the two flexible chains that connect on each side of the symmetry axis the fixed rigid link at the top to the mobile rigid link at the bottom, the mobile link displacement is larger than the input displacement along the actuation direction. The symmetry enables analyzing only half of the mechanism, as shown in Figure 5.26b, where the bottom link's vertical translation is generated by means of the guided support at its end. All definitions that have been introduced in the section analyzing the double-symmetry displacement-amplification mechanism are also valid for this mechanism. As a consequence, the relationships connecting the input and output displacements (which are u_{Cx} and u_{Oy} as per Figure 5.26b) to the actuation and external loads (identified as f_a and f_l in the same Figure 5.26b) need to be formulated.

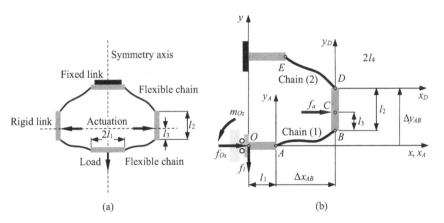

(a) (b)

FIGURE 5.26 Generic displacement-amplification flexible device with one symmetry axis: (a) skeleton representation; (b) half-mechanism.

The series mechanism of Figure 5.26b is formed of two flexible chains: chain (1), which is comprised between points A and B, and chain (2), which is located between points D and E. The two chains are connected by the rigid link BD. The serial chain of Figure 5.26b also comprises two end rigid links. The reactions at the guided end need to be determined first. The in-plane displacement vector at O is expressed in terms of the load vectors at O and C as:

$$[u_O] = \left(\left[C_O^{(1)} \right] + \left[C_O^{(2)} \right] \right)[f_O] + \left[C_{OC}^{(2)} \right][f_C] = \left[C_O^{(1,2)} \right][f_O] + \left[C_{OC}^{(2)} \right][f_C]$$

(5.176)

with: $\left[C_O^{(1,2)} \right] = \left[C_O^{(1)} \right] + \left[C_O^{(2)} \right],$

where

$$[u_O] = \begin{bmatrix} 0 & u_{Oy} & 0 \end{bmatrix}^T;$$

(5.177)

$$[f_O] = \begin{bmatrix} f_{Ox} & -f_l & m_{Oz} \end{bmatrix}^T; \quad [f_C] = \begin{bmatrix} f_a & 0 & 0 \end{bmatrix}^T.$$

The compliance matrix $\left[C_O^{(1,2)} \right]$, as shown in Eq. (5.176), adds the compliance matrices of the two flexible chains of Figure 5.26b, namely, $\left[C_O^{(1)} \right]$ and $\left[C_O^{(2)} \right]$, that are calculated with respect to O. $\left[C_{OC}^{(2)} \right]$ is the compliance matrix of chain 2 calculated at O and related to the load $[f_C]$, which is applied at C. The reactions at O are determined by rearranging the two scalar equations of Eqs. (5.176) and (5.177) that express zero displacements at O as follows:

$$\begin{bmatrix} f_{Ox} \\ m_{Oz} \end{bmatrix} = \begin{bmatrix} C_O^{(1,2)}(1,1) & C_O^{(1,2)}(1,3) \\ C_O^{(1,2)}(1,3) & C_O^{(1,2)}(3,3) \end{bmatrix}^{-1} \begin{bmatrix} -C_{OC}^{(2)}(1,1) & C_{OC}^{(2)}(1,2) \\ -C_{OC}^{(2)}(1,3) & C_{OC}^{(2)}(2,3) \end{bmatrix} \begin{bmatrix} f_a \\ f_l \end{bmatrix}$$

$$= \begin{bmatrix} a_{11} & a_{12} \\ a_{21} & a_{22} \end{bmatrix} \begin{bmatrix} f_a \\ f_l \end{bmatrix}.$$

(5.178)

The two reactions are substituted into the equation expressing the y displacement at O in Eqs. (5.176) and (5.177), which yields:

$$u_{Oy} = b_1 \cdot f_a + b_2 \cdot f_l$$

with

(5.179)

$$\begin{cases} b_1 = a_{11} \cdot C_O^{(1,2)}(1,2) + a_{21} \cdot C_O^{(1,2)}(2,3) + C_{OC}^{(2)}(1,2) \\ b_2 = a_{12} \cdot C_O^{(1,2)}(1,2) + a_{22} \cdot C_O^{(1,2)}(2,3) - C_O^{(1,2)}(2,2) \end{cases}$$

The in-plane displacement vector at C is calculated as:

$$[u_C] = \left[C_{CO}^{(2)} \right][f_O] + \left[C_C^{(2)} \right][f_C],$$

(5.180)

where

$$[u_C] = \begin{bmatrix} u_{Cx} & u_{Cy} & \theta_{Cz} \end{bmatrix}^T.$$

(5.181)

The two compliance matrices of Eq. (5.180) correspond to the flexible chain 2, which is related to the load vectors at O and at C and produces the displacement at point C. The input displacement is found by combining Eqs. (5.178), (5.180) and (5.181):

$$u_{Cx} = c_1 \cdot f_a + c_2 \cdot f_l$$

with （5.182)

$$\begin{cases} c_1 = a_{11} \cdot C_{CO}^{(2)}(1,1) + a_{21} \cdot C_{CO}^{(2)}(1,3) + C_C^{(2)}(1,1) \\ c_2 = a_{12} \cdot C_{CO}^{(2)}(1,1) + a_{22} \cdot C_{CO}^{(2)}(1,3) - C_{CO}^{(2)}(1,2) \end{cases}$$

Equations (5.179) and (5.182) are instrumental in evaluating the performance qualifiers that have been introduced in relation to the displacement-amplification mechanism with two symmetry axes. Thus, when considering that the load force is zero, $f_l = 0$, Eqs. (5.179) and (5.182) simplify to:

$$u_{Oy} = b_1 \cdot f_a; \quad u_{Cx} = c_1 \cdot f_a. \tag{5.183}$$

As a consequence, the displacement amplification $(d.a.)$, input stiffness (k_{in}) and cross-stiffness (k_c) are calculated as:

$$d.a. = \frac{|u_{Oy}|}{|u_{Cx}|} = \left|\frac{b_1}{c_1}\right|; \quad k_{in} = \frac{f_a}{|u_{Cx}|} = \frac{1}{|c_1|}; \quad k_c = \frac{f_a}{|u_{Oy}|} = \frac{1}{|b_1|}. \tag{5.184}$$

The block load is evaluated by annulling the output displacement in Eq. (5.179), namely:

$$f_{l,max} = -\frac{b_1}{b_2} \cdot f_a. \tag{5.185}$$

The output stiffness is calculated using $f_a = 0$ in Eq. (5.179):

$$k_{out} = \frac{f_l}{|u_{Oy}|} = \frac{1}{|b_2|}. \tag{5.186}$$

The cross-stiffness is evaluated from Eqs. (5.179) and (5.182) with $f_a = 0$ as:

$$k_c = \frac{f_l}{|u_{Cx}|} = \frac{1}{|c_2|}. \tag{5.187}$$

Comparing the two expressions of the cross-stiffness given in Eqs. (5.184) and (5.187), it follows that $b_1 = c_2$. This is indeed so because due to the reciprocal-displacement theorem – see Gere and Timoshenko [39] – the displacement produced by f_a along the y-direction at O (which is u_{Oy}) is equal to the displacement produced by f_l along the direction of f_a at C (which is u_{Cx}) when the two forces are equal, i.e., $f_a = f_l$.

Note that Eqs. (5.166), which allow relating the displacement amplification to the input and cross-stiffnesses and the block force to the output and cross-stiffnesses for a double-symmetry displacement-amplification mechanism, are also retrieved from Eqs. (5.184) through (5.187), (5.189) and (5.190).

Example 5.11

Study the in-plane performance of the displacement-amplification mechanism sketched in Figure 5.27, which possesses one symmetry axis in terms of both geometry and loading, by formulating its displacement amplification, input, output and cross-stiffnesses, and block load. Consider that all the straight-axis flexible hinges are identical. Calculate these performance qualifiers for: hinges of a constant rectangular cross-section with an in-plane thickness $t = 0.001$ m, an out-of-plane width $w = 0.008$ m, a length $l = 0.016$ m and a material Young's modulus $E = 2 \cdot 10^{11}$ N/m². Known are also the angle $\delta = 8°$ and the lengths $l_1 = 0.01$ m, $l_2 = 0.03$ m, $l_3 = 0.04$ m and $l_4 = 0.018$ m.

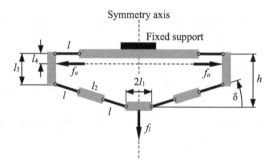

FIGURE 5.27 Skeleton representation of a displacement-amplification device with one symmetry axis for both geometry and load.

Solution:

The half-model of the mechanism depicted in Figure 5.27 is shown in Figure 5.28 with geometric parameters, boundary conditions and loads. The flexible chain (1) of Figure 5.28 comprises two identical hinges and the connecting rigid link (all links are collinear and inclined at an angle δ with respect to the global x_O axis), whereas the chain (2) of the same Figure 5.28 consists of the horizontal flexible hinge DE. Note that:

$$h = (2l + l_2) \cdot \sin\delta + l_3. \tag{5.188}$$

In order to calculate the two reactions at O and the output displacement at the same point, the following compliance matrices need to be first evaluated based on Figure 5.28:

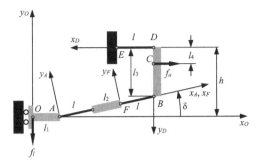

FIGURE 5.28 Half-model of displacement-amplification device with one symmetry axis.

$$\begin{cases} \left[C_O^{(1,2)}\right] = \left[C_O^{(1)}\right] + \left[C_O^{(2)}\right] = \left[T_{OA}\right]^T \left[C_A^{(1)}\right]\left[T_{OA}\right] + \left[T_{OD}\right]^T \left[C_D^{(2)}\right]\left[T_{OD}\right] \\ \left[C_{OC}^{(2)}\right] = \left[T_{OD}\right]^T \left[C_D^{(2)}\right]\left[T_{CD}\right], \end{cases}$$ (5.189)

where $\left[C_A^{(1)}\right]$ is the compliance of chain (1) and $\left[C_D^{(2)}\right]$ is the compliance of chain (2), both evaluated in their local frames located at A and D, respectively. The translation matrices of Eqs. (5.189) are:

$$[T_{OA}] = \begin{bmatrix} 1 & 0 & 0 \\ 0 & 1 & 0 \\ 0 & -l_1 & 1 \end{bmatrix}; \quad [T_{OD}] = \begin{bmatrix} 1 & 0 & 0 \\ 0 & 1 & 0 \\ \Delta y_{OD} & -\Delta x_{OD} & 1 \end{bmatrix};$$

$$[T_{CD}] = \begin{bmatrix} 1 & 0 & 0 \\ 0 & 1 & 0 \\ l_4 & 0 & 1 \end{bmatrix};$$ (5.190)

where $\Delta x_{OD} = (2 \cdot l + l_2) \cdot \cos\delta + l_1$ and $\Delta y_{OD} = h = (2 \cdot l + l_2) \cdot \sin\delta + l_3$. The compliance matrices $\left[C_A^{(1)}\right]$ and $\left[C_D^{(2)}\right]$ of Eq. (5.189) are calculated as:

$$\begin{cases} \left[C_A^{(1)}\right] = [R_A]^T \left([C] + [T_{AF}]^T [C][T_{AF}]\right)[R_A] \\ \left[C_D^{(2)}\right] = [R_D]^T [C][R_D] \end{cases},$$ (5.191)

with $[C]$ being the native-frame, in-plane compliance matrix of the flexible hinge. The translation and rotation matrices of Eq. (5.191) are:

$$[T_{AF}] = \begin{bmatrix} 1 & 0 & 0 \\ 0 & 1 & 0 \\ 0 & -\Delta x_{AF} & 1 \end{bmatrix} = \begin{bmatrix} 1 & 0 & 0 \\ 0 & 1 & 0 \\ 0 & -(l + l_2) & 1 \end{bmatrix};$$

$$[R_A] = \begin{bmatrix} \cos\delta & \sin\delta & 0 \\ -\sin\delta & \cos\delta & 0 \\ 0 & 0 & 1 \end{bmatrix};$$

$$[R_D] = \begin{bmatrix} \cos(\pi) & \sin(\pi) & 0 \\ -\sin(\pi) & \cos(\pi) & 0 \\ 0 & 0 & 1 \end{bmatrix} = \begin{bmatrix} 1 & 0 & 0 \\ 0 & 1 & 0 \\ 0 & 0 & 1 \end{bmatrix}. \quad (5.192)$$

The compliance matrices of Eq. (5.180) that enable expressing the output displacement vector at C in terms of the load vectors at O and at C are calculated as:

$$\left[C_{CO}^{(2)} \right] = [T_{OD}]^T \left[C_D^{(2)} \right][T_{CD}]; \quad \left[C_C^{(2)} \right] = [T_{CD}]^T \left[C_D^{(2)} \right][T_{CD}], \quad (5.193)$$

with the two translation matrices defined in Eqs. (5.190). The following numerical results are obtained: $d.a. = 6.738$, $k_{in} = 4{,}459.8\,\text{N/m}$, $k_c = 661.8\,\text{N/m}$, $k_{out} = 6{,}814.4\,\text{N/m}$ and $f_{l,max} = 1.326 \cdot f_a$.

5.2 SPATIAL (3D) MECHANISMS

This section is a brief introduction to the quasi-static response of serial mechanisms of spatial (three-dimensional, 3D) configuration. It includes three-dimensional rigid-body transformations (translation and rotation), the compliance matrix method for a generic, fixed-free spatial compliant chain and an example that studies a Cartesian serial robot.

5.2.1 RIGID-LINK LOAD AND DISPLACEMENT TRANSFORMATION

Similar to the planar case, loads and displacements can be transferred through translation and rotation in a three-dimensional rigid link. The load vector $[f_1] = [f_{x1}\, f_{y1}\, m_{z1}\, m_{x1}\, m_{y1}\, f_{z1}]^T$ of Figure 5.29 is transferred from point O_1 and the reference frame $O_1 x_1 y_1 z_1$ to point O_2 where it is referred to the $O_2 x_5 y_5 z_5$ frame as $[f_5] = [f_{x5}\, f_{y5}\, m_{z5}\, m_{x5}\, m_{y5}\, f_{z5}]^T$. As discussed shortly, this transfer can be achieved by means of one translation and three rotations.

Likewise, the displacement vector $[u_5] = [u_{x5}\, u_{y5}\, \theta_{z5}\, \theta_{x5}\, \theta_{y5}\, u_{z5}]^T$ is transferred back to point O_1 in the $O_1 x_1 y_1 z_1$ from O_2 as $[u_1] = [u_{x1}\, u_{y1}\, \theta_{z1}\, \theta_{x1}\, \theta_{y1}\, u_{z1}]^T$ by means of the reverse transformation sequence formed of the three rotations and one translation.

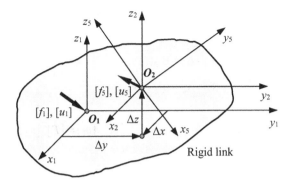

FIGURE 5.29 Displaced reference frames in a rigid link for load/displacement three-dimensional transformation.

5.2.1.1 Translation

The forces and moments applied at point O_1 in the $O_1x_1y_1z_1$ reference frame need to be transferred to point O_2 in the $O_2x_2y_2z_2$ reference frame whose axes are, respectively, parallel to the axes of the $O_1x_1y_1z_1$ reference frame. The following relationships can be formulated based on Figure 5.29, which identifies the offsets Δx, Δy and Δz, and should be used in conjunction with Figure 5.2 that shows the in-plane and out-of-plane forces/moments and displacements:

$$\begin{cases} f_{x2} = f_{x1} \\ f_{y2} = f_{y1} \\ m_{z2} = m_{z1} + f_{x1} \cdot \Delta y - f_{y1} \cdot \Delta x \\ m_{x2} = m_{x1} + f_{y1} \cdot \Delta z - f_{z1} \cdot \Delta y \\ m_{y2} = m_{y1} - f_{x1} \cdot \Delta z + f_{z1} \cdot \Delta x \\ f_{z2} = f_{z1} \end{cases} \quad \text{or} \quad [f_2] = [T][f_1], \tag{5.194}$$

where the load vectors $[f_1]$, $[f_2]$ and the translation matrix $[T]$ are:

$$[f_1] = \begin{Bmatrix} f_{x1} \\ f_{y1} \\ m_{z1} \\ m_{x1} \\ m_{y1} \\ f_{z1} \end{Bmatrix}; [f_2] = \begin{Bmatrix} f_{x2} \\ f_{x2} \\ m_{z2} \\ m_{x2} \\ m_{y2} \\ f_{z2} \end{Bmatrix}; [T] = \begin{bmatrix} 1 & 0 & 0 & 0 & 0 & 0 \\ 0 & 1 & 0 & 0 & 0 & 0 \\ \Delta y & -\Delta x & 1 & 0 & 0 & 0 \\ 0 & \Delta z & 0 & 1 & 0 & -\Delta y \\ -\Delta z & 0 & 0 & 0 & 1 & \Delta x \\ 0 & 0 & 0 & 0 & 0 & 1 \end{bmatrix}.$$

$$\tag{5.195}$$

The displacements recorded in the frame located at O_2 are translated similarly to the frame placed at O_1 as:

$$
\begin{cases}
u_{x1} = u_{x2} - \theta_{y2} \cdot \Delta z + \theta_{z2} \cdot \Delta y \\
u_{y1} = u_{y2} + \theta_{x2} \cdot \Delta z - \theta_{z2} \cdot \Delta x \\
\theta_{z1} = \theta_{z2} \\
\theta_{x1} = \theta_{x2} \\
\theta_{y1} = \theta_{y2} \\
u_{z1} = u_{z2} - \theta_{x2} \cdot \Delta y + \theta_{y2} \cdot \Delta x
\end{cases}
\quad \text{or} \quad [u_1] = [T]^T [u_2],
\qquad (5.196)
$$

where $[u_2] = [u_{x2}\, u_{y2}\, \theta_{z2}\, \theta_{x2}\, \theta_{y2}\, u_{z2}]^T$, and $[T]$ is expressed in Eq. (5.195).

5.2.1.2   Rotation

The frame $O_2 x_5 y_5 z_5$ has its axes rotated from the axes of the translated frame $O_2 x_2 y_2 z_2$. The position of the $O_2 x_5 y_5 z_5$ frame can be achieved by means of three rotations. The particular orientation of this frame is achieved using the following rotation sequence involving the Euler angles φ, θ and ψ:

- Rotation φ around the z_2 axis, which results in the $O_2 x_3 y_3 z_3$ frame;
- Rotation θ around the x_3 axis, which results in the $O_2 x_4 y_4 z_4$ frame;
- Rotation ψ around the z_4 axis, which results in the $O_2 x_5 y_5 z_5$ frame.

This three-angle rotation sequence is depicted in Figure 5.30.

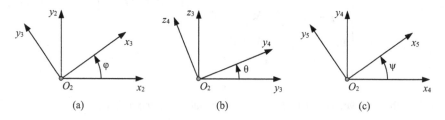

FIGURE 5.30   Spatial frame rotation realized by a sequence formed of the Euler angles: (a) rotation of angle φ around the z_2 axis; (b) rotation of angle θ around the x_3 axis; (c) rotation of angle ψ around the z_4 axis.

Let us start with these rotations using the reverse sequence; as such, the rotation around the z_4 axis, which is pictured in Figure 5.30c, produces the load vector $[f_5] = [f_{x5}\, f_{y5}\, m_{z5}\, m_{x5}\, m_{y5}\, f_{z5}]^T$, from the load vector $[f_4] = [f_{x4}\, f_{y4}\, m_{z4}\, m_{x4}\, m_{y4}\, f_{z4}]^T$, by means of a planar rotation of angle ψ. As discussed in Section 5.1.1.1, the two load vectors are connected by means of a rotation matrix $[R_{4-5}]$ as:

$$[f_5] = [R_{4-5}][f_4] \quad \text{with} \quad [R_{4-5}] = \begin{bmatrix} [R_\psi]_{3\times3} & [0]_{3\times3} \\ [0]_{3\times3} & [R_\psi]_{3\times3} \end{bmatrix};$$

$$[R_\psi]_{3\times3} = \begin{bmatrix} \cos\psi & \sin\psi & 0 \\ -\sin\psi & \cos\psi & 0 \\ 0 & 0 & 1 \end{bmatrix}. \tag{5.197}$$

The load vector $[f_4]$ is obtained from the $[f_3] = [f_{x3}\, f_{y3}\, m_{z3}\, m_{x3}\, m_{y3}\, f_{z3}]^T$ vector by a rotation θ around the x_3 axis, as illustrated in Figure 5.30b. In keeping with the separation between in-plane and out-of-plane components of this rotation, the following relationship exists between the components of $[f_4]$ and $[f_3]$:

$$\begin{Bmatrix} f_{y4} \\ f_{z4} \\ m_{x4} \\ m_{y4} \\ m_{z4} \\ f_{x4} \end{Bmatrix} = \begin{bmatrix} \cos\theta & \sin\theta & 0 & 0 & 0 & 0 \\ -\sin\theta & \cos\theta & 0 & 0 & 0 & 0 \\ 0 & 0 & 1 & 0 & 0 & 0 \\ 0 & 0 & 0 & \cos\theta & \sin\theta & 0 \\ 0 & 0 & 0 & -\sin\theta & \cos\theta & 0 \\ 0 & 0 & 0 & 0 & 0 & 1 \end{bmatrix} \begin{Bmatrix} f_{y3} \\ f_{z3} \\ m_{x3} \\ m_{y3} \\ m_{z3} \\ f_{x3} \end{Bmatrix}. \tag{5.198}$$

This relationship is rearranged to retrieve the component sequence provided in $[f_3]$ and $[f_4]$ as:

$$[f_4] = [R_{3-4}][f_3] \text{ with } [R_{3-4}] = \begin{bmatrix} 1 & 0 & 0 & 0 & 0 & 0 \\ 0 & \cos\theta & 0 & 0 & 0 & \sin\theta \\ 0 & 0 & \cos\theta & 0 & -\sin\theta & 0 \\ 0 & 0 & 0 & 1 & 0 & 0 \\ 0 & 0 & \sin\theta & 0 & \cos\theta & 0 \\ 0 & -\sin\theta & 0 & 0 & 0 & \cos\theta \end{bmatrix}. \tag{5.199}$$

Lastly, the rotation φ illustrated in Figure 5.30a generates $[f_3]$ from $[f_2] = [f_{x2}\, f_{y2}\, m_{z2}\, m_{x2}\, m_{y2}\, f_{z2}]^T$ as:

$$[f_3] = [R_{2-3}][f_2] \quad \text{with} \quad [R_{2-3}] = \begin{bmatrix} [R_\varphi]_{3\times3} & [0]_{3\times3} \\ [0]_{3\times3} & [R_\varphi]_{3\times3} \end{bmatrix};$$

$$[R_\varphi]_{3\times3} = \begin{bmatrix} \cos\varphi & \sin\varphi & 0 \\ -\sin\varphi & \cos\varphi & 0 \\ 0 & 0 & 1 \end{bmatrix}. \tag{5.200}$$

Combining now Eqs. (5.197), (5.199) and (5.200) yields:

$$[f_5] = [R][f_2] \quad \text{with} \quad [R] = [R_{4-5}][R_{3-4}][R_{2-3}], \tag{5.201}$$

where $[R]$ is the rotation matrix realizing the angular change from the $O_2x_2y_2z_2$ frame to the $O_2x_5y_5z_5$ frame.

The displacement vector $[u_5] = [u_{x5} \; u_{y5} \; \theta_{z5} \; \theta_{x5} \; \theta_{y5} \; u_{z5}]^T$ is transformed into the vector $[u_2] = [u_{x2} \; u_{y2} \; \theta_{z2} \; \theta_{x2} \; \theta_{y2} \; u_{z2}]^T$ by following a similar sequence of rotations as:

$$[u_2] = [R]^T [u_5], \tag{5.202}$$

with $[R]$ of Eq. (5.201). The demonstration of Eq. (5.202) is rather straightforward and is not included here.

Combining now Eqs. (5.194) and (5.201), as well as Eqs. (5.196) and (5.202) results in:

$$\begin{cases} [f_5] = [R][T][f_1] \\ [u_1] = [T]^T [R]^T [u_5] \quad \text{or} \quad [u_5] = \left([R]^T\right)^{-1} \left([T]^T\right)^{-1} [u_1]. \end{cases} \tag{5.203}$$

5.2.2 FIXED-FREE SERIAL CHAIN

Consider the 3D serial chain of Figure 5.31 comprising a rigid link and a flexible hinge, which is represented as a straight-axis member for the sake of simplicity. A load vector $[f_1]$ acts on the rigid link at F. This vector is transferred to the root H of the flexible hinges through a translation and a combined rotation and becomes the vector $[f_5]$. The 3D displacements at point P on the rigid link need to be evaluated. The relative position of F with respect to H is defined by the offsets Δx_{FH}, Δy_{FH} and Δz_{FH}; similarly, point P is positioned at Δx_{PH}, Δy_{PH} and Δz_{PH} from point H. Figure 5.31 identifies these six offsets by means of their resultant vectors $[\Delta r_{PH}] = [\Delta x_{PH}, \; \Delta y_{PH}, \; \Delta z_{PH}]^T$ and $[\Delta r_{FH}] = [\Delta x_{FH}, \; \Delta y_{FH}, \; \Delta z_{FH}]^T$. Similarly, only the relevant load and displacement vectors are indicated in the same Figure 5.31.

The displacement vector $[u_5] = [u_H]$ and the load vector $[f_5] = [f_H]$ are related by means of the compliance matrix $[C]$ of the flexible hinge as:

$$[u_H] = [C][f_H] \quad \text{with}$$

$$[C] = \begin{bmatrix} C_{u_x-f_x} & 0 & 0 & 0 & 0 & 0 \\ 0 & C_{u_y-f_y} & C_{u_y-m_z} & 0 & 0 & 0 \\ 0 & C_{u_y-m_z} & C_{\theta_z-m_z} & 0 & 0 & 0 \\ 0 & 0 & 0 & C_{\theta_x-m_x} & 0 & 0 \\ 0 & 0 & 0 & 0 & C_{\theta_y-m_y} & C_{\theta_y-f_z} \\ 0 & 0 & 0 & 0 & C_{\theta_y-f_z} & C_{u_z-f_z} \end{bmatrix}. \tag{5.204}$$

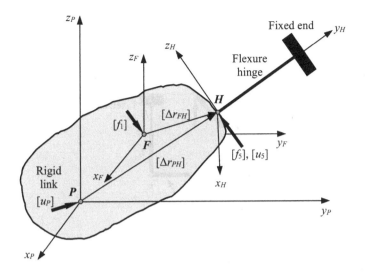

FIGURE 5.31 Schematic representation of a 3D serial chain with one rigid link, one flexible hinge, one point of spatial loading (F) and a point of displacements of interest (P).

The displacement and load vectors of Eqs. (5.203) are substituted into Eq. (5.204), which yields, after some algebraic manipulations, the displacement vector at P, $[u_1] = [u_P]$:

$$[u_1] = [C_{PFH}][f_1] \quad \text{with} \quad [C_{PFH}] = [T_{PH}]^T [R]^T [C][R][T_{FH}]. \quad (5.205)$$

The matrix $[C_{PFH}]$ of Eq. (5.205) is the global compliance matrix, which is evaluated in the frame $O_P x_P y_P z_P$ from the local-frame compliance matrix $[C]$ with the aid of the rotation matrix $[R]$ and the translation matrix $[T_{FH}]$, which transfers the load from F to H, and $[T_{PH}]$, which moves the displacements from H to P.

In a manner similar to the one utilized for planar serial compliant mechanisms, Eq. (5.205) can be applied to 3D mechanisms comprising multiple flexible hinges and loads.

Example 5.12

The Cartesian serial mechanism of Figure 5.32 is formed of three identical straight-axis flexible hinges of a constant circular cross-section (diameter is d) and several rigid connecting links. The spatial chain is fixed at one end and acted upon at the free end by three Cartesian forces. Calculate the three displacements at the free end knowing: $l = 0.01$ m, $d = 0.003$ m, $l_x = l_y = l_z = 0.05$ m, $f_x = 150$ N, $f_y = 100$ N and $f_z = 80$ N; the material properties of the flexible hinges are $E = 2 \cdot 10^{11}$ N/m^2 and $\mu = 0.3$.

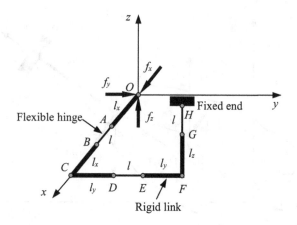

FIGURE 5.32   Spatial Cartesian serial flexible-hinge mechanism.

Solution:

The 3D leg's global compliance matrix is evaluated in the $Oxyz$ reference frame as:

$$\left[C^{(l)}\right]=\left[T_1\right]^T[C][T_1]+\left[T_2\right]^T\left[R_2\right]^T[C]\left[R_2\right]\left[T_2\right]+\left[T_3\right]^T\left[R_3\right]^T[C]\left[R_3\right]\left[T_3\right], \quad (5.206)$$

where "l" identifies the 3D *leg* and the hinge local-frame compliance matrix [C] is expressed in Eq. (5.204); the subscripts 1, 2 and 3 denote the three flexible hinges shown in Figure 5.32: hinge AB is #1, hinge DE is #2, and hinge GH is #3. The translation matrices of Eq. (5.206) are calculated with Eq. (5.195), while the rotation matrices are evaluated by means of Eqs. (5.197), (5.199)–(5.201). The following particular parameters are used to evaluate these matrices:

$$\begin{cases} \Delta x_1 = l_x; \quad \Delta y_1 = \Delta z_1 = 0; \\ \begin{cases} \Delta x_2 = l_x + l + l_x = l + 2l_x; \quad \Delta y_2 = l_y; \quad \Delta z_2 = 0; \\ \varphi_2 = 90°; \quad \theta_2 = \psi_2 = 0; \end{cases} \\ \begin{cases} \Delta x_3 = \Delta x_2; \quad \Delta y_3 = l_y + l + l_y = l + 2l_y; \quad \Delta z_3 = l_z; \\ \varphi_3 = 180°; \quad \theta_3 = \psi_3 = 90° \end{cases} \end{cases} \quad . \quad (5.207)$$

The load vector applied at the free end is $[f] = [f_x\, f_y\, 0\, 0\, 0\, f_z]^T$. As per Eq. (5.205), the translation displacement at the free end are: $u_x = 3.5 \cdot 10^{-7}\,\text{m}$, $u_y = 2.73 \cdot 10^{-3}\,\text{m}$ and $u_z = 3.3 \cdot 10^{-3}\,\text{m}$.

REFERENCES

1. Smith, S.T., *Flexures – Elements of Elastic Mechanisms*, Gordon and Breach Science Publishers, Amsterdam, 2000.
2. Xu, Q., *Design and Implementation of Large-Range Compliant Micropositioning Systems*, Wiley, London, 2017.
3. Henin, S. et al., *The Art of Flexure Mechanism Design*, EPFL Press, Lausanne, 2017.
4. Zhang, X. and Zhu, B., *Topology Optimization of Compliant Mechanisms*, Springer, Berlin, 2018.
5. Zentner, L. and Linβ, S., *Compliant Mechanisms: Mechanics of Elastically Deformable Mechanisms, Actuators and Sensors*, De Gruiter Oldenbourg, Berlin, 2019.
6. Boronkay, T.G. and Mei, C., Analysis and design of multiple input flexible link mechanisms, *Journal of Mechanisms*, 5 (1), 29, 1970.
7. Williams, D.W. and Turcic, D.A., An inverse kinematic analysis procedure for flexible open-loop mechanisms, *Mechanism and Machine Theory*, 27, 701, 1992.
8. Jouaneh, M. and Yang, R., Modeling of flexure-hinge type lever mechanisms, *Precision Engineering*, 27, 407, 2003.
9. Lobontiu, N., Compliance-based matrix method for modeling the quasi-static response of planar serial flexure-hinge mechanisms, *Precision Engineering*, 38, 639, 2014.
10. Du, H. et al., Topological optimization of mechanical amplifiers for piezoelectric actuators under dynamic motion, *Smart Materials and Structures*, 9, 788, 2000.
11. Kim, J.H., Kim, S.H., and Kwak, Y.K., Development of a piezoelectric actuator using a three-dimensional bridge-type hinge mechanism, *Review of Scientific Instruments*, 74 (5), 2918, 2003.
12. Quyang, P.R., Zhang, W.J., and Gupta, M.M., Design of a new compliant mechanical amplifier, *Proceedings of IDETC/CIE 2005, ASME 2005 International Design Engineering Technical Conferences & Computers and Information in Engineering Conference*, Long Beach, CA, September 24–28, 2005.
13. Choi, S.B., Han, S.S., and Lee, Y.S., Fine motion control of a moving stage using a piezoactuator associated with a displacement amplifier, *Smart Materials and Structures*, 14, 222, 2005.
14. Lobontiu, N. and Garcia, E., Static response of planar compliant devices with small-deformation flexure hinges, *Mechanics Based Design of Structures and Machines*, 32 (4), 459, 2004.
15. Lobontiu, N. and Garcia, E., Analytical model of displacement amplification and stiffness optimization for a class of flexure-based compliant mechanisms, *Computers & Structures*, 81, 2797, 2003.
16. Ma, H.-W. et al., Analysis of the displacement amplification ratio of bridge-type flexure hinge, *Sensors and Actuators A*, 132, 730, 2006.
17. Mottard, P. and St-Amant, Y., Analysis of flexural hinge orientation for amplified piezo-driven actuators, *Smart Materials and Structures*, 18, 1, 2009.
18. Scire, F.E. and Teague, E.C., Piezodriven 50-μm range stage with subnanometer resolution, *Review of Scientific Instruments*, 49, 1735, 1978.
19. Her, I. and Chang, J.C., A linear scheme for displacement analysis of micropositioning stages with flexure hinges, *ASME Journal of Mechanical Design*, 116, 770, 1994.
20. Park, S.R. and Yang, S.H., A mathematical approach for analyzing ultra precision positioning system with compliant mechanism, *Material Process Technology*, 164–165, 1584, 2005.
21. Mukhopadhyay, D. et al., A SOI-MEMS-based planar parallel-kinematics nanopositioning stage, *Sensors and Actuators A*, 147, 340, 2008.

22. Zettl, B., Szyszkowski, W., and Zhang, W.J., Accurate low DOF modeling of a planar compliant mechanism with flexure hinges: the equivalent beam methodology, *Precision Engineering*, 29, 237, 2005.
23. Furukawa, E. and Mizuno, M., Displacement amplification and reduction by means of a linkage, *Japan Society of Precision Engineering*, 24, 285, 1990.
24. Kim, J.H., Kim, S.H., and Kwak, Y.K., Development and optimization of 3-D bridge-type hinge mechanisms, *Sensors and Actuators A*, 116, 530, 2004.
25. Chang, S.H. and Du, B.C., A precision piezodriven micropositioner mechanism with large travel range, *Review of Scientific Instruments*, 69, 1785, 1998.
26. Tian, Y. et al., Development and dynamic modeling of a flexure-based Scott-Russell mechanism for nano-manipulation, *Mechanical System Signal Processing*, 23, 957, 2009.
27. Choi, S.B. et al., A magnification device for precision mechanisms featuring piezo-actuators and flexure hinges: design and experimental validation, *Mechanism and Machine Theory*, 42, 1184, 2007.
28. Choi, K.B., Lee, J.J., and Hata, S., A piezo-driven compliant stage with double mechanical amplification mechanisms arranged in parallel, *Sensors and Actuators A*, 161, 173, 2010.
29. Tian, Y. et al., Design and forward kinematics of the compliant micro-manipulator with lever mechanisms, *Precision Engineering*, 23, 466, 2009.
30. Wu, M. et al., Compliance analysis for flow effector actuator based on serial compliant mechanism with double guide ends, *IEEE Access*, 7, 28913, 2019.
31. Pham, H.H. and Chen, I.M., Stiffness modeling of flexure parallel mechanism, *Precision Engineering*, 29, 467, 2005.
32. Ryu, J.W. et al., Inverse kinematic modeling of a coupled flexure hinge mechanism, *IEEE ASME Transactions on Mechatronics*, 9, 657, 1999.
33. Yong, Y.K. and Lu, T.F., Kinetostatic modeling of 3-RRR compliant micro-motion stages with flexure hinges, *Mechanism and Machine Theory*, 44, 1156, 2009.
34. Ryu, J.W., Gweon, D.G., and Moon, K.S., Optimal design of a flexure hinge based XYθ wafer stage, *Precision Engineering*, 21, 18, 1997.
35. Hua, W. and Xianmin, Z., Input coupling analysis and optimal design of a 3-DOF compliant micro-positioning stage, *Mechanism and Machine Theory*, 43, 400, 2008.
36. Yue, Y. et al., Relationship among input-force, payload, stiffness and displacement of a 3-DOF perpendicular parallel micro-manipulator, *Mechanism and Machine Theory*, 45, 756, 2010.
37. Wang, J. et al., On the validity of compliance-based matrix method in output compliance modeling of flexure-hinge mechanism, *Precision Engineering*, 56, 485, 2019.
38. Lobontiu, N., Out-of-plane (diaphragm) compliances of circular-axis notch flexible hinges with midpoint radial symmetry, *Mechanics Based Design of Structures and Machines*, 42, 517, 2014.
39. Gere, J.M. and Timoshenko, S.P., *Mechanics of Materials*, 4th ed., PWS Publishing, Boston, MA, 1996.

6 Quasi-Static Response of Parallel Flexible-Hinge Mechanisms

Parallel mechanisms with conventional rotation/translation joints and their application in robotics and manipulation are a rich, mature field of research and engineering development – related information is provided by Merlet [1], Gosselin and Wang [2], Li [3], Khan et al. [4], Geike and McPhee [5], Liu et al. [6,7], Wu et al. [8], Gosselin and Angeles [9], Staicu [10], Huang and Thebert [11], Bonev et al. [12], Staicu [13], Wang and Mills [14], Briot and Bonev [15], Briot et al. [16], Merlet et al. [17], Yang and O'Brien [18], Choi [19], Yu et al. [20], Binaud et al. [21] and Ebrahimi et al. [22]. Parallel-connection flexible-hinge mechanisms have also been developed, particularly for precision-motion/positioning applications, as presented by Smith [23], Xu [24], Henin et al. [25], Zhang and Zhu [26], Zentner and Linβ [27], Su et al. [28], Li and Xu [29], Lu et al. [30], Yong and Lu [31], Tian etal. [32], Hua and Xianmin [33], Zettl et al. [34], Yi et al. [35], Pham and Chen [36], Oetomo et al. [37], Pham and Chen [38], Mukhopadyay et al. [39], Hesselbach et al. [40], Her and Chang [41], Ryu et al. [42,43], Awtar et al. [44], Huang and Schimmels [45], Hopkins and Culpepper [46], Hopkins et al. [47], Awtar et al. [48] – to cite just a few of the numerous publications in this area.

Using the compliance matrix modeling approach, this chapter concentrates on the quasi-static behavior of compliant mechanisms that are formed of flexible hinges or chains connected in parallel. While the focus of this chapter falls on planar (two-dimensional, 2D) flexible-hinge designs, spatial (three-dimensional, 3D) mechanisms are also discussed. Different loading, boundary conditions and symmetry configurations are thoroughly analyzed. Particular designs include straight-axis suspensions, folded-chain compliant devices, self-similar flexible mechanisms, trapeze and parallelogram mechanisms, displacement-amplification devices, planar stages with radial symmetry and grippers.

6.1 PLANAR (2D) MECHANISMS

This section utilizes the compliance matrix approach to connect loads to displacements for a variety of planar (2D) parallel flexible-hinge mechanisms that can formally be placed in two categories: in the first category, there is a single load acting on the rigid link, which is common to all flexible chains that are connected to it; the second category comprises mechanisms with multiple loads acting on both the common rigid link and some of the component flexible chains. A substructuring approach is also presented that enables to model, analyze and design the quasi-static response of flexible-hinge mechanisms of branched (complex) architecture.

6.1.1 Load on Rigid Link Common to All Flexible Chains

Consider m flexible chains (each formed of several flexible segments and rigid links coupled in series) that connect two adjacent rigid links to form a planar mechanism. A load vector, generically denoted by $[f]$, acts at a point O on the mobile rigid link, as illustrated in Figure 6.1a. The load vector can be partitioned into an in-plane component $[f_{ip}]$ and an out-of-plane sub-vector $[f_{op}]$.

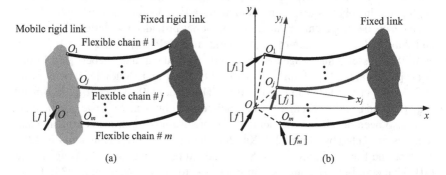

(a)   (b)

FIGURE 6.1 Planar mechanism with m flexible chains connected in parallel and acted upon by a single load applied to the common mobile rigid link: (a) schematic representation; (b) load distribution on individual chains.

Because the m flexible chains are connected in parallel, they all undergo the same motion/displacement at point O on the common rigid link, namely:

$$[u_O] = \left[u_O^{(1)}\right] = \cdots = \left[u_O^{(j)}\right] = \cdots = \left[u_O^{(m)}\right]. \tag{6.1}$$

As indicated in Figure 6.1b, the load $[f]$ is distributed to the m links due to static equilibrium such that the sum of all loads applied to the m points O_j equals the original load at O, namely:

$$[f] = [f_1] + \cdots + [f_j] + \cdots + [f_m]. \tag{6.2}$$

With respect to the same point O, the individual-chain force loads are connected to displacements by means of corresponding compliance matrices as:

$$\left[u_O^{(j)}\right] = \left[C_{OO_j}^{(j)}\right][f_j] \quad \text{or} \quad [f_j] = \left[C_{OO_j}^{(j)}\right]^{-1}\left[u_O^{(j)}\right] = \left[K_{OO_j}^{(j)}\right][u_O]; \quad j = 1,2,\ldots,m, \tag{6.3}$$

where $\left[C_{OO_j}^{(j)}\right]$ and $\left[K_{OO_j}^{(j)}\right]$ are the compliance and stiffness matrices corresponding to the chain j, both evaluated in the global frame $Oxyz$. Assume that the rigid links are connected by a single flexible chain, which is statically equivalent to the actual m parallel-connection chains. The load and displacement at O are related by means of the compliance matrix $[C_O]$ that characterizes the single, equivalent chain, namely:

$$[u_O] = [C_O][f] \quad \text{or} \quad [f] = [C_O]^{-1}[u_O] = [K_O][u_O],$$ (6.4)

with $[K_O]$ being the overall parallel-connection stiffness matrix. Combining now Eqs. (6.1) – (6.4) yields:

$$[C_O] = \left(\sum_{j=1}^{m} \left[C_{OO_j}^{(j)} \right]^{-1} \right)^{-1} \quad \text{or} \quad [K_O] = [C_O]^{-1} = \sum_{j=1}^{m} \left[C_{OO_j}^{(j)} \right]^{-1}.$$ (6.5)

Equation (6.5) can be expressed separately for in-plane and out-of-plane loads/displacements, thus resulting in corresponding compliance/stiffness matrices as:

$$\begin{cases} \left[C_{O,ip} \right] = \left(\sum_{j=1}^{m} \left[C_{OO_j,ip}^{(j)} \right]^{-1} \right)^{-1} ; \quad \left[K_{O,ip} \right] = \left[C_{O,ip} \right]^{-1} = \sum_{j=1}^{m} \left[C_{OO_j,ip}^{(j)} \right]^{-1} ; \\ \\ \left[C_{O,op} \right] = \left(\sum_{j=1}^{m} \left[C_{OO_j,op}^{(j)} \right]^{-1} \right)^{-1} ; \quad \left[K_{O,op} \right] = \left[C_{O,op} \right]^{-1} = \sum_{j=1}^{m} \left[C_{OO_j,op}^{(j)} \right]^{-1} \end{cases}$$. (6.6)

This section studies suspension mechanisms with straight-axis and folded flexible chains, axially or radially symmetric compliant mechanisms, self-similar flexible-hinge designs, as well as trapeze and parallelogram flexure-based configurations.

6.1.1.1 Suspension Mechanism with Two Identical, Symmetrical and Collinear, Straight-Axis Flexible Chains

Several parallel-connection planar mechanisms are composed of two identical, straight-axis and collinear flexible portions that are symmetric with respect to the connecting rigid link. The flexible chains can be formed of flexible hinges and rigid links or solely of flexible hinges.

The mechanism whose skeleton representation is shown in Figure 6.2 comprises a central rigid link connected to two identical and symmetrical flexible chains that are collinear. A load vector acts on the rigid link at point O, which is placed on the symmetry axis. This load may include actuation components, as well as external load components, which can be reduced to a single load vector at O. Assuming that the in-plane and out-of-plane compliance matrices of the two identical chains (with one end free – the end closest to the rigid link – and the other end fixed) are known with respect to the central $Oxyz$ reference frame, the in-plane and out-of-plane compliance matrices of the full suspension mechanism can be calculated with respect to the same reference frame.

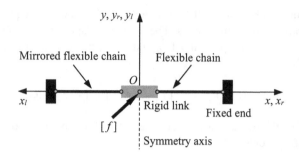

FIGURE 6.2 Parallel suspension mechanism formed of two identical, collinear and symmetric straight-axis flexible chains.

Consider that the in-plane compliance matrix $\left[C_{O,ip}^{(r)} \right]$ of the flexible segment to the right of the central rigid link (the *right* segment) is given/known. Because of symmetry with respect to the y axis, the in-plane compliance matrix $\left[C_{O,ip}^{(l)} \right]$ of the identical and mirrored left chain is also known. In the $Oxyz$ reference frame, these matrices are expressed as:

$$\left[C_{O,ip}^{(r)} \right] = \begin{bmatrix} C_{u_x-f_x}^{(r)} & 0 & 0 \\ 0 & C_{u_y-f_y}^{(r)} & C_{u_y-m_z}^{(r)} \\ 0 & C_{u_y-m_z}^{(r)} & C_{\theta_z-m_z}^{(r)} \end{bmatrix}; \quad \left[C_{O,ip}^{(l)} \right] = \begin{bmatrix} C_{u_x-f_x}^{(r)} & 0 & 0 \\ 0 & C_{u_y-f_y}^{(r)} & -C_{u_y-m_z}^{(r)} \\ 0 & -C_{u_y-m_z}^{(r)} & C_{\theta_z-m_z}^{(r)} \end{bmatrix}.$$

$$(6.7)$$

The particular form of $\left[C_{O,ip}^{(r)} \right]$ is derived in Eq. (3.35), while the minus signs in $\left[C_{O,ip}^{(l)} \right]$ indicate mirroring with respect to the y axis. The full-mechanism compliance matrix is calculated based on Eqs. (6.6) and (6.7) as:

$$\left[C_{O,ip} \right] = \left(\left[C_{O,ip}^{(r)} \right]^{-1} + \left[C_{O,ip}^{(l)} \right]^{-1} \right)^{-1} = \frac{1}{2} \cdot \begin{bmatrix} C_{u_x-f_x} & 0 & 0 \\ 0 & C_{u_y-f_y} & 0 \\ 0 & 0 & C_{\theta_z-m_z} \end{bmatrix}, \quad (6.8)$$

with

$$C_{u_x-f_x} = C_{u_x-f_x}^{(r)}; \quad C_{u_y-f_y} = C_{u_y-f_y}^{(r)} - \frac{\left(C_{u_y-m_z}^{(r)} \right)^2}{C_{\theta_z-m_z}^{(r)}}; \quad C_{\theta_z-m_z} = C_{\theta_z-m_z}^{(r)} - \frac{\left(C_{u_y-m_z}^{(r)} \right)^2}{C_{u_y-f_y}^{(r)}}. \quad (6.9)$$

A similar approach is used to express the out-of-plane compliance matrix of the full mechanism. The out-of-plane compliance matrices of the right and left flexible chains are expressed in the central frame $Oxyz$ as:

$$\left[C_{O,op}^{(r)}\right] = \begin{bmatrix} C_{\theta_x-m_x}^{(r)} & 0 & 0 \\ 0 & C_{\theta_y-m_y}^{(r)} & C_{\theta_y-f_z}^{(r)} \\ 0 & C_{\theta_y-f_z}^{(r)} & C_{u_z-f_z}^{(r)} \end{bmatrix}; \quad \left[C_{O,op}^{(l)}\right] = \begin{bmatrix} C_{\theta_x-m_x}^{(r)} & 0 & 0 \\ 0 & C_{\theta_y-m_y}^{(r)} & -C_{\theta_y-f_z}^{(r)} \\ 0 & -C_{\theta_y-f_z}^{(r)} & C_{u_z-f_z}^{(r)} \end{bmatrix}.$$

$$(6.10)$$

The structure of $\left[C_{O,op}^{(r)}\right]$ is also presented in Eq. (3.40); the minus signs in $\left[C_{O,op}^{(l)}\right]$ is consistent with mirroring about the symmetry axis. The full-mechanism out-of-plane compliance matrix is calculated based on the second Eq. (6.6) with the two matrices of Eq. (6.10) as:

$$\left[C_{O,op}\right] = \left(\left[C_{O,op}^{(r)}\right]^{-1} + \left[C_{O,op}^{(l)}\right]^{-1}\right)^{-1} = \frac{1}{2} \cdot \begin{bmatrix} C_{\theta_x-m_x} & 0 & 0 \\ 0 & C_{\theta_y-m_y} & 0 \\ 0 & 0 & C_{u_z-f_z} \end{bmatrix}, \quad (6.11)$$

where

$$C_{\theta_x-m_x} = C_{\theta_x-m_x}^{(r)}; \quad C_{\theta_y-m_y} = C_{\theta_y-m_y}^{(r)} - \frac{\left(C_{\theta_y-f_z}^{(r)}\right)^2}{C_{u_z-f_z}^{(r)}}; \quad C_{u_z-f_z} = C_{u_z-f_z}^{(r)} - \frac{\left(C_{\theta_y-f_z}^{(r)}\right)^2}{C_{\theta_y-m_y}^{(r)}}.$$

$$(6.12)$$

The two compliance matrices of Eqs. (6.8) and (6.11) are diagonal, which indicates full decoupling of the loads acting along the x, y and z reference axes.

Example 6.1

The mechanism of Figure 6.3 is a particular design of the generic configuration of Figure 6.2. The right half of the mechanism comprises two identical straight-axis flexible hinges of a constant circular cross-section of diameter $d = 0.0015$ m. The following parameters are also given: $l = 0.02$ m, $l_1 = 0.03$ m, $l_2 = 0.05$ m and $E = 2 \cdot 10^{11}$ N/m^2. Knowing that the in-plane displacements at the center O are $u_{Ox} = 10$ μm and $u_{Oy} = 8$ mm, calculate the in-plane force f and its angle δ.

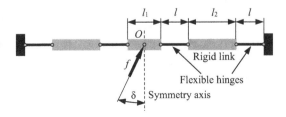

FIGURE 6.3 A symmetric suspension mechanism with straight-axis collinear links.

Solution:

The in-plane compliance matrix of the right half of mechanism of Figure 6.3 is calculated as:

$$\left[C_{O,ip}^{(r)} \right] = \left[T_{OA,ip} \right]^{T} \left[C_{ip} \right] \left[T_{OA,ip} \right] + \left[T_{OC,ip} \right]^{T} \left[C_{ip} \right] \left[T_{OC,ip} \right]$$

$$\text{with} \quad \left[T_{OA,ip} \right] = \begin{bmatrix} 1 & 0 & 0 \\ 0 & 1 & 0 \\ 0 & -l_1/2 & 1 \end{bmatrix}; \quad \left[T_{OC,ip} \right] = \begin{bmatrix} 1 & 0 & 0 \\ 0 & 1 & 0 \\ 0 & -(l_1/2 + l + l_2) & 1 \end{bmatrix}. \tag{6.13}$$

$[C_{ip}]$ is the local-frame, in-plane compliance matrix of a straight-axis flexible hinge – see Eqs. (2.13) and (2.33). The displacements at O are expressed using Eq. (6.8) as:

$$\begin{bmatrix} u_{O_x} \\ u_{O_y} \\ \theta_{O_x} \end{bmatrix} = \frac{1}{2} \cdot \begin{bmatrix} C_{u_x - f_x} & 0 & 0 \\ 0 & C_{u_y - f_y} & 0 \\ 0 & 0 & C_{\theta_z - m_z} \end{bmatrix} \begin{bmatrix} f \cdot \sin \delta \\ f \cdot \cos \delta \\ 0 \end{bmatrix}. \tag{6.14}$$

The compliances of Eq. (6.14) are provided in Eq. (6.9) in terms of the right half-compliances, which are found from Eq. (6.13). After expressing u_{O_x} and u_{O_y} from Eq. (6.14) and some algebraic manipulations, the force f and angle δ are found:

$$f = 2\sqrt{\left(\frac{u_{O_x}}{C_{u_x - f_x}} \right)^2 + \left(\frac{u_{O_y}}{C_{u_y - f_y}} \right)^2}; \quad \delta = \tan^{-1} \left(\frac{u_{O_x} \cdot C_{u_y - f_y}}{u_{O_y} \cdot C_{u_x - f_x}} \right). \tag{6.15}$$

The numerical values of these parameters are $f = 177.419\,\text{N}$ and $\delta = 84.9°$.

6.1.1.2   Folded Chains with Curvilinear-Axis Hinges

Figure 6.4a shows a mechanism that uses folded hinges and is also illustrative from a conceptual standpoint of the generic configuration of Figure 6.2. The in-plane, quasi-static response of this mechanism is of interest – more details of this

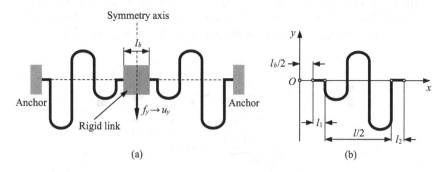

(a)   (b)

FIGURE 6.4   Skeleton representation of: (a) symmetric stage with two identical and mirrored folded flexible hinges; (b) half-folded hinge segment.

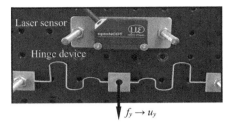

FIGURE 6.5 Experimental setup for measuring the stiffness $K_{fy\text{-}uy}$ of the stage represented in *Figure 6.4*.

configuration are provided in Lobontiu et al. [49]. The mechanism uses the folded hinge of Figure 4.28b as its half-hinge, which is defined by a circular envelope and segments with elliptical tops. In addition to the basic, folded half-chain of Figure 4.28, the half-architecture shown in Figure 6.4b adds two straight-axis rigid links at its ends to connect to the middle rigid link and the anchor point. As discussed in Lobontiu et al. [49], several prototypes of the type shown in Figure 6.4a were fabricated out of aluminum and plastic materials; they were tested experimentally by applying a force f_y (along the symmetry axis) on the middle rigid link. The resulting displacement u_y along the same axis was read by a laser displacement sensor, and the resulting stiffness $K_{fy\text{-}uy} = f_y/u_y$ was determined; a photograph of the experimental setup is shown in Figure 6.5. A similar path was followed to obtain the stiffness $K_{fy\text{-}uy}$ by using finite element simulation.

The basic half-folded hinge of Figure 4.28b, as well as the half-flexible segment of Figure 6.4b, is not symmetric with respect to the x axis. Details are offered in Lobontiu et al. [49] on how to calculate the in-plane compliance matrix of the half-portion of Figure 6.4a by adding all the translated compliance matrices corresponding to the folded segment and the two end segments. The in-plane compliance matrices of the right half of the segment and the left (mirrored) half are expressed in the $Oxyz$ reference frame as:

$$
\left[C_{O,ip}^{(r)}\right] = \begin{bmatrix} C_{u_x-f_x}^{(r)} & C_{u_x-f_y}^{(r)} & C_{u_x-m_z}^{(r)} \\ C_{u_x-f_y}^{(r)} & C_{u_y-f_y}^{(r)} & C_{u_y-m_z}^{(r)} \\ C_{u_x-m_z}^{(r)} & C_{u_y-m_z}^{(r)} & C_{\theta_z-m_z}^{(r)} \end{bmatrix}; \quad
\left[C_{O,ip}^{(l)}\right] = \begin{bmatrix} C_{u_x-f_x}^{(r)} & -C_{u_x-f_y}^{(r)} & C_{u_x-m_z}^{(r)} \\ -C_{u_x-f_y}^{(r)} & C_{u_y-f_y}^{(r)} & -C_{u_y-m_z}^{(r)} \\ C_{u_x-m_z}^{(r)} & -C_{u_y-m_z}^{(r)} & C_{\theta_z-m_z}^{(r)} \end{bmatrix},
$$

$$(6.16)$$

where the superscripts "r" and "l" denote right and left, respectively. The in-plane compliance matrix of the full device is calculated as:

$$
\left[C_{O,ip}\right] = \left(\left[C_{O,ip}^{(r)}\right]^{-1} + \left[C_{O,ip}^{(l)}\right]^{-1}\right)^{-1} = \frac{1}{2} \cdot \begin{bmatrix} C_{u_x-f_x} & 0 & C_{u_x-m_z} \\ 0 & C_{u_y-f_y} & 0 \\ C_{u_x-m_z} & 0 & C_{\theta_z-m_z} \end{bmatrix}, \quad (6.17)
$$

with

$$
\left\{
\begin{array}{l}
C_{u_x-f_x} = C_{u_x-f_x}^{(r)} - \dfrac{\left(C_{u_x-f_y}^{(r)}\right)^2}{C_{u_y-f_y}^{(r)}} ; \\[4mm]
C_{u_x-m_z} = C_{u_x-m_z}^{(r)} - \dfrac{C_{u_x-f_y}^{(r)} \cdot C_{u_y-m_z}^{(r)}}{C_{u_y-f_y}^{(r)}} ; \\[4mm]
C_{u_y-f_y} = C_{u_y-f_y}^{(r)} + \dfrac{2C_{u_x-f_y}^{(r)} \cdot C_{u_x-m_z}^{(r)} \cdot C_{u_y-m_z}^{(r)} - C_{u_x-f_x}^{(r)} \cdot \left(C_{u_y-m_z}^{(r)}\right)^2 - C_{\theta_z-m_z}^{(r)} \cdot \left(C_{u_x-f_y}^{(r)}\right)^2}{C_{\theta_z-m_z}^{(r)} \cdot C_{u_x-f_x}^{(r)} - \left(C_{u_x-m_z}^{(r)}\right)^2} ; \\[4mm]
C_{\theta_z-m_z} = C_{\theta_z-m_z}^{(r)} - \dfrac{\left(C_{u_y-m_z}^{(r)}\right)^2}{C_{u_y-f_y}^{(u)}}
\end{array}
\right.
$$

$$(6.18)$$

Note in Eq. (6.17) that there is decoupling between the x and y axes, as well as between the y and z axes, as illustrated by the zeros in the respective positions.

The vector relationship between the loads and displacements at the center O is:

$$
\left\{
\begin{array}{c}
u_x \\ u_y \\ \theta_z
\end{array}
\right\}
= \frac{1}{2} \cdot
\begin{bmatrix}
C_{u_x-f_x} & 0 & C_{u_x-m_z} \\
0 & C_{u_y-f_y} & 0 \\
C_{u_x-m_z} & 0 & C_{\theta_z-m_z}
\end{bmatrix}
\left\{
\begin{array}{c}
0 \\ f_y \\ 0
\end{array}
\right\},
\qquad (6.19)
$$

and therefore,

$$
\left\{
\begin{array}{l}
u_x = 0 \\[2mm]
u_y = \dfrac{1}{2} \cdot C_{u_y-f_y} \cdot f_y \rightarrow K_{f_y-u_y} = \dfrac{f_y}{u_y} = \dfrac{2}{C_{u_y-f_y}} . \\[3mm]
\theta_z = 0
\end{array}
\right.
\qquad (6.20)
$$

The experimental, finite element and analytical stiffness results (the latter ones based on Eqs. (6.16) through (6.20)) generated small deviations – see Lobontiu et al. [49].

6.1.1.3　Mechanisms with Identical, Symmetrical and Non-Collinear Flexible Chains

The compliance matrix approach is applied here to characterize the quasi-static response of planar flexible mechanisms that are symmetric with respect to either one axis or two perpendicular axes. Mechanisms with one symmetry axis are formed of two identical flexible chains that are mirrored with respect to the symmetry axis. Mechanisms with two perpendicular symmetry axes comprise four identical flexible chains that are mirrored with respect to the two axes and their intersection point, which is the symmetry center. In all cases studied here, the identical flexible chains connect to a moving rigid link.

6.1.1.3.1 Mechanisms with One Symmetry Axis

Planar flexible-hinge mechanisms that are symmetric with respect to either the y axis or the x axis are analyzed in this section.

Symmetry line is the y axis:

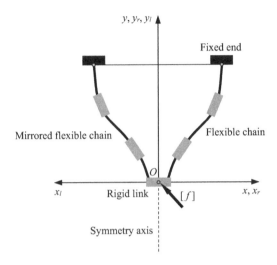

FIGURE 6.6 Parallel mechanism formed of two identical serial flexible chains that are symmetric with respect to the y axis.

The mechanism of Figure 6.6 is formed of two identical flexible chains that are symmetric with respect to the global y axis and may comprise straight- and curvilinear-axis flexible hinges.

Using "r" to denote the right flexible chain in Figure 6.6 and "l" to identify the identical left segment, which is mirrored with respect to the symmetry y axis, the in-plane compliance matrices of these two segments in the global frame $Oxyz$ are expressed in Eqs. (6.16); furthermore, the in-plane compliance matrix of the full mechanism is the one provided in Eqs. (6.17) and (6.18).

A similar approach is followed to express the out-of-plane compliance matrix of the mechanism shown in Figure 6.6. The compliance matrices of the two identical and mirrored chains are:

$$\left[C_{O,op}^{(r)}\right] = \begin{bmatrix} C_{\theta_x-m_x}^{(r)} & C_{\theta_x-m_y}^{(r)} & C_{\theta_x-f_z}^{(r)} \\ C_{\theta_x-m_y}^{(r)} & C_{\theta_y-m_y}^{(r)} & C_{\theta_y-f_z}^{(r)} \\ C_{\theta_x-f_z}^{(r)} & C_{\theta_y-f_z}^{(r)} & C_{u_z-f_z}^{(r)} \end{bmatrix}; \quad \left[C_{O,op}^{(l)}\right] = \begin{bmatrix} C_{\theta_x-m_x}^{(r)} & -C_{\theta_x-m_y}^{(r)} & C_{\theta_x-f_z}^{(r)} \\ -C_{\theta_x-m_y}^{(r)} & C_{\theta_y-m_y}^{(r)} & -C_{\theta_y-f_z}^{(r)} \\ C_{\theta_x-f_z}^{(r)} & -C_{\theta_y-f_z}^{(r)} & C_{u_z-f_z}^{(r)} \end{bmatrix}.$$

$$(6.21)$$

The out-of-plane compliance matrix of the full mechanism is calculated based on those of the two parallel-connection chains of Eq. (6.21) as:

$$\left[C_{O,op}\right]=\left(\left[C_{O,op}^{(r)}\right]^{-1}+\left[C_{O,op}^{(l)}\right]^{-1}\right)^{-1}=\frac{1}{2}\cdot\begin{bmatrix} C_{\theta_x-m_x} & 0 & C_{\theta_x-f_z} \\ 0 & C_{\theta_y-m_y} & 0 \\ C_{\theta_x-f_z} & 0 & C_{u_z-f_z} \end{bmatrix}, \quad (6.22)$$

with

$$\begin{cases} C_{\theta_x-m_x}=C_{\theta_x-m_x}^{(r)}-\dfrac{\left(C_{\theta_x-m_y}^{(r)}\right)^2}{C_{\theta_y-m_y}^{(r)}}; \\[4mm] C_{\theta_x-f_z}=C_{\theta_x-f_z}^{(r)}-\dfrac{C_{\theta_x-m_y}^{(r)}\cdot C_{\theta_y-f_z}^{(r)}}{C_{\theta_y-m_y}^{(r)}}; \\[4mm] C_{\theta_y-m_y}=C_{\theta_y-m_y}^{(r)}+\dfrac{C_{u_z-f_z}^{(r)}\cdot\left(C_{\theta_x-m_y}^{(r)}\right)^2+C_{\theta_y-f_z}^{(y)}\cdot\left(C_{\theta_y-f_z}^{(r)}\cdot C_{\theta_x-m_x}^{(r)}-2\cdot C_{\theta_x-m_y}^{(r)}\cdot C_{\theta_x-f_z}^{(r)}\right)}{\left(C_{\theta_x-f_z}^{(r)}\right)^2-C_{u_z-f_z}^{(r)}\cdot C_{\theta_x-m_x}^{(r)}}; \\[4mm] C_{u_z-f_z}=C_{u_z-f_z}^{(r)}-\dfrac{\left(C_{\theta_y-f_z}^{(r)}\right)^2}{C_{\theta_y-m_y}^{(r)}}. \end{cases}$$

$$(6.23)$$

Like in the in-plane situation, there is decoupling between the x and y axes, as well as between the y and z axes, which is illustrated by the zeros in the respective positions in the matrix of Eq. (6.22).

Out-of-plane stage with fractal (self-similar), circular-axis flexible segments: Figure 6.7a shows the top-view skeleton representation of a planar stage for piston-type (z axis), out-of-plane motion – see Lobontiu et al. [50] for more details. The device consists of n several concentric, circular segments that are

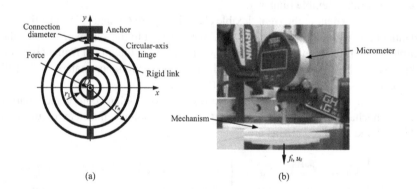

(a)   (b)

FIGURE 6.7 (a) Top-view skeleton representation of fractal (self-similar) mechanism with serially connected circular-axis flexible hinges for out-of-plane motion; (b) experimental setup.

interconnected serially in a folded manner along a radial direction. A force applied along the z axis (perpendicular to the mechanism plane) at the center O on a rigid link will cause all circular segments to deform out of the plane. The total motion at the center is large as it adds the relatively small deformations of all circular segments – with the outer rim being fixed. This mechanism has a fractal (or self-similar) geometry with a basic pattern (the circle) repeated on a scaled manner.

An arbitrary segment i (with $i = 1, 3, 5, \ldots$) of radius r_i is sketched in Figure 6.8a.

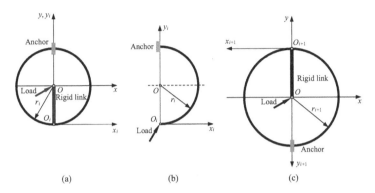

(a) (b) (c)

FIGURE 6.8 (a) Generic full-circle segment i ($i = 1, 3, 5, \ldots$); (b) right half-circular segment; (c) generic full-circle segment $i + 1$.

This segment is formed by combining the right half-circular segment in parallel with the left half-circular segment, which is structurally identical to the right half and mirrored with respect to the y axis. In the local frame $O_i x_i y_i z_i$, the out-of-plane compliance matrices of the left and right half-circular segments are expressed in Eq. (6.21) and the full-circular segment compliance matrix, denoted here by $\left[C^{(i)}_{O_i, op} \right]$, is determined as per Eq. (6.22). This compliance matrix is translated from its local frame to the global frame $Oxyz$ based on the equation:

$$\left[C^{(i)}_{O, op} \right] = \left[T^{(i)}_{OO_i, op} \right]^T \left[C^{(i)}_{O_i, op} \right] \left[T^{(i)}_{OO_i, op} \right] \quad \text{with} \quad \left[T^{(i)}_{OO_i, op} \right] = \begin{bmatrix} 1 & 0 & r_i \\ 0 & 1 & 0 \\ 0 & 0 & 1 \end{bmatrix}.$$

$$(6.24)$$

The compliance matrix of the segment $i + 1$, which is sketched together with its virtual connection to the center and the boundary conditions in Figure 6.8c, is similarly calculated as:

$$\left[C^{(i+1)}_{O, op} \right] = \left[T^{(i+1)}_{OO_{i+1}, op} \right]^T \left[R^{(i+1)} \right]^T \left[C^{(i+1)}_{O_{i+1}, op} \right] \left[R^{(i+1)} \right] \left[T^{(i+1)}_{OO_{i+1}, op} \right] \quad \text{with}$$

$$\left[T^{(i)}_{OO_i, op} \right] = \begin{bmatrix} 1 & 0 & -r_{i+1} \\ 0 & 1 & 0 \\ 0 & 0 & 1 \end{bmatrix}; \quad \left[R^{(i+1)} \right] = \begin{bmatrix} \cos \pi & \sin \pi & 0 \\ -\sin \pi & \cos \pi & 0 \\ 0 & 0 & 1 \end{bmatrix}.$$

$$(6.25)$$

Using the compliances of a half-circular segment based on Eqs. (2.109), (2.110) and Eqs. (A2.27) through (A2.32) for a center angle $\alpha = 180°$, the out-of-plane compliance of any full-circular segment is diagonal:

$$
\left[C_{O,op}^{(i)} \right] = \begin{bmatrix} C_{\theta_x - m_x}^{(i)} & 0 & 0 \\ 0 & C_{\theta_y - m_y}^{(i)} & 0 \\ 0 & 0 & C_{u_z - f_z}^{(i)} \end{bmatrix}; \quad i = 1, 2, ..., n, \quad (6.26)
$$

with

$$
\begin{cases} C_{\theta_x - m_x}^{(i)} = \dfrac{\pi}{4} \cdot \left(\dfrac{1}{EI_{ni}} + \dfrac{1}{GI_{ti}} \right) \cdot r_i \\[2ex] C_{\theta_y - m_y}^{(i)} = \dfrac{1}{4\pi} \cdot \left(\dfrac{\pi^2}{EI_{ni}} + \dfrac{\pi^2 - 8}{GI_{ti}} \right) \cdot r_i \\[2ex] C_{u_z - f_z}^{(i)} = \dfrac{\left(\pi^2 - 8 \right) EI_{n,i} + \pi^2 GI_{t,i}}{2\pi GI_{ti} \left(EI_{ni} + GI_{ti} \right)} \cdot r_i^3 = a_i \cdot r_i^3 \end{cases} \quad , \quad (6.27)
$$

as demonstrated in Lobontiu et al. [50]. In Eqs. (6.27), E and G are the material Young's and shear moduli, I_{ni} is the cross-sectional moment of area related to out-of-plane bending, and I_{ti} is the similar moment of area related to torsion. The out-of-plane compliance matrix of the full device illustrated in Figure 6.7a is calculated by adding the compliances of the n fractal serially connected circular segments using Eq. (6.26), namely:

$$
\left[C_{O,op} \right] = \sum_{i=1}^{n} \left[C_{O,op}^{(i)} \right] = \begin{bmatrix} C_{\theta_x - m_x} & 0 & 0 \\ 0 & C_{\theta_y - m_y} & 0 \\ 0 & 0 & C_{u_z - f_z} \end{bmatrix}
$$

$$
= \begin{bmatrix} \displaystyle\sum_{i=1}^{n} C_{\theta_x - m_x}^{(i)} & 0 & 0 \\ 0 & \displaystyle\sum_{i=1}^{n} C_{\theta_y - m_y}^{(i)} & 0 \\ 0 & 0 & \displaystyle\sum_{i=1}^{n} C_{u_z - f_z}^{(i)} \end{bmatrix}. \quad (6.28)
$$

Based on this generic model, the paper by Lobontiu et al. [50] discusses several particular designs of constant rectangular cross-sections, such as configurations having the same cross-sections for all circular segments and identical radial segment gap or designs with all circular segments deforming identically. Finite element analysis and experimental testing of a prototype – see photograph of experimental setup in Figure 6.7b – confirmed the analytical model predictions.

Symmetry line is the x axis:

Consider the generic planar mechanism of Figure 6.9, which is formed of two identical flexible chains that are mirrored/symmetric with respect to the x axis.

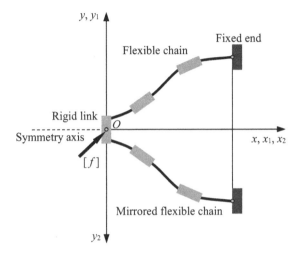

FIGURE 6.9 Parallel mechanism formed of two identical serial flexible chains that are symmetric with respect to the x axis.

Assuming the load is applied on the rigid link at point O on the symmetry line, the in-plane and out-of-plane compliance matrices of the full mechanism can be formulated in terms of those defining the half-mechanism. Denote by "u" the upper flexible chain in Figure 6.8 that is comprised between point O and its fixed end, whereas its mirrored lower counterpart is denoted by "l". The in-plane compliance matrices of the upper chain $\left[C_{O,ip}^{(u)} \right]$ and the lower chain $\left[C_{O,ip}^{(l)} \right]$ have the following generic equations in the reference frame $Oxyz$:

$$\left[C_{O,ip}^{(u)} \right] = \begin{bmatrix} C_{u_x-f_x}^{(u)} & C_{u_x-f_y}^{(u)} & C_{u_x-m_z}^{(u)} \\ C_{u_x-f_y}^{(u)} & C_{u_y-f_y}^{(u)} & C_{u_y-m_z}^{(u)} \\ C_{u_x-m_z}^{(u)} & C_{u_y-m_z}^{(u)} & C_{\theta_z-m_z}^{(u)} \end{bmatrix};$$

$$(6.29)$$

$$\left[C_{O,ip}^{(l)} \right] = \begin{bmatrix} C_{u_x-f_x}^{(u)} & -C_{u_x-f_y}^{(u)} & -C_{u_x-m_z}^{(u)} \\ -C_{u_x-f_y}^{(u)} & C_{u_y-f_y}^{(u)} & C_{u_y-m_z}^{(u)} \\ -C_{u_x-m_z}^{(u)} & C_{u_y-m_z}^{(u)} & C_{\theta_z-m_z}^{(u)} \end{bmatrix}.$$

Because the two chains are connected in parallel, the in-plane compliance matrix of the entire mechanism of Figure 6.9 is determined with respect to the common point O as:

$$\left[C_{O,ip} \right] = \left(\left[C_{O,ip}^{(u)} \right]^{-1} + \left[C_{O,ip}^{(l)} \right]^{-1} \right)^{-1} = \frac{1}{2} \cdot \begin{bmatrix} C_{u_x-f_x} & 0 & 0 \\ 0 & C_{u_y-f_y} & C_{u_y-m_z} \\ 0 & C_{u_y-m_z} & C_{\theta_z-m_z} \end{bmatrix}, \quad (6.30)$$

with

$$\begin{cases} C_{u_x-f_x} = C_{u_x-f_x}^{(u)} + \dfrac{C_{\theta_z-m_z}^{(u)} \cdot \left(C_{u_x-f_y}^{(u)} \right)^2 + C_{u_x-m_z}^{(u)} \cdot \left(C_{u_x-m_z}^{(u)} \cdot C_{u_y-f_y}^{(u)} - 2 \cdot C_{u_x-f_y}^{(u)} \cdot C_{u_y-m_z}^{(u)} \right)}{\left(C_{u_y-m_z}^{(u)} \right)^2 - C_{\theta_z-m_z}^{(u)} \cdot C_{u_y-f_y}^{(u)}}; \\[4mm] C_{u_y-f_y} = C_{u_y-f_y}^{(u)} - \dfrac{\left(C_{u_x-f_y}^{(u)} \right)^2}{C_{u_x-f_x}^{(u)}}; \\[4mm] C_{u_y-m_z} = C_{u_y-m_z}^{(u)} - \dfrac{C_{u_x-f_y}^{(u)} \cdot C_{u_x-m_z}^{(u)}}{C_{u_x-f_x}^{(u)}}; \\[4mm] C_{\theta_z-m_z} = C_{\theta_z-m_z}^{(u)} - \dfrac{\left(C_{u_x-m_z}^{(u)} \right)^2}{C_{u_x-f_x}^{(u)}}. \end{cases}$$

$$(6.31)$$

Note from Eq. (6.30) that the mechanism of Figure 6.9 behaves as a straight-axis flexible hinge whose in-plane compliance matrix indicates no interaction (coupling) between the x and y axes and the x and z axes.

A similar procedure is applied to determine the out-of-plane compliance matrix of the parallel mechanism. The compliance matrices of the upper and lower serial chains are formulated as:

$$\left[C_{O,op}^{(u)} \right] = \begin{bmatrix} C_{\theta_x-m_x}^{(u)} & C_{\theta_x-m_y}^{(u)} & C_{\theta_x-f_z}^{(u)} \\ C_{\theta_x-m_y}^{(u)} & C_{\theta_y-m_y}^{(u)} & C_{\theta_y-f_z}^{(u)} \\ C_{\theta_x-f_z}^{(u)} & C_{\theta_y-f_z}^{(u)} & C_{u_z-f_z}^{(u)} \end{bmatrix};$$

$$(6.32)$$

$$\left[C_{O,op}^{(l)} \right] = \begin{bmatrix} C_{\theta_x-m_x}^{(u)} & -C_{\theta_x-m_y}^{(u)} & -C_{\theta_x-f_z}^{(u)} \\ -C_{\theta_x-m_y}^{(u)} & C_{\theta_y-m_y}^{(u)} & C_{\theta_y-f_z}^{(u)} \\ -C_{\theta_x-f_z}^{(u)} & C_{\theta_y-f_z}^{(u)} & C_{u_z-f_z}^{(u)} \end{bmatrix}.$$

The out-of-plane compliance matrix of the mechanism is calculated based on those of the two parallel-connection chains of Eq. (6.32) as:

$$
\left[C_{O,op}\right]=\left(\left[C_{O,op}^{(u)}\right]^{-1}+\left[C_{O,op}^{(l)}\right]^{-1}\right)^{-1}=\frac{1}{2}\cdot\begin{bmatrix} C_{\theta_x-m_x} & 0 & 0 \\ 0 & C_{\theta_y-m_y} & C_{\theta_y-f_z} \\ 0 & C_{\theta_y-f_z} & C_{u_z-f_z} \end{bmatrix},\quad(6.33)
$$

with

$$
\begin{cases}
C_{\theta_x-m_x}=C_{\theta_x-m_x}^{(u)}+\dfrac{C_{u_z-f_z}^{(u)}\cdot\left(C_{\theta_x-m_y}^{(u)}\right)^2+C_{\theta_x-f_z}^{(u)}\cdot\left(C_{\theta_x-f_z}^{(u)}\cdot C_{\theta_y-m_y}^{(u)}-2\cdot C_{\theta_x-m_y}^{(u)}\cdot C_{\theta_y-f_z}^{(u)}\right)}{\left(C_{\theta_y-f_z}^{(u)}\right)^2-C_{u_z-f_z}^{(u)}\cdot C_{\theta_y-m_y}^{(u)}};\\[3ex]
C_{\theta_y-m_y}=C_{\theta_y-m_y}^{(u)}-\dfrac{\left(C_{\theta_x-m_y}^{(u)}\right)^2}{C_{\theta_x-m_x}^{(u)}};\\[3ex]
C_{\theta_y-f_z}=C_{\theta_y-f_z}^{(u)}-\dfrac{C_{\theta_x-m_y}^{(u)}\cdot C_{\theta_x-f_z}^{(u)}}{C_{\theta_x-m_x}^{(u)}};\\[3ex]
C_{u_z-f_z}=C_{u_z-f_z}^{(u)}-\dfrac{\left(C_{\theta_x-f_z}^{(u)}\right)^2}{C_{\theta_x-m_x}^{(u)}}.
\end{cases}
$$

$$(6.34)$$

Note, again, that the out-of-plane compliance matrix of Eq. (6.33) is similar to that of a straight-axis flexible hinge with no coupling between the x and y axes and x and z axes, as indicated by the zero components in the respective positions of the matrix.

Trapeze and parallelogram mechanisms are examples of parallel-connection devices with one symmetry axis that are formed of two identical (and mirrored) serial chains – these configurations are analyzed in this section.

Trapeze mechanisms: A conventional rotation-joint trapeze mechanism is shown schematically in Figure 6.10a. In a displaced configuration (indicated with dotted lines in the same figure), the link, which is originally parallel to the fixed link, assumes a position that is no longer parallel to the fixed link. A generic planar, flexible-hinge trapeze mechanism is sketched in Figure 6.10b that uses two identical and mirrored flexible chains. Its upper half-portion is a generalization of the serial mechanism studied in *Example 5.3*, which derived the in-plane and out-of-plane compliance matrices for a configuration where the inclined chain is formed of three collinear links: two flexible hinges and an intermediate rigid segment. Using Eqs. (6.30), (6.31) and (6.33), (6.34) in conjunction with the corresponding half-mechanism compliance matrices derived in *Example 5.3* results in the full-mechanism in-plane and out-of-plane compliances, which are not included here.

Parallelogram mechanisms: When $\delta=0$, the trapeze mechanism of Figure 6.10b becomes a parallelogram mechanism, like the one with two parallel flexible chains sketched in Figure 6.11a.

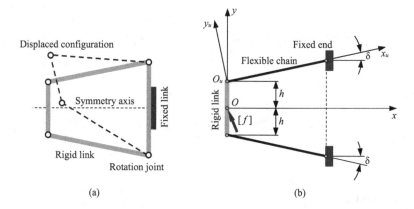

FIGURE 6.10 Trapeze mechanism with symmetry axis: (a) device with rotary joints; (b) configuration with straight-axis flexible chains.

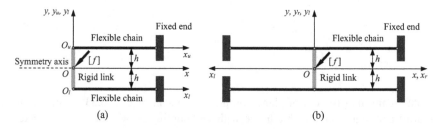

FIGURE 6.11 Parallelogram mechanisms with symmetry axis and identical straight-axis flexible chains formed of (a) two chains; (b) four chains.

The in-plane compliances that form the matrix of Eq. (6.31) become:

$$
\begin{cases}
C_{u_x-f_x} = C^{(u)}_{u_x-f_x}; \\[2mm]
C_{u_y-f_y} = \dfrac{C^{(u)}_{u_x-f_x} \cdot C^{(u)}_{u_y-f_y} + h^2 \cdot \left[C^{(u)}_{\theta_z-m_z} \cdot C^{(u)}_{u_y-f_y} - \left(C^{(u)}_{u_y-m_z} \right)^2 \right]}{C^{(u)}_{u_x-f_x} + h^2 \cdot C^{(u)}_{\theta_z-m_z}}; \\[4mm]
C_{u_y-m_z} = \dfrac{C^{(u)}_{u_x-f_x} \cdot C^{(u)}_{u_y-m_z}}{C^{(u)}_{u_x-f_x} + h^2 \cdot C^{(u)}_{\theta_z-m_z}}; \\[4mm]
C_{\theta_z-m_z} = \dfrac{C^{(u)}_{u_x-f_x} \cdot C^{(u)}_{\theta_z-m_z}}{C^{(u)}_{u_x-f_x} + h^2 \cdot C^{(u)}_{\theta_z-m_z}}.
\end{cases}
\qquad (6.35)
$$

Equations (6.35) were obtained by translating the in-plane compliance matrix of a straight-axis flexible hinge – Eq. (2.13); the offsets in the translation matrix are: $\Delta x_{OO_u} = 0$, $\Delta y_{OO_u} = h$.

The out-of-plane compliances of Eq. (6.34) change to:

$$
\begin{cases}
C_{\theta_x - m_x} = \dfrac{C_{\theta_x - m_x}^{(u)} \cdot \left[\left(C_{\theta_y - f_z}^{(u)} \right)^2 - C_{\theta_y - m_y}^{(u)} \cdot C_{u_z - f_z}^{(u)} \right]}{\left(C_{\theta_y - f_z}^{(u)} \right)^2 - C_{\theta_y - m_y}^{(u)} \cdot \left(C_{u_z - f_z}^{(u)} + h^2 \cdot C_{\theta_x - m_x}^{(u)} \right)} ; \\[4mm]
C_{\theta_y - m_y} = C_{\theta_y - m_y}^{(u)} ; \\[2mm]
C_{\theta_y - f_z} = C_{\theta_y - f_z}^{(u)} ; \\[2mm]
C_{u_z - f_z} = C_{u_z - f_z}^{(u)} .
\end{cases}
\tag{6.36}
$$

Equations (6.36) were obtained from the out-of-plane compliance matrix of a straight-axis flexible hinge – Eq. (2.16) – in combination with a corresponding translation matrix defined by the offsets: $\Delta x_{OO_u} = 0$, $\Delta y_{OO_u} = h$.

The mechanism shown in Figure 6.11b is a parallel connection of the mechanism sketched in Figure 6.11a, which is mirrored with respect to the y axis in terms of the central rigid-link motion. As per Eq. (6.30), the in-plane compliance matrices of the half-mechanism situated to the right of the y axis in Figure 6.11b and of its mirrored left half are:

$$
\left[C_{O,ip}^{(r)} \right] = \frac{1}{2} \cdot
\begin{bmatrix}
C_{u_x - f_x} & 0 & 0 \\
0 & C_{u_y - f_y} & C_{u_y - m_z} \\
0 & C_{u_y - m_z} & C_{\theta_z - m_z}
\end{bmatrix} ;
$$

$$
\left[C_{O,ip}^{(l)} \right] = \frac{1}{2} \cdot
\begin{bmatrix}
C_{u_x - f_x} & 0 & 0 \\
0 & C_{u_y - f_y} & -C_{u_y - m_z} \\
0 & -C_{u_y - m_z} & C_{\theta_z - m_z}
\end{bmatrix} ,
\tag{6.37}
$$

with the particular compliances of one flexible chain being expressed in Eq. (6.35). The full-mechanism compliance matrix is:

$$
\left[C_{O,ip} \right] = \left(\left[C_{O,ip}^{(r)} \right]^{-1} + \left[C_{O,ip}^{(l)} \right]^{-1} \right)^{-1}
$$

$$
= \frac{1}{4} \cdot
\begin{bmatrix}
C_{u_x - f_x} & 0 & 0 \\[3mm]
0 & C_{u_y - f_y} - \dfrac{\left(C_{u_y - m_z} \right)^2}{C_{\theta_z - m_z}} & 0 \\[5mm]
0 & 0 & C_{\theta_z - m_z} - \dfrac{\left(C_{u_y - m_z} \right)^2}{C_{u_y - f_y}}
\end{bmatrix} .
\tag{6.38}
$$

Similarly, the out-of-plane compliance matrices of the right and half-portions of the mechanism of Figure 6.11b are expressed based on Eq. (6.33) as:

$$\left[C_{O,op}^{(r)}\right]=\frac{1}{2}\cdot\begin{bmatrix} C_{\theta_x-f_x} & 0 & 0 \\ 0 & C_{\theta_y-m_y} & C_{\theta_y-f_z} \\ 0 & C_{\theta_y-f_z} & C_{u_z-f_z} \end{bmatrix};$$

$$(6.39)$$

$$\left[C_{O,op}^{(l)}\right]=\frac{1}{2}\cdot\begin{bmatrix} C_{\theta_x-f_x} & 0 & 0 \\ 0 & C_{\theta_y-m_y} & -C_{\theta_y-f_z} \\ 0 & -C_{\theta_y-f_z} & C_{u_z-f_z} \end{bmatrix},$$

with the individual compliances of Eqs. (6.36). The full-mechanism out-of-plane compliance matrix is:

$$\left[C_{O,op}\right]=\left(\left[C_{O,op}^{(r)}\right]^{-1}+\left[C_{O,op}^{(l)}\right]^{-1}\right)^{-1}$$

$$=\frac{1}{4}\cdot\begin{bmatrix} C_{\theta_x-m_x} & 0 & 0 \\ 0 & C_{\theta_y-m_y}-\dfrac{\left(C_{\theta_y-f_z}\right)^2}{C_{u_z-f_z}} & 0 \\ 0 & 0 & C_{u_z-f_z}-\dfrac{\left(C_{\theta_y-f_z}\right)^2}{C_{\theta_y-m_y}} \end{bmatrix}.$$

$$(6.40)$$

Note that both compliance matrices of Eqs. (6.38) and (6.40) are diagonal – this is actually a mechanism with two symmetry axes and a resulting symmetry center, which is separately studied in Section 6.1.1.3.2. The diagonal form of the compliance matrices indicates full load decoupling between the x, y and z axes for both in-plane motion and out-of-plane motion.

Consider now the mechanism shown in skeleton form in Figure 6.12a. It is a series combination of two dissimilar parallelogram mechanisms, such as the one of Figure 6.11a. Instead of using a regular, cascading physical connection, the two parallelogram stages are coupled in a folded and more compact manner to reduce the planar footprint.

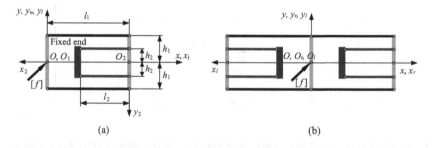

(a)   (b)

FIGURE 6.12 Folded parallelogram mechanisms: (a) simple (single); (b) double.

The main parameters of the longer, outer parallelogram mechanism are l_1 and h_1, while the shorter, inner mechanism is defined by l_2 and h_2. With respect to the reference frame $Oxyz$, the in-plane compliance matrix is calculated as:

$$\left[C_{O,ip} \right] = \left[C_{O_1,ip}^{(1)} \right] + \left[T_{OO_2,ip}^{(2)} \right]^T \left[R^{(2)} \right]^T \left[C_{O_2,ip}^{(2)} \right] \left[R^{(2)} \right] \left[T_{OO_2,ip}^{(2)} \right], \quad (6.41)$$

where

$$\left[T_{OO_2,ip}^{(2)} \right] = \begin{bmatrix} 1 & 0 & 0 \\ 0 & 1 & 0 \\ \Delta y_2 & -\Delta x_2 & 1 \end{bmatrix} = \begin{bmatrix} 1 & 0 & 0 \\ 0 & 1 & 0 \\ 0 & -l_1 & 1 \end{bmatrix};$$

$$\left[R^{(2)} \right] = \begin{bmatrix} \cos \pi & \sin \pi & 0 \\ -\sin \pi & \cos \pi & 0 \\ 0 & 0 & 1 \end{bmatrix}.$$

$$(6.42)$$

The compliance matrices $\left[C_{O,ip}^{(1)} \right], \left[C_{O,ip}^{(2)} \right]$ are calculated by means of Eqs. (6.30) and (6.35). Because the mechanism is symmetric with respect to the x axis, there is decoupling between the x and y axes, as well as between the x and z axes in the matrix $\left[C_{O,ip} \right]$. The nonzero components of the in-plane compliance matrix $\left[C_{O,ip} \right]$ of Eq. (6.41) are not included here being too complex.

Similarly, the out-of-plane compliance matrix of the mechanism of Figure 6.12a is:

$$\left[C_{O,op} \right] = \left[C_{O_1,op}^{(1)} \right] + \left[T_{OO_2,op}^{(2)} \right]^T \left[R^{(2)} \right]^T \left[C_{O_2,op}^{(2)} \right] \left[R^{(2)} \right] \left[T_{OO_2,op}^{(2)} \right], \quad (6.43)$$

with

$$\left[T_{OO_2,op}^{(2)} \right] = \begin{bmatrix} 1 & 0 & -\Delta y_2 \\ 0 & 1 & \Delta x_2 \\ 0 & 0 & 1 \end{bmatrix} = \begin{bmatrix} 1 & 0 & 0 \\ 0 & 1 & l_1 \\ 0 & 0 & 1 \end{bmatrix}; \quad (6.44)$$

and the compliance matrices $\left[C_{O,ip}^{(1)} \right]$ and $\left[C_{O,ip}^{(2)} \right]$ are calculated by means of Eqs. (6.33) and (6.36). The matrix $\left[C_{O,op} \right]$ of Eq. (6.43) shows no coupling between the x axis, on the one hand, and the y and z axes, on the other hand, due to mechanism symmetry about the x axis; as a result, the matrix components that model xy and xz interaction are zero – the nonzero components of $\left[C_{O,op} \right]$ are too involved and are not provided explicitly here.

The double-folded mechanism of Figure 6.12b consists of two identical and mirrored simple folded mechanisms, like the one of Figure 6.12a, which are connected in parallel. If we denote the mechanism to the right of the middle rigid link by the superscript "r" and the mirrored mechanism to the left of the middle rigid link by "l", the in-plane and out-of-plane compliance matrices of the mechanism shown in Figure 6.12b are expressed with respect to the central frame $Oxyz$ as:

$$\left[C_{O,ip}\right] = \left(\left[C_{O,ip}^{(r)}\right]^{-1} + \left[C_{O,ip}^{(l)}\right]^{-1}\right)^{-1}; \quad \left[C_{O,op}\right] = \left(\left[C_{O,op}^{(r)}\right]^{-1} + \left[C_{O,op}^{(l)}\right]^{-1}\right)^{-1}. \quad (6.45)$$

Both the in-plane and out-of-plane compliance matrices of Eq. (6.45) are diagonal, but their nonzero diagonal terms are not given here being too complex. The diagonal form is attributable to the mechanism being symmetric about the x and y axes.

Example 6.2

The chains of the double-folded parallelogram of Figure 6.12b are all single, straight-axis flexible hinges of constant, rectangular cross-section with an in-plane thickness t and an out-of-plane width w. An in-plane load f is applied at O, which is formed of f_x and f_y. Graphically study the variation of the displacement ratio $r_u = u_{Oy}/u_{Ox}$ in terms of h_2/h_1, l_2/l_1, and the angle α between f and the x axis. Known are $t = 0.001$ m, $l_1 = 0.05$ m and $h_1 = 0.015$ m.

Solution:

Because the matrix $[C_{O,ip}]$ of Eq. (6.45) is diagonal, the following relationships result:

$$\begin{bmatrix} u_{Ox} \\ u_{Oy} \\ \theta_{Oz} \end{bmatrix} = \begin{bmatrix} C_{O,ip}(1,1) & 0 & 0 \\ 0 & C_{O,ip}(2,2) & 0 \\ 0 & 0 & C_{O,ip}(3,3) \end{bmatrix} \begin{bmatrix} f_{Ox} \\ f_{Oy} \\ m_{Oz} \end{bmatrix} \rightarrow r_u$$

$$= \frac{u_{Oy}}{u_{Ox}} = \frac{C_{O,ip}(2,2) \cdot f_{Oy}}{C_{O,ip}(1,1) \cdot f_{Ox}} = \frac{C_{O,ip}(2,2)}{C_{O,ip}(1,1)} \cdot \tan\alpha. \quad (6.46)$$

The in-plane compliance matrix of the right half mechanism of Figure 6.12a is:

$$\left[C_{O,ip}^{(r)}\right] = \left[C_{O_1,ip}^{(1)}\right] + \left[T_{OO_2,ip}^{(2)}\right]^T \left[R^{(2)}\right]^T \left[C_{O_2,ip}^{(2)}\right]\left[R^{(2)}\right]\left[T_{OO_2,ip}^{(2)}\right], \quad (6.47)$$

with the translation and rotation matrices of Eqs. (6.42). The compliance matrices of the two serially connected pairs of flexures are those given in Eqs. (6.30) and (6.35), namely:

$$\left[C_{O,ip}^{(i)}\right] = \frac{1}{2} \cdot \begin{bmatrix} C_{u_x-f_x}^{(i)} & 0 & 0 \\ 0 & C_{u_y-f_y}^{(i)} & C_{u_y-m_z}^{(i)} \\ 0 & C_{u_y-m_z}^{(i)} & C_{\theta_z-m_z}^{(i)} \end{bmatrix}, \quad i = 1,2, \quad (6.48)$$

with

$$\begin{cases} C_{u_x-f_x}^{(i)} = C_{u_x-f_x}^{(hi)}; \\[2mm] C_{u_y-f_y}^{(i)} = \dfrac{C_{u_x-f_x}^{(hi)} \cdot C_{u_y-f_y}^{(hi)} + \left[C_{\theta_z-m_z}^{(hi)} \cdot C_{u_y-f_y}^{(hi)} - \left(C_{u_y-m_z}^{(hi)} \right)^2 \right]}{C_{u_x-f_x}^{(hi)} + h_i^2 \cdot C_{\theta_z-m_z}^{(hi)}}; \\[4mm] C_{u_y-m_z}^{(i)} = \dfrac{C_{u_x-f_x}^{(hi)} \cdot C_{u_y-m_z}^{(hi)}}{C_{u_x-f_x}^{(hi)} + h_i^2 \cdot C_{\theta_z-m_z}^{(hi)}}; \\[4mm] C_{\theta_z-m_z}^{(i)} = \dfrac{C_{u_x-f_x}^{(hi)} \cdot C_{\theta_z-m_z}^{(hi)}}{C_{u_x-f_x}^{(hi)} + h_i^2 \cdot C_{\theta_z-m_z}^{(hi)}} \end{cases} \qquad i = 1,2. \quad (6.49)$$

In Eqs. (6.48) and (6.49), "i" denotes either of the two pairs of parallel-connection hinges, while "hi" stands for the individual hinge 1 or 2 – these compliances are:

$$C_{u_x-f_x}^{(hi)} = \frac{l_i}{EA}; \quad C_{u_y-f_y}^{(hi)} = \frac{l_i^3}{3EI_z}; \quad C_{u_y-m_z}^{(hi)} = -\frac{l_i^2}{2EI_z}; := C_{\theta_z-m_z}^{(hi)} = \frac{l_i}{EI_z}, \qquad (6.50)$$

where $A = w \cdot t$ and $I_z = w \cdot t^3/12$.

As seen in Eq. (6.46), the displacement ratio varies linearly with $\tan\alpha$ when the compliance ratio of the same equation is constant. Figure 6.13a and b shows the displacement ratio in terms of the nondimensional variables $r_l = l_2/l_1$ and $r_h = h_2/h_1$. While the displacement ratio r_u is almost insensitive to variations in r_h, as seen in Figure 6.13b, the same ratio reaches a minimum value for approximately $r_l = 0.5$, as illustrated in Figure 6.13a.

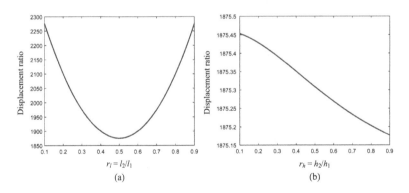

FIGURE 6.13 Plots of displacement ratio r_u in terms of (a) $r_l = l_2/l_1$ ratio; (b) $r_h = h_2/h_1$ ratio.

6.1.1.3.2 Mechanisms with Two Perpendicular Symmetry Axes

The skeleton-form planar mechanism of Figure 6.14 has two perpendicular symmetry axes, the x and y axes; as a consequence, the mechanism has a symmetry center – the intersection point O of the two symmetry axes. It can be considered that this mechanism combines in parallel the y-symmetric, two-chain upper

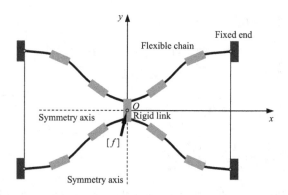

FIGURE 6.14 Planar parallel mechanism formed of four identical flexible chains that are symmetric with respect to the x and y axes.

portion with the identical lower segment that is the mirrored upper part with respect to the x axis.

The in-plane compliance of this mechanism is calculated as:

$$\left[C_{O,ip} \right] = \left(\left[C_{O,ip}^{(uh)} \right]^{-1} + \left[C_{O,ip}^{(lh)} \right]^{-1} \right)^{-1}, \tag{6.51}$$

where the superscript "uh" denotes the upper half-portion of the mechanism sketched in Figure 6.14, and "lh" identifies the lower half-section of the mechanism. The compliance matrix $\left[C_{O,ip}^{(uh)} \right]$ can be expressed by considering that the upper half-mechanism is the parallel combination of its mirrored right and left portions (quarters); as a result:

$$\left[C_{O,ip}^{(uh)} \right] = \frac{1}{2} \cdot \begin{bmatrix} C_{u_x-f_x} & 0 & C_{u_x-m_z} \\ 0 & C_{u_y-f_y} & 0 \\ C_{u_x-m_z} & 0 & C_{\theta_z-m_z} \end{bmatrix};$$

$$\left[C_{O,ip}^{(lh)} \right] = \frac{1}{2} \cdot \begin{bmatrix} C_{u_x-f_x} & 0 & -C_{u_x-m_z} \\ 0 & C_{u_y-f_y} & 0 \\ -C_{u_x-m_z} & 0 & C_{\theta_z-m_z} \end{bmatrix}, \tag{6.52}$$

where the individual compliances of Eq. (6.52) are explicitly given in Eqs. (6.18) and are based on the individual compliances of the upper-right flexible quarter mechanism of Figure 6.14. Combining Eqs. (6.51) and (6.52) yields:

$$\left[C_{O,ip} \right] = \frac{1}{4} \cdot \begin{bmatrix} C_{u_x-f_x} - \dfrac{\left(C_{u_x-m_z} \right)^2}{C_{\theta_z-m_z}} & 0 & 0 \\ 0 & C_{u_y-f_y} & 0 \\ 0 & 0 & C_{\theta_z-m_z} - \dfrac{\left(C_{u_x-m_z} \right)^2}{C_{u_x-f_x}} \end{bmatrix}. \tag{6.53}$$

It can be seen that the in-plane compliance matrix of the mechanism that has four identical flexible chains coupled in parallel and two symmetry axes is diagonal in a reference frame placed at the symmetry center O where the x and y axes are symmetry lines.

A similar procedure is followed in order to determine the out-of-plane compliance matrix of the parallel mechanism of Figure 6.14. The compliance matrices of the upper and lower halves are obtained from Eq. (6.22) as:

$$
\left[C_{O,op}^{(uh)} \right] = \frac{1}{2} \cdot \begin{bmatrix} C_{\theta_x-m_x} & 0 & C_{\theta_x-f_z} \\ 0 & C_{\theta_y-m_y} & 0 \\ C_{\theta_x-f_z} & 0 & C_{u_z-f_z} \end{bmatrix};
$$

$$
\left[C_{O,op}^{(lh)} \right] = \frac{1}{2} \cdot \begin{bmatrix} C_{\theta_x-m_x} & 0 & -C_{\theta_x-f_z} \\ 0 & C_{\theta_y-m_y} & 0 \\ -C_{\theta_x-f_z} & 0 & C_{u_z-f_z} \end{bmatrix},
$$

(6.54)

with the individual compliance of Eq. (6.54) being expressed in Eqs. (6.23) in terms of the known compliances of the upper-right flexible chain of Figure 6.14. The out-of-plane compliance of the full mechanism is:

$$
\left[C_{O,ip} \right] = \left(\left[C_{O,op}^{(uh)} \right]^{-1} + \left[C_{O,op}^{(lh)} \right]^{-1} \right)^{-1}
$$

$$
= \frac{1}{4} \cdot \begin{bmatrix} C_{\theta_x-m_x} - \dfrac{\left(C_{\theta_x-f_z} \right)^2}{C_{u_z-f_z}} & 0 & 0 \\ 0 & C_{\theta_y-m_y} & 0 \\ 0 & 0 & C_{u_z-f_z} - \dfrac{\left(C_{\theta_x-f_z} \right)^2}{C_{\theta_x-m_x}} \end{bmatrix}
$$

(6.55)

Note, again, the diagonal form of this matrix when calculated with respect to the reference frame located at the symmetry center O and with its x and y axes being the mechanism symmetry axes.

6.1.1.3.3   Mechanisms with Radial Symmetry

The planar mechanism whose top-view skeleton architecture is depicted in Figure 6.15 has three identical flexible chains (arms) converging radially to a central rigid link of center O; the arms are placed equidistantly at a center angle of 120° from one another.

Assume that the in- and out-of-plane compliance matrices of the three identical flexible chains are known and determined in their local frames $Ox_1y_1z_1$ (identical to the global frame $Oxyz$), $Ox_2y_2z_2$ and $Ox_3y_3z_3$, respectively, as:

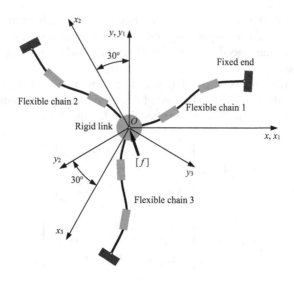

FIGURE 6.15 Planar parallel mechanism formed of three identical serial flexible chains that are radially symmetric.

$$
\left[C_{O,ip}^{(1)} \right] = \begin{bmatrix}
C_{u_x-f_x}^{(1)} & C_{u_x-f_y}^{(1)} & C_{u_x-m_z}^{(1)} \\
C_{u_x-f_y}^{(1)} & C_{u_y-f_y}^{(1)} & C_{u_y-m_z}^{(1)} \\
C_{u_x-m_z}^{(1)} & C_{u_y-m_z}^{(1)} & C_{\theta_z-m_z}^{(1)}
\end{bmatrix};
$$

$$
\left[C_{O,op}^{(1)} \right] = \begin{bmatrix}
C_{\theta_x-m_x}^{(1)} & C_{\theta_x-m_y}^{(1)} & C_{\theta_x-f_z}^{(1)} \\
C_{\theta_x-m_y}^{(1)} & C_{\theta_y-m_y}^{(1)} & C_{\theta_y-f_z}^{(1)} \\
C_{\theta_x-f_z}^{(1)} & C_{\theta_y-f_z}^{(1)} & C_{u_z-f_z}^{(1)}
\end{bmatrix},
$$

(6.56)

The chains 2 and 3 are rotated by the angles $2\pi/3$ and $4\pi/3$ with respect to chain 1. The three chains are connected in parallel to the central rigid link, and therefore, the in-plane compliance matrix of the full mechanism in the central frame $Oxyz$ is calculated as:

$$
\left[C_{O,ip} \right] = \left(\left[C_{O,ip}^{(1)} \right]^{-1} + \left(\left[R^{(2)} \right]^T \left[C_{O,ip}^{(1)} \right] \left[R^{(2)} \right] \right)^{-1} + \left(\left[R^{(3)} \right]^T \left[C_{O,ip}^{(1)} \right] \left[R^{(3)} \right] \right)^{-1} \right)^{-1}
$$

$$
= \begin{bmatrix}
C_{u_x-f_x} & 0 & 0 \\
0 & C_{u_y-f_y} & 0 \\
0 & 0 & C_{\theta_z-m_z}
\end{bmatrix}.
$$

(6.57)

Similarly, the out-of-plane compliance of the full mechanism is:

$$
\left[C_{O,op} \right] = \left(\left[C_{O,op}^{(1)} \right]^{-1} + \left(\left[R^{(2)} \right]^{T} \left[C_{O,op}^{(1)} \right] \left[R^{(2)} \right] \right)^{-1} + \left(\left[R^{(3)} \right]^{T} \left[C_{O,op}^{(1)} \right] \left[R^{(3)} \right] \right)^{-1} \right)^{-1}
$$

$$
= \begin{bmatrix} C_{\theta_x - m_x} & 0 & 0 \\ 0 & C_{\theta_y - m_y} & 0 \\ 0 & 0 & C_{u_z - f_z} \end{bmatrix}.
$$

(6.58)

The two rotation matrices in Eqs. (6.57) and (6.58) are:

$$
\left[R^{(2)} \right] = \begin{bmatrix} \cos(2\pi/3) & \sin(2\pi/3) & 0 \\ -\sin(2\pi/3) & \cos(2\pi/3) & 0 \\ 0 & 0 & 1 \end{bmatrix};
$$

$$
\left[R^{(3)} \right] = \begin{bmatrix} \cos(4\pi/3) & \sin(4\pi/3) & 0 \\ -\sin(4\pi/3) & \cos(4\pi/3) & 0 \\ 0 & 0 & 1 \end{bmatrix},
$$

(6.59)

The in-plane and out-of-plane compliance matrices are both diagonal, as seen in Eqs. (6.57) and (6.58), but their components are too complex and are not explicitly given here.

It can be demonstrated that any other stage configuration with an odd number m of identical chains disposed equidistantly radially, provided one chain (the first one, usually) is aligned with one of the global axes (like x), results in in-plane and out-of-plane compliance matrices that are diagonal when calculated in a central reference frame. For m identical chains, the in-plane compliance matrix is calculated by generalizing Eq. (6.57):

$$
\left[C_{O,ip} \right] = \left(\left[C_{O,ip}^{(1)} \right]^{-1} + \sum_{j=1}^{m-1} \left(\left[R^{(j+1)} \right]^{T} \left[C_{O,ip}^{(1)} \right] \left[R^{(j+1)} \right] \right)^{-1} \right)^{-1}
$$

$$
= \begin{bmatrix} C_{u_x - f_x} & 0 & 0 \\ 0 & C_{u_y - f_y} & 0 \\ 0 & 0 & C_{\theta_z - m_z} \end{bmatrix}.
$$

(6.60)

Similarly, Eq. (6.58) becomes:

$$
\begin{aligned}
\left[C_{O,op}\right] &= \left(\left[C_{O,op}^{(1)}\right]^{-1} + \sum_{j=1}^{m-1} \left(\left[R^{(j+1)}\right]^{T} \left[C_{O,op}^{(1)}\right] \left[R^{(j+1)}\right]\right)^{-1}\right)^{-1} \\
&= \begin{bmatrix} C_{\theta_x - m_x} & 0 & 0 \\ 0 & C_{\theta_y - m_y} & 0 \\ 0 & 0 & C_{u_z - f_z} \end{bmatrix},
\end{aligned}
\tag{6.61}
$$

with

$$
\left[R^{(j+1)}\right] = \begin{bmatrix} \cos\left(2j\cdot\pi/m\right) & \sin\left(2j\cdot\pi/m\right) & 0 \\ \sin\left(2j\cdot\pi/m\right) & \cos\left(2j\cdot\pi/m\right) & 0 \\ 0 & 0 & 1 \end{bmatrix}, \quad j = 1,2,...,m-1. \tag{6.62}
$$

While not demonstrated here, it can also be shown that stages with an even number of identical flexible chains ($m = 4, 6, ...$) that are connected in parallel in a radial, angularly equidistant manner produce in-plane and out-of-plane compliance matrices that are diagonal.

Stages with identical radial-symmetry serial folded chains:

The planar mechanism shown in Figure 6.16a consists of a central, rigid link and m identical, serial compliant chains (arms) that are placed radially and equidistantly in the same plane with the central link. Each flexible arm has a folded configuration comprising circular-axis flexible segments interconnected with rigid, straight-axis links of radial direction – see Figures 4.30 and 6.16b. The in- and out-of-plane compliance matrices of these arms are provided in Section 4.2.4.3.

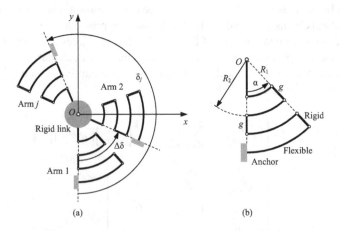

(a)   (b)

FIGURE 6.16 Planar stage with identical radial-symmetry flexible arms: (a) skeleton generic representation; (b) arm representation.

Because the identical arms are connected in parallel with respect to the central link, the in-plane stiffness matrix in the global frame $Oxyz$ is the sum of individual-arm stiffness matrices after pertinent rotations that align the local frames to the global frame $Oxyz$, which is expressed in compliance form as:

$$\left[K_{O,ip}\right]=\left[C_{O,ip}\right]^{-1}=\sum_{j=1}^{m}\left[C_{O,ip}^{(j)}\right]^{-1}$$

$$\text{with } \left[C_{O,ip}^{(j)}\right]=\left[R^{(j)}\right]^{T}\left[C_{O_j,ip}^{(j)}\right]\left[R^{(j)}\right]; \quad \left[R^{(j)}\right]=\begin{bmatrix} \cos\delta_j & \sin\delta_j & 0 \\ -\sin\delta_j & \cos\delta_j & 0 \\ 0 & 0 & 1 \end{bmatrix}.$$

(6.63)

where $\left[C_{O_j,ip}^{(j)}\right]$ is the in-plane compliance matrix of a folded arm in its local frame – see Eq. (4.66). Similarly, the out-of-plane stiffness matrix at the mechanism center is:

$$\left[K_{O,op}\right]=\left[C_{O,op}\right]^{-1}=\sum_{j=1}^{m}\left[C_{O,op}^{(j)}\right]^{-1} \quad \text{with } \left[C_{O,op}^{(j)}\right]=\left[R^{(j)}\right]^{T}\left[C_{O_j,op}^{(j)}\right]\left[R^{(j)}\right],$$

(6.64)

where $\left[C_{O_j,op}^{(j)}\right]$ is the out-of-plane, local-frame compliance matrix of a folded arm. Note that δ_j is measured from the radial direction of arm 1, which is consistent with the arm 1 being positioned by $\delta_1 = 0$. For m identical radial arms, the center angle between two consecutive arms $\Delta\delta$, and the position angle δ_i of arm i with respect to the global axis x are:

$$\Delta\delta = \frac{2\pi}{m}; \quad \delta_j = (j-1)\cdot\Delta\delta = \frac{2\pi(j-1)}{m}.$$

(6.65)

Example 6.3

Consider the flexible-hinge mechanism of Figure 6.16. The circular segments have a constant rectangular cross-section with an in-plane width w and an out-of-plane thickness t.

    i. For an arm defined by $n = 6$ radial segments, $w = 0.001$ m, $t = 0.005$ m, $R_1 = 0.03$ m, $g = 0.015$ m, $\alpha = 50°$, $E = 2\cdot10^{11}$ N/m^2 and $\mu = 0.3$, evaluate the mechanism compliance $C_{\theta_z-m_z}$ as well the ratio $C_{\theta_z-m_z}/C_{\theta_x-m_x}$ for $m = 2, 3, 4, 5, 6$ arms;

    ii. For all the parameters provided at (i), except for $w = 0.005$ m and $t = 0.001$ m, evaluate the compliance $C_{u_z-f_z}$ as well as the ratio $C_{u_z-f_z}/C_{u_x-f_x}$ in terms of the same values of m.

Solution:

    i. These configurations are all thick designs, with the constant out-of-plane thickness t larger than the circular segments in-plane width

w; as a consequence, the compliance $C_{\theta_z-m_z}$ is an active compliance, whereas $C_{\theta_x-m_x}$ is a parasitic one, which needs to be minimized. Table 6.1 lists the values of $C_{\theta_z-m_z}$ and the compliance ratio. As expected, the active compliance decreases with the increasing number of parallel-connection arms. The compliance ratio $C_{\theta_z-m_z}/C_{\theta_x-m_x}$, on the other hand, remains constant for m equal or larger than 3; this indicates that both the active and the parasitic compliances have the same sensitivity to variations in the number of arms when m ranges from 3 to 6. Also, note that the active compliance is larger than the parasitic one for the given geometrical parameters.

ii. These are all thin designs since the constant out-of-plane thickness t is smaller than the circular segments in-plane width w; as a consequence, the compliance $C_{u_z-f_z}$ is an active compliance, while $C_{u_x-f_x}$ is a parasitic compliance. Table 6.2 displays the values of $C_{u_z-f_z}$ and $C_{u_z-f_z}/C_{u_x-f_x}$ in terms of the number of arms m. Again, the active compliance decreases when the number of parallel-connection arms increases. Like at (i), the compliance ratio $C_{u_z-f_z}/C_{u_x-f_x}$ remains constant for m ranging from 3 to 6, which signals that both the active compliance and the parasitic one have the same sensitivity to the variation in the number of arms m. However, the active compliance is larger than the parasitic one, as illustrated by the compliance ratio.

TABLE 6.1

In-Plane Compliance $C_{\theta_z-m_z}$ and Compliance Ratio $C_{\theta_z-m_z}/C_{\theta_x-m_x}$ in Terms of the Arm Number m

	m				
	2	3	4	5	6
$C_{\theta_z-m_z}$ [m^{-1} N^{-1}]	0.184	0.123	0.092	0.074	0.061
$C_{\theta_z-m_z}/C_{\theta_x-m_x}$	9.03	13.69	13.69	13.69	13.69

TABLE 6.2

Out-of-Plane Compliance $C_{u_z-f_z}$ and Compliance Ratio $C_{u_z-f_z}/C_{u_x-f_x}$ in Terms of the Arm Number m

	m				
	2	3	4	5	6
$C_{u_z-f_z}$ [m N^{-1} $\cdot$ 10^{-3}]	0.713	0.476	0.357	0.285	0.238
$C_{u_z-f_z}/C_{u_x-f_x}$	16.49	18.42	18.42	18.42	18.42

6.1.2 LOAD ON RIGID LINK COMMON TO ALL FLEXIBLE CHAINS AND LOADS ON FLEXIBLE CHAINS

A more complex loading condition is encountered when additional loads are applied to all or just some of the parallel chains in addition to the common rigid-link load. This section derives a compliance matrix model for planar mechanisms with multiple loads applied to parallel-connection mechanisms. The model is subsequently applied to planar, displacement-amplification and branched flexible mechanisms.

6.1.2.1 Generic Compliance-Matrix Model

The compliance-matrix model developed here applies to both in-plane and out-of-plane flexible-hinge mechanisms, and consequently, the regular identifiers "*ip*" and "*op*" are not utilized. Before generalizing to a configuration with m flexible chains that are loaded, the simpler design with only two chains connected in parallel and acted upon by two force vectors $[f_{A1}]$ and $[f_{A2}]$ is analyzed, based on Figure 6.17.

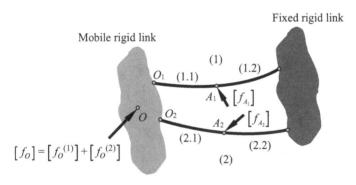

FIGURE 6.17 Planar mechanism with two loaded flexible chains connected in parallel to a rigid link that is acted upon by an external load.

Consider that each of the two chains is separated into two portions by the point of application of the external load. As such, chain (1) is formed of the flexible segments (1.1) and (1.2), whereas the chain (2) is divided into the segments (2.1) and (2.2). The displacement at point O on the common rigid link is also shared by the two flexible chains when their displacements are transferred at O from their roots O_1 and O_2, which is expressed as:

$$[u_O] = \left[C_O^{(1)} \right]\left[f_O^{(1)} \right] + \left[C_{OA_1}^{(1.2)} \right][f_{A_1}] = \left[C_O^{(2)} \right]\left[f_O^{(2)} \right] + \left[C_{OA_2}^{(2.2)} \right][f_{A_2}], \quad (6.66)$$

where $\left[C_O^{(1)} \right]$ represents the compliance of the entire chain 1 transferred at O, while $\left[C_{OA_1}^{(1.2)} \right]$ is the compliance of the portion 1.2 of chain 1, also transferred at O; the compliances $\left[C_O^{(2)} \right]$ and $\left[C_{OA_2}^{(2.2)} \right]$ define the similar portions of the chain 2.

The external force at O is divided into the two components used in Eq. (6.66), each distributed to either of the two chains, namely:

$$\left[f_O^{(1)}\right]+\left[f_O^{(2)}\right]=\left[f_O\right].\tag{6.67}$$

Equations (6.66) and (6.67) are solved for $\left[f_O^{(1)}\right]$ and $\left[f_O^{(2)}\right]$, resulting in:

$$\left[\begin{array}{c}\left[f_O^{(1)}\right]\\\left[f_O^{(2)}\right]\end{array}\right]=[A]\left[\begin{array}{c}\left[f_{A_1}\right]\\\left[f_{A_2}\right]\\\left[f_O\right]\end{array}\right];\ [A]=[A_1]^{-1}[A_2],\left\{\begin{array}{l}[A_1]=\left[\begin{array}{cc}\left[C_O^{(1)}\right]&-\left[C_O^{(2)}\right]\\{}[I]&[I]\end{array}\right]\\[A_2]=\left[\begin{array}{ccc}-\left[C_{OA_1}^{(1.2)}\right]&\left[C_{OA_2}^{(2.2)}\right]&[0]\\{}[0]&[0]&[I]\end{array}\right]\end{array}\right.,$$

$$\tag{6.68}$$

where $[I]$ and $[0]$ are the 3×3 identity and zero matrices, respectively. With the load components expressed in Eqs. (6.68), the displacement at O can be determined from Eq. (6.66). The displacements at A_1 and A_2 can also be calculated when expressing the following displacement differences corresponding to chains 1 and 2, respectively:

$$\begin{array}{ll}\left[u_O\right]-\left[u_{A_1}\right]=\left[C_O^{(1.1)}\right]\left[f_O^{(1)}\right]&\text{or}\quad\left[u_{A_1}\right]=\left[u_O\right]-\left[C_O^{(1.1)}\right]\left[f_O^{(1)}\right]\\\left[u_O\right]-\left[u_{A_2}\right]=\left[C_O^{(2.1)}\right]\left[f_O^{(2)}\right]&\text{or}\quad\left[u_{A_2}\right]=\left[u_O\right]-\left[C_O^{(2.1)}\right]\left[f_O^{(2)}\right]\end{array},\tag{6.69}$$

with the aid of Eqs. (6.66) and (6.68).

Consider now the general design with m parallel-connection flexible chains, each acted upon by a load vector in addition to the load applied to the common rigid link, as illustrated in Figure 6.18. It is necessary to first evaluate the load

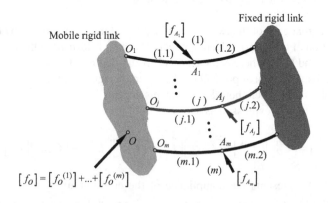

FIGURE 6.18 Planar mechanism with m loaded flexible chains connected in parallel and acted upon by a load applied to the common rigid link.

components $\left[f_O^{(j)} \right]$ ($j = 1, 2, \ldots, m$) that are applied to the m parallel-connection chains and resulting from $[f_O]$.

The displacement $[u_O]$ is shared by all the flexible chains at O because they are connected in parallel. Similar to Eq. (6.66), the following $m-1$ equations, each expressing the displacement $[u_O]$, are formulated:

$$
\begin{cases}
\left[C_O^{(1)} \right]\left[f_O^{(1)} \right] - \left[C_O^{(2)} \right]\left[f_O^{(2)} \right] = \left[C_{OA_2}^{(2.2)} \right]\left[f_{A_2} \right] - \left[C_{OA_1}^{(1.2)} \right]\left[f_{A_1} \right] \\
\ldots \\
\left[C_O^{(j-1)} \right]\left[f_O^{(j-1)} \right] - \left[C_O^{(j)} \right]\left[f_O^{(j)} \right] = \left[C_{OA_j}^{(j.2)} \right]\left[f_{A_j} \right] - \left[C_{OA_{j-1}}^{(j-1.2)} \right]\left[f_{A_{j-1}} \right] \\
\ldots \\
\left[C_O^{(m-1)} \right]\left[f_O^{(m-1)} \right] - \left[C_O^{(m)} \right]\left[f_O^{(m)} \right] = \left[C_{OA_m}^{(m.2)} \right]\left[f_{A_m} \right] - \left[C_{OA_{m-1}}^{(m-1.2)} \right]\left[f_{A_{m-1}} \right] \\
\left[f_O^{(1)} \right] + \cdots + \left[f_O^{(j)} \right] + \cdots + \left[f_O^{(m)} \right] = [f_O]
\end{cases}
\tag{6.70}
$$

The last Eq. (6.70) shows that the sum of the m load components that are transferred to the m chains is equal to the external load applied at O. The m Eqs. (6.70) are solved for the m unknown forces as:

$$
\begin{bmatrix}
\left[f_O^{(1)} \right] \\
\left[f_O^{(2)} \right] \\
\ldots \\
\left[f_O^{(j-1)} \right] \\
\left[f_O^{(j)} \right] \\
\ldots \\
\left[f_O^{(m-1)} \right] \\
\left[f_O^{(m)} \right]
\end{bmatrix}
= [A]_{m\times(m+1)}
\begin{bmatrix}
\left[f_{A_1} \right] \\
\left[f_{A_2} \right] \\
\ldots \\
\left[f_{A_{j-1}} \right] \\
\left[f_{A_j} \right] \\
\ldots \\
\left[f_{A_m} \right] \\
[f_O]
\end{bmatrix}
; \quad [A] = [A_1]^{-1}[A_2],
\tag{6.71}
$$

where

$$
[A_1]_{m\times m} =
\begin{bmatrix}
\left[C_O^{(1)} \right] & -\left[C_O^{(2)} \right] & \cdots & [0] & [0] & \cdots & [0] & [0] \\
[0] & \left[C_O^{(2)} \right] & \cdots & [0] & [0] & \cdots & [0] & [0] \\
\cdots & \cdots & \cdots & \cdots & \cdots & & & \\
[0] & [0] & \cdots & \left[C_O^{(j-1)} \right] & -\left[C_O^{(j)} \right] & \cdots & [0] & [0] \\
[0] & [0] & \cdots & [0] & \left[C_O^{(j)} \right] & \cdots & [0] & [0] \\
\cdots & \cdots & \cdots & \cdots & \cdots & \cdots & \cdots & \cdots \\
[0] & [0] & \cdots & [0] & [0] & \cdots & \left[C_O^{(m-1)} \right] & -\left[C_O^{(m)} \right] \\
[I] & [I] & \cdots & [I] & [I] & \cdots & [I] & [I]
\end{bmatrix}
; \tag{6.72a}
$$

$$
[A_2]_{m \times (m+1)} =
\begin{bmatrix}
-\left[C_{OA_1}^{(1.2)}\right] & \left[C_{OA_2}^{(2.2)}\right] & [0] & \cdots & [0] & \cdots & [0] & [0] & [0] \\
[0] & -\left[C_{OA_2}^{(2.2)}\right] & \left[C_{OA_3}^{(3.2)}\right] & \cdots & [0] & \cdots & [0] & [0] & [0] \\
\cdots & \cdots & \cdots & \cdots & \cdots & \cdots & \cdots & \cdots & \cdots \\
[0] & [0] & [0] & \cdots & [0] & \cdots & -\left[C_{OA_{m-1}}^{(m-1.2)}\right] & \left[C_{OA_m}^{(m.2)}\right] & [0] \\
[0] & [0] & [0] & \cdots & [0] & \cdots & [0] & [0] & [I]
\end{bmatrix}.
$$

$$(6.72b)$$

The displacement vectors at O and at the m points A_j where the external forces are applied are calculated as:

$$
\begin{cases}
[u_O] = \left[C_O^{(j)}\right]\left[f_O^{(j)}\right] + \left[C_{OA_j}^{(j.2)}\right]\left[f_{A_j}\right] \\
\left[u_{A_j}\right] = [u_O] - \left[C_O^{(j.1)}\right]\left[f_O^{(j)}\right]
\end{cases}; \quad j = 1,2\ldots,m, \qquad (6.73)
$$

where $\left[f_O^{(j)}\right]$ are determined by means of Eqs. (6.71) and (6.72).

6.1.2.2 Displacement-Amplification Mechanisms

Displacement-amplification mechanisms with flexible hinges are regularly utilized to augment relatively low displacement levels provided by piezoelectric actuation in macroscale applications or by various other actuation means in micro/nanosystems – see, for instance, Mottard and St-Amant [51], Ma et al. [52], Du et al. [53], Jouaneh and Yang [54], Kim et al. [55], Ni et al. [56], Choi et al. [57] and Lobontiu and Garcia [58]. Displacement-amplification planar mechanisms with one or two symmetry axes and with load components that are not applied along the structural symmetry axes need to be modeled as parallel combinations of two symmetric halves of the overall mechanism, as studied in this section.

6.1.2.2.1 Mechanisms with Two Symmetry Axes

In Chapter 5, the quasi-static response of displacement-amplification mechanisms with two symmetry axes is reduced to the analysis of a quarter-mechanism serial chain with two guided ends along perpendicular directions when the external and actuation loads are applied at points on the symmetry axes and along these axes. When the external loads are not applied along the symmetry axes, as shown in Figure 6.19a, it is no longer possible to analyze one-quarter mechanism only to reflect the full-mechanism behavior; instead, the mechanism should be considered as being structured from two halves that are coupled in parallel at the point on the symmetry axes where a non-symmetric load is applied. The right half of the mechanism is shown in Figure 6.19b with its portion of the total external load applied at point O.

The general Eq. (6.68), which expresses the loads pertaining to each of the two individual chains that are coupled in parallel at O (see Figure 6.17), also applies to the two halves of the Figure 6.19a mechanism. The matrix $[A]$ of Eq. (6.68) is:

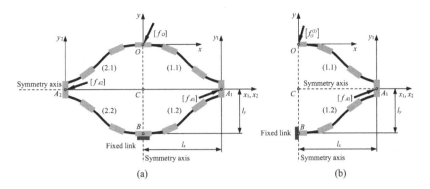

FIGURE 6.19 Generic displacement-amplification flexible mechanism with two symmetry axes under non-symmetric external load: (a) full mechanism; (b) half-mechanism.

$$[A] = \begin{bmatrix} [A_{11}] & [A_{12}] & [A_{13}] \\ [A_{21}] & [A_{22}] & [A_{23}] \end{bmatrix} = \begin{bmatrix} \left[C_O^{(1)} \right] & -\left[C_O^{(2)} \right] \\ [I] & [I] \end{bmatrix}^{-1}$$

$$\begin{bmatrix} -\left[C_{OA_1}^{(1.2)} \right] & \left[C_{OA_2}^{(2.2)} \right] & [0] \\ [0] & [0] & [I] \end{bmatrix}, \quad (6.74)$$

where the superscript (1) denotes the right half mechanism – the one of Figure 6.19b – and (2) indicates the left portion of the full mechanism depicted in Figure 6.19a. As shown in Figure 6.19, the two quarters at the bottom of the mechanism are denoted by (1.2) and (2.2), and they are also included in Eq. (6.74).

As discussed in the section dedicated to the series connection of two identical and mirrored chains, the compliance matrix of the right half-chain of Figure 6.19b is expressed with respect to the reference frame $Oxyz$ as:

$$\left[C_O^{(rh)} \right] = \left[C_O^{(1)} \right] = \left[T_{OA_1} \right]^T \left[C_{A_1}^{(1)} \right] \left[T_{OA_1} \right]. \quad (6.75)$$

The compliance matrix $\left[C_{A_1}^{(1)} \right]$ of the half-segment at A_1 and the corresponding translation matrix from A_1 to O are calculated as per Eqs. (5.105) and (5.106), namely:

$$\left[C_{A_1}^{(rh)} \right] = \left[C_{A_1}^{(1)} \right] = 2 \cdot \begin{bmatrix} C_{u_x - f_x}^{(q)} & 0 & 0 \\ 0 & C_{u_y - f_y}^{(q)} & C_{u_y - m_z}^{(q)} \\ 0 & C_{u_y - m_z}^{(q)} & C_{\theta_z - m_z}^{(q)} \end{bmatrix} = 2 \cdot \begin{bmatrix} C_{u_x - f_x}^{(1.1)} & 0 & 0 \\ 0 & C_{u_y - f_y}^{(1.1)} & C_{u_y - m_z}^{(1.1)} \\ 0 & C_{u_y - m_z}^{(1.1)} & C_{\theta_z - m_z}^{(1.1)} \end{bmatrix};$$

$$\left[T_{OA_1} \right] = \begin{bmatrix} 1 & 0 & 0 \\ 0 & 1 & 0 \\ \Delta y_{OA_1} & -\Delta x_{OA_1} & 1 \end{bmatrix} \quad \text{with} \quad \Delta x_{OA_1} = l_x; \quad \Delta y_{OA_1} = -l_y$$

$$(6.76)$$

The superscript "q" denotes the basic quarter chain identified as (1.1) in the frame $A_1x_1y_1$ of Figure 6.19. The final expression of the in-plane compliance matrix of Eq. (6.75) is of the form:

$$\left[C_O^{(1)}\right] = \begin{bmatrix} C_O^{(1)}(1,1) & C_O^{(1)}(1,2) & C_O^{(1)}(1,3) \\ C_O^{(1)}(1,2) & C_O^{(1)}(2,2) & C_O^{(1)}(2,3) \\ C_O^{(1)}(1,3) & C_O^{(1)}(2,3) & C_O^{(1)}(3,3) \end{bmatrix}. \tag{6.77}$$

The in-plane compliance of the left half-chain of Figure 6.19a is found from the right half-compliance matrix of Eq. (6.77) by using the y-axis symmetry of this segment, which effects in changing the signs of the components involving x-y and y-z axes coupling, namely:

$$\left[C_O^{(lh)}\right] = \left[C_O^{(2)}\right] = \begin{bmatrix} C_O^{(1)}(1,1) & -C_O^{(1)}(1,2) & C_O^{(1)}(1,3) \\ -C_O^{(1)}(1,2) & C_O^{(1)}(2,2) & -C_O^{(1)}(2,3) \\ C_O^{(1)}(1,3) & -C_O^{(1)}(2,3) & C_O^{(1)}(3,3) \end{bmatrix}. \tag{6.78}$$

The compliance matrix $\left[C_{OA_1}^{(1.2)}\right]$ of Eq. (6.74) is determined as:

$$\left[C_{OA_1}^{(1.2)}\right] = \left[T_{OA_1}\right]^T \left[C_{A_1}^{(1.2)}\right] = \left[T_{OA_1}\right]^T \begin{bmatrix} C_{u_x-f_x}^{(q)} & -C_{u_x-f_y}^{(q)} & -C_{u_x-m_z}^{(q)} \\ -C_{u_x-f_y}^{(q)} & C_{u_y-f_y}^{(q)} & C_{u_y-m_z}^{(q)} \\ -C_{u_x-m_z}^{(q)} & C_{u_y-m_z}^{(q)} & C_{\theta_z-m_z}^{(q)} \end{bmatrix}$$

$$= \left[T_{OA_1}\right]^T \begin{bmatrix} C_{u_x-f_x}^{(1.1)} & -C_{u_x-f_y}^{(1.1)} & -C_{u_x-m_z}^{(1.1)} \\ -C_{u_x-f_y}^{(1.1)} & C_{u_y-f_y}^{(1.1)} & C_{u_y-m_z}^{(1.1)} \\ -C_{u_x-m_z}^{(1.1)} & C_{u_y-m_z}^{(1.1)} & C_{\theta_z-m_z}^{(1.1)} \end{bmatrix}. \tag{6.79}$$

Note that the compliance matrix multiplying the translation matrix in Eq. (6.79) results from the following compliance matrix:

$$\left[C_{A_1}^{(q)}\right] = \left[C_{A_1}^{(1.1)}\right] = \begin{bmatrix} C_{u_x-f_x}^{(q)} & C_{u_x-f_y}^{(q)} & C_{u_x-m_z}^{(q)} \\ C_{u_x-f_y}^{(q)} & C_{u_y-f_y}^{(q)} & C_{u_y-m_z}^{(q)} \\ C_{u_x-m_z}^{(q)} & C_{u_y-m_z}^{(q)} & C_{\theta_z-m_z}^{(q)} \end{bmatrix}$$

$$= \begin{bmatrix} C_{u_x-f_x}^{(1.1)} & C_{u_x-f_y}^{(1.1)} & C_{u_x-m_z}^{(1.1)} \\ C_{u_x-f_y}^{(1.1)} & C_{u_y-f_y}^{(1.1)} & C_{u_y-m_z}^{(1.1)} \\ C_{u_x-m_z}^{(1.1)} & C_{u_y-m_z}^{(1.1)} & C_{\theta_z-m_z}^{(1.1)} \end{bmatrix} \tag{6.80}$$

that defines the segment 1.1, the basic quarter, with respect to the reference frame centered at A_1 by changing the signs of the matrix elements in Eq. (6.80) that reflect x-y and x-z coupling. Similarly, the in-plane compliance matrix $\left[C_{OA_2}^{(2,2)} \right]$ of Eq. (6.74) is calculated as:

$$\left[C_{OA_2}^{(2,2)} \right] = \left[T_{OA_2} \right]^T \left[C_{A_2}^{(2,2)} \right];$$

$$\left[C_{A_2}^{(2,2)} \right] = \left[T_{A_2 A_1} \right]^T \begin{bmatrix} C_{u_x - f_x}^{(q)} & C_{u_x - f_y}^{(q)} & -C_{u_x - m_z}^{(q)} \\ C_{u_x - f_y}^{(q)} & C_{u_y - f_y}^{(q)} & -C_{u_y - m_z}^{(q)} \\ -C_{u_x - m_z}^{(q)} & -C_{u_y - m_z}^{(q)} & C_{\theta_z - m_z}^{(q)} \end{bmatrix} \left[T_{A_2 A_1} \right]. \tag{6.81}$$

where

$$\left[T_{OA_2} \right] = \begin{bmatrix} 1 & 0 & 0 \\ 0 & 1 & 0 \\ -l_y & l_x & 1 \end{bmatrix}; \quad \left[T_{A_2 A_1} \right] = \begin{bmatrix} 1 & 0 & 0 \\ 0 & 1 & 0 \\ 0 & -2l_x & 1 \end{bmatrix}. \tag{6.82}$$

The particular signs of the compliance matrix that multiplies the translation matrix of Eq. (6.81) were obtained by starting from the compliance matrix of the basic segment 1.1 given in Eq. (6.80): a mirroring was first performed with respect to the y axis (which changes the signs of the components indicating coupling between the x-y and the y-z axes) followed by a reflection with respect to the x axis (which changes the signs in the newly obtained matrix of the components that indicate coupling between the x-y and x-z axes). The compliance matrices of Eqs. (6.69), which enable expressing the displacements at points A_1 and A_2, are calculated as:

$$\left[C_O^{(1,1)} \right] = \left[T_{OA_1} \right]^T \left[C_{A_1}^{(1,1)} \right] \left[T_{OA_1} \right];$$

$$\left[C_O^{(2,1)} \right] = \begin{bmatrix} C_O^{(1,1)}(1,1) & -C_O^{(1,1)}(1,2) & C_O^{(1,1)}(1,3) \\ -C_O^{(1,1)}(2,1) & C_O^{(1,1)}(2,2) & -C_O^{(1,1)}(2,3) \\ C_O^{(1,1)}(3,1) & -C_O^{(1,1)}(3,2) & C_O^{(1,1)}(3,3) \end{bmatrix}. \tag{6.83}$$

Relevant displacements and performance qualifiers:
    The displacements at the points of interest and the mechanism performance qualifiers can be evaluated once the in-plane and out-of-plane loads on the two parallel chains have been determined. The displacements at point O (the output

port) are calculated by using either the right half-chain or the left half-one as per Eq. (6.66). Utilizing Eqs. (6.66) and (6.74) allows us to express $[u_O]$ as a linear combination of the displacements produced by the three loads:

$$[u_O] = [C_{OA_1}][f_{A_1}] + [C_{OA_2}][f_{A_2}] + [C_O][f_O] \quad \text{with}:$$

$$\begin{cases} [C_{OA_1}] = [C_O^{(1)}][A_{11}] + [C_{OA_1}^{(1.2)}] = [C_O^{(2)}][A_{21}] \\ [C_{OA_2}] = [C_O^{(1)}][A_{12}] = [C_O^{(2)}][A_{22}] + [C_{OA_2}^{(2.2)}] \\ [C_O] = [C_O^{(1)}][A_{13}] = [C_O^{(2)}][A_{23}] \end{cases} \quad . \tag{6.84}$$

Similarly, the displacements at A_1 and A_2, which are originally expressed as in Eq. (6.69), become:

$$\begin{cases} [u_{A_1}] = [C_{A_1}][f_{A_1}] + [C_{A_1 A_2}][f_{A_2}] + [C_{A_1 O}][f_O] \\ \text{with} \quad [C_{A_1}] = [C_{OA_1}] - [C_O^{(1.1)}][A_{11}]; \\ [C_{A_1 A_2}] = [C_{OA_2}] - [C_O^{(1.1)}][A_{12}]; \quad [C_{A_1 O}] = [C_O] - [C_O^{(1.1)}][A_{13}]; \\ [u_{A_2}] = [C_{A_2 A_1}][f_{A_1}] + [C_{A_2}][f_{A_2}] + [C_{A_2 O}][f_O] \\ \text{with} \quad [C_{A_2 A_1}] = [C_{OA_1}] - [C_O^{(2.1)}][A_{21}]; \\ [C_{A_2}] = [C_{OA_2}] - [C_O^{(2.1)}][A_{22}]; \quad [C_{A_2 O}] = [C_O] - [C_O^{(2.1)}][A_{23}] \end{cases} \tag{6.85}$$

after substituting $[f_O^{(1)}]$ and $[f_O^{(2)}]$ of Eqs. (6.74) into Eqs. (6.69).

For a freely actuated system (with no external load, $[f_O] = [0]$), Eqs. (6.84) and (6.85) simplify to:

$$\begin{cases} [u_O] = [C_{OA_1}][f_{A_1}] + [C_{OA_2}][f_{A_2}] \\ [u_{A_1}] = [C_{A_1}][f_{A_1}] + [C_{A_1 A_2}][f_{A_2}] \\ [u_{A_2}] = [C_{A_2 A_1}][f_{A_1}] + [C_{A_2}][f_{A_2}] \end{cases} \tag{6.86}$$

The last two Eqs. (6.86) result in the following load/displacement relationship:

$$[f_A] = [K_{in}] = [u_A] \quad \text{with}:$$

$$[f_A] = \begin{bmatrix} [f_{A_1}] \\ [f_{A_2}] \end{bmatrix}; \quad [u_A] = \begin{bmatrix} [u_{A_1}] \\ [u_{A_2}] \end{bmatrix}; \tag{6.87}$$

$$[K_{in}] = \begin{bmatrix} [K_{in,11}] & [K_{in,12}] \\ [K_{in,21}] & K_{in,22} \end{bmatrix} = \begin{bmatrix} [C_{A_1}] & [C_{A_1 A_2}] \\ [C_{A_2 A_1}] & [C_{A_2}] \end{bmatrix}^{-1},$$

which identifies the *input stiffness matrix* $[K_{in}]$. Combining Eq. (6.87) with the first Eq. (6.86) yields the displacement at O as a linear combination of the displacements at A_1 and A_2, namely:

$$[u_O] = \left([C_{OA_1}][K_{in,11}] + [C_{OA_2}][K_{in,21}]\right)[u_{A_1}]$$
$$+ \left([C_{OA_1}][K_{in,12}] + [C_{OA_2}][K_{in,22}]\right)[u_{A_2}] \qquad (6.88)$$

Equation (6.88) may serve at qualifying the displacement amplification/reduction of a specific mechanism.

Another particular load case is when the two actuation vectors are zero, which simplifies Eqs. (6.84) and (6.85) to:

$$\begin{cases} [u_O] = [C_O][f_O] \quad \text{or} \quad [f_O] = [C_O]^{-1}[u_O] = [K_{out}][u_O] \\ [u_{A_1}] = [C_{A_1O}][f_O]; \quad [u_{A_2}] = [C_{A_2O}][f_O] \end{cases} \qquad (6.89)$$

where the output stiffness matrix $[K_{out}]$ is formulated. Equations (6.89) also enable expressing the following relationships between the displacement vectors:

$$\begin{cases} [u_O] = [C_O][C_{A_1O}]^{-1}[u_{A_1}] = [C_O][C_{A_2O}]^{-1}[u_{A_2}]; \\ [u_{A_1}] = [C_{A_1O}][C_{A_2O}]^{-1}[u_{A_2}]; \\ [u_{A_2}] = [C_{A_2O}][C_{A_1O}]^{-1}[u_{A_1}]. \end{cases} \qquad (6.90)$$

Example 6.4

The two-axis symmetric mechanism shown in skeleton form in Figure 6.20a is actuated by the horizontal forces $f_{A1} = f_{A2}$. Known are the lengths of the rigid links: $l_1 = 0.01$ m, $l_2 = 0.06$ m and $l_3 = 0.03$ m (see Figure 6.20b). The identical flexible hinges are all horizontal and of constant rectangular cross-section with an in-plane thickness $t = 0.001$ m and an out-of-plane width $w = 0.006$ m; their length is $l = 0.015$ m. The material Young's modulus is $E = 2 \cdot 10^{11}$ N/m².

i. Study the mechanism displacement amplification in terms of the angle δ;

ii. For $\delta = 10°$ and assuming the two horizontal actuation forces are not identical, express f_{A1} and f_{A2} in terms of specified displacements u_{Ox} and u_{Oy} at point O. Calculate the particular values of f_{A1} and f_{A2} when $u_{Ox} = 0.005$ m and $u_{Oy} = 0.01$ m.

Solution:

i. The compliance matrix $[C_O^{(1)}]$ is calculated with the aid of Eqs. (6.75)–(6.77), while the compliance matrix $[C_O^{(2)}]$ is evaluated based on Eq. (6.78) with:

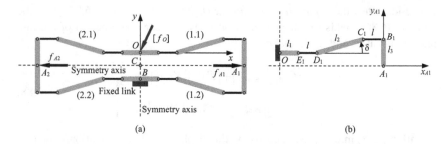

FIGURE 6.20 Displacement-amplification flexible mechanism with two symmetry axes: (a) full mechanism with no external load and symmetric actuation; (b) quarter-mechanism geometry.

$$\begin{cases} \Delta x_{OA_1} = l_x = l_1 + l_2 \cos\delta + 2l \\ \Delta y_{OA_1} = -l_y = -\left(l_3 - l_2 \sin\delta\right) \end{cases} \tag{6.91}$$

The compliance matrix of the quarter mechanism, which is sketched in Figure 6.20b and is denoted by (1.1), is:

$$\left[C_{A_1}^{(1.1)}\right] = \left[T_{A_1B_1}\right]^T [C]\left[T_{A_1B_1}\right] + \left[T_{A_1D_1}\right]^T [C]\left[T_{A_1D_1}\right], \tag{6.92}$$

where the flexible-hinge in-plane compliance matrix $[C]$ and the translation matrices are:

$$[C] = \begin{bmatrix} \dfrac{l}{EA} & 0 & 0 \\[2mm] 0 & \dfrac{l^3}{3EI_z} & -\dfrac{l^2}{2EI_z} \\[2mm] 0 & -\dfrac{l^2}{2EI_z} & \dfrac{l}{EI_z} \end{bmatrix}; \quad [T_{A_1B_1}] = \begin{bmatrix} 1 & 0 & 0 \\ 0 & 1 & 0 \\ l_3 & 0 & 1 \end{bmatrix};$$

$$[T_{A_1D_1}] = \begin{bmatrix} 1 & 0 & 0 \\ 0 & 1 & 0 \\ l_3 - l_2\sin\delta & l + l_2\cos\delta & 1 \end{bmatrix}. \tag{6.93}$$

The cross-sectional properties of Eq. (6.93) are: $A = w \cdot t$ and $I_z = w \cdot t^3/12$. The compliance matrices $\left[C_{OA_1,ip}^{(1.2)}\right]$ and $\left[C_{OA_2,ip}^{(2.2)}\right]$ are calculated by means of Eqs. (6.79) and (6.81), respectively. Ultimately, the displacement vectors at O and A_1 are determined by means of the compliance matrices required in Eqs. (6.84)–(6.86), namely:

$$[u_O] = ([C_{OA_1}] - [C_{OA_2}]) \begin{bmatrix} f_A \\ 0 \\ 0 \end{bmatrix}; \quad [u_{A_1}] = ([C_{A_1}] - [C_{A_1A_2}]) \begin{bmatrix} f_A \\ 0 \\ 0 \end{bmatrix}. \quad (6.94)$$

The displacement amplification is calculated as $d.a. = u_{Oy}/u_{A_1x}$, and Figure 6.21 plots this ratio in terms of δ varying in the $[8°; 30°]$ interval. It can be seen that the displacement amplification decreases nonlinearly with δ increasing from a value of 18 at $8°$ to a value of 4.5 when $\delta = 30°$.

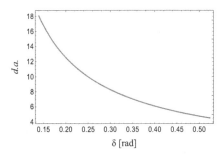

FIGURE 6.21 Plot of displacement amplification in terms of the rigid lever inclination angle δ.

ii. The first Eq. (6.86) is written in the explicit form:

$$\begin{bmatrix} u_{Ox} \\ u_{Oy} \\ \theta_{Oz} \end{bmatrix} = [C_{OA_1}] \begin{bmatrix} f_{A_1} \\ 0 \\ 0 \end{bmatrix} + [C_{OA_2}] \begin{bmatrix} f_{A_2} \\ 0 \\ 0 \end{bmatrix}. \quad (6.95)$$

By selecting only the x and y components, Eqs. (6.95) are reformulated as:

$$\begin{bmatrix} f_{A_1} \\ f_{A_2} \end{bmatrix} = \begin{bmatrix} C_{OA_1}(1,1) & C_{OA_2}(1,1) \\ C_{OA_1}(2,1) & C_{OA_2}(2,1) \end{bmatrix}^{-1} \begin{bmatrix} u_{Ox} \\ u_{Oy} \end{bmatrix}, \quad (6.96)$$

where the matrix is formed with components of the two matrices $[C_{OA_1}]$ and $[C_{OA_2}]$, which are evaluated as discussed at (i). For the given numerical parameters, the following numerical results are obtained: $f_{A_1} = 41.13$ N and $f_{A_2} = -11.50$ N.

6.1.2.2.2 Mechanisms with One Symmetry Axis

The planar flexible mechanism of Figure 6.22a has one symmetry axis and is acted upon by the loads $[f_O]$, $[f_{A_1}]$ and $[f_{A_2}]$, which are not parallel with or perpendicular to the symmetry axis. Figure 6.22b shows the mechanism right half, which is denoted by (1), and which is formed of two dissimilar portions (1.1) and (1.2).

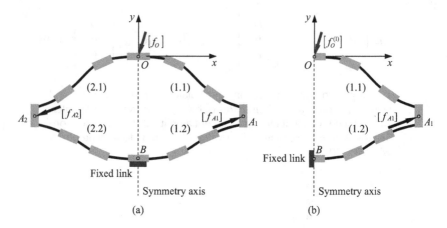

FIGURE 6.22 Generic displacement-amplification flexible mechanism with one symmetry axis under non-symmetric external loads: (a) full-mechanism configuration; (b) half-mechanism.

The procedure of calculating the displacements at points O, A_1 and A_2 is almost identical to the one presented in Section 6.1.2.2.1, which studied mechanisms with two symmetry axes. The only difference is that the compliance matrices of the right half mechanism of Figure 6.22b can no longer be calculated in terms of a quarter mechanism because this mechanism does not have a second symmetry axis.

The following *Example* details the steps that are necessary to evaluate the relevant matrices and displacements of a displacement-amplification mechanism with one symmetry axis for non-symmetric external loading.

Example 6.5

The mechanism of Figure 6.23a has one symmetry axis; it consists of six identical flexible hinges of length $l = 0.015$ m and a constant rectangular cross-section with an in-plane thickness $t = 0.0015$ m and an out-of-plane width

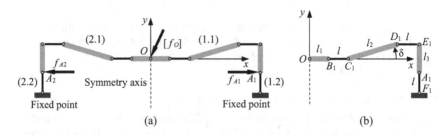

FIGURE 6.23 Displacement-amplification flexible mechanism with one symmetry axis under non-symmetric external load and symmetric actuation: (a) full-mechanism configuration; (b) half-mechanism.

$w = 0.006$ m. The material Young's modulus is $E = 2 \cdot 10^{11}$ N/m². All other links are rigid, and the following lengths (identified in Figure 6.23b) are given: $l_1 = 0.01$ m and $l_3 = 0.04$ m. The horizontal forces f_{A1} and f_{A2} produce the actuation, while the force vector $[f_O]$ is the external load.

i. Study the displacement amplification in terms of the inclined link length l_2 considering that the two horizontal actuation forces are identical in magnitude, $f_{A1} = f_{A2}$, and there is no external load at O;

ii. For $l_2 = 0.03$ m, $f_{A1} = 150$ N, $f_{A2} = -150$ N, $f_{Ox} = 0$ and $f_{Oy} = -10$ N, calculate the displacements u_{Ox} and u_{Oy} at O.

Solution:

i. The in-plane compliance matrix of the half-mechanism illustrated in Figure 6.23b and denoted by (1) is:

$$\left[C_O^{(1)}\right] = \left[C_O^{(1.1)}\right] + \left[T_{OA_1}\right]^T [R]^T [C][R]\left[T_{OA_1}\right];$$

$$\text{with} \quad \left[C_O^{(1.1)}\right] = \left[T_{OB_1}\right]^T [C]\left[T_{OB_1}\right] + \left[T_{OD_1}\right]^T [C]\left[T_{OD_1}\right] \tag{6.97}$$

The compliance matrix $\left[C_O^{(1.1)}\right]$ corresponds to the portion 1.1 of the half-mechanism, which comprises the links spanning the chain from O to A_1. The flexible-hinge compliance matrix $[C]$ is given in Eq. (6.93), while the translation and rotation matrices of Eq. (6.97) are:

$$[T_{OB_1}] = \begin{bmatrix} 1 & 0 & 0 \\ 0 & 1 & 0 \\ \Delta y_{OB_1} & -\Delta x_{OB_1} & 1 \end{bmatrix}; \quad [T_{OD_1}] = \begin{bmatrix} 1 & 0 & 0 \\ 0 & 1 & 0 \\ \Delta y_{OD_1} & -\Delta x_{OD_1} & 1 \end{bmatrix};$$

$$[T_{OA_1}] = \begin{bmatrix} 1 & 0 & 0 \\ 0 & 1 & 0 \\ \Delta y_{OA_1} & -\Delta x_{OA_1} & 1 \end{bmatrix}; \quad [R] = \begin{bmatrix} \cos(-\pi/2) & \sin(-\pi/2) & 0 \\ -\sin(-\pi/2) & \cos(-\pi/2) & 0 \\ 0 & 0 & 1 \end{bmatrix},$$

$$\tag{6.98}$$

with

$$\begin{cases} \Delta x_{OB_1} = l_1; \quad \Delta y_{OB_1} = 0; \\ \Delta x_{OD_1} = l_1 + l_2 \cos\delta + l; \quad \Delta y_{OD_1} = l_2 \sin\delta; \\ \Delta x_{OA_1} = l_1 + l_2 \cos\delta + 2l; \quad \Delta y_{OA_1} = -(l_3 - l_2 \sin\delta) \end{cases} \tag{6.99}$$

Due to symmetry about the y axis, the compliance matrices $\left[C_O^{(2.1)}\right]$ and $\left[C_O^{(2)}\right]$ are obtained from the source matrices $\left[C_O^{(1.1)}\right]$ and $\left[C_O^{(1)}\right]$ of Eqs. (6.97) as:

$$\left[C_O^{(2,1)}\right] = \begin{bmatrix} C_O^{(1,1)}(1,1) & -C_O^{(1,1)}(1,2) & C_O^{(1,1)}(1,3) \\ -C_O^{(1,1)}(2,1) & C_O^{(1,1)}(2,2) & -C_O^{(1,1)}(2,3) \\ C_O^{(1,1)}(3,1) & -C_O^{(1,1)}(3,2) & C_O^{(1,1)}(3,3) \end{bmatrix};$$

$$\left[C_O^{(2)}\right] = \begin{bmatrix} C_O^{(1)}(1,1) & -C_O^{(1)}(1,2) & C_O^{(1)}(1,3) \\ -C_O^{(1)}(2,1) & C_O^{(1)}(2,2) & -C_O^{(1)}(2,3) \\ C_O^{(1)}(3,1) & -C_O^{(1)}(3,2) & C_O^{(1)}(3,3) \end{bmatrix}.$$

(6.100)

The portion 1.2 of the half-mechanism consists of the flexure A_1F_1 in Figure 6.23b. The compliance matrix of the chain 1.2 with respect to O is:

$$\left[C_{OA_1}^{(1,2)}\right] = \left[T_{OA_1}\right]^T [R]^T [C],$$ (6.101)

where $\left[T_{OA_1}\right]$ and $[R]$ are provided in Eq. (6.98). The compliance matrix $\left[C_{OA_2}^{(2,2)}\right]$ is calculated as:

$$\left[C_{OA_2}^{(2,2)}\right] = \left[T_{OA_2}\right]^T [R]^T [C]; \quad \left[T_{OA_2}\right] = \begin{bmatrix} 1 & 0 & 0 \\ 0 & 1 & 0 \\ \Delta y_{OA_1} & -\Delta x_{OA_1} & 1 \end{bmatrix},$$ (6.102)

with Δy_{OA_1} and Δx_{OA_1} provided in Eq. (6.99).

Equations (6.97) through (6.102) enable to calculate the matrix $[A]$ of Eq. (6.74), which permits to evaluate the displacements $[u_O]$, $[u_{A_1}]$ and $[u_{A_2}]$ of Eqs. (6.84) and (6.85). The displacement amplification is calculated as $d.a. = u_{Oy}/u_{A_1x}$; the plot of Figure 6.24 graphs the displacement amplification in terms of the inclined rigid lever length l_2. It can be seen that the displacement amplification increases slightly nonlinearly with l_2 increasing.

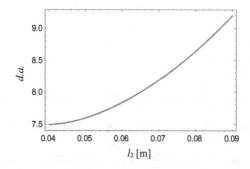

FIGURE 6.24 Plot of displacement amplification in terms of the inclined rigid lever length l_2.

ii. For the particular actuation and external loads of this *Example*, Eqs. (6.84) become:

$$\begin{bmatrix} u_{Ox} \\ u_{Oy} \\ \theta_{Oz} \end{bmatrix} = \begin{bmatrix} C_{OA_1} \end{bmatrix} \begin{bmatrix} f_{A_1} \\ 0 \\ 0 \end{bmatrix} + \begin{bmatrix} C_{OA_2} \end{bmatrix} \begin{bmatrix} f_{A_2} \\ 0 \\ 0 \end{bmatrix} + \begin{bmatrix} C_O \end{bmatrix} \begin{bmatrix} 0 \\ f_{Oy} \\ 0 \end{bmatrix}. \qquad (6.103)$$

From Eq. (6.103), the two displacements at O are expressed as:

$$\begin{cases} u_{Ox} = C_{OA_1}(1,1) \cdot f_{A_1} + C_{OA_2}(1,1) \cdot f_{A_2} + C_O(1,2) \cdot f_{Oy} \\ u_{Oy} = C_{OA_1}(2,1) \cdot f_{A_1} + C_{OA_2}(2,1) \cdot f_{A_2} + C_O(2,2) \cdot f_{Oy} \end{cases}. \qquad (6.104)$$

The matrices $[C_O], [C_{OA_1}]$ and $[C_{OA_2}]$ are calculated by means of Eqs. (6.84). For the given numerical values, Eq. (6.104) yields $u_{Ox} = 0$ and $u_{Oy} = -0.00026\,\text{m}$.

6.1.2.3 Mechanisms with Radial Symmetry – $xy\theta$ Stages

The planar mechanism sketched in Figure 6.25 is formed of a central rigid link and three identical flexible chains that are disposed symmetrically in a radial manner with respect to the rigid link's center O, 120° from one another. On each flexible chain, an external load is applied at an intermediate point, while another external load acts at the rigid-link center – all these loads are considered planar here, but they can also comprise out-of-plane components. The chain loads $[f_{A_1}], [f_{A_2}]$ and $[f_{A_3}]$ may comprise actuation forces/moments, while the

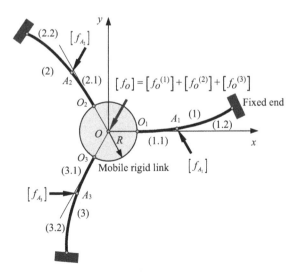

FIGURE 6.25 Planar mechanism with three loaded, radially-symmetric, identical flexible chains connected in parallel to a rigid link that is acted upon by an external load.

load $[f_O]$ is the external load (payload). The chain loads are applied at points A_1, A_2 and A_3, which subdivide each chain into two segments, (1.1)-(1.2), (2.1)-(2.2) and (3.1)-(3.2). The central rigid link does not have to be of circular shape; the only condition is that the three chains be placed at the same distance (R) from the center O.

Under in-plane loading, the central rigid link can translate along the x and y axes, as well as rotate around the z axis – because of these motion capabilities, the generic architecture of Figure 6.25 is known as a $xy\theta$ *mechanism*.

The main objective is to evaluate the relevant displacements of the central rigid link produced by the external loads. It is first necessary to determine the portions of the external load $[f_O]$ that are allotted to each of the three chains. As per the generic Eqs. (6.71) and (6.72), these loads are:

$$
\begin{bmatrix} [f_O^{(1)}] \\ [f_O^{(2)}] \\ [f_O^{(3)}] \end{bmatrix} = [A] \begin{bmatrix} [f_{A_1}] \\ [f_{A_2}] \\ [f_{A_3}] \\ [f_O] \end{bmatrix};
$$

(6.105)

$$
[A] = [A_1]^{-1}[A_2] = \begin{bmatrix} [A_{11}] & [A_{12}] & [A_{13}] & [A_{14}] \\ [A_{21}] & [A_{22}] & [A_{23}] & [A_{24}] \\ [A_{31}] & [A_{32}] & [A_{33}] & [A_{34}] \end{bmatrix},
$$

with

$$
[A_1] = \begin{bmatrix} [C_O^{(1)}] & -[C_O^{(2)}] & [0] \\ [0] & [C_O^{(2)}] & [C_O^{(3)}] \\ [I] & [I] & [I] \end{bmatrix};
$$

(6.106)

$$
[A_2] = \begin{bmatrix} -[C_{OA_1}^{(1.2)}] & [C_{OA_2}^{(2.2)}] & [0] & [0] \\ [0] & -[C_{OA_2}^{(2.2)}] & [C_{OA_3}^{(3.2)}] & [0] \\ [0] & [0] & [0] & [I] \end{bmatrix},
$$

where $[I]$ and $[0]$ are the 3×3 identity and zero matrices.

Because the three flexible chains are identical (in terms of both geometry and external load position), it is sufficient to know the compliances $[C_O^{(1)}]$ and $[C_{OA_1}^{(1.2)}]$ defining chain 1 in order to calculate the similar compliances of the chains 2 and 3. Due to radial symmetry, the chains 2 and 3 compliance matrices of Eq. (6.106) are obtained as:

$$\begin{cases} \left[C_O^{(2)} \right] = \left[R^{(2)} \right]^T \left[C_O^{(1)} \right] \left[R^{(2)} \right]; & \left[C_O^{(3)} \right] = \left[R^{(3)} \right]^T \left[C_O^{(1)} \right] \left[R^{(3)} \right]; \\ \left[C_{OA_2}^{(2.2)} \right] = \left[R^{(2)} \right]^T \left[C_{OA_1}^{(1.2)} \right] \left[R^{(2)} \right]; & \left[C_{OA_3}^{(3.2)} \right] = \left[R^{(3)} \right]^T \left[C_{OA_1}^{(1.2)} \right] \left[R^{(3)} \right] \end{cases} \tag{6.107}$$

The rotation matrices of Eqs. (6.107) are:

$$\left[R^{(2)} \right] = \begin{bmatrix} \cos(2\pi/3) & \sin(2\pi/3) & 0 \\ -\sin(2\pi/3) & \cos(2\pi/3) & 0 \\ 0 & 0 & 1 \end{bmatrix};$$

$$\left[R^{(3)} \right] = \begin{bmatrix} \cos(4\pi/3) & \sin(4\pi/3) & 0 \\ -\sin(4\pi/3) & \cos(4\pi/3) & 0 \\ 0 & 0 & 1 \end{bmatrix}. \tag{6.108}$$

The displacement vector at O is evaluated by means of Eq. (6.73) as:

$$[u_O] = \left[C_O^{(1)} \right] \left[f_O^{(1)} \right] + \left[C_{OA_1}^{(1.2)} \right] \left[f_{A_1} \right] = \left[C_O^{(2)} \right] \left[f_O^{(2)} \right]$$

$$+ \left[C_{OA_2}^{(2.2)} \right] \left[f_{A_2} \right] = \left[C_O^{(3)} \right] \left[f_O^{(3)} \right] + \left[C_{OA_3}^{(3.2)} \right] \left[f_{A_3} \right]. \tag{6.109}$$

Combining Eqs. (6.105), (6.106) and (6.109) results in the following displacement vector:

$$[u_O] = \left[C_{OA_1} \right] \left[f_{A_1} \right] + \left[C_{OA_2} \right] \left[f_{A_2} \right] + \left[C_{OA_3} \right] \left[f_{A_3} \right] + \left[C_O \right] \left[f_O \right], \tag{6.110}$$

where

$$\left[C_{OA_1} \right] = \left[C_O^{(1)} \right] \left[A_{11} \right] + \left[C_{OA_1}^{(1.2)} \right] = \left[C_O^{(2)} \right] \left[A_{21} \right] = \left[C_O^{(3)} \right] \left[A_{31} \right]$$

$$\left[C_{OA_2} \right] = \left[C_O^{(1)} \right] \left[A_{12} \right] = \left[C_O^{(2)} \right] \left[A_{22} \right] + \left[C_{OA_2}^{(2.2)} \right] = \left[C_O^{(3)} \right] \left[A_{32} \right]$$

$$\left[C_{OA_3} \right] = \left[C_O^{(1)} \right] \left[A_{13} \right] = \left[C_O^{(2)} \right] \left[A_{23} \right] = \left[C_O^{(3)} \right] \left[A_{33} \right] + \left[C_{OA_3}^{(3.2)} \right] \tag{6.111}$$

$$\left[C_O \right] = \left[C_O^{(1)} \right] \left[A_{14} \right] = \left[C_O^{(2)} \right] \left[A_{24} \right] = \left[C_O^{(3)} \right] \left[A_{34} \right]$$

The displacements at points A_1, A_2 and A_3, which may not be of primary relevance, are calculated as:

$$\left[u_{A_1}\right]=\left[u_O\right]-\left[C_O^{(1.1)}\right]\left[f_O^{(1)}\right]; \quad \left[u_{A_2}\right]=\left[u_O\right]-\left[C_O^{(2.1)}\right]\left[f_O^{(2)}\right];$$
$$\left[u_{A_3}\right]=\left[u_O\right]-\left[C_O^{(3.1)}\right]\left[f_O^{(3)}\right] \tag{6.112}$$

The chains 2 and 3 compliance matrices of Eqs. (6.112) are determined through rotations from the similar compliance defining chain 1, namely:

$$\left[C_O^{(2.1)}\right]=\left[R^{(2)}\right]^T\left[C_O^{(1.1)}\right]\left[R^{(2)}\right]; \quad \left[C_O^{(3.1)}\right]=\left[R^{(3)}\right]^T\left[C_O^{(1.1)}\right]\left[R^{(3)}\right], \tag{6.113}$$

with the rotation matrices of Eq. (6.108). The explicit form of the displacements expressed in Eqs. (6.112) is not included here.

Example 6.6

The $xy\theta$ skeleton representation mechanism of Figure 6.26a comprises three identical chains, one of which is illustrated in Figure 6.26b. Each chain is formed of three identical straight-axis right circular flexible hinges and two rigid links. The flexible hinges are defined by $r = 0.008$ m, in-plane minimum thickness $t = 0.001$ m, out-of-plane width $w = 0.006$ m and $E = 2 \cdot 10^{11}$ N/m². The other geometric parameters of the mechanism are: $R = 0.04$ m, $l_1 = 0.02$ m, $l_2 = 0.06$ m and $l_{21} = 0.055$ m ($l_2 = l_{21} + l_{22}$).

i. For a moment $m_{Oz} = 20$ Nm applied at O and no actuation, calculate the rotation angle at the same point;

ii. Calculate the external loads f_{Ox}, f_{Oy} and m_{Oz} to be applied at O in order to produce zero displacements at O when the following actuation forces are applied $f_{A1} = f_{A2} = f_{A3} = 500$ N. Assume the three actuation forces are perpendicular to the rigid links they act on.

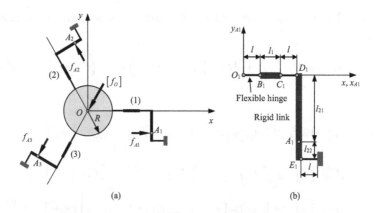

(a)   (b)

FIGURE 6.26 Skeleton representation of a planar-stage mechanism with three identical flexible chains: (a) top view of $xy\theta$ stage; (b) chain configuration with rigid links and identical flexible hinges.

Solution:

The classical counterpart (based on regular joints) of the hinge mechanism sketched in Figure 6.26a is the triple-rotation-joint (*RRR*) device shown in Figure 6.27 – see Li [3], Khan et al. [4], Geike and McPhee [5], Liu et al. [6,7], Wu et al. [8], Gosselin and Angeles [9], Staicu [10], Huang and Thebert [11], Bonev et al. [12] or Staicu [13], for instance.

　i. The configuration of chain (1) of Figure 6.26b indicates that the compliance matrix $\left[C_O^{(1.1)}\right]$ is a serial combination of two flexible-hinge compliances, matrices [C] namely:

$$\left[C_O^{(1.1)}\right]=\left[T_{OO_1}\right]^T[C]\left[T_{OO_1}\right]+\left[T_{OC_1}\right]^T[C]\left[T_{OC_1}\right] \quad \text{with}:$$

$$\left[T_{OO_1}\right]=\begin{bmatrix} 1 & 0 & 0 \\ 0 & 1 & 0 \\ 0 & -R & 1 \end{bmatrix}; \quad \left[T_{OC_1}\right]=\begin{bmatrix} 1 & 0 & 0 \\ 0 & 1 & 0 \\ 0 & -(R+l+l_1) & 1 \end{bmatrix}. \qquad (6.114)$$

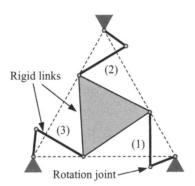

Rigid links

Rotation joint

FIGURE 6.27　Conventional $xy\theta$ stage with three RRR chains.

The compliance matrix $\left[C_O^{(1)}\right]$ is expressed as:

$$\left[C_O^{(1)}\right]=\left[C_O^{(1.1)}\right]+\left[T_{OE_1}\right]^T[C]\left[T_{OE_1}\right] \quad \text{with} \quad \left[T_{OE_1}\right]=\begin{bmatrix} 1 & 0 & 0 \\ 0 & 1 & 0 \\ -l_2 & -(R+2l+l_1) & 1 \end{bmatrix}.$$

$$(6.115)$$

The compliance matrix $\left[C_{A_1}^{(1.2)}\right]$ is determined as:

$$\left[C_{A_1}^{(1.2)}\right]=\left[T_{A_1E_1}\right]^T[C]\left[T_{A_1E_1}\right] \quad \text{with} \quad \left[T_{A_1E_1}\right]=\begin{bmatrix} 1 & 0 & 0 \\ 0 & 1 & 0 \\ -l_{22} & 0 & 1 \end{bmatrix}. \qquad (6.116)$$

Based on it, the following compliance matrix is evaluated:

$$\left[C_{OA_1}^{(1.2)}\right]=\left[T_{OA_1}\right]^T\left[C_{A_1}^{(1.2)}\right] \quad \text{with} \quad \left[T_{OA_1}\right]=\begin{bmatrix} 1 & 0 & 0 \\ 0 & 1 & 0 \\ -l_{21} & -(R+2l+l_1) & 1 \end{bmatrix}.$$

(6.117)

The matrices of Eqs. (6.115) and (6.117), which define chain 1, are used in conjunction with the corresponding rotation matrices to determine the similar matrices of chains 2 and 3 as per Eqs. (6.107) and (6.108). The matrix [A] is calculated based on Eqs. (6.105) and (6.106). The displacements at O can now be evaluated with Eqs. (6.110) and (6.111) by taking into account that the load vectors at A_1, A_2 and A_3 are zero, whereas the load at O is $[u_O] = [0\ 0\ m_{Oz}]^T$. The resulting rotation angle at the same point is $\theta_{Oz} = 1.75°$.

ii. The three actuation vectors are:

$$[f_{A1}]=\begin{bmatrix} f_{A1} \\ 0 \\ 0 \end{bmatrix}; \quad [f_{A2}]=\begin{bmatrix} -f_{A2}\cdot\sin(\pi/6) \\ f_{A2}\cdot\cos(\pi/6) \\ 0 \end{bmatrix}; \quad [f_{A3}]=\begin{bmatrix} -f_{A3}\cdot\sin(\pi/6) \\ -f_{A3}\cdot\cos(\pi/6) \\ 0 \end{bmatrix}.$$

(6.118)

Because $[u_O] = [0\ 0\ 0]^T$, Eq. (6.110) is solved for the external load vector as:

$$[f_O]=-[C_O]^{-1}\left([C_{OA_1}][f_{A1}]+[C_{OA_2}][f_{A2}]+[C_{OA_3}][f_{A3}]\right).$$ (6.119)

With the numerical values of this *Example*, the moment at O is $m_{Oz} = 0.009$ Nm, whereas the x and y forces are negligible (of the order of 10^{-15}), which is to be expected given the symmetry of the actuation loading.

6.1.3 BRANCHED FLEXIBLE PLANAR MECHANISMS AND SUBSTRUCTURING

The structure, external loading and support boundary conditions of flexible-hinge mechanisms are oftentimes quite complex. In order to evaluate overall mechanism performance qualifiers, such as relevant displacements or stiffnesses, it is necessary to first study various mechanism portions that are simpler, either structurally or in terms of loading/boundary conditions, before being able to model the full mechanism. This process is known as *substructuring* and is similar to the finite element procedure having the same name. Structural complexity that requires substructuring is mainly related to *branched* compliant mechanisms – they consist of three or more flexible chains that connect at a point with or without loads (external or support reactions) acting on them. The branched links can be *passive* or *structurally branched* if there are no internal loads acting on them;

conversely, if external loads are applied internally to at least one of the component links, the branched link is *active* or *kinematically branched* – see Noveanu et al. [59] for a similar study. The process of substructuring leads to *reduction* whereby branched portions can be combined in series and/or in parallel to produce simpler, equivalent sections whose compliances can be evaluated from the original branched links compliances.

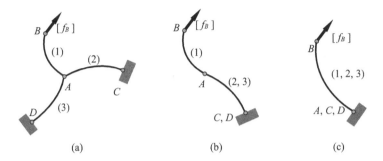

(a)   (b)   (c)

FIGURE 6.28 Reduction of structurally branched flexible mechanism with one end-point external load: (a) original mechanism; (b) equivalent, intermediate series mechanism; (c) final, equivalent single-link mechanism.

The planar mechanism sketched in Figure 6.28a, for instance, is structurally branched as its three flexible links/chains (1), (2) and (3) are rigidly connected at A. There are no loads applied internally on the three flexible links, and the only load, $[f_B]$, acts at the free end of chain (1); the other two chains are fixed at their opposite ends. The links (2) and (3) are connected in parallel with respect to point A, and they conceptually result in a single, equivalent link – denoted by (2, 3) in Figure 6.28b. This link is serially coupled to link (1) at A, and the final equivalent configuration is the one-link mechanism denoted by (1, 2, 3) in Figure 6.28c. Because of the parallel/series connection of the original links, the compliance matrix of the equivalent, single-link mechanism can be calculated in terms of the three original links' compliance matrices. This enables to determine the displacement at B, as well as the displacements at other internal points, such as at the junction point A.

The configuration of Figure 6.29a is similar to the one of Figure 6.28a and represents a kinematically branched mechanism because 3 is an active link (it has an external load $[f_E]$ applied to it). The original three-link mechanism can be decomposed into two equivalent, two-link mechanisms, as illustrated in Figure 6.29b, where the force at A is a reaction introduced by the mechanism portion separated from the analyzed segment. The reactions at A and D are found by means of a continuity vector equation requiring that the displacement at A be the same for the two two-link mechanisms of Figure 6.29b plus another boundary equation reflecting the zero displacements at D. With the reaction at A determined and the external load at B provided, all relevant displacements (such as at points A, B or E) can simply be calculated by using the two serial component mechanisms of Figure 6.29b.

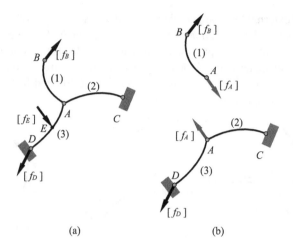

FIGURE 6.29 Reduction of kinematically branched flexible mechanism: (a) original mechanism; (b) equivalent component serial chains with internal reaction load.

Substructuring a branched flexible-hinge mechanism with both passive and active chains regularly includes the following procedural steps:

- Identification of passive links (structurally branched portions), formation of simpler equivalent chains through structural parallel/series-connection reduction and evaluation of their compliance matrices from component compliance matrices;
- Identification of passive links with special boundary conditions (such as guided ends) and of active links; calculation of relevant internal and support reactions and displacements at internally and externally constrained points of interest by using displacement continuity and zero boundary-condition equations;
- Calculation of relevant displacements at other points (usually unconstrained) of the simpler/component chains that resulted through structural reduction;
- Evaluation of the full-mechanism performance (such as output/input displacement/force ratios or input/output stiffness).

Example 6.7

The planar mechanism shown in the skeleton form of Figure 6.30a is formed of two rigid links and three identical constant, rectangular cross-section flexible hinges. An external load $[f_E]$ is applied at point E. Calculate the x and y components of an actuation force $[f_C]$ at C that will result in zero x and y displacements at E. Known are the properties of the flexible hinge: $l = 0.008\,\text{m}$, $w = 0.004\,\text{m}$ (out-of-plane width), $t = 0.001\,\text{m}$ (in-plane thickness) and $E = 2\cdot10^{11}\,\text{N/}$ m^2, the lengths $l_1 = 0.01\,\text{m}$, $l_2 = 0.06\,\text{m}$ and $l_3 = 0.006\,\text{m}$, and the global-frame force components $f_{Ex} = -1\,\text{N}$ and $f_{Ey} = -4\,\text{N}$.

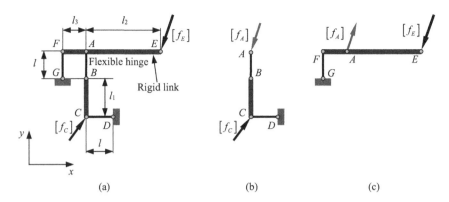

(a)　　　　　　　　(b)　　　　　　　　(c)

FIGURE 6.30 Skeleton representation of a planar branched flexible-hinge mechanism: (a) mechanism structure and loading; (b) first chain with actuation load and internal reaction; (c) second chain with external load and internal reaction.

Solution:

The mechanism is decomposed into chain (1) – which is shown in Figure 6.30b – and chain (2) – represented in Figure 6.30c. The internal reaction force $[f_A]$ is applied on both chains at point A. The two chains are structurally connected in parallel at A, and therefore, the displacement $[u_A]$ is the same for both chains. In terms of chain 1, this displacement is calculated as:

$$[u_A] = \left[C_A^{(1)} \right] [f_A] + \left[C_{AC}^{(f)} \right] [f_C], \tag{6.120}$$

where

$$\left[C_A^{(1)} \right] = [R]^T \left[C^{(f)} \right] [R] + [T_{AC}]^T \left[C^{(f)} \right] [T_{AC}]; \quad \left[C_{AC}^{(f)} \right] = [T_{AC}]^T \left[C^{(f)} \right] \tag{6.121}$$

with

$$[R] = \begin{bmatrix} \cos(-\pi/2) & \sin(-\pi/2) & 0 \\ -\sin(-\pi/2) & \cos(-\pi/2) & 0 \\ 0 & 0 & 1 \end{bmatrix}; \quad [T_{AC}] = \begin{bmatrix} 1 & 0 & 0 \\ 0 & 1 & 0 \\ -(l+l_1) & 0 & 1 \end{bmatrix}. \tag{6.122}$$

The compliance matrix $[C^{(f)}]$ of the flexible hinge is given in Eq. (6.93).

The displacement $[u_A]$ can also be expressed in connection with chain 2 as:

$$[u_A] = \left[C_{AEF}^{(f)} \right] [f_E] + \left[C_{AF}^{(f)} \right] [f_A], \tag{6.123}$$

where

$$\left[C_{AEF}^{(f)} \right] = [T_{AF}]^T [R]^T \left[C^{(f)} \right] [R] [T_{EF}]; \quad \left[C_{AF}^{(f)} \right] = [T_{AF}]^T [R]^T \left[C^{(f)} \right] [R] [T_{AF}] \tag{6.124}$$

with

$$[T_{AF}]=\begin{bmatrix} 1 & 0 & 0 \\ 0 & 1 & 0 \\ 0 & l_3 & 1 \end{bmatrix}; \quad [T_{EF}]=\begin{bmatrix} 1 & 0 & 0 \\ 0 & 1 & 0 \\ 0 & l_2+l_3 & 1 \end{bmatrix}. \quad (6.125)$$

Equations (6.120) and (6.123) allow us to solve for the internal reaction $[f_A]$ as:

$$[f_A]=[A_E][f_E]+[A_C][f_C]. \quad (6.126)$$

The two new matrices of Eq. (6.126) are:

$$[A_E]=\left(\left[C_A^{(1)}\right]-\left[C_{AF}^{(f)}\right]\right)^{-1}\left[C_{AEF}^{(f)}\right]; \quad [A_C]=-\left(\left[C_A^{(1)}\right]-\left[C_{AF}^{(f)}\right]\right)^{-1}\left[C_{AC}^{(f)}\right]. \quad (6.127)$$

The displacement at E is determined by analyzing chain 2 and is expressed as:

$$[u_E]=\left[C_{EF}^{(f)}\right][f_E]+\left[C_{EAF}^{(f)}\right][f_A]. \quad (6.128)$$

Combining Eqs. (6.126) and (6.128) results in:

$$[u_E]=[C_E][f_E]+[C_{EC}][f_C], \quad (6.129)$$

with

$$[C_E]=\left[C_{EF}^{(f)}\right]+\left[C_{EAF}^{(f)}\right][A_E]; \quad [C_{EC}]=\left[C_{EAF}^{(f)}\right][A_C]. \quad (6.130)$$

Considering that $u_{Ex}=0$ and $u_{Ey}=0$, the following relationship results from Eq. (6.129):

$$\begin{bmatrix} f_{Cx} \\ f_{Cy} \end{bmatrix}=[A_f]\begin{bmatrix} f_{Ex} \\ f_{Ey} \end{bmatrix}, \quad (6.131)$$

where

$$[A_f]=\begin{bmatrix} C_{EC}(1,1) & C_{EC}(1,2) \\ C_{EC}(2,1) & C_{EC}(2,2) \end{bmatrix}^{-1}\begin{bmatrix} C_E(1,1) & C_E(1,2) \\ C_E(2,1) & C_E(2,2) \end{bmatrix}. \quad (6.132)$$

The components of the matrices defining $[A_f]$ in Eq. (6.132) are collected from the respective 3 × 3 matrices expressed in Eqs. Eq. (6.129). For the numerical values of this *Example*, the following actuation forces are obtained from Eq. (6.131): $f_{Cx}=-346\,\text{N}$ and $f_{Cy}=-46\,\text{N}$.

A flexible-hinge planar gripper:

A flexible-hinge, piezoelectric-actuated planar gripper of branched configuration with amplified and nonparallel output motion of its gripping jaws is studied in Noveanu et al. [59] by means of the analytical substructuring method, by finite element analysis and through experimental testing of a specimen. The photograph of a gripper prototype is shown in Figure 6.31a. The gripper has a symmetry axis, and the planar structure of the upper half is illustrated in Figure 6.31b. Also shown in Figure 6.31b is a cylindrical specimen in contact with the end effector. The gripper uses identical right circularly corner-filleted flexible hinges.

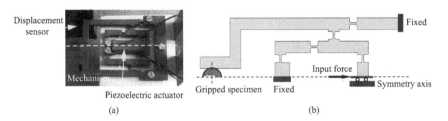

FIGURE 6.31   Flexible-hinge, piezoelectric-actuated compliant gripper: (a) prototype photograph; (b) geometry of the half-gripper elastic frame.

The piezoelectric actuator produces a horizontal force and displacement of the guided link; this motion is amplified twice resulting in a symmetric tilt of the two external lever arms, which produces the gripping motion.

This mechanism can be simplified to the substructured configuration of Figure 6.32, which uses five serial chains denoted as (1) to (5). The five chains have their individual global frames centered at points A (chain (1)), B (chains (2) and (4)) and C (chains (3) and (5)), respectively. All these reference frames were chosen with their axes parallel to the mechanism global reference frame xyz of Figure 6.32. Noveanu et al. [59] provide full details on how to calculate the in-plane compliance matrices $\left[C_A^{(1)}\right], \left[C_B^{(2)}\right], \left[C_C^{(3)}\right], \left[C_B^{(4)}\right]$ and $\left[C_C^{(5)}\right]$ of the five chains. The matrix subscripts indicate the point with respect to which a compliance matrix of a given chain was evaluated in its own reference frame.

The half-elastic frame of Figure 6.32 is modeled to evaluate several gripper performance qualifiers, for instance, its displacement amplification, which

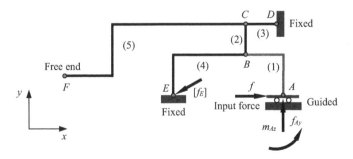

FIGURE 6.32   Skeleton representation of half-gripper with basic flexible chains, actuation force, reactions and boundary conditions.

is $d.a. = u_{Fy}/u_{Ax}$, when the actuation force f is applied at the guided end A. The compliance matrix at B that couples chains 2 and 3 serially is calculated as:

$$\left[C_B^{(2,3)} \right] = \left[C_B^{(2)} \right] + \left[C_B^{(3)} \right] = \left[C_B^{(2)} \right] + \left[T_{BC} \right]^T \left[C_C^{(3)} \right] \left[T_{BC} \right]. \qquad (6.133)$$

The series connection of chains 2 and 3 is based on a valid assumption since chain 5, which also converges at B, is a free chain. The compliance matrix at B results from the parallel connection between the chains 2, 3 and 4 and is expressed as:

$$\left[C_B^{(2,3,4)} \right] = \left(\left[C_B^{(2,3)} \right]^{-1} + \left[C_B^{(4)} \right]^{-1} \right)^{-1}. \qquad (6.134)$$

The compliance matrix $[C_A]$ corresponding to the serial combination of chains 1 and 2, 3, 4 is determined as:

$$\left[C_A \right] = \left[C_A^{(1)} \right] + \left[C_A^{(2,3,4)} \right] = \left[C_A^{(1)} \right] + \left[T_{AB} \right]^T \left[C_B^{(2,3,4)} \right] \left[T_{AB} \right]. \qquad (6.135)$$

The displacement u_{Ax} is a component of $[u_A]$, which is calculated as:

$$[u_A] = [C_A] \cdot [f_A]$$

$$\text{with} \quad [u_A] = \left[\begin{array}{ccc} u_{Ax} & u_{Ay} = 0 & \theta_{Az} = 0 \end{array} \right]^T; \quad [f_A] = \left[\begin{array}{ccc} f & f_{Ay} & m_{Az} \end{array} \right]^{T.}$$

$$(6.136)$$

After rearranging it, Eq. (6.136) is solved for the unknown displacements and reaction forces at A as:

$$\left[\begin{array}{c} u_{Ax} \\ f_{Ay} \\ m_{Az} \end{array} \right] = \left[\begin{array}{ccc} 1 & -C_A(1,2) & -C_A(1,3) \\ 0 & -C_A(2,2) & -C_A(2,3) \\ 0 & -C_A(3,2) & -C_A(3,3) \end{array} \right]^{-1} \left[\begin{array}{c} C_A(1,1) \\ C_A(2,1) \\ C_A(3,1) \end{array} \right] \cdot f, \qquad (6.137)$$

where $C_A(i,j)$ are the elements of the 3×3 symmetric matrix $[C_A]$ evaluated with Eq. (6.135). Equation (6.137) results in:

$$u_{Ax} = a \cdot f; \quad f_{Ay} = b \cdot f; \quad m_{Az} = c \cdot f, \qquad (6.138)$$

with a, b and c known factors.

The displacement vector at point B produced by the force vector applied at A when considering the combined chain 2, 3, 4 is expressed as:

$$[u_B] = \left[C_B^{(2,3,4)} \right] \left[T_{BA} \right] \left[f_A \right]. \qquad (6.139)$$

The equivalent load to be applied at B in order to produce the displacement calculated in Eq. (6.139) by considering the series chain 2, 3 is:

$$[f_B] = \left[C_B^{(2,3)} \right]^{-1} \left[u_B \right]. \qquad (6.140)$$

The displacement produced at C by the equivalent load at B determined in Eq. (6.140) is calculated as:

$$[u_C] = \left[C_C^{(3)}\right][T_{CB}][f_B].$$

(6.141)

The displacement at F is evaluated by transferring the displacement at C in Eq. (6.141):

$$[u_F] = [T_{FC}]^T[u_C].$$

(6.142)

Equations (6.139) through (6.142) are combined into:

$$[u_F] = [C_{FA}][f_A];$$
$$[C_{FA}] = [T_{FC}]^T\left[C_C^{(3)}\right][T_{CB}]\left[C_B^{(2,3)}\right]^{-1}\left[C_B^{(2,3,4)}\right][T_{BA}]$$

(6.143)

The displacement at F is $[u_F] = [\ u_{Fx}\ u_{Fy}\ \theta_{Fz}\ ]^T$, whereas the load at A is $[f_A] = [f\ b\cdot f\ c\cdot f]^T$; as a consequence, the y displacement at F results from Eqs. (6.143):

$$u_{Fy} = [C_{FA}(2,1) + b\cdot C_{FA}(2,2) + c\cdot C_{FA}(2,3)]\cdot f,$$

(6.144)

where C_{FA} $(2, j)$ – with $j = 1, 2, 3$ – are the components on the second row of the 3×3 matrix $[C_{FA}]$.

The displacement amplification is calculated with the displacement u_{Ax} of Eq. (6.138) and the displacement u_{Fy} expressed in Eq. (6.144) as:

$$d.a. = \frac{1}{a}\cdot[C_{FA}(2,1) + b\cdot C_{FA}(2,2) + c\cdot C_{FA}(2,3)].$$

(6.145)

6.2 SPATIAL (3D) MECHANISMS

This section is a brief study of the quasi-static response of three-dimensional mechanisms that are formed of a rigid link (platform) acted upon by external loads; the link connects to m flexible chains, as illustrated in Figure 6.33.

Assuming the external load (which may consist of several components applied at different points on the rigid link) is reduced to a unique load vector $[f]$ at point F in Figure 6.33, the displacement $[u_P]$ at another point P on the rigid link is evaluated as:

$$[u_P] = [C_P][f] \quad \text{with} \quad \begin{cases} [C_P] = \left(\sum_{j=1}^{m}\left[C_{PFH_j}\right]^{-1}\right)^{-1} \\ \\ \left[C_{PFH_j}\right] = \left[T_{PH_j}\right]^T\left[R_{PH_j}\right]^T\left[C_{H_j}^{(j)}\right]\left[R_{FH_j}\right]\left[T_{FH_j}\right] \end{cases}.$$

(6.146)

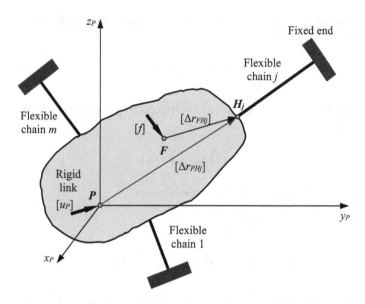

FIGURE 6.33 Schematic representation of a 3D mechanism formed of a rigid link with actuation and external load, and several parallel-connection flexible chains.

The compliance matrix $\left[C_{H_j}^{(j)} \right]$ corresponds to the generic flexible chain denoted by j in Figure 6.33 ($j = 1, 2, \ldots, m$) and is calculated in the local frame of each flexible chain. The matrices $\left[T_{PH_j} \right]^T$ and $\left[R_{PH_j} \right]^T$ of Eq. (6.147) realize the displacement transfer from point H_j to point P, while the matrices $\left[R_{FH_j} \right]$ and $\left[T_{FH_j} \right]$ transfer the load from point F to the flexible chain root H_j.

Similar to the spatial application discussed in Chapter 5, we study here a three-dimensional Cartesian mechanism to illustrate the generic Eq. (6.146) that allows the calculation of the displacement at a point on the rigid link.

Example 6.8

The spatial mechanism of Figure 6.34 is formed of a central rigid platform (link) that is supported by three identical 3D flexible chains of the type shown in Figure 5.32 and analyzed in *Example 5.12*. A flexible leg configuration is shown in Figure 6.35. It comprises three identical straight-axis flexible hinges of length *l* and constant circular cross-section (diameter *d*), and three rigid links. As shown in Figure 6.34, the legs are distributed in an equidistant manner radially and are offset from the rigid platform along the z axis by means of identical rigid links of length *h*.

A force vector [*f*] acts at the center *O* of the rigid platform. Calculate the displacements at *O* knowing all the geometric and material parameters defining the identical flexible legs: $l = 0.015\,\mathrm{m}$, $d = 0.002\,\mathrm{m}$, $l_x = l_y = l_z = 0.04\,\mathrm{m}$, $E = 2 \cdot 10^{11}\,\mathrm{N/m^2}$ and $\mu = 0.3$. Known are also: $r = 0.04\,\mathrm{m}$, $h = 0.02\,\mathrm{m}$, $f_x = 20\,\mathrm{N}$, $f_y = 30\,\mathrm{N}$ and $f_z = -100\,\mathrm{N}$.

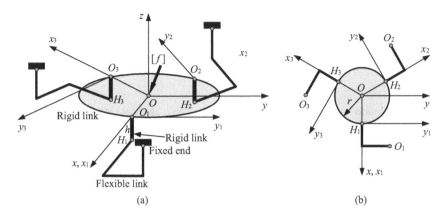

FIGURE 6.34 Spatial Cartesian parallel flexible-hinge mechanism: (a) three-dimensional rendering; (b) top view.

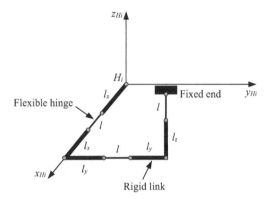

FIGURE 6.35 Spatial Cartesian flexible chain supporting the mechanism of *Figure 6.34*.

Solution:

The compliance matrix of the mechanism is calculated in the $Oxyz$ reference frame as:

$$[C_O] = \left(\left[C_O^{(1)} \right]^{-1} + \left[C_O^{(2)} \right]^{-1} + \left[C_O^{(3)} \right]^{-1} \right)^{-1}. \tag{6.147}$$

The individual flexible legs compliances of Eq. (6.147) are:

$$\left[C_O^{(j)} \right] = \left[T_{OH_j}^{(j)} \right]^T \left[R^{(j)} \right]^T [C] \left[R^{(j)} \right] \left[T_{OH_j}^{(j)} \right], \tag{6.148}$$

with $j = 1, 2, 3$. The flexible chain, local-frame compliance matrix $[C]$ is calculated by means of Eqs. (5.206) and (5.207). The translation matrices of Eq. (6.148) are calculated as in Eq. (5.195) with the following offsets:

$$\begin{cases} \Delta x_{OH_1} = 0, \Delta y_{OH_1} = -r, \Delta z_{OH_1} = -h; \\ \Delta x_{OH_2} = r \cdot \cos(\pi/6), \Delta y_{OH_2} = r \cdot \sin(\pi/6), \Delta z_{OH_2} = -h; \\ \Delta x_{OH_3} = -r \cdot \cos(\pi/6), \Delta y_{OH_3} = r \cdot \sin(\pi/6), \Delta z_{OH_3} = -h; \end{cases} \quad (6.149)$$

The rotation matrices of Eq. (6.148) are calculated as in Eqs. (5.197), (5.199)–(5.201):

$$\left[R^{(j)} \right] = \left[R^{(j)}_{4-5} \right]\left[R^{(j)}_{3-4} \right]\left[R^{(j)}_{2-3} \right] \qquad (6.150)$$

with

$$\begin{cases} \left[R^{(1)} \right] = [I] \\ \left[R^{(2)} \right] = \left[R^{(2)}_{4-5} \right] \text{and } \psi_2 = 2\pi/3, \\ \left[R^{(3)} \right] = \left[R^{(3)}_{4-5} \right] \text{and } \psi_3 = 4\pi/3 \end{cases} \qquad (6.151)$$

where [I] is the identity matrix. The displacement vector at O is calculated as:

$$[u_O] = [C_O][f_O] \quad \text{with} \quad \begin{cases} [u_O] = \begin{bmatrix} u_{Ox} & u_{Oy} & \theta_{Oz} & \theta_{Ox} & \theta_{Oy} & u_{Oz} \end{bmatrix}^T \\ [u_O] = \begin{bmatrix} f_{Ox} & f_{Oy} & 0 & 0 & 0 & f_{Oz} \end{bmatrix}^T \end{cases},$$

$$(6.152)$$

and the following numerical results are obtained: $u_{Ox} = 8 \cdot 10^{-8}$ m, $u_{Oy} = 0.00032$ m and $u_{Oz} = -0.0005$ m.

REFERENCES

1. Merlet, J.P., *Parallel Robots*, Kluwer, Dordrecht, 2000.
2. Gosselin, C.M. and Wang, J., Singularity loci of planar parallel manipulators with revolute actuators, *Robotics and Autonomous Systems*, 21, 377, 1997.
3. Li, Z., Study of planar three-degree-of-freedom 2-RRR parallel manipulators, *Mechanism and Machine Theory*, 38, 409, 2003.
4. Khan, W.A. et al., Modular and recursive kinematics and dynamics for parallel manipulators, *Multibody System Dynamics*, 14, 419, 2005.
5. Geike, T. and McPhee, J., Inverse dynamic analysis of parallel manipulators with full mobility, *Mechanism and Machine Theory*, 38, 549, 2003.
6. Liu, X.J., Wang, J., and Pritschow, G., Performance atlases and optimum design of planar 5R symmetrical parallel mechanisms, *Mechanism and Machine Theory*, 41, 119, 2006.
7. Liu, X.J., Wang, J., and Pritschow, G., Kinematics, singularity and workspace of planar 5R symmetrical parallel mechanisms, *Mechanism and Machine Theory*, 41, 145, 2006.

8. Wu, J. et al., Performance comparison of three planar 3-DOF manipulators with 4-RRR, 3-RRR and 2-RRR structures, *Mechatronics*, 20, 510, 2010.

9. Gosselin, C.M. and Angeles, J., Optimum kinematic design of a planar three-degree-of-freedom parallel manipulator, *ASME Journal of Mechanical Design*, 110, 34, 1998.

10. Staicu, S., Matrix modeling of inverse dynamics of spatial and planar parallel robots, *Multibody System Dynamics*, 27, 239, 2012.

11. Huang, M.Z. and Thebert, J.L., A study of workspace and singularity characteristics for design of 3-DOF planar parallel robots, *International Journal of Advanced Manufacturing Technology*, 51, 789, 2010.

12. Bonev, I.A., Zlatanov, D., and Gosselin, C.M., Singularity analysis of 3 DOF planar parallel mechanisms via screw theory, *ASME Journal of Mechanical Design*, 25, 573, 2003.

13. Staicu, S., Inverse dynamics of 3-PRR planar parallel robot, *Robotics and Autonomous Systems*, 57, 556, 2009.

14. Wang, X. and Mills, J.K., Dynamic modeling of a flexible-link planar parallel platform using a substructuring approach, *Mechanism and Machine Theory*, 41, 671, 2006.

15. Briot, S. and Bonev, I.A., Accuracy analysis of 3-DOF planar parallel robots, *Mechanism and Machine Theory*, 43, 445, 2008.

16. Briot, S. et al., Self-motions of general 3-RPR planar parallel robots, *International Journal of Robotics Research*, 27, 855, 2008.

17. Merlet, J.P., Gosselin, C.M., and Mouly, N., Workspaces of planar parallel manipulators, *Mechanism and Machine Theory*, 33, 7, 1998.

18. Yang, Y. and O'Brien, J.F., A case study of planar 3-RPR parallel robot singularity free workspace design, *Proceedings of the 2007 IEEE International Conference on Mechatronics and Automation*, Harbin, China, 1834, 2007.

19. Choi, K.B., Kinematic analysis and optimal design of 3-PPR planar parallel manipulator, *KSME International Journal*, 17, 528, 2003.

20. Yu, A., Bonev, I.A., and Zsombor-Murray, P., Geometric approach to the accuracy analysis of a class of 30DOF planar parallel robots, *Mechanism and Machine Theory*, 43, 364, 2008.

21. Binaud, M. et al., Comparison of 3-PPR parallel manipulators based on their sensitivity to joint clearances, *Proceedings of the 2010 IEEE/RSJ International Conference on Intelligent Robotic Systems*, Taipei, Taiwan, 2778, 2010.

22. Ebrahimi, I., Carretero, J.A., and Boudreau, R., 3-PRRR redundant planar parallel manipulator: inverse displacement, workspace and singularity analyses, *Mechanism and Machine Theory*, 42, 1007, 2007.

23. Smith, S.T., *Flexures – Elements of Elastic Mechanisms*, Gordon and Breach Science Publishers, Amsterdam, 2000.

24. Xu, Q., *Design and Implementation of Large-Range Compliant Micropositioning Systems*, Wiley, London, 2017.

25. Henin, S. et al., *The Art of Flexure Mechanism Design*, EPFL Press, Lausanne, 2017.

26. Zhang, X. and Zhu, B., *Topology Optimization of Compliant Mechanisms*, Springer, Berlin, 2018.

27. Zentner, L. and Linβ, S., *Compliant Mechanisms: Mechanics of Elastically Deformable Mechanisms, Actuators and Sensors*, De Gruiter Oldenbourg, Berlin, 2019.

28. Su, H.J., Shi, H., and Yu, J.J., A symbolic formulation for analytical compliance analysis and synthesis of flexure mechanisms, *ASME Journal of Mechanical Design*, 134, 051009, 2012.

29. Li, Y. and Xu, Q., Optimal design of a novel 2-DOF compliant parallel micro-manipulator for nanomanipulation, *Proceedings of the 2005 IEEE International Conference on Automation Science and Engineering*, Edmonton, Canada, 118, 2005.

30. Lu, T.F. et al., A three-DOF compliant micromotion stage with flexure hinges, *Industrial Robot*, 31, 355, 2004.

31. Yong, Y.K. and Lu, T.F., Kinetostatic modeling of 3-RRR compliant micro-motion stages with flexure hinges, *Mechanism and Machine Theory*, 44, 1156, 2009.

32. Tian, Y., Shirinzadeh, B., and Zhang, D., Design and dynamics of a 3-DOF flexure-based mechanism for micro/nano manipulation, *Microelectronics Engineering*, 87, 230, 2010.

33. Hua, W. and Xianmin, Z., Input coupling analysis and optimal design of a 3-DOF compliant micro-positioning stage, *Mechanism and Machine Theory*, 43, 400, 2008.

34. Zettl, B., Szyszkowski, W., and Zhang, W.J., Accurate low DOF modeling of a planar compliant mechanism with flexure hinges: the equivalent beam methodology, *Journal of Precision Engineering*, 29, 237, 2005.

35. Yi, B.J. et al., Design and experiment of a 3-DOF parallel micromechanism utilizing flexure hinges, *IEEE Transactions on Robotics and Automation*, 4, 604, 2003.

36. Pham, H.H. and Chen, I.M., Stiffness modeling of flexure parallel mechanism, *Journal of Precision Engineering*, 29, 467, 2005.

37. Oetomo, D. et al., An interval-based method for workspace analysis of planar flexure-jointed mechanism, *ASME Journal of Mechanical Design*, 131, 011014-1, 2009.

38. Pham, H.H. and Chen, I.M., Evaluation of resolution of flexure parallel mechanisms for ultraprecision manipulation, *Review of Scientific Instruments*, 75, 3016, 2004.

39. Mukhopadyay, D. et al., A SOI-MEMS-based 3-DOF planar parallel-kinematics nanopositioning stage, *Sensors and Actuators A*, 147, 340, 2008.

40. Hesselbach, J. et al., Aspects on design of high precision parallel robots, *Assembly Automation*, 24, 49, 2004.

41. Her, I. and Chang, J.C., A linear scheme for the displacement analysis of micropositioning stages with flexure hinges, *ASME Journal of Mechanical Design*, 116, 770, 1994.

42. Ryu, J.W., Gweon, D.G., and Moon, K.S., Optimal design of a flexure hinge based XYθ wafer stage, *Journal of Precision Engineering*, 21, 18, 1997.

43. Ryu, J.W. et al., Inverse kinematic modeling of a coupled flexure hinge mechanism, *Mechatronics*, 9, 657, 1999.

44. Awtar, S., Slocum, A.H., and Sevincer, E., Characteristics of beam-based flexure modules, *ASME Journal of Mechanical Design*, 129, 625, 2007.

45. Huang, S. and Schimmels, J.M., The bounds and realization of spatial stiffness achieved with simple springs connected in parallel, *IEEE Transactions on Robotics and Automation*, 14 (3), 466, 1998.

46. Hopkins, J.B. and Culpepper, M.L., Synthesis of multi-degree of freedom parallel flexure systems via freedom and constraint topology (FACT) – part I: principles, *Journal of Precision Engineering*, 34, 259, 2010.

47. Hopkins, J.B., Vericella, J.J., and Harvey, C.D., Modeling and generating parallel flexure elements, *Journal of Precision Engineering*, 38, 525, 2014.

48. Awtar, S., Shimotsu, K., and Sen, S., Elastic averaging in flexure mechanisms: a three-beam parallelogram flexure case study, *ASME Journal of Mechanical Robotics*, 2, 041006-1, 2010.

49. Lobontiu, N., Wight-Crask, J., and Kawagley, C., Straight-axis folded flexure hinges: in-plane elastic response, *Journal of Precision Engineering*, 57C, 54, 2019.

50. Lobontiu, N. et al., Stiffness design of circular-axis hinge, self-similar mechanism with large out-of-plane motion, *ASME Journal of Mechanical Design*, 141, 092302-12019.
51. Mottard, P. and St-Amant, Y., Analysis of flexural hinge orientation for amplified piezo-driven actuators, *Smart Materials and Structures*, 18, 1, 2009.
52. Ma, H.-W. et al., Analysis of the displacement amplification ratio of bridge-type flexure hinge, *Sensors and Actuators A*, 132, 730, 2006.
53. Du, H. et al., Topological optimization of mechanical amplifiers for piezoelectric actuators under dynamic motion, *Smart Materials and Structures*, 9, 788, 2000.
54. Jouaneh, M. and Yang, R., Modeling of flexure-hinge type lever mechanisms, *Precision Engineering*, 27, 407, 2003.
55. Kim, J.H., Kim, S.H., and Kwak, Y.K., Development of a piezoelectric actuator using a three-dimensional bridge-type hinge mechanism, *Review of Scientific Instruments*, 74 (5), 2918, 2003.
56. Ni, Y. et al., Modeling and analysis of an overconstrained flexure-based compliant mechanism, *Measurement*, 50, 270, 2014.
57. Choi, S.B., Han, S.S., and Lee, Y.S., Fine motion control of a moving stage using a piezoactuator associated with a displacement amplifier, *Smart Materials and Structures*, 14, 222, 2005.
58. Lobontiu, N. and Garcia, E., Analytical model of displacement amplification and stiffness optimization for a class of flexure-based compliant mechanisms, *Computers & Structures*, 81, 2797, 2003.
59. Noveanu, S. et al., Substructure compliance matrix model of planar branched flexure-hinge mechanisms: design, testing and characterization of a gripper, *Mechanism and Machine Theory*, 91, 1, 2015.

7 Dynamics of Flexible-Hinge Mechanisms

The dynamics of flexible-hinge mechanisms regarded as flexible multibody systems can be studied by means of distributed-parameter or lumped-parameter (including finite element) approaches – Shabana [1]. Most often, the dynamic modeling of these mechanisms addresses the natural-frequency, modal, transient, steady-state and frequency-domain responses. Methods of formulating the dynamic mathematical model include d'Alembert principle – Boyet and Coiffet [2], Lagrange's equations – Vibet [3], Surdilovic and Vukobratovic [4], Hamilton's variational principle – Yu [5] or Lyapunov analysis – Indri and Tornambe [6]. The resulting dynamic models are derived by utilizing either local frames – Boutaghou and Erdman [7] – or global coordinates – see Simo and Vu-Quoc [8]. Methods and algorithms aimed at reducing the number of coordinates have been developed in order to simplify the mathematical models, such as those proposed by McPhee [9], Liew et al. [10], Engstler and Kaps [11], Cui and Haque [12], Smith [13] or Rosner et al. [14]. Dynamic models of flexible-hinge mechanisms have mainly been utilized in applications of manipulation (including the micro-/nano-domain), robotics – as discussed by Gao [15], Bagci and Streit [16] or Tian et al. [17] – precision positioning, sensing and actuation – see Ling [18], Du et al. [19], Wang et al. [20], Polit and Dong [21] or Lin et al. [22].

This chapter studies the dynamic response of serial and parallel flexible-hinge mechanisms by using a lumped-parameter compliance-based approach. The majority of applications are planar structurally, but spatial configurations are also analyzed. The degrees of freedom (DOF) are the displacements of the mobile rigid links' centers of mass. Mechanisms where the inertia of flexible hinges can be neglected are studied, as well as mechanisms that include the inertia of hinges. The compliance matrix of flexible-hinge mechanisms is formed based on information from Chapters 5 and 6 in direct relation to the DOF. The damping results predominantly from the viscous interaction of the rigid links with the surrounding fluid, but similar contributions from the flexible hinges are also incorporated, specifically in the proportional damping model. The free undamped response of flexible-hinge mechanisms, together with the natural frequencies, is analyzed, as well as the free damped and the forced responses. The steady-state and the frequency-domain behavior of flexible mechanisms subjected to sinusoidal excitation are also studied here.

7.1 DYNAMIC MODELING

This section presents a general overview of the lumped-parameter mathematical models that describe the linear dynamic response of flexible-hinge mechanisms. The basic matrix equations are presented corresponding to the free (undamped and damped) response and the forced response of these mechanisms. While a derivation of these models is not included here, vibrations texts provide the respective derivation – see, for instance, Shabana [1] and Meirovitch [23]. Discussed are also the steady-state and frequency-response mathematical models for flexible-hinge mechanisms under harmonic (sinusoidal) input.

7.1.1 DEGREES OF FREEDOM (DOF)

The number of differential equations forming the mathematical model of a dynamic flexible-mechanism system is identical to the number of independent parameters fully defining the state of that mechanism, which are the DOF. The mechanisms studied in this book consist of series/parallel combinations of flexible hinges and rigid links. As demonstrated by concrete examples later in this chapter, the DOF of a flexible-hinge mechanism are the translation and rotary motions of the mobile rigid links' mass centers.

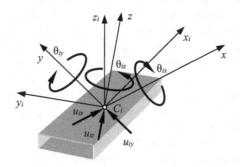

FIGURE 7.1   Rigid body in three-dimensional space with six DOF in local and global reference frames.

The rigid body of Figure 7.1 has six DOF because its spatial position is determined by the six displacements at the center C_i, which are collected in the vector:

$$[u_i]_{6\times1} = \begin{bmatrix} u_{ix} & u_{iy} & \theta_{iz} & \theta_{ix} & \theta_{iy} & u_{iz} \end{bmatrix}^T \tag{7.1}$$

The six displacement components of Eq. (7.1) are expressed in the reference frame $C_i xyz$ whose axes are parallel to the axes of the global frame – Figure 7.1 also displays the local frame $C_i x_i y_i z_i$. Therefore, a three-dimensional (3D) compliant mechanism with n rigid links (and the corresponding flexible hinges that connect the rigid links) undergoing spatial motion has $6n$ DOF, which are arranged in the overall displacement vector:

$$[u]_{6n\times1} = \begin{bmatrix} [u_1]_{6\times1} & [u_2]_{6\times1} & \cdots & [u_i]_{6\times1} & \cdots & [u_n]_{6\times1} \end{bmatrix}^T. \tag{7.2}$$

A planar (2D) rigid link can move in its plane or out of its plane; in both cases, the link has three DOF in either of these motions, which are collected in a related displacement vector as:

$$\left[u_{i,ip} \right]_{3\times1} = \left[\begin{array}{ccc} u_{ix} & u_{iy} & \theta_{iz} \end{array} \right]^{T} ; \ \left[u_{i,op} \right]_{3\times1} = \left[\begin{array}{ccc} \theta_{ix} & \theta_{iy} & u_{iz} \end{array} \right]^{T} \ (7.3)$$

where "ip" stands for in-plane and "op" symbolizes out-of-plane. As a result, a planar mechanism with n rigid links and flexible hinges undergoing either in-plane or out-of-plane motion has $3n$ DOF, which can be organized as:

$$\left[u_{ip} \right]_{3n\times1} = \left[\begin{array}{ccccc} \left[u_{1,ip} \right]_{3\times1} & \left[u_{2,ip} \right]_{3\times1} & \cdots & \left[u_{i,ip} \right]_{3\times1} & \cdots & \left[u_{n,ip} \right]_{3\times1} \end{array} \right]^{T}$$

$$\left[u_{op} \right]_{3n\times1} = \left[\begin{array}{ccccc} \left[u_{1,op} \right]_{3\times1} & \left[u_{2,op} \right]_{3\times1} & \cdots & \left[u_{i,op} \right]_{3\times1} & \cdots & \left[u_{n,op} \right]_{3\times1} \end{array} \right]^{T}.$$

$$(7.4)$$

A compliant mechanism's number of DOF sets the dimensions of the matrices defining the system and of the load and displacement vectors. A spatial mechanism with $6n$ DOF has square inertia, compliance/stiffness and damping matrices of dimension $6n \times 6n$, while the load/displacement is a $6n \times 1$ vector. Similarly, the dimension of these matrices and the load/displacement vector is $3n$ for a planar compliant mechanism moving either in-plane or out-of-plane.

The mathematical models presented in the subsequent Sections 7.1.2–7.1.4 can be derived by means of general dynamics methods such as Newton's second law of motion or Lagrange's equations.

7.1.2 TIME-DOMAIN FREE UNDAMPED RESPONSE

When only the inertia and the elastic features of a flexible-hinge mechanism are taken into account and there is no external load, the resulting mathematical model and its solution/response are free undamped or natural.

Consider a multiple-DOF flexible mechanism whose state is defined by a general displacement vector $[u]$. The free, undamped mathematical model is the matrix differential equation:

$$[M][\ddot{u}] + [K][u] = [0] \quad \text{with} \quad [K] = [C]^{-1}, \tag{7.5}$$

where $[M]$ is the inertia (mass) matrix, $[K]$ is the stiffness matrix and $[C]$ is the compliance matrix. As per Den Hartog [24], Thompson and Dillon Dahleh [25], Rao [26] or Inman [27], the solution to Eq. (7.5) is:

$$[u] = [U]\sin(\omega_{n}t + \varphi). \tag{7.6}$$

Combining Eqs. (7.5) and (7.6) leads to the eigenvalue problem:

$$\left(-\omega_n^2[M]+[K]\right)[U]=[0] \quad \text{or} \quad [D][U]=\lambda_n[U], \tag{7.7}$$

where $[U]$ is the eigenvector corresponding to a natural frequency ω_n. The dynamic matrix $[D]$ and the eigenvalue λ_n are:

$$[D]=[M]^{-1}[K]=[M]^{-1}[C]^{-1}=\left([C][M]\right)^{-1}; \quad \lambda_n=\omega_n^2. \tag{7.8}$$

There are several methods available for solving Eq. (7.5), such as time-stepping algorithms (Runge–Kutta, Newmark), modal analysis (superposition), Laplace transform or transfer function procedures. The Laplace transform method can be utilized for systems having a relatively small number of DOF and is also applied in this chapter to find the free undamped response of flexible-hinge mechanisms. Consider that the initial displacement vector $[u(0)]$ and the initial velocity vector $[v(0)]$ are known. Applying the Laplace transform to Eq. (7.5), the following Laplace-domain solution is obtained:

$$[U(s)]=\left(s^2[M]+[K]\right)^{-1}[M]\left(s[u(0)]+[v(0)]\right), \tag{7.9}$$

where $[U(s)]$ is the Laplace transform of $[u(t)]$. The solution $[u(t)]$ is found as the inverse Laplace transform of $[U(s)]$ of Eq. (7.9).

As indicated in Eq. (7.5), the stiffness matrix $[K]$ is the inverse of the compliance matrix $[C]$. Details will be provided throughout this chapter on how to assemble the matrix $[C]$ consistent with the DOF resulting from the independent motions of the moving rigid links for several serial and parallel compliant mechanisms. Section 7.2 discusses the formulation of the inertia matrix $[M]$.

7.1.3 Time-Domain Free Damped Response

The free damped dynamic model of a flexible-hinge mechanism adds damping (usually of viscous nature) to the inertia and compliance/stiffness properties of a free undamped system.

The lumped-parameter mathematical model of a mechanical system that includes viscous damping properties in addition to inertia and stiffness is the equation:

$$[M][\ddot{u}]+[B][\dot{u}]+[K][u]=[0], \tag{7.10}$$

where $[B]$ is the damping matrix. Applying the Laplace transform to Eq. (7.10) with the initial conditions $[u(0)]$ and $[v(0)]$, the following Laplace-domain solution is obtained:

$$[U(s)]=\left(s^2[M]+s[B]+[K]\right)^{-1}\left((s[M]+[B])[u(0)]+[M][v(0)]\right). \tag{7.11}$$

The time-domain solution $[u(t)]$ is found by inverse Laplace transforming $[U(s)]$ of Eq. (7.11).

While the damping matrix $[B]$ can be formulated independently, an alternative model uses the proportional damping (or Rayleigh damping), which assumes that $[B]$ is a linear combination of $[M]$ and $[K]$ in the form:

$$[B] = \alpha[M] + \beta[K], \tag{7.12}$$

where the positive coefficients α and β weigh the contributions of $[M]$ and $[K]$. For $[B]$ defined as in Eq. (7.12), the solution of Eq. (7.11) changes to:

$$[U(s)] = \left(\left(s^2 + \alpha s\right)[M] + \left(\beta s + 1\right)[K]\right)^{-1}\left(\left((s+\alpha)[M] + \beta[K]\right)[u(0)] + [M][v(0)]\right). \tag{7.13}$$

7.1.4 Time-Domain Forced Response

When external forcing is applied to a flexible-hinge mechanism, the following time-domain mathematical model results:

$$[M][\ddot{u}] + [B][\dot{u}] + [K][u] = [f], \tag{7.14}$$

where $[f]$ is the time-dependent external forcing vector. Laplace transforming Eq. (7.14) by using the initial-condition vectors $[u(0)]$ and $[v(0)]$, the resulting Laplace-domain solution is:

$$[U(s)] = \left(s^2[M] + s[B] + [K]\right)^{-1}\left(\left(s[M] + [B]\right)[u(0)] + [M][v(0)] + [F(s)]\right), \tag{7.15}$$

where $[F(s)]$ is the Laplace transform of $[f(t)]$. The inverse Laplace transform is applied to $[U(s)]$ of Eq. (7.15) to obtain the time-domain displacement $[u(t)]$.

Equation (7.15) can also be formulated in terms of the transfer function matrix $[G(s)]$ and an equivalent forcing vector $[F_{eq}(s)]$ as:

$$[U(s)] = [G(s)][F_{eq}(s)] \quad \text{with} \quad \begin{cases} [G(s)] = \left(s^2[M] + s[B] + [K]\right)^{-1} \\ [F_{eq}(s)] = \left(s[M] + [B]\right)[u(0)] \\ \quad + [M][v(0)] + [F(s)]. \end{cases} \tag{7.16}$$

For zero initial conditions, Eq. (7.16) simplifies to:

$$[U(s)] = [G(s)][F(s)]. \tag{7.17}$$

7.1.5 STEADY-STATE AND FREQUENCY-DOMAIN RESPONSES FOR HARMONIC FORCING

When the external loading applied to an n-DOF flexible-hinge mechanism is harmonic, for instance, sinusoidal:

$$[f(t)] = \begin{bmatrix} f_1(t) & f_2(t) & \cdots & f_i(t) & \cdots & f_n(t) \end{bmatrix}^T \quad \text{with}$$

$$f_i(t) = F_i \sin(\omega_i t), \quad i = 1, 2, \ldots, n. \tag{7.18}$$

where F_i are amplitudes and ω_i are frequencies ($i = 1$ to n), the steady-state, time-domain response $[u(t)]$ of the compliant mechanism is also sinusoidal and of the form:

$$[u(t)] = \begin{bmatrix} u_1(t) & u_2(t) & \cdots & u_i(t) & \cdots & u_n(t) \end{bmatrix}^T \quad \text{with}:$$

$$u_i(t) = \sum_{k=1}^{n} |G_{ik}(\omega_k \cdot j)| F_k \sin(\omega_k t + \measuredangle G_{ik}(\omega_k \cdot j)), \quad i = 1, 2, \ldots, n. \tag{7.19}$$

The complex numbers $G_{ik}(\omega_k j)$ of Eq. (7.19) are components resulting from the transfer function matrix $[G(s)]$ of Eq. (7.16) where the substitution $s = \omega \cdot j$ ($j^2 = -1$) was used. The complex number $G_{ik}(\omega_k j)$ is defined by its modulus $|G_{ik}(\omega_k \cdot j)|$ and its phase angle $\measuredangle G_{ik}(\omega_k \cdot j)$. The variation of these two amounts in terms of the frequency ω_k is known as frequency response, and the resulting plots are the Bode plots.

7.2 INERTIA (MASS) MATRIX

For the majority of compliant mechanisms whose rigid links are substantially heavier than the flexible hinges, the inertia (mass) matrix results solely from the rigid links. There are, however, devices where the dimensions and mass properties of the rigid links and the flexible hinges are of the same order of magnitude. These applications require consideration and inclusion of flexible-hinge inertia fractions into the overall inertia matrix; both categories are characterized in this section.

7.2.1 HEAVY RIGID LINKS AND MASSLESS FLEXIBLE HINGES

The inertia matrix of the 3D link shown in Figure 7.1 is expressed in connection to the nodal displacement vector of Eq. (7.1) as a diagonal matrix in the local frame $C_i x_i y_i z_i$:

$$\left[M_i^{(l)} \right]_{6 \times 6} = \text{diag}\left(m_i \quad m_i \quad J_{iz_i} \quad J_{ix_i} \quad J_{iy_i} \quad m_i \right), \tag{7.20}$$

where m_i is the mass of the rigid link i, and J_{ix_i}, J_{iy_i} and J_{iz_i} are the link mass moments of inertia with respect to the central local reference frame $C_i x_i y_i z_i$ axes. The superscript "l" in the matrix of Eq. (7.20) denotes the local frame.

For the in-plane or out-of-plane motions of planar mechanisms, the local-frame inertia matrices are related to the displacements defined in Eqs. (7.3) and are expressed similarly to Eq. (7.20):

$$\left[M_{i,ip}^{(l)} \right]_{3\times3} = \text{diag}\left(\begin{array}{ccc} m_i & m_i & J_{izi} \end{array} \right); \quad \left[M_{i,op}^{(l)} \right]_{3\times3} = \text{diag}\left(\begin{array}{ccc} J_{ixi} & J_{iyi} & m_i \end{array} \right).$$

(7.21)

7.2.2 HEAVY RIGID LINKS AND HEAVY FLEXIBLE HINGES

When the dimensions and masses of the rigid links and flexible hinges are comparable, the inertia of the flexible hinges should be combined with that of the rigid links. This approach is particularly relevant to micro-/nanoscale compliant mechanisms. Two modeling approaches are presented here, and they are both similar to finite element methods that derive inertia matrices, as discussed in Chapter 8. The simplest approach is the *lumped-inertia* model, which places fractions of the total mass of the flexible hinge at mobile end(s) corresponding to the relevant DOF and resulting in a diagonal inertia matrix. The other approach evaluates mass fractions to be placed at the flexible-hinge end(s) to account for all possible motions and hinge deformations (see, for instance, Lobontiu et al. [28], Lobontiu [29], Lobontiu and Garcia [30], Lobontiu [31] or [32]) and results in *consistent-inertia* matrices comprising off-diagonal components. Two separate cases are considered here: in one case, the flexible hinge is fixed at one and connects to a mobile rigid link at the other end; in the other case, the flexible hinge connects to adjacent mobile rigid links at both ends.

This chapter only covers the inertia of straight-axis flexible hinges, and as it is shown, the consistent-model results are identical to the ones obtained by the finite element approach – see Chapter 8, which also covers the inertia of curvilinear-axis flexible hinges.

7.2.2.1 Fixed-Free Flexible Hinges

The flexible hinge of Figure 7.2 is fixed at one and connects to a mobile rigid link at the other end. Inertia matrices are derived here corresponding to the in-plane and out-of-plane vibrations of the hinge "free" end H (which is actually connected to a moving rigid link) using both a consistent formulation and a lumped-parameter model. The resulting inertia matrices are then transferred (through rotation and translation) at the center C of the rigid link, and the resulting transformed hinge inertia matrices are added to the corresponding rigid-link inertia matrices.

7.2.2.1.1 Consistent-Inertia Model

The kinetic energy T of a lumped-parameter dynamic system is expressed in matrix form as:

$$T = \frac{1}{2} \cdot [\dot{u}]^T [M][\dot{u}],$$

(7.22)

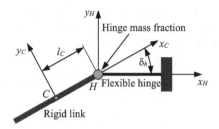

FIGURE 7.2   Skeleton representation of mechanism with one heavy flexible hinge, which is fixed at one end and connected to a moving heavy rigid link at the other end.

where $[\dot{u}]$ is the velocity vector corresponding to the displacement vector $[u]$ and $[M]$ is the inertia matrix. The aim in this section is to derive the in-plane and out-of-plane inertia matrices of a fixed-free flexible hinge by using the equality between the kinetic energy of the actual (distributed-parameter) hinge and an equivalent system with inertia fractions located at the hinge free (moving) end.

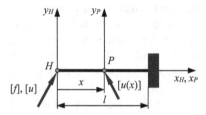

FIGURE 7.3   Skeleton representation of a fixed-free flexible hinge with end-point load.

Consider the skeleton representation of the fixed-free (cantilever-type), straight-axis flexible hinge of Figure 7.3, which is acted upon at its free end H by a spatial load vector $[f]$. Assume that the cross-section of the flexible hinge is variable. The first goal is to establish a relationship between the displacements $[u]$ at the free end and the displacement $[u(x)]$ at an arbitrary point P of position x. Axial, torsional and bending vibrations of the hinge are analyzed separately.

Axial vibrations:

Under an axial force f_x applied at point H in Figure 7.3, the displacements u_x at H and $u_x(x)$ at P are calculated as:

$$
\begin{aligned}
u_x &= C_{u_x-f_x} \cdot f_x \\
u_x(x) &= C_{u_x-f_x}(x) \cdot f_x
\end{aligned}
\;\rightarrow\; u_x(x) = \frac{C_{u_x-f_x}(x)}{C_{u_x-f_x}} \cdot u_x \text{ or } u_x(x,t) = \frac{C_{u_x-f_x}(x)}{C_{u_x-f_x}} \cdot u_x(t),
$$

$$(7.23)$$

where $C_{u_x-f_x}$ is the axial compliance of the full flexible hinge and $C_{u_x-f_x}(x)$ is the axial compliance of the segment comprised between P and the fixed end of the hinge. While $C_{u_x-f_x}$ is provided in Eqs. (2.6), the x-dependent compliance of Eqs. (7.23) is calculated as:

$$C_{u_x - f_x}(x) = \frac{1}{E} \cdot \int_x^l \frac{dx}{A(x)}. \tag{7.24}$$

For a constant cross-section hinge with $A(x) = A$, Eq. (7.23) becomes:

$$u_x(x,t) = \left(1 - \frac{x}{l}\right) \cdot u_x(t). \tag{7.25}$$

Equations (7.23) and (7.25) indicate that the axial displacement $u_x(x, t)$ at P is the product of a space-dependent function and a time-dependent function, which is the axial displacement at the free end u_x. Therefore, Eq. (7.23) results in:

$$\frac{\partial u_x(x,t)}{\partial t} = \frac{C_{u_x - f_x}(x)}{C_{u_x - f_x}} \cdot \frac{du_x(t)}{dt}. \tag{7.26}$$

The kinetic energy of the flexible hinge during axial vibrations (along the x axis) is calculated as indicated in Thompson and Dillon Dahleh [25], Rao [26] and Inman [27]:

$$T_a = \frac{1}{2} \cdot \rho \cdot \int_0^l A(x) \cdot \left(\frac{\partial u_x(x,t)}{\partial t}\right)^2 dx = \frac{1}{2} \cdot \rho \cdot \left(\int_0^l A(x) \cdot \left(\frac{C_{u_x - f_x}(x)}{C_{u_x - f_x}}\right)^2 dx\right) \cdot \left(\frac{du_x}{dt}\right)^2. \tag{7.27}$$

If a single particle of mass M_a is placed at the free end of the hinge, the kinetic energy of this particle during its motion along the x axis is:

$$T_a = \frac{1}{2} \cdot M_a \cdot \left(\frac{du_{Hx}}{dt}\right)^2. \tag{7.28}$$

Assuming that the actual, distributed-inertia flexible hinge and the particle of mass M_a have the same kinetic energy and identical x-axis motions at point H, Eqs. (7.27) and (7.28) yield the value of the inertia-equivalent particle mass:

$$M_a = \rho \cdot \int_0^l A(x) \cdot \left(\frac{C_{u_x - f_x}(x)}{C_{u_x - f_x}}\right)^2 dx. \tag{7.29}$$

For a constant cross-section flexible hinge of area A, Eqs. (7.25) and (7.29) result in:

$$M_a = \rho \cdot A \cdot \int_0^l \left(1 - \frac{x}{l}\right)^2 dx = \frac{1}{3} \cdot (\rho \cdot A \cdot l) = \frac{1}{3} \cdot m_h, \tag{7.30}$$

where m_h is the mass of the full flexible hinge.

Torsional vibrations:

Based on Figure 7.3 and applying a reasoning similar to the one used to evaluate the axial vibrations of the fixed-free flexible hinge, it can be shown that the end-point rotation θ_x and the rotation $\theta_x(x)$ at the arbitrary point P that are produced when a moment m_x is applied at point H along the longitudinal (x) axis are related as:

$$\begin{aligned} \theta_x &= C_{\theta_x-m_x} \cdot m_x \\ \theta_x(x) &= C_{\theta_x-m_x}(x) \cdot m_x \end{aligned} \quad \rightarrow \theta_x(x) = \frac{C_{\theta_x-m_x}(x)}{C_{\theta_x-m_x}} \cdot \theta_x$$

$$\text{or} \quad \begin{cases} \theta_x(x,t) = \dfrac{C_{\theta_x-m_x}(x)}{C_{\theta_x-m_x}} \cdot \theta_x(t) \\[4mm] \dfrac{\partial \theta_x(x,t)}{\partial t} = \dfrac{C_{\theta_x-m_x}(x)}{C_{\theta_x-m_x}} \cdot \dfrac{d\theta_x(t)}{dt} \end{cases} \quad (7.31)$$

Equation (7.31) also includes the time derivative of $\theta_x(x, t)$. $C_{\theta_x-m_x}$ is defined in Eq. (2.6), and the x-dependent compliance of Eq. (7.31) is defined as:

$$C_{\theta_x-m_x}(x) = \frac{1}{G} \cdot \int_x^l \frac{dx}{I_t(x)}. \quad (7.32)$$

For a constant cross-section hinge, the torsion moment of area is $I_t(x) = I_t$ and Eq. (7.31) simplifies to:

$$\theta_x(x,t) = \left(1 - \frac{x}{l}\right) \cdot \theta_x(t). \quad (7.33)$$

The kinetic energy of the distributed-inertia flexible hinge during torsional vibrations around the x axis is expressed as in Thompson and Dillon Dahleh [25], Rao [26] and Inman [27]:

$$T_t = \frac{1}{2} \cdot \rho \cdot \int_0^l I_x(x) \cdot \left(\frac{\partial \theta_x(x,t)}{\partial t}\right)^2 dx = \frac{1}{2} \cdot \rho \cdot \left(\int_0^l I_t(x) \cdot \left(\frac{C_{\theta_x-m_x}(x)}{C_{\theta_x-m_x}}\right)^2 dx\right) \cdot \left(\frac{d\theta_x}{dt}\right)^2. \quad (7.34)$$

The same kinetic energy of Eq. (7.34) can be produced by a body placed at the free end of the flexible hinge; if the x-axis mass moment of inertia of this body is J_t, then its kinetic energy is:

$$T_t = \frac{1}{2} \cdot J_t \cdot \left(\frac{d\theta_x}{dt}\right)^2. \quad (7.35)$$

Equating T_t of Eqs. (7.34) and (7.35) yields the equivalent mass moment of inertia:

$$J_t = \rho \cdot \int_0^l I_t(x) \cdot \left(\frac{C_{\theta_x - m_x}(x)}{C_{\theta_x - m_x}} \right)^2 dx \qquad (7.36)$$

For a constant cross-section hinge with torsional moment of area I_t, Eqs. (7.33) and (7.36) yield:

$$J_t = \rho \cdot I_t \cdot \int_0^l \left(1 - \frac{x}{l} \right)^2 dx = \frac{1}{3} \cdot (\rho \cdot I_t \cdot l) = \frac{1}{3} \cdot J_{h,x}, \qquad (7.37)$$

where $J_{h,x}$ is the torsional moment of inertia of the full hinge with respect to the x axis.

Bending vibrations:

In-plane and out-of-plane bending vibrations are treated separately, and the corresponding inertia matrices will be derived next.

Consider that a load vector $[f_{bz}] = [f_y \ m_z]^T$ is applied at the free end H of the flexible hinge shown in Figure 7.3, where "bz" denotes bending about the z axis. This load generates a displacement vector $[u_{bz}] = [u_y \ \theta_z]^T$ at the same point and a displacement vector $[u_{bz}(x)] = [u_y(x) \ \theta_z(x)]^T$ at the arbitrary point P. The two displacement vectors are expressed as:

$$[u_{bz}] = [C_{H,bz}][f_{bz}]; \quad [u_{bz}(x)] = [C_{PH,bz}(x)][f_{bz}], \qquad (7.38)$$

where $[C_{H,bz}]$ is the bending compliance matrix of the full flexible hinge with respect to the reference frame positioned at H and related to the load at H. $[C_{PH,bz}(x)]$ is the compliance matrix of the segment comprised between the arbitrary point P and the fixed end, calculated at point P in the local frame $Px_Py_Pz_P$, and produced by the same load applied at H. This latter matrix is calculated as:

$$[C_{PH,bz}(x)] = [C_{P,bz}(x)][T_{HP,bz}], \qquad (7.39)$$

where $[C_{P,bz}(x)]$ is the compliance matrix of the segment comprised between P and the fixed point when evaluated with respect to the local frame placed at P, and considering the load is applied at P, as well. $[T_{HP,bz}]$ is the in-plane translation matrix moving the load from H to P. For a variable cross-section flexible hinge, these matrices are evaluated as:

$$[C_{H,bz}] = \begin{bmatrix} C_{u_y - f_y} & C_{u_y - m_z} \\ C_{u_y - m_z} & C_{\theta_z - m_z} \end{bmatrix}; \quad [C_{P,bz}(x)] = \begin{bmatrix} C_{u_y - f_y}(x) & C_{u_y - m_z}(x) \\ C_{u_y - m_z}(x) & C_{\theta_z - m_z}(x) \end{bmatrix};$$

$$[T_{HP,bz}] = \begin{bmatrix} 1 & 0 \\ -x & 1 \end{bmatrix}. \qquad (7.40)$$

The compliances of $[C_{H,bz}]$ are defined in Eqs. (2.6), while the compliances depending on x in Eq. (7.40) are:

$$C_{u_y-f_y}(x) = \frac{1}{E} \cdot \int_x^l \frac{x^2}{I_z(x)} dx; \quad C_{u_y-m_z}(x) = -\frac{1}{E} \cdot \int_x^l \frac{x}{I_z(x)} dx; \quad C_{\theta_z-m_z}(x) = \frac{1}{E} \cdot \int_x^l \frac{dx}{I_z(x)},$$

$$(7.41)$$

where $I_z(x)$ is the z-axis area moment of the variable cross-section. Eliminating the load vector between the two Eqs. (7.38) results in:

$$[u_{bz}(x)] = [S_b(x)][u_{bz}] \text{ with } [S_b(x)] = [C_{PH,bz}(x)][C_{H,bz}]^{-1} = \begin{bmatrix} S_{b1}(x) & S_{b2}(x) \\ S_{b3}(x) & S_{b4}(x) \end{bmatrix}.$$

$$(7.42)$$

The bending *shape-function matrix* $[S_b(x)]$, which is the conventional name used in finite element analysis (see also Chapter 8), is a square matrix of dimension 2. The deflection at P is expressed by means of the first-row elements of $[S_b(x)]$ of Eq. (7.42), namely:

$$u_y(x) = S_{b1}(x) \cdot u_y + S_{b2}(x) \cdot \theta_z \quad \text{or} \quad u_y(x,t) = [S_{b,1}(x)]^T [u_{bz}(t)]$$

$$\text{with} \quad [S_{b,1}(x)] = \begin{bmatrix} S_{b1}(x) & S_{b2}(x) \end{bmatrix}^T.$$

$$(7.43)$$

For a constant cross-section, the shape functions of Eq. (7.43) become:

$$S_{b1}(x) = 1 - 3 \cdot \frac{x^2}{l^2} + 2 \cdot \frac{x^3}{l^3}; \quad S_{b2}(x) = x - 2 \cdot \frac{x^2}{l} + \frac{x^3}{l^2}. \quad (7.44)$$

Equation (7.43) indicates that a similar relationship exists between the velocity fields corresponding to the two displacement vectors of that equation, namely:

$$\frac{\partial u_y(x,t)}{\partial t} = [S_{b,1}(x)]^T \left[\frac{du_{bz}(t)}{dt} \right]. \quad (7.45)$$

As per Thompson and Dillon Dahleh [25], Rao [26] and Inman [27], the bending-related kinetic energy of the distributed-inertia flexible hinge is calculated as:

$$T_b = \frac{1}{2} \cdot \rho \cdot \int_0^l A(x) \cdot \left(\frac{\partial u_y(x,t)}{\partial t} \right)^2 dx$$

$$= \frac{1}{2} \cdot \rho \cdot \left[\frac{du_{bz}(t)}{dt} \right]^T \left(\int_0^l A(x) \cdot [S_{b,1}(x)][S_{b,1}(x)]^T dx \right) \cdot \left[\frac{du_{bz}(t)}{dt} \right]. \quad (7.46)$$

Equation (7.46) used Eq. (7.45), as well as the identity:

$$a = [V_1]^T [V_2] \rightarrow a^2 = a^T \cdot a = [V_2]^T [V_1][V_1]^T [V_2], \qquad (7.47)$$

where a is a scalar that can be written as the product of two equal-dimension vectors $[V_1]$ and $[V_2]$. In Eqs. (7.45) and (7.46):

$$a = \frac{\partial u_y(x,t)}{\partial t}; \quad [V_1] = [S_{b,1}(x)]; \quad [V_2] = \left[\frac{du_{bz}(t)}{dt}\right]. \qquad (7.48)$$

Comparing Eqs. (7.46) and (7.22) shows that the in-plane, bending-related symmetric inertia matrix is:

$$[M_b] = \rho \cdot \int_0^l A(x) \cdot [S_{b,1}(x)][S_{b,1}(x)]^T \, dx = \begin{bmatrix} M_b(1,1) & M_b(1,2) \\ M_b(1,2) & M_b(2,2) \end{bmatrix}. \qquad (7.49)$$

For a constant cross-section flexible hinge, Eq. (7.49) reduces to:

$$[M_b] = \frac{1}{35} \cdot m_h \cdot \begin{bmatrix} 13 & \frac{11}{6} \cdot l \\ \frac{11}{6} \cdot l & \frac{1}{3} \cdot l^2 \end{bmatrix}, \qquad (7.50)$$

where m_h is the mass of the flexible hinge.

Note on inertia matrix corresponding to out-of-plane bending:

The development and equations obtained for in-plane bending (namely, Eqs. (7.49) and (7.50)) are also valid for out-of-plane bending in relation to a displacement vector $[u_{by}] = [u_z \ \theta_y]^T$ at the end H in Figure 7.3 and a displacement vector $[u_{by}(x)] = [u_z(x) \ \theta_y(x)]^T$ at the arbitrary point P – "by" denotes bending around the y axis.

Example 7.1

A straight-axis, right elliptical flexible hinge of rectangular cross-section is defined by a constant out-of-plane width $w = 0.008$ m, ellipse semi-axis length $b = 0.006$ m, material Young's modulus $E = 2 \cdot 10^{11} \, \text{N/m}^2$ and mass density $\rho = 7,800 \, \text{kg/m}^3$. Consider the other ellipse semi-axis a and the minimum thickness t are variable in the intervals $0.005 \le a \le 0.015$ and $0.0007 \le t \le 0.012$ (both measured in meters). Study the variation of the in-plane, direct-bending inertia fraction $M_b(1,1)$ when calculating it via the shape functions of a constant cross-section hinge vs. the shape functions defined based on compliances and the actual variable cross-section.

Solution:

The compliance matrix $[C_{H,bz}]$ of Eq. (7.40) is calculated by means of Eqs. (A3.3), whereas the compliance matrix $[C_{P,bz}(x)]$ of the same Eq. (7.40) is calculated with the following components, which are originally provided in Eq. (2.53):

$$
\begin{cases}
C_{u_y-f_y}(x) = C_{u_y-f_y}(\varphi) = \dfrac{12a^3}{Ew} \cdot \displaystyle\int_{\varphi}^{\pi/2} \dfrac{\sin^2 \varphi \cdot \cos \varphi}{t(\varphi)^3} \cdot d\varphi; \\[4mm]
C_{u_y-m_z}(x) = C_{u_y-m_z}(\varphi) = -\dfrac{12a^2}{Ew} \cdot \displaystyle\int_{\varphi}^{\pi/2} \dfrac{\sin \varphi \cdot \cos \varphi}{t(\varphi)^3} \cdot d\varphi; \\[4mm]
C_{\theta_z-m_z}(x) = C_{\theta_z-m_z}(\varphi) = \dfrac{12a}{Ew} \cdot \displaystyle\int_{\varphi}^{\pi/2} \dfrac{\cos \varphi}{t(\varphi)^3} \cdot d\varphi;,
\end{cases}
\tag{7.51}
$$

where $t(\varphi) = t + 2b(1 - \cos\varphi)$ is the variable, in-plane thickness of the hinge, as per Eq. (2.50). The shape-function matrix of Eq. (7.42) provides the components $S_{b1}(x)$ and $S_{b2}(x)$ that form the vector $[S_{b,1}(x)]$ of Eq. (7.43). Using $x = a \cdot \sin\varphi$, the shape functions of Eqs. (7.44), which consider the cross-section as constant, can be evaluated in terms of the variable angle φ, and the corresponding shape-function vector $[S_{b,1}(x)]$ is formed. The inertia matrix is calculated for both sets of shape functions based on Eq. (7.49), which is conditioned as:

$$
[M_b] = \rho \cdot w \cdot a \cdot \int_{-\pi/2}^{\pi/2} [S_b(\varphi)][S_b(\varphi)]^T t(\varphi) \cdot \cos \varphi \cdot d\varphi.
\tag{7.52}
$$

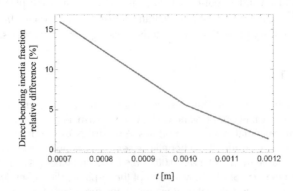

FIGURE 7.4 Plot of relative difference between inertia fractions in terms of the minimum thickness t for a right elliptical flexible hinge of rectangular cross-section.

The relative difference is evaluated between the component $M_b(1,1)$ of the inertia matrix that uses shape functions based on the variable cross-section and the same component of the inertia matrix that is calculated based on the constant cross-section shape functions of Eq. (7.44). The relative error is constant and equal to 5.58% when the parameter a varies (and t assumes a value of 0.001 m). However, for t ranging in the given interval (and $a = 0.01$ m), the relative error in the direct-bending inertia fraction varies as illustrated in Figure 7.4. As the figure shows, the relative errors between the two models' predictions decrease with t increasing.

In-plane and out-of-plane inertia matrices:
The in-plane and out-of-plane inertia matrices of the flexible hinge can now be formed. For an in-plane displacement vector $[u_{ip}] = [u_x\ u_y\ \theta_z]^T$, the inertia matrix is formally assembled from the axially-related Eq. (7.29) and the bending-related Eq. (7.49), namely:

$$\left[M_{ip}^{(h)}\right] = \begin{bmatrix} M_a & 0 & 0 \\ 0 & M_b(1,1) & M_b(1,2) \\ 0 & M_b(1,2) & M_b(2,2) \end{bmatrix};$$

$$\left[M_{ip}^{(h)}\right] = m_h \cdot \begin{bmatrix} \dfrac{1}{3} & 0 & 0 \\ 0 & \dfrac{13}{35} & \dfrac{11}{210} \cdot l \\ 0 & \dfrac{11}{210} \cdot l & \dfrac{1}{105} \cdot l^2 \end{bmatrix}. \tag{7.53}$$

The second Eq. (7.53) expresses the inertia matrix of a constant cross-section, fixed-free flexible hinge.

The out-of-plane inertia matrix combines torsion and bending, and is correspondingly assembled from Eqs. (7.36) and (7.49) based on the free-end displacement vector $[u_{op}] = [\theta_x\ \theta_y\ u_z]^T$:

$$\left[M_{op}^{(h)}\right] = \begin{bmatrix} J_t & 0 & 0 \\ 0 & M_b(2,2) & M_b(1,2) \\ 0 & M_b(1,2) & M_b(1,1) \end{bmatrix};$$

$$\left[M_{op}^{(h)}\right] = \begin{bmatrix} \dfrac{1}{3} \cdot J_{h,x} & 0 & 0 \\ 0 & \dfrac{1}{105} \cdot m_h \cdot l^2 & \dfrac{11}{210} \cdot m_h \cdot l \\ 0 & \dfrac{11}{210} \cdot m_h \cdot l & \dfrac{13}{35} \cdot m_h \end{bmatrix}. \tag{7.54}$$

Again, the second Eq. (7.54) provides the inertia matrix of a constant cross-section, fixed-free flexible hinge. Note that both the in-plane and out-of-plane inertia matrices of Eqs. (7.53) and (7.54) are symmetric and evaluated in their local reference frames. The superscript "h" in Eqs. (7.53) and (7.54) stands for hinge. For a hinge with variable cross-section, the flexure hinge mass and the x-axis moment of inertia are calculated as:

$$m_h = \rho \cdot \int_0^l A(x)dx; \quad J_{h,x} = \rho \cdot \int_0^l I_t(x)dx. \quad (7.55)$$

7.2.2.1.2 Lumped-Parameter Inertia Model

The distributed-inertia of a fixed-free flexible hinge can be equivalently transformed in lumped-inertia fractions placed at the free, moving end in order to account for axial, torsional and bending vibrations of the hinge, as pictured in Figure 7.5.

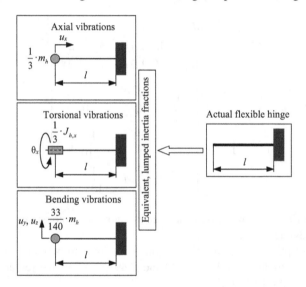

FIGURE 7.5 Lumped-parameter inertia fractions resulting from a distributed-inertia fixed-free flexible hinge.

The mass fraction accounting for the axial (x axis) vibration is equal to 1/3 of the flexible-hinge mass m_h. This is a value that corresponds to a constant cross-section flexible hinge – see Inman [27], Lobontiu and Garcia [30], Lobontiu [31] and [32] for more details – but can also be used as a reasonable approximation for variable cross-section hinges. In a similar manner, the lumped-parameter torsional moment of inertia (accounting for rotary vibrations along the x axis) that is located at the hinge end H is equal to 1/3 of the full hinge mass moment of inertia $J_{h,x}$ for a constant cross-section member, but is a fair approximation for variable cross-section hinges, as well. These values were also obtained through the consistent approach in this section. The flexible-hinge inertia representing the translation vibrations along the y or z axes at the end H is equal to 33/140 of m_h, which corresponds to a constant cross-section flexible hinge and is also used to approximate the similar feature of a variable cross-section hinge. Details on how to derive the bending-related mass fraction are provided in Thompson and Dillon Dahleh [25], Inman [27] and Lobontiu [32].

As in the consistent method, the in-plane DOF of the flexible hinge are collected in the vector $[u_x \ u_y \ \theta_z]^T$, whereas the out-of-plane DOF are the vector's components $[\theta_x \ \theta_y \ u_z]^T$ – both vectors are defined in the hinge local reference frame $Hx_H y_H z_H$ – see Figure 7.3. Corresponding to these local DOF, the following in-plane and out-of-plane lumped-parameter inertia matrices are formed:

$$
\left[M_{ip}^{(h)} \right] = \begin{bmatrix} \dfrac{1}{3} \cdot m_h & 0 & 0 \\ 0 & \dfrac{33}{140} \cdot m_h & 0 \\ 0 & 0 & 0 \end{bmatrix}; \quad \left[M_{op}^{(h)} \right] = \begin{bmatrix} \dfrac{1}{3} \cdot J_{h,x} & 0 & 0 \\ 0 & 0 & 0 \\ 0 & 0 & \dfrac{33}{140} \cdot m_h \end{bmatrix}.
$$

(7.56)

7.2.2.1.3 Assembling the Inertia Matrix Resulting from Rigid Links and Fixed-Free Flexible Hinges

Consider, again, the simple chain formed of the rigid link and the flexible hinge shown in Figure 7.2. An in-plane inertia load $\left[f_{ip} \right] = \begin{bmatrix} f_x & f_y & m_z \end{bmatrix}^T$ acting at H results in a displacement vector $\left[u_{ip} \right] = \begin{bmatrix} u_x & u_y & \theta_z \end{bmatrix}^T$ at the same point. These vectors are transferred from the hinge local frame $Hx_H y_H z_H$ to the global frame $Cx_C y_C z_C$ (which is located at the rigid-link center of mass C) where they become $\left[f_{C,ip} \right]$ and $\left[u_{C,ip} \right]$, respectively. Chapters 3 and 4 detail the procedure of transferring loads and displacements between reference frames that are both offset (non-collocated) and rotated. In a similar manner, the local-frame load and displacement (also acceleration) vectors are connected to their global-frame counterparts as:

$$
\begin{cases} \left[f_{ip} \right] = \left[T_{CH,ip} \right] \left[R_{CH} \right] \left[f_{C,ip} \right]; \\ \left[u_{ip} \right] = \left[T_{CH,ip} \right] \left[R_{CH} \right] \left[u_{C,ip} \right] \rightarrow \left[\ddot{u}_{ip} \right] = \left[T_{CH,ip} \right] \left[R_{CH} \right] \left[\ddot{u}_{C,ip} \right] \end{cases},
$$

(7.57)

where the translation and rotation matrices are:

$$
\left[T_{CH,ip} \right] = \begin{bmatrix} 1 & 0 & 0 \\ 0 & 1 & 0 \\ 0 & -l_C & 1 \end{bmatrix}; \left[R_{CH} \right] = \begin{bmatrix} \cos(-\delta_h) & \sin(-\delta_h) & 0 \\ -\sin(-\delta_h) & \cos(-\delta_h) & 0 \\ 0 & 0 & 1 \end{bmatrix}.
$$

(7.58)

The in-plane inertia load and acceleration vectors are connected by inertia matrices in the two reference frames as:

$$\left[f_{ip} \right] = \left[M_{ip}^{(h)} \right]\left[\ddot{u}_{ip} \right]; \quad \left[f_{C,ip} \right] = \left[M_{C,ip}^{(h)} \right]\left[\ddot{u}_{C,ip} \right]. \tag{7.59}$$

Substituting the local-frame vectors of Eqs. (7.57) into the first Eq. (7.59) results in an equation similar to the second Eq. (7.59) where the global-frame inertia matrix of the flexible hinge is expressed in terms of the local-frame one as:

$$\left[M_{C,ip}^{(h)} \right] = \left[T_{CH,ip} \right]^{T}\left[R_{CH} \right]^{T}\left[M_{ip}^{(h)} \right]\left[R_{CH} \right]\left[T_{CH,ip} \right]. \tag{7.60}$$

With a similar reasoning, it can be demonstrated that the out-of-plane, global-frame inertia matrix of the flexible hinge can be calculated in terms of the local-frame inertia matrix as:

$$\left[M_{C,op}^{(h)} \right] = \left[T_{CH,op} \right]^{T}\left[R_{CH} \right]^{T}\left[M_{op}^{(h)} \right]\left[R_{CH} \right]\left[T_{CH,op} \right] \text{ with } \left[T_{CH,op} \right] = \begin{bmatrix} 1 & 0 & 0 \\ 0 & 1 & l_{C} \\ 0 & 0 & 1 \end{bmatrix}. \tag{7.61}$$

The overall in-plane and out-of-plane inertia matrices of the two-member mechanism shown in Figure 7.2 add the corresponding inertia matrices of the rigid link and the flexible hinge as:

$$\left[M_{C,ip} \right] = \left[M_{C,ip}^{(r)} \right] + \left[M_{C,ip}^{(h)} \right]; \quad \left[M_{C,op} \right] = \left[M_{C,op}^{(r)} \right] + \left[M_{C,op}^{(h)} \right], \tag{7.62}$$

where the superscript "r" denotes the rigid link. The inertia matrices of the rigid link are:

$$\left[M_{C,ip}^{(r)} \right] = \begin{bmatrix} m_{r} & 0 & 0 \\ 0 & m_{r} & 0 \\ 0 & 0 & J_{r,z} \end{bmatrix}; \quad \left[M_{C,op}^{(r)} \right] = \begin{bmatrix} J_{r,x} & 0 & 0 \\ 0 & J_{r,y} & 0 \\ 0 & 0 & m_{r} \end{bmatrix}, \tag{7.63}$$

where m_{r} is the mass of the rigid link and $J_{r,x}$, $J_{r,y}$ and $J_{r,z}$ are the mass moments of inertia of the rigid link with respect to the global axes centered at C in Figure 7.2.

7.2.2.2 Free-Free Flexible Hinges

Compliant mechanisms may include flexible hinges that are connected at both ends to moving rigid links – Figure 7.6 shows the skeleton representation of such a configuration. These hinges are "free-free" in the sense that both their ends are movable, unlike the previous hinge configuration of Figure 7.2, which is "fixed-free" because one of its ends is fixed.

When the flexible hinge is relatively heavy, hinge inertia fractions can be formulated to be placed at both ends H_1 and H_2 based on either a consistent approach or a lumped-parameter model, as discussed in this section.

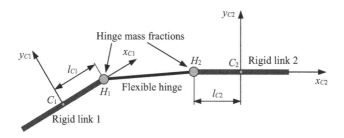

FIGURE 7.6 Skeleton representation of planar mechanism with one heavy flexible hinge connected at both ends to moving heavy rigid links.

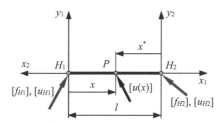

FIGURE 7.7 Skeleton representation of a free-free flexible hinge with end-point loads.

7.2.2.2.1 Consistent-Inertia Model

Figure 7.7 shows the skeleton representation of a free-free, straight-axis flexible hinge, which is acted upon at its ends H_1 and H_2 by two spatial load vectors. The first goal is to establish a relationship between the displacement $[u(x)]$ at an arbitrary point P of position x and the displacements $[u_{H1}]$ and $[u_{H2}]$ at the free ends. Axial, torsional and bending inertia are addressed separately.

Axial vibrations:

Consider that the flexible hinge of Figure 7.7 can only deform axially under the action of forces aligned with the hinge and applied at the two ends. Under an axial force f_{1x} applied at point H_1 in Figure 7.6, the displacement u_{1x} at H_1 and the displacement $u_x(x)$ at P, both relative to the displacement u_{2x} at H_2, are calculated as:

$$
\begin{aligned}
u_{1x} - u_{2x} &= C_{u_x-f_x} \cdot f_{1x}; \\
u_x(x) - u_{2x} &= C_{u_x-f_x}(x) \cdot f_{1x}
\end{aligned}
\quad \rightarrow u_x(x,t) = S_{a1}(x) \cdot u_{1x}(t) + S_{a2}(x) \cdot u_{2x}(t)
$$

$$(7.64)$$

$$
\text{with} \quad S_{a1}(x) = \frac{C_{u_x-f_x}(x)}{C_{u_x-f_x}}; \quad S_{a2}(x) = 1 - \frac{C_{u_x-f_x}(x)}{C_{u_x-f_x}},
$$

where the compliance $C_{u_x-f_x}$ is the axial compliance of the full hinge and is the compliance of the segment comprised between P and H_2 – it is formulated in Eq. (7.24).

The functions $S_{a1}(x)$ and $S_{a2}(x)$ are the axial shape functions – they are similar to the bending shape functions introduced in Section 7.2.2.1.1. Equation (7.64) can also be written in vector form as:

$$u_x(x,t) = [S_a(x)]^T \cdot [u_a(t)] \rightarrow \frac{\partial u_x(x,t)}{\partial t} = [S_a(x)]^T \cdot \left[\frac{du_a(t)}{dt} \right]$$

with
$$\begin{cases} [S_a(x)] = \begin{bmatrix} S_{a1}(x) & S_{a2}(x) \end{bmatrix}^T \\ [u_a(t)] = \begin{bmatrix} u_{1x}(t) & u_{2x}(t) \end{bmatrix}^T. \end{cases}$$
(7.65)

The kinetic energy of the flexible hinge with regard to its axial (x axis) vibrations is calculated as:

$$T_a = \frac{1}{2} \cdot \rho \cdot \int_0^l A(x) \cdot \left(\frac{\partial u_x(x,t)}{\partial t} \right)^2 dx$$

$$= \frac{1}{2} \cdot \rho \cdot \left[\frac{du_a(t)}{dt} \right]^T \left(\int_0^l A(x) \cdot [S_a(x)] \cdot [S_a(x)]^T dx \right) \cdot \left[\frac{du_a(t)}{dt} \right].$$
(7.66)

Comparing Eq. (7.66) to the generic expression of the kinetic energy of Eq. (7.22) yields the symmetric inertia matrix of the free-free flexible hinge owing to axial effects:

$$[M_a] = \rho \cdot \int_0^l A(x) \cdot [S_a(x)] \cdot [S_a(x)]^T dx = \begin{bmatrix} M_a(1,1) & M_a(1,2) \\ M_a(1,2) & M_a(2,2) \end{bmatrix}.$$
(7.67)

For a constant cross-section hinge, the shape functions of Eqs. (7.64) become:

$$S_{a1}(x) = 1 - \frac{x}{l}; \quad S_{a2}(x) = \frac{x}{l},$$
(7.68)

and the corresponding inertia matrix is:

$$[M_a] = \begin{bmatrix} \frac{1}{3} \cdot m_h & \frac{1}{6} \cdot m_h \\ \frac{1}{6} \cdot m_h & \frac{1}{3} \cdot m_h \end{bmatrix} = \frac{1}{6} \cdot m_h \cdot \begin{bmatrix} 2 & 1 \\ 1 & 2 \end{bmatrix}.$$
(7.69)

Note that adding all the mass fractions in the matrix of Eq. (7.69) results in m_h, the mass of the flexible hinge.

Torsional vibrations:

Torsion vibration is approached similarly to axial vibration. Under an axial moment m_{1x} applied at point H_1 in Figure 7.7, the rotations θ_{1x} at H_1 and $\theta_x(x)$ at P are calculated relative to the rotation θ_{2x} at H_2 as:

$$\begin{aligned} \theta_{1x} - \theta_{2x} &= C_{\theta_x - m_x} \cdot m_{1x}; \\ \theta_x(x,t) - \theta_{2x} &= C_{\theta_x - m_x}(x) \cdot m_{1x} \end{aligned} \rightarrow \theta_x(x,t) = S_{t1}(x) \cdot \theta_{1x}(t) + S_{t2}(x) \cdot \theta_{2x}(t)$$

(7.70)

$$\text{with} \quad S_{t1}(x) = \frac{C_{\theta_x - m_x}(x)}{C_{\theta_x - m_x}}; \quad S_{t2}(x) = 1 - \frac{C_{\theta_x - m_x}(x)}{C_{\theta_x - m_x}},$$

where $S_{t1}(x)$ and $S_{t2}(x)$ are the torsional shape functions. The rotation at point P can be expressed in vector form as:

$$\theta_x(x,t) = [S_t(x)]^T \cdot [\theta_t(t)] \rightarrow \frac{\partial \theta_x(x,t)}{\partial t} = [S_t(x)]^T \cdot \left[\frac{d\theta_t(t)}{dt} \right]$$

$$\text{with} \quad \begin{cases} [S_t(x)] = \begin{bmatrix} S_{t1}(x) & S_{t2}(x) \end{bmatrix}^T \\ [\theta_t(t)] = \begin{bmatrix} \theta_{1x}(t) & \theta_{2x}(t) \end{bmatrix}^T \end{cases} . \quad (7.71)$$

The kinetic energy of the flexible hinge during torsional vibrations around the x axis is:

$$\begin{aligned} T_t &= \frac{1}{2} \cdot \rho \cdot \int_0^l I_t(x) \cdot \left(\frac{\partial \theta_x(x,t)}{\partial t} \right)^2 dx \\ &= \frac{1}{2} \cdot \rho \cdot \left[\frac{d\theta_t(t)}{dt} \right]^T \left(\int_0^l I_t(x) \cdot [S_t(x)] \cdot [S_t(x)]^T dx \right) \cdot \left[\frac{d\theta_t(t)}{dt} \right]. \end{aligned} \quad (7.72)$$

Comparing Eq. (7.72) to the generic kinetic energy given in Eq. (7.22) indicates that the symmetric inertia matrix of the free-free flexible hinge due to torsion is:

$$[M_t] = \rho \cdot \int_0^l I_t(x) \cdot [S_t(x)] \cdot [S_t(x)]^T dx = \begin{bmatrix} M_t(1,1) & M_t(1,2) \\ M_t(1,2) & M_t(2,2) \end{bmatrix}. \quad (7.73)$$

For a constant cross-section hinge, the torsion shape functions of Eqs. (7.70) are identical to the axial shape functions of Eq. (7.68), which is $[S_t(x)] = [S_a(x)]$, and the inertia matrix of Eq. (7.73) becomes:

$$[M_t] = \begin{bmatrix} \dfrac{1}{3} \cdot J_{h,x} & \dfrac{1}{6} \cdot J_{h,x} \\ \dfrac{1}{6} \cdot J_{h,x} & \dfrac{1}{3} \cdot J_{f,x} \end{bmatrix} = \dfrac{1}{6} \cdot J_{h,x} \cdot \begin{bmatrix} 2 & 1 \\ 1 & 2 \end{bmatrix}, \qquad (7.74)$$

where $J_{h,x} = \rho \cdot I_t \cdot l$ is the mass moment of inertia of the flexible hinge with respect to the x axis.

Bending vibrations:

A process similar to the one used to identify the bending shape functions for a fixed-free flexible hinge is applied to express the deflection $u_y(x)$ at the arbitrary point P in terms of the bending-related displacement vector $[u_{bz}] = [u_{1y}\ \theta_{1z}\ u_{2y}\ \theta_{2z}]^T$, where u_{1y}, θ_{1z}, u_{2y} and θ_{2z} are deflections and rotations at the ends H_1 and H_2 in Figure 7.7. The deflection $u_y(x)$ is obtained as a superposition of the case studied when a load is applied at H_1 and H_2 is assumed fixed (see the bending of the fixed-free flexible hinge) with the reversed case where a load is applied at H_2 and H_1 is assumed fixed.

For H_2 fixed, a load vector $[f_{1,bz}] = [f_{1y}\ \mathrm{m}_{1z}]^T$ is applied at the end H_1 of the flexible hinge. This load generates a displacement vector $[u_{1,bz}] = [u_{1y}\ \theta_{1z}]^T$ at the same point and a displacement vector $[u_{1,bz}(x)] = [u_{1y}(x)\ \theta_{1z}(x)]^T$ at the arbitrary point P. Applying a reasoning similar to the one used for the fixed-free flexible hinge, the following equation results:

$$[u_{1,bz}(x)] = [S_{b1}(x)][u_{1,bz}] \quad \text{with} \quad [S_{b1}(x)] = [C_{PH_1,bz}(x)][C_{H_1,bz}]^{-1}, \quad (7.75)$$

where

$$[C_{PH_1,bz}(x)] = [C_{P,bz}(x)][T_{H_1P,bz}]. \qquad (7.76)$$

The shape-function matrix $[S_{b1}(x)]$ of Eq. (7.75) is identical to the shape-function matrix $[S_b(x)]$ expressed in Eq. (7.42). The compliance matrix $[C_{P,bz}(x)]$ is expressed in Eq. (7.40). The compliance matrix $[C_{H_1,bz}]$ of Eq. (7.75) and the translation matrix of Eq. (7.76) are:

$$[C_{H_1,bz}] = \begin{bmatrix} C_{H_1,u_y-f_y} & C_{H_1,u_y-m_z} \\ C_{H_1,u_y-m_z} & C_{H_1,\theta_z-m_z} \end{bmatrix}; \quad [T_{H_1P,bz}] = \begin{bmatrix} 1 & 0 \\ -x & 1 \end{bmatrix}. \quad (7.77)$$

As in Eq. (7.43), the deflection at P is:

$$u_{1y}(x) = S_{b1}(x) \cdot u_{1y} + S_{b2}(x) \cdot \theta_{1z}, \qquad (7.78)$$

where $S_{b1}(x)$ and $S_{b2}(x)$ are the elements on the first row of the 2×2 shape-function matrix $[S_{b1}(x)]$ of Eq. (7.75), which are identical to the similar elements of $[S_b(x)]$ in Eq. (7.42). For a constant cross-section flexible hinge, the two shape functions of Eq. (7.78) become those of Eq. (7.44).

Assume now that a forcing vector $[f_{2,bz}] = [f_{2y}\ \mathrm{m}_{2z}]^T$ is applied at the end H_2 of the flexible hinge while the other end H_1 is fixed – see Figure 7.7. An approach similar to the one used previously is used to formulate the displacement vector $[u_{2,bz}] = [u_{2y}\ \theta_{2z}]^T$ at H_2 and $[u_{2,bz}(x^*)] = [u_{2y}(x^*)\ \theta_{2z}(x^*)]^T$ at P. Note that $x^* = l - x$ is the coordinate in the reference frame centered at H_2 in Figure 7.7. These two displacement vectors are related as:

$$\left[u_{2,bz}(x)\right] = \left[S_{b2}(x^*)\right]^T \left[u_{2,bz}\right] \quad \text{with} \quad \left[S_{b2}(x^*)\right] = \left[C_{PH2,bz}(x^*)\right]\left[C_{H2,bz}\right]^{-1}. \quad (7.79)$$

Because the reference frame $H_2x_2y_2z_2$ has its x_2 and z_2 axes flipped compared to the right reference frame $H_1x_1y_1z_1$, the actual compliance matrix $[S_{b2}(x)]$ has minus signs in front of its off-diagonal components of Eq. (7.79), namely:

$$[S_{b2}(x)] = \begin{bmatrix} S_{b2,1}(x) & S_{b2,2}(x) \\ S_{b2,3}(x) & S_{b2,4}(x) \end{bmatrix} = \begin{bmatrix} S_{b2,1}(x^*) & -S_{b2,2}(x^*) \\ -S_{b2,3}(x^*) & S_{b2,4}(x^*) \end{bmatrix}. \quad (7.80)$$

The x^*-dependent compliance matrix of Eq. (7.79) is:

$$\left[C_{PH2,bz}(x^*)\right] = \left[C_{P,bz}(x^*)\right]\left[T_{H2P,bz}\right]. \quad (7.81)$$

The other compliance matrix of Eq. (7.79) and the new matrices of Eq. (7.81) are:

$$[C_{H2,bz}] = \begin{bmatrix} C_{H2,u_y-f_y} & C_{H2,u_y-m_z} \\ C_{H2,u_y-m_z} & C_{H2,\theta_z-m_z} \end{bmatrix};$$

$$[C_{P,bz}(x^*)] = \begin{bmatrix} C_{u_y-f_y}(x^*) & C_{u_y-m_z}(x^*) \\ C_{u_y-m_z}(x^*) & C_{\theta_z-m_z}(x^*) \end{bmatrix}; \quad [T_{H2P,bz}] = \begin{bmatrix} 1 & 0 \\ -x^* & 1 \end{bmatrix}. \quad (7.82)$$

The components of the $[C_{P,bz}(x^*)]$ matrix are calculated as:

$$\begin{cases} C_{u_y-f_y}(x^*) = \int_{x^*}^{l} \dfrac{(x^*)^2}{EI_z(x^*)} \cdot dx^* = \int_{0}^{l-x} \dfrac{(l-x)^2}{EI_z(x)} \cdot dx; \\[3mm] C_{u_y-m_z}(x^*) = -\int_{x^*}^{l} \dfrac{x^*}{EI_z(x^*)} \cdot dx^* = -\int_{0}^{l-x} \dfrac{(l-x)}{EI_z(x)} \cdot dx; \\[3mm] C_{\theta_z-m_z}(x^*) = \int_{x^*}^{l} \dfrac{dx^*}{EI_z(x^*)} = \int_{0}^{l-x} \dfrac{dx}{EI_z(x)}, \end{cases} \quad (7.83)$$

which used $x^* = l - x$.

From Eq. (7.79), the deflection at P is expressed in terms of the first-row elements of the shape-function matrix expressed in Eq. (7.80), namely:

$$u_{2y}(x) = S_{b2,1}(x) \cdot u_{2y} + S_{b2,2}(x) \cdot \theta_{2z} = S_{b3}(x) \cdot u_{2y} + S_{b4}(x) \cdot \theta_{2z}, \quad (7.84)$$

where the "*" superscript notation was dropped. When the flexible-hinge cross-section is constant, the shape functions of Eq. (7.84) become:

$$S_{b2,1}(x) = S_{b3}(x) = 3 \cdot \frac{x^2}{l^2} - 2 \cdot \frac{x^3}{l^3}; \quad S_{b2,2}(x) = S_{b4}(x) = -\frac{x^2}{l} + \frac{x^3}{l^2}. \quad (7.85)$$

For a free-free flexible hinge, the deflections $u_{1y}(x)$ of Eq. (7.78) and $u_{2y}(x)$ of Eq. (7.84) need to be superimposed to obtain the total deflection $u_y(x)$ at P:

$$u_y(x) = u_{1y}(x) + u_{2y}(x) = S_{b1}(x) \cdot u_{1y} + S_{b2}(x) \cdot \theta_{1z} + S_{b3}(x) \cdot u_{2y} + S_{b4}(x) \cdot \theta_{2z} \quad (7.86)$$

$$\text{or} \quad u_y(x,t) = \left[S_b(x)\right]^T \left[u_{bz}(t)\right]$$

with

$$\begin{cases} \left[S_b(x)\right] = \left[S_{b1}(x) \quad S_{b2}(x) \quad S_{b3}(x) \quad S_{b4}(x)\right]^T \\ \left[u_{bz}\right] = \left[u_{1y} \quad \theta_{1z} \quad u_{2y} \quad \theta_{2z}\right]^T. \end{cases} \quad (7.87)$$

With these distribution functions and following a development similar to the one used in the case of the fixed-free flexible hinge, the following bending-related symmetric inertia matrix is obtained:

$$\left[M_b^{(h)}\right] = \rho \cdot \int_0^l A(x) \cdot \left[S_b(x)\right]\left[S_b(x)\right]^T dx$$

$$= \begin{bmatrix} M_b(1,1) & M_b(1,2) & M_b(1,3) & M_b(1,4) \\ M_b(1,2) & M_b(2,2) & M_b(2,3) & M_b(2,4) \\ M_b(1,3) & M_b(2,3) & M_b(3,3) & M_b(3,4) \\ M_b(1,4) & M_b(2,4) & M_b(3,4) & M_b(4,4) \end{bmatrix}. \quad (7.88)$$

For a constant cross-section flexible hinge, Eq. (7.88) becomes:

$$\left[M_b^{(h)}\right] = \frac{1}{420} \cdot m_h \cdot \begin{bmatrix} 156 & 22l & 54 & -13l \\ 22l & 4l^2 & 13l & -3l^2 \\ 54 & 13l & 156 & -22l \\ -13l & -3l^2 & -22l & 4l^2 \end{bmatrix}. \quad (7.89)$$

where m_h is the mass of the flexible hinge.

Note on inertia matrix corresponding to out-of-plane bending:

The inertia matrix that captures the bending about the y axis is not given explicitly here, but is incorporated in the out-of-plane inertia matrix shortly in this section. The components of the y-axis bending inertia matrix are the ones of Eqs. (7.88) and (7.89), but they are rearranged to reflect the component sequence in the out-of-plane displacement vector $\left[u_{by}\right]=\left[\theta_{1y} \quad u_{1z} \quad \theta_{2y} \quad u_{2z}\right]^T$.

In-plane and out-of-plane inertia matrices:

The in-plane inertia matrix of a free-free flexible hinge can now be assembled by considering axial fractions from Eq. (7.67) and bending contributions as per Eq. (7.88) in relation to the in-plane displacement vector $[u_{ip}] = [u_{1x}\, u_{1y}\, \theta_{1z}\, u_{2x}\, u_{2y}\, \theta_{2z}]^T$:

$$\left[M_{ip}^{(h)}\right]=\begin{bmatrix} M_a(1,1) & 0 & 0 & M_a(1,2) & 0 & 0 \\ 0 & M_b(1,1) & M_b(1,2) & 0 & M_b(1,3) & M_b(1,4) \\ 0 & M_b(1,2) & M_b(2,2) & 0 & M_b(2,3) & M_b(2,4) \\ M_a(1,2) & 0 & 0 & M_a(2,2) & 0 & 0 \\ 0 & M_b(1,3) & M_b(2,3) & 0 & M_b(3,3) & M_b(3,4) \\ 0 & M_b(1,4) & M_b(2,4) & 0 & M_b(3,4) & M_b(4,4) \end{bmatrix}.$$

$$(7.90)$$

For a constant cross-section flexible hinge, combining Eq. (7.90) with Eqs. (7.69) and (7.89) results in:

$$\left[M_{ip}^{(h)}\right]=\frac{1}{420}\cdot m_h \cdot \begin{bmatrix} 140 & 0 & 0 & 70 & 0 & 0 \\ 0 & 156 & 22l & 0 & 54 & -13l \\ 0 & 22l & 4l^2 & 0 & 13l & -3l^2 \\ 70 & 0 & 0 & 140 & 0 & 0 \\ 0 & 54 & 13l & 0 & 156 & -22l \\ 0 & -13l & -3l^2 & 0 & -22l & 4l^2 \end{bmatrix}.$$

$$(7.91)$$

The out-of-plane inertia matrix is based on the displacement vector $[u_{op}] = [\theta_{1x}\, \theta_{1y}\, u_{1z}\, \theta_{2x}\, \theta_{2y}\, u_{2z}]^T$ and assembles torsion contributions from Eq. (7.73) and bending fractions from Eq. (7.88) in the form:

$$\left[M_{op}^{(h)}\right]=\begin{bmatrix} M_t(1,1) & 0 & 0 & M_t(1,2) & 0 & 0 \\ 0 & M_b(2,2) & M_b(1,2) & 0 & M_b(2,4) & M_b(2,3) \\ 0 & M_b(1,2) & M_b(1,1) & 0 & M_b(1,4) & M_b(1,3) \\ M_t(1,2) & 0 & 0 & M_t(2,2) & 0 & 0 \\ 0 & M_b(2,4) & M_b(1,4) & 0 & M_b(4,4) & M_b(3,4) \\ 0 & M_b(2,3) & M_b(1,3) & 0 & M_b(3,4) & M_b(3,3) \end{bmatrix}.$$

$$(7.92)$$

Equations (7.92), (7.74) and (7.89) are combined when the flexible-hinge cross-section is constant and yield:

$$\left[M_{op}^{(h)}\right]=\frac{1}{420}\cdot$$

$$\begin{bmatrix} 140\cdot J_{h,x} & 0 & 0 & 70\cdot J_{h,x} & 0 & 0 \\ 0 & 4\cdot m_h\cdot l^2 & 22\cdot m_h\cdot l & 0 & -3\cdot m_h\cdot l^2 & 13\cdot m_h\cdot l \\ 0 & 22\cdot m_h\cdot l & 156\cdot m_h & 0 & -13\cdot m_h\cdot l & 54\cdot m_h \\ 70\cdot J_{h,x} & 0 & 0 & 140\cdot J_{h,x} & 0 & 0 \\ 0 & -3\cdot m_h\cdot l^2 & -13\cdot m_h\cdot l & 0 & 4\cdot m_h\cdot l^2 & -22\cdot m_h\cdot l \\ 0 & 13\cdot m_h\cdot l & 54\cdot m_h & 0 & -22\cdot m_h\cdot l & 156\cdot m_h \end{bmatrix}.$$

$$(7.93)$$

Note, again, that the in-plane and out-of-plane inertia matrices of Eqs. (7.90) through (7.93) are symmetric.

7.2.2.2.2 Lumped-Parameter Inertia Model

The lumped-parameter model distributes inertia fractions to the two ends of the flexible hinge in a diagonal inertia matrix. Figure 7.8 illustrates this process for axial, torsional and bending vibrations.

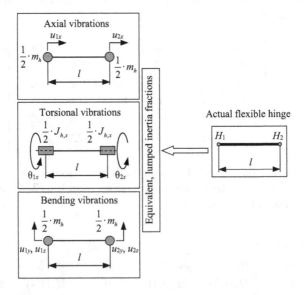

FIGURE 7.8 Lumped-parameter inertia fractions resulting from a distributed-inertia free-free flexible hinge.

As shown in Figure 7.8, identical fractions that are equal to half the mass of the flexible hinge are associated with the axial and bending translations; similarly, half the flexible-hinge mass moment of inertia about the x axis is apportioned at each of the two ends to account for torsion. As a result, the in-plane, lumped-parameter inertia matrix, which is associated with the displacement vector $[u_{ip}] = [u_{1x}\ u_{1y}\ \theta_{1z}\ u_{2x}\ u_{2y}\ \theta_{2z}]^T$, is of diagonal form, namely:

$$\left[M_{ip}^{(h)} \right] = \text{diag}\left(\frac{1}{2} \cdot m_h \quad \frac{1}{2} \cdot m_h \quad 0 \quad \frac{1}{2} \cdot m_h \quad \frac{1}{2} \cdot m_h \quad 0 \right). \tag{7.94}$$

The out-of-plane, lumped-parameter inertia matrix, which is associated with the displacement vector $[u_{op}] = [\theta_{1x}\ \theta_{1y}\ u_{1z}\ \theta_{2x}\ \theta_{2y}\ u_{2z}]^T$, is also of diagonal form, namely:

$$\left[M_{op}^{(h)} \right] = \text{diag}\left(\frac{1}{2} \cdot J_{h,x} \quad 0 \quad \frac{1}{2} \cdot m_h \quad \frac{1}{2} \cdot J_{h,x} \quad 0 \quad \frac{1}{2} \cdot m_h \right). \tag{7.95}$$

When a straight-axis flexible hinge is part of a three-dimensional mechanism, its inertia matrix conforms to the displacement vector $[u] = [[u_{ip}]\ [u_{op}]]^T$ and is diagonally formed of the in-plane and out-of-plane inertia matrices as:

$$\left[M^{(h)} \right] = \begin{bmatrix} \left[M_{ip}^{(h)} \right] & [0] \\ [0] & \left[M_{op}^{(h)} \right] \end{bmatrix}. \tag{7.96}$$

7.2.2.2.3 Assembling the Inertia Matrix Resulting from Rigid Links and Free-Free Flexible Hinges

As seen in the previous section, the inertia matrix of a heavy, mobile flexible hinge consists of inertia fractions that correspond to the hinge end-point DOF. In either in-plane or out-of-plane vibrations, the square inertia matrix is a 6×6 matrix. On the other hand, the in-plane or out-of-plane inertia matrix of a rigid link is defined at its mass center as a 3×3 matrix. Due to this dimensional discrepancy, it is not straightforward to combine the inertia matrices of a flexible link and an adjacent moving rigid link. The following procedure eliminates this discrepancy and comprises the following steps:

Step 1: Calculate the inertia matrices of all rigid links in their local frames originally and then in the global frame through appropriate rotation;

Step2: Assemble all rigid-link inertia matrices into a single matrix, which is consistent with the DOF of the flexible-hinge mechanism;

Step 3: Similarly calculate inertia matrices of all the heavy flexible hinges in their local frames originally and then in the global frame through appropriate rotation;

Step 4: Transform (actually reduce) the inertia matrices of all flexible hinges from the DOF corresponding to their moving end points to the adjacent rigid-link centers DOF, which are also the mechanism DOF;

Step 5: Assemble all the mobile flexible hinges transformed inertia matrices obtained at *Step 4* into a single inertia matrix, which is consistent with the mechanism DOF;

Step 6: Add the transformed flexible-hinge inertia matrix to the rigid-link inertia matrix to obtain the mechanism inertia matrix.

While *Steps 1, 2, 3, 5* and *6* involve regular matrix operations, *Step 4* is based on expressing the displacements at the ends of a rigid link (where heavy flexible links are connected) in terms of the displacements at the rigid-link center. These transformations are obtained considering the small-displacement rigid-body motion of the rigid link. Consider a mobile rigid link that is connected at its ends to two flexible hinges, as illustrated in Figure 7.9. This portion is part of a planar mechanism, which performs small in-plane or out-of-plane motions.

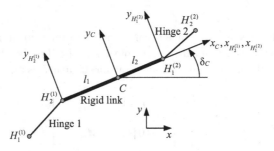

FIGURE 7.9 Mobile rigid link connected to two flexible hinges in planar configuration.

In the local frame $Cx_Cy_Cz_C$, the in-plane displacements of the two end points of the rigid link and those of the center C are connected as:

$$
\begin{cases}
\left[u_{C,ip}^{(l)} \right] = \left[T_{CH_2^{(1)},ip} \right]^T \left[u_{H_2^{(1)},ip}^{(l)} \right] \\
\left[u_{C,ip}^{(l)} \right] = \left[T_{CH_1^{(2)},ip} \right]^T \left[u_{H_1^{(2)},ip}^{(l)} \right]
\end{cases}
\text{ or }
\begin{cases}
\left[u_{H_2^{(1)},ip}^{(l)} \right] = \left(\left[T_{CH_2^{(1)},ip} \right]^T \right)^{-1} \left[u_{C,ip}^{(l)} \right] \\
\left[u_{H_1^{(2)},ip}^{(l)} \right] = \left(\left[T_{CH_1^{(2)},ip} \right]^T \right)^{-1} \left[u_{C,ip}^{(l)} \right]
\end{cases}
\tag{7.97}
$$

$$
\text{with } \left[T_{CH_2^{(1)},ip} \right] =
\begin{bmatrix}
1 & 0 & 0 \\
0 & 1 & 0 \\
0 & l_1 & 0
\end{bmatrix};
\quad
\left[T_{CH_1^{(2)},ip} \right] =
\begin{bmatrix}
1 & 0 & 0 \\
0 & 1 & 0 \\
0 & -l_2 & 0
\end{bmatrix},
$$

where the superscript "*l*" denotes the local frames centered at $H_2^{(1)}, H_1^{(2)}, C$, which share their x axes, as shown in Figure 7.9. The three local-frame displacement vectors of Eq. (7.97) are each formed of the regular components denoting x-axis translation, y-axis translation and z-axis rotation. These local-frame vectors are rotated into the global frame xyz (identified in Figure 7.9) to produce the following global-frame vectors:

$$\begin{cases} \left[u_{H_2^{(1)},ip}\right]=[R_C]^T\left[u_{H_2^{(1)},ip}^{(l)}\right] \\[2mm] \left[u_{H_1^{(2)},ip}\right]=[R_C]^T\left[u_{H_1^{(2)},ip}^{(l)}\right] \quad \text{with} \quad [R_C]=\begin{bmatrix} \cos\delta_C & \sin\delta_C & 0 \\ -\sin\delta_C & \cos\delta_C & 0 \\ 0 & 0 & 1 \end{bmatrix}. \\[2mm] \left[u_{C,ip}\right]=[R_C]^T\left[u_{C,ip}^{(l)}\right] \end{cases} \tag{7.98}$$

Equations (7.97) and (7.98) are combined to express the end-point displacement vectors in terms of the center-point displacement vectors in the global frame as:

$$\begin{cases} \left[u_{H_2^{(1)},ip}\right]=\left[Tr_{CH_2^{(1)},ip}\right]\left[u_{C,ip}\right] \\[2mm] \left[u_{H_1^{(2)},ip}\right]=\left[Tr_{CH_1^{(2)},ip}\right]\left[u_{C,ip}\right] \end{cases} \text{with} \begin{cases} \left[Tr_{CH_2^{(1)},ip}\right]=[R_C]^T\left(\left[T_{CH_2^{(1)},ip}\right]^T\right)^{-1}[R_C] \\[2mm] \left[Tr_{CH_1^{(2)},ip}\right]=[R_C]^T\left(\left[T_{CH_1^{(2)},ip}\right]^T\right)^{-1}[R_C] \end{cases}.$$

$$\tag{7.99}$$

Similarly, the out-of-plane displacements at the two end points of the rigid link are related to those of the center C as:

$$\begin{cases} \left[u_{H_2^{(1)},op}^{(l)}\right]=\left(\left[T_{CH_2^{(1)},op}\right]^T\right)^{-1}\left[u_{C,op}^{(l)}\right] \\[2mm] \left[u_{H_1^{(2)},op}^{(l)}\right]=\left(\left[T_{CH_1^{(2)},op}\right]^T\right)^{-1}\left[u_{C,op}^{(l)}\right] \end{cases} \text{with} \begin{cases} \left[T_{CH_2^{(1)},op}\right]=\begin{bmatrix} 1 & 0 & 0 \\ 0 & 1 & -l_1 \\ 0 & 0 & 0 \end{bmatrix} \\[4mm] \left[T_{CH_1^{(2)},op}\right]=\begin{bmatrix} 1 & 0 & 0 \\ 0 & 1 & l_2 \\ 0 & 0 & 0 \end{bmatrix} \end{cases}.$$

$$\tag{7.100}$$

The local-frame components of the three vectors in Eq. (7.100) are the regular x-axis rotation, y-axis rotation and z-axis translation – all determined in the local frames located at the points $H_2^{(1)}, H_1^{(2)}$ and C. The following global-frame displacement vectors are obtained by rotating the local-frame vectors of Eq. (7.100):

$$\left[u_{H_2^{(1)},op}\right]=[R_C]^T\left[u_{H_2^{(1)},op}^{(l)}\right]; \quad \left[u_{H_1^{(2)},op}\right]=[R_C]^T\left[u_{H_1^{(2)},op}^{(l)}\right]; \quad \left[u_{C,op}\right]=[R_C]^T\left[u_{C,op}^{(l)}\right],$$

$$\tag{7.101}$$

with $[R_C]$ of Eq. (7.98). Combination of Eqs. (7.100) and (7.101) yields:

$$\begin{cases} \left[u_{H_2^{(1)},op}\right]=\left[Tr_{CH_2^{(1)},op}\right]\left[u_{C,op}\right] \\[2mm] \left[u_{H_1^{(2)},op}\right]=\left[Tr_{CH_1^{(2)},op}\right]\left[u_{C,op}\right] \end{cases} \text{with} \begin{cases} \left[Tr_{CH_2^{(1)},op}\right]=[R_C]^T\left(\left[T_{CH_2^{(1)},op}\right]^T\right)^{-1}[R_C] \\[2mm] \left[Tr_{CH_1^{(2)},op}\right]=[R_C]^T\left(\left[T_{CH_1^{(2)},op}\right]^T\right)^{-1}[R_C] \end{cases}.$$

$$\tag{7.102}$$

Steps 3–5 are detailed in the following *Example 7.2*, while *Example 7.7* follows up on *Example 7.2* to apply the full algorithm of incorporating flexible-hinge inertia in an actual planar mechanism.

Example 7.2

The generic planar mechanism shown in skeleton form in Figure 7.10 is formed of two flexible hinges and two rigid links. Assemble the overall in-plane and out-of-plane inertia matrices with respect to the mass centers of the two rigid links that are due to the flexible hinges only. Consider that known are the local-frame inertia matrices of the two flexible hinges, as well as all geometric parameters identified in Figure 7.10.

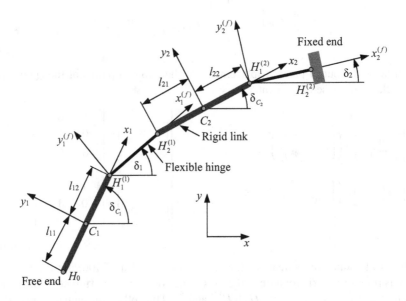

FIGURE 7.10 Skeleton representation of a planar fixed-free mechanism with two flexible hinges and two rigid links.

Solution:

In-plane inertia matrix:
    Based on Eqs. (7.97)–(7.99), the connections between the in-plane displacement vectors $\left[u_{H_1^{(1)},ip} \right]$, $\left[u_{H_2^{(1)},ip} \right]$ and $\left[u_{H_1^{(2)},ip} \right]$ at the moving points $H_1^{(1)}$, $H_2^{(1)}$ and $H_1^{(2)}$, which are the end points of the two flexible hinges, and the similar displacement vectors $\left[u_{C_1,ip} \right]$ and $\left[u_{C_2,ip} \right]$ at the centers C_1 and C_2 of the two rigid links are expressed as:

$$\begin{cases} \left[u_{H_1^{(1)},ip} \right] = \left[Tr_{C_1H_1^{(1)},ip} \right] \left[u_{C_1,ip} \right]; \\ \left[u_{H_2^{(1)},ip} \right] = \left[Tr_{C_2H_2^{(1)},ip} \right] \left[u_{C_2,ip} \right]; \\ \left[u_{H_1^{(2)},ip} \right] = \left[Tr_{C_2H_1^{(2)},ip} \right] \left[u_{C_2,ip} \right] \end{cases} \tag{7.103}$$

where the transformation matrices are:

$$\begin{cases} \left[Tr_{C_1H_1^{(1)},ip} \right] = \left[R_{C_1} \right]^T \left[T_{C_1H_1^{(1)},ip} \right] \left[R_{C_1} \right]; \\ \left[Tr_{C_2H_2^{(1)},ip} \right] = \left[R_{C_2} \right]^T \left[T_{C_2H_2^{(1)},ip} \right] \left[R_{C_2} \right]; \\ \left[Tr_{C_2H_1^{(2)},ip} \right] = \left[R_{C_2} \right]^T \left[T_{C_2H_1^{(2)},ip} \right] \left[R_{C_2} \right]. \end{cases} \tag{7.104}$$

The translation matrices of Eq. (7.104) are:

$$\left[T_{C_1H_1^{(1)},ip} \right] = \begin{bmatrix} 1 & 0 & 0 \\ 0 & 1 & 0 \\ 0 & -l_{12} & 1 \end{bmatrix}; \quad \left[T_{C_2H_2^{(1)},ip} \right] = \begin{bmatrix} 1 & 0 & 0 \\ 0 & 1 & 0 \\ 0 & l_{21} & 1 \end{bmatrix};$$

$$\left[T_{C_2H_1^{(2)},ip} \right] = \begin{bmatrix} 1 & 0 & 0 \\ 0 & 1 & 0 \\ 0 & -l_{22} & 1 \end{bmatrix}. \tag{7.105}$$

The rotation matrices of Eq. (7.104) are calculated as in Eq. (7.98) with the rotation angles δ_{C_1} and δ_{C_2} of Figure 7.10.

Equations (7.103) are assembled in the following form:

$$\left[u_{H,ip} \right]_{9\times1} = \left[Tr_{CH,ip} \right]_{9\times6} \left[u_{C,ip} \right]_{6\times1}, \tag{7.106}$$

where

$$\begin{cases} \left[u_{H,ip} \right]_{9\times1} = \left[\ \left[u_{H_1^{(1)},ip} \right]_{3\times1} \ \ \left[u_{H_2^{(1)},ip} \right]_{3\times1} \ \ \left[u_{H_1^{(2)},ip} \right]_{3\times1} \ \right]^T; \\ \left[u_{C,ip} \right]_{6\times1} = \left[\ \left[u_{C_1,ip} \right]_{3\times1} \ \ \left[u_{C_2,ip} \right]_{3\times1} \ \right]^T \end{cases} \tag{7.107}$$

and

$$\left[Tr_{CH,ip} \right]_{9\times6} = \begin{bmatrix} \left[Tr_{C_1H_1^{(1)},ip} \right]_{3\times3} & [0]_{3\times3} \\ [0]_{3\times3} & \left[Tr_{C_2H_2^{(1)},ip} \right]_{3\times3} \\ [0]_{3\times3} & \left[Tr_{C_2H_1^{(2)},ip} \right]_{3\times3} \end{bmatrix}. \tag{7.108}$$

The total kinetic energy resulting from the two heavy flexible hinges is expressed as:

$$T = \frac{1}{2}\left[\dot{u}_{H,ip}\right]_{1\times9}^{T}\left[M_{H,ip}\right]_{9\times9}\left[\dot{u}_{H,ip}\right]_{9\times1},$$ (7.109)

where the inertia matrix pertaining to the two flexible hinges of Figure 7.10 is:

$$\left[M_{H,ip}\right]_{9\times9} = \begin{bmatrix} \left[M_{ip}^{(1)}\right]_{6\times6} & [0]_{6\times3} \\ [0]_{3\times6} & \left[M_{ip}^{(2)}\right]_{3\times3} \end{bmatrix}$$ (7.110)

with $\left[M_{ip}^{(1)}\right]$ representing the inertia matrix of flexible hinge 1 and $\left[M_{ip}^{(2)}\right]$ repre-

senting the inertia matrix of hinge 2 – both are expressed in the global frame. They are obtained from the corresponding local-frame inertia matrices by means of rotations with the angles δ_1 and δ_2 – see Figure 7.10 – as:

$$\left[M_{ip}^{(1)}\right]_{6\times6} = \left[R^{(1)}\right]_{6\times6}^{T}\left[M_{ip}^{(1h)}\right]_{6\times6}\left[R^{(1)}\right]_{6\times6}$$
$$\left[M_{ip}^{(2)}\right]_{3\times3} = \left[R^{(2)}\right]_{3\times3}^{T}\left[M_{ip}^{(2h)}\right]_{3\times3}\left[R^{(2)}\right]_{3\times3}$$ (7.111)

where the inertia matrices in the right-hand sides of Eqs. (7.111) are determined in the local frames of the two hinges; they can be calculated by means of either the consistent formulation or the lumped-parameter model. The rotation matrices of Eq. (7.111) are:

$$\left[R^{(1)}\right]_{6\times6} = \begin{bmatrix} \left[R^{(1)}\right]_{3\times3} & [0]_{3\times3} \\ [0]_{3\times3} & \left[R^{(1)}\right]_{3\times3} \end{bmatrix}$$

$$\left[R^{(1)}\right]_{3\times3} = \begin{bmatrix} \cos\delta_1 & \sin\delta_1 & 0 \\ -\sin\delta_1 & \cos\delta_1 & 0 \\ 0 & 0 & 1 \end{bmatrix}; \quad \left[R^{(2)}\right]_{3\times3} = \begin{bmatrix} \cos\delta_2 & \sin\delta_2 & 0 \\ -\sin\delta_2 & \cos\delta_2 & 0 \\ 0 & 0 & 1 \end{bmatrix}$$ (7.112)

Note that the in-plane, local-frame inertia matrices of the two flexible hinges of Figure 7.10 are of 6×6 dimension (hinge 1) and 3×3 dimension (hinge 2) – the reduced dimension of the latter hinge is because it has one end fixed. Combining now Eqs. (7.106) and (7.109) yields the inertia matrix of the two flexible hinges that corresponds to the displacement vector, $\left[u_{C,ip}\right]$ which is:

$$T = \frac{1}{2}\left[\dot{u}_{C,ip}\right]_{1\times6}^{T}\left[M_{C,ip}\right]_{6\times6}\left[\dot{u}_{C,ip}\right]_{6\times1} \text{ with}$$
$$\left[M_{C,ip}\right]_{6\times6} = \left[Tr_{CH,ip}\right]_{6\times9}^{T}\left[M_{H,ip}\right]_{9\times9}\left[Tr_{CH,ip}\right]_{9\times6}.$$ (7.113)

Out-of-plane inertia matrix:
An equation similar to Eq. (7.103) is obtained that connects the out-of-plane displacement vectors at the three moving hinge ends $\left[u_{H_1^{(1)},op} \right], \left[u_{H_2^{(1)},op} \right]$ and $\left[u_{H_1^{(2)},op} \right]$ to the similar displacement vectors at the centers C_1 and C_2 of the two rigid links: $\left[u_{C_1,op} \right], \left[u_{C_2,op} \right]$ – simply use "*op*" instead of "*ip*" in Eqs. (7.103); the out-of-plane translation matrices are:

$$\left[T_{C_1 H_1^{(1)},op} \right] = \begin{bmatrix} 1 & 0 & 0 \\ 0 & 1 & l_{12} \\ 0 & 0 & 1 \end{bmatrix}; \quad \left[T_{C_2 H_2^{(1)},op} \right] = \begin{bmatrix} 1 & 0 & 0 \\ 0 & 1 & -l_{21} \\ 0 & 0 & 1 \end{bmatrix};$$

$$\left[T_{C_2 H_1^{(2)},op} \right] = \begin{bmatrix} 1 & 0 & 0 \\ 0 & 1 & l_{22} \\ 0 & 0 & 1 \end{bmatrix} \qquad (7.114)$$

Equations similar to Eqs. (7.107) through (7.113) are derived for the out-of-plane motion – the only formal difference being the subscript "*op*" that needs to be used instead of "*ip*"; eventually, the reduced out-of-plane inertia matrix is:

$$\left[M_{C,op} \right]_{6\times6} = \left[Tr_{CH,op} \right]_{6\times9}^T \left[M_{H,op} \right]_{9\times9} \left[Tr_{CH,op} \right]_{9\times6}. \qquad (7.115)$$

7.3 DAMPING MATRIX

Similar to the formulation of the inertia matrix [*M*], the damping matrix [*B*] of a flexible-hinge mechanism mainly results from the external viscous friction between the more massive rigid links and the surrounding fluid. It can thus be considered that [*B*] is solely associated with the motion of the rigid links, and this model is preponderantly used in the related literature. However, when the rigid links and flexible hinges have comparable dimensions, [*B*] should also include contributions from the flexible hinges. It is particularly advantageous in this latter approach to utilize the proportional damping model, according to which the damping matrix is a linear combination of the inertia matrix and the stiffness matrix (the inverse of the compliance matrix) – see also Section 7.1.3. These damping models are utilized in this chapter.

The loss mechanisms through damping are, however, very complex, and there are multiple contributing sources, such as internal losses, particularly in the flexible hinges that deform during cyclic vibration when a relaxation process due to thermoelastic phenomena occurs – see Lazan [33], Nashif et al. [34] and Rivin [35], for instance. Vibratory microscale compliant mechanisms are additionally subjected to damping losses through mechanisms such as squeeze- or slide-film damping, substrate, volume and/or thermoelastic losses – as discussed in Lobontiu [32]. These damping mechanisms are not included in the book.

7.3.1 Viscous Damping from Rigid Links

The viscous damping matrix of the 3D rigid link sketched in Figure 7.1 is connected to the six DOF and the corresponding nodal displacement vector of Eq. (7.1); it is the following local-frame diagonal matrix:

$$\left[B_i^{(l)}\right]_{6\times6} = \text{diag}\left(\begin{array}{cccccc} b_{t,x_i} & b_{t,y_i} & b_{r,z_i} & b_{r,x_i} & b_{r,y_i} & b_{t,z_i} \end{array}\right), \quad (7.116)$$

where b_{t,x_i}, b_{t,y_i} and b_{t,z_i} are damping coefficients associated with the translation of rigid link i along the local directions x_i, y_i and z_i, whereas b_{r,x_i}, b_{r,y_i} and b_{r,z_i} are the damping coefficients characterizing the rotations of the rigid link i around the local axes.

The local-frame damping matrices defining the in-plane or out-of-plane motions are expressed as:

$$\left[B_{i,ip}^{(l)}\right]_{3\times3} = \text{diag}\left(\begin{array}{ccc} b_{t,x_i} & b_{t,y_i} & b_{r,z_i} \end{array}\right); \quad \left[B_{i,op}^{(l)}\right]_{3\times3}$$

$$= \text{diag}\left(\begin{array}{ccc} b_{r,x_i} & b_{r,y_i} & b_{t,z_i} \end{array}\right). \quad (7.117)$$

7.3.2 Proportional Viscous Damping from Rigid Links and Flexible Hinges

When the dimensions of the rigid links and those of the flexible hinges are comparable, viscous damping contributions from the flexible hinges should be added to those of the rigid links. Separate evaluation of the damping due to flexible hinges can be pursued as discussed, for instance, in Lobontiu [32]. However, the proportional viscous damping model allows for the natural incorporation of the hinge damping addition; according to this model, the viscous damping matrix is a linear combination of the inertia matrix and the stiffness matrix as given in Eq. (7.12), where α and β are coefficients weighing the inertia and stiffness matrices. The rigid-link and flexible-hinge damping matrices are:

$$\left[B^{(r)}\right] = \alpha_r\left[M^{(r)}\right] + \beta[K]; \quad \left[B^{(h)}\right] = \alpha_h\left[M^{(h)}\right] + \beta[K], \quad (7.118)$$

where all the matrices in the right-hand sides of Eqs. (7.118) are expressed in the mechanism global frame. The total damping matrix is therefore:

$$\begin{aligned} [B] &= \left[B^{(r)}\right] + \left[B^{(h)}\right] = \alpha_r\left[M^{(r)}\right] + \alpha_h\left[M^{(h)}\right] + 2\beta[K] \\ &= \alpha_r\left[M^{(r)}\right] + \alpha_h\left[M^{(h)}\right] + 2\beta[C]^{-1}, \end{aligned} \quad (7.119)$$

where "r" and "h" stand for rigid link and flexible hinges, respectively.

Formation of the compliance matrix $[C]$ can be achieved to conform with the dimensionality of the compliant mechanism, which is also the dimensionality of the inertia matrix $[M]$ as dictated by the mechanism DOF. Subsequent sections in this chapter will separately discuss the derivation of $[C]$ for serial and parallel flexible-hinge mechanisms.

7.4 SERIAL MECHANISMS

This section studies the dynamics of mechanisms that are formed by serially connecting flexible hinges and rigid links. Mechanisms with fixed-free ends, as well as mechanisms with other boundary conditions (overconstrained), are analyzed here. Examples include devices of planar structure that undergo in-plane or out-of-plane motion, as well as spatial mechanisms that move three dimensionally.

7.4.1 PLANAR (2D) MECHANISMS

Fixed-free and overconstrained mechanisms are included here, which can be of simple configurations, such as devices formed of one rigid link and one flexible hinge. More complex mechanism designs with multiple rigid links and flexible hinges are also studied.

7.4.1.1 Fixed-Free Chain with One Flexible Hinge and One Rigid Link

Consider the mechanism sketched in Figure 7.11; it is formed of a single hinge connected to a moving rigid link at one end and fixed at the other end – the two links are collinear in this section. The mathematical models of the free undamped and the free damped responses are derived here, and related examples are analyzed.

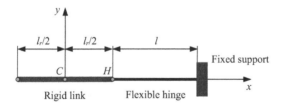

FIGURE 7.11 Skeleton representation of a fixed-free collinear chain formed of a flexible hinge and a rigid link.

7.4.1.1.1 Free Undamped Response

The mathematical models of the free undamped response for both in-plane and the out-of-plane vibrations are formulated in this section. While the rigid link is inherently heavy, the flexible hinge may be either massless or heavy.

In-plane motion:

The planar motion of the rigid link (and of the mechanism) is fully defined by three parameters related to the link center of mass (and symmetry) C: the horizontal and vertical translations u_{Cx} and u_{Cy}, and the link rotation θ_{Cz} around the z axis. As a result, the system sketched in Figure 7.11 has three DOF. The equation defining the in-plane free undamped vibrations of the system is:

$$[M_{C,ip}][\ddot{u}_{C,ip}] + [K_{C,ip}][u_{C,ip}] = [0] \quad \text{with}$$

$$[u_{C,ip}] = \begin{bmatrix} u_{Cx} & u_{Cx} & \theta_{Cz} \end{bmatrix}^T; \quad [K_{C,ip}] = [C_{C,ip}]^{-1}. \tag{7.120}$$

The inertia matrix $[M_{C,ip}]$ of Eq. (7.120) adds inertia contributions from the rigid link and the flexible hinge as in Eq. (7.62). The rigid-link inertia matrix is expressed in Eq. (7.63), whereas the flexible-hinge inertia matrix is transferred to C as in Eq. (7.60), namely:

$$[M_{C,ip}^{(h)}] = [T_{CH,ip}]^T [M_{H,ip}^{(h)}][T_{CH,ip}] \quad \text{with} \quad [T_{CH,ip}] = \begin{bmatrix} 1 & 0 & 0 \\ 0 & 1 & 0 \\ 0 & -l_r/2 & 1 \end{bmatrix}. \tag{7.121}$$

Equation (7.121) resulted from Eq. (7.60) by using the identity matrix instead of the rotation matrix in that equation. The flexible-hinge inertia matrix $\left[M_{H,ip}^{(h)} \right]$ is provided generically in the first Eq. (7.53) based on the consistent model and in Eq. (7.56) for the lumped model. The rigid-link mass moment of inertia in Eq. (7.63) is $J_{Cz} = m_r \cdot l_r^2 / 12$, with m_r being the mass of the rigid link.

The stiffness matrix $[K_{C,ip}]$ of Eq. (7.120) results from inverting the compliance matrix $[C_{C,ip}]$; the latter is evaluated at C, by means of a translation matrix that transfers the flexible-hinge native compliance matrix from H to C:

$$[C_{C,ip}] = [T_{CH,ip}]^T [C_{H,ip}][T_{CH,ip}], \tag{7.122}$$

with $[T_{CH,ip}]$ provided in Eq. (7.121). The in-plane dynamic matrix, which allows calculating the natural frequencies, is determined as:

$$[D_{C,ip}] = [M_{C,ip}]^{-1}[K_{C,ip}] = [M_{C,ip}]^{-1}[C_{C,ip}]^{-1} = ([C_{C,ip}][M_{C,ip}])^{-1}. \tag{7.123}$$

Out-of-plane motion:

A similar approach is followed to derive the mathematical model of the out-of-plane free vibrations of the system sketched in Figure 7.11. The rigid link's center-relevant motions are the rotations θ_{Cx} and θ_{Cy} around the x and y axes, and the translation u_{Cz} along the z axis. The equation of motion is:

$$[M_{C,op}][\ddot{u}_{C,op}] + [K_{C,op}][u_{C,op}] = [0] \quad \text{with}$$

$$[u_{C,op}] = \begin{bmatrix} \theta_{Cx} & \theta_{Cx} & u_{Cz} \end{bmatrix}^T; \quad [K_{C,op}] = [C_{C,op}]^{-1}. \tag{7.124}$$

The out-of-plane inertia matrix of Eq. (7.124) combines inertia fractions resulting from the rigid link and the flexible hinge as per Eq. (7.62). The rigid-link inertia matrix is expressed in Eq. (7.63), where $J_{Cx} = \rho \cdot I_t \cdot l_r, J_{Cy} = m_r \cdot l_r^2 / 12$ and I_t is the rigid-link torsional moment of area. The out-of-plane inertia matrix of the flexible hinge is translated at the center C based on the modified Eq. (7.61) as:

$$\left[M_{C,op}^{(h)} \right] = \left[T_{CH,op} \right]^T \left[M_{H,op}^{(h)} \right] \left[T_{CH,op} \right] \quad \text{with} \quad \left[T_{CH,op} \right] = \begin{bmatrix} 1 & 0 & 0 \\ 0 & 1 & l_r / 2 \\ 0 & 0 & 1 \end{bmatrix}.$$

$$(7.125)$$

The compliance matrix $[C_{C,op}]$ is translated at C from the hinge end point H as:

$$\left[C_{C,op} \right] = \left[T_{CH,op} \right]^T \left[C_{H,op} \right] \left[T_{CH,op} \right], \tag{7.126}$$

with $[T_{CH,op}]$ given in Eq. (7.125). The out-of-plane dynamic matrix is calculated as:

$$\left[D_{C,op} \right] = \left[M_{C,op} \right]^{-1} \left[K_{C,op} \right] = \left[M_{C,op} \right]^{-1} \left[C_{C,op} \right]^{-1} = \left(\left[C_{C,op} \right] \left[M_{C,op} \right] \right)^{-1}, \tag{7.127}$$

and it enables to evaluate of the corresponding natural frequencies.

Example 7.3

The flexible hinge of the mechanism sketched in Figure 7.11 is right circularly corner-filleted of rectangular cross-section; its length is $l = 0.015$ m, the fillet radius is $r = l/6$, the minimum thickness is $t = 0.001$ m, the constant out-of-plane width is $w = 0.007$ m, and Young's modulus is $E = 2 \cdot 10^{11}$ N/m². The rigid link has a length $l_r = 0.02$ m. The mass density of both the flexible hinge and the rigid link is $\rho = 7,800$ kg/m³. Consider that the mass of the flexible hinge is a fraction c of the rigid-link mass, i.e., $m_h = c \cdot m_r$. Using the mass ratio c as an independent variable, compare the in-plane natural frequencies of the mechanism with massless hinge to the corresponding natural frequencies obtained when also including the flexible-hinge inertia. Use both the consistent model and the lumped model for the hinge inertia.

Solution:

The in-plane compliance matrix of this mechanism is calculated by means of Eqs. (7.122). The corresponding compliance matrix of the right circularly corner-filleted flexure hinge is evaluated as explained in Section 3.2.2.2.3. The inertia matrix when only the rigid link is accounted for is calculated as per Eq. (7.63), and the resulting natural frequencies are the square roots of the eigenvalues of the dynamic matrix expressed in Eq. (7.123).

When the inertia of the flexure hinge is added to the inertia of the rigid link, the resulting in-plane inertia matrix becomes the one expressed in Eq. (7.62). The global-frame inertia matrix of the flexible hinge is calculated as in Eq. (7.121). The consistent-model hinge inertia matrix is given in Eq. (7.53), whereas the one based on the lumped model is provided in Eq. (7.56) – both are calculated with respect to the reference frame located at H in Figure 7.11. The mass of the flexure hinge is $m_h = \rho \cdot w \cdot [\pi r^2 + (l - 2r) \cdot t] = 0.00162$ kg. Using now the inertia matrix of Eq. (7.62) in conjunction with the same compliance matrix results in altered natural frequencies. By varying the mass ratio c from 0.02 to 0.5, the following relative variations in the natural frequencies:

$$\Delta\omega_{ni} = \frac{\omega_{ni} - \omega_{ni}^*}{\omega_{ni}} \cdot 100 \qquad (7.128)$$

are plotted in Figure 7.12, where ω_{ni} ($i = 1, 2, 3$) are the natural frequencies of the mechanism without flexible-hinge inertia, while ω_{ni}^* are the natural frequencies of the mechanism with hinge inertia included. Figure 7.12a uses the consistent-inertia model for flexible hinges, whereas Figure 7.12b employs the lumped-parameter model of flexible-hinge inertia. According to both models, the largest natural frequency ω_{n1} is the most affected by the inclusion of the flexible-hinge inertia; the natural frequency ω_{n3} follows in terms of deviation magnitude, while the second natural frequency ω_{n2} is the least affected by the inclusion of the flexible-hinge inertia. It can also be seen that the relative errors of the two models yield identical predictions for the largest natural frequencies.

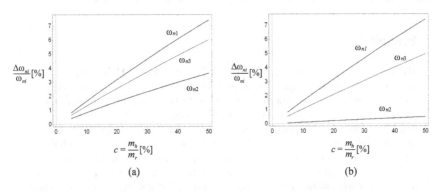

FIGURE 7.12 Plots of relative variation of natural frequencies in terms of the flexible-hinge mass fraction when the hinge inertia is expressed with the: (a) consistent model; (b) lumped-parameter model.

7.4.1.1.2 Free Damped Response

With the three DOF defined in Eq. (7.120), the mathematical model of the in-plane free damped vibrations of the Figure 7.11 mechanism is:

$$\left[M_{C,ip}\right]\left[\ddot{u}_{C,ip}\right] + \left[B_{C,ip}\right]\left[\dot{u}_{C,ip}\right] + \left[C_{C,ip}\right]^{-1}\left[u_{C,ip}\right] = [0]. \qquad (7.129)$$

For proportional damping resulting only from the rigid link, the damping matrix of Eq. (7.129) is calculated as:

$$\left[B_{C,ip}\right] = \alpha_r\left[M_{C,ip}^{(r)}\right] + \beta\left[C_{C,ip}\right]^{-1}, \qquad (7.130)$$

whereas for proportional damping produced by both the rigid link and the flexible hinge, the damping matrix $[B_{C,ip}]$ is:

$$\left[B_{C,ip} \right] = \left[B_{C,ip}^{(r)} \right] + \left[B_{C,ip}^{(h)} \right] = \alpha_r \left[M_{C,ip}^{(r)} \right] + \alpha_h \left[M_{C,ip}^{(h)} \right] + 2\beta \left[C_{C,ip} \right]^{-1}. \qquad (7.131)$$

An equation similar to Eq. (7.129) is the mathematical model governing the out-of-plane free damped response of the mechanism shown in Figure 7.11, namely:

$$\left[M_{C,op} \right]\left[\ddot{u}_{C,op} \right] + \left[B_{C,op} \right]\left[\dot{u}_{C,op} \right] + \left[C_{C,op} \right]^{-1}\left[u_{C,op} \right] = [0] \qquad (7.132)$$

for $[u_{C,op}]$ of Eq. (7.124). The proportional damping matrix produced by the rigid link solely is:

$$\left[B_{C,op} \right] = \alpha_r \left[M_{C,op}^{(r)} \right] + \beta \left[C_{C,op} \right]^{-1}. \qquad (7.133)$$

When proportional damping results from both the rigid link and the flexible hinge, the damping matrix of Eq. (7.132) is calculated as:

$$\left[B_{C,op} \right] = \left[B_{C,op}^{(r)} \right] + \left[B_{C,op}^{(h)} \right] = \alpha_r \left[M_{C,op}^{(r)} \right] + \alpha_h \left[M_{C,op}^{(h)} \right] + 2\beta \left[C_{C,op} \right]^{-1}. \qquad (7.134)$$

Example 7.4

The mechanism of Figure 7.11 is subjected to proportional viscous damping and an initial displacement $[u_{C,ip}(0)] = [u_{Cx}(0)\ u_{Cy}(0)\ \theta_{Cz}(0)]^T = [0\ 0.01\ 0]^T$ at C (nonzero displacement is expressed in m – meters). Use all the geometric and material parameters of *Example 7.3*, and plot the time responses $u_{Cy}(t)$ for proportional damping related to the rigid link only and defined by $\alpha_r = 0.25$ and $\beta = 0.3$; also, plot $u_{Cy}(t)$ when there is additional damping associated with the motion of the flexible hinges – defined by $\alpha_h = 0.15$. Apply the lumped-parameter inertia model for the flexible hinge, and assume that the mass fraction is $c = m_h/m_r = 0.1$.

Solution:

Based on Eq. (7.11), the in-plane, Laplace-domain response is:

$$\left[U_{C,ip}(s) \right] = \left[\ U_{Cx}(s)\quad U_{Cy}(s)\quad \Theta_{Cz}(s)\ \right]^T$$

$$= \left(s^2 \left[M_{C,ip} \right] + s\left[B_{C,ip} \right] + \left[K_{C,ip} \right] \right)^{-1}\left(s\left[M_{C,ip} \right] + \left[B_{C,ip} \right] \right)\left[u_{C,ip}(0) \right]. \qquad (7.135)$$

The inertia matrix of Eq. (7.135) is calculated by means of Eqs. (7.56), (7.62), (7.63) and (7.121). The compliance matrix $\left[C_{C,ip} \right]$ is calculated as in *Example 7.3*, whereas the damping matrix is evaluated with Eqs. (7.130) and (7.131). Figure 7.13 plots the time response of this mechanism based on the two damping models. The more pronounced effect of the flexible-hinge-added damping is observed in the time response that converges to zero faster than the one corresponding to a massless hinge.

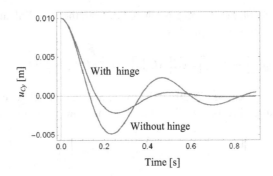

FIGURE 7.13 Time plot of free damped response u_{Cy} for the compliant mechanism as shown in *Figure 7.11*.

7.4.1.2 Fixed-Free Chain with Multiple Flexible Hinges and Rigid Links

Figure 7.14a shows a planar serial chain with one end free and the opposite end fixed. It consists of two rigid links and two flexible hinges. While this configuration is specific, it illustrates the general case of a fixed-free serial chain with multiple rigid links and flexible hinges.

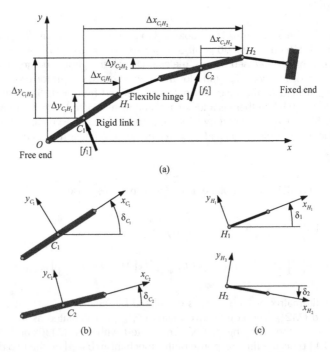

FIGURE 7.14 (a) Skeleton representation of a planar fixed-free serial chain formed of two flexible hinges and two rigid links; (b) local frames of rigid links; (c) local frames of flexible hinges.

The in-plane state of this mechanism is fully defined by the positions of the two rigid links; specifically, the in-plane positions of the rigid links are determined when the x- and y-translations of the two centers C_1 and C_2 are specified, together with the z-axis rotations of these links at their mass centers. As a result, the in-plane motion of the compliant mechanism is defined by six DOF in terms of its planar motion. A similar reasoning shows that the out-of-plane motion of the same mechanism is characterized by six DOF: x-axis and y-axis rotations alongside z-axis translations at C_1 and C_2. Formulated here are the inertia matrix pertaining to the two heavy rigid links and the compliance matrix associated with the two flexible hinges for both the in-plane and out-of-plane motions.

7.4.1.2.1 In-Plane Motion

The global-frame, independent coordinates of motion (DOF) for the in-plane motion are:

$$
\left[u_{ip} \right] = \left[\begin{array}{c} \left[u_{ip}^{(1)} \right] \\ \left[u_{ip}^{(2)} \right] \end{array} \right] \quad \text{with} \quad \begin{cases} \left[u_{ip}^{(1)} \right] = \left[\begin{array}{ccc} u_{C_1 x} & u_{C_1 y} & \theta_{C_1 z} \end{array} \right]^T \\ \left[u_{ip}^{(2)} \right] = \left[\begin{array}{ccc} u_{C_2 x} & u_{C_2 y} & \theta_{C_2 z} \end{array} \right]^T \end{cases} . \quad (7.136)
$$

They are recorded in the global frame $Oxyz$, and the superscripts (1) and (2) identify the two rigid links. Corresponding to these DOF, the two rigid links of Figure 7.14a have inertia matrices that are formulated in their local frames $C_1 x_{C_1} y_{C_1} z_{C_1}$ and $C_2 x_{C_2} y_{C_2} z_{C_2}$ as:

$$
\left[M_{ip}^{(1,l)} \right]_{3 \times 3} = \left[\begin{array}{ccc} m_1 & 0 & 0 \\ 0 & m_1 & 0 \\ 0 & 0 & J_{1z} \end{array} \right]; \quad \left[M_{ip}^{(2,l)} \right]_{3 \times 3} = \left[\begin{array}{ccc} m_2 & 0 & 0 \\ 0 & m_2 & 0 \\ 0 & 0 & J_{2z} \end{array} \right]. \quad (7.137)
$$

The local-frame displacement vectors are defined similarly to the global ones of Eqs. (7.136), namely:

$$
\left[u_{ip}^{(l)} \right] = \left[\begin{array}{c} \left[u_{ip}^{(1,l)} \right] \\ \left[u_{ip}^{(2,l)} \right] \end{array} \right] \quad \text{with} \quad \begin{cases} \left[u_{ip}^{(1,l)} \right] = \left[\begin{array}{ccc} u_{C_1 x C_1} & u_{C_1 y C_1} & \theta_{C_1 z C_1} \end{array} \right]^T \\ \left[u_{ip}^{(2,l)} \right] = \left[\begin{array}{ccc} u_{C_2 x C_2} & u_{C_2 y C_2} & \theta_{C_2 z C_2} \end{array} \right]^T \end{cases} . \quad (7.138)
$$

The superscript "l" indicates the local frames in Eqs. (7.137) and (7.138). Two inertia-force vectors are formed based on the local-frame inertia matrices of Eqs. (7.137) and the displacement vectors of Eqs. (7.138):

$$\left[f_{ip}^{(1,l)} \right] = \left[M_{ip}^{(1,l)} \right]\left[\ddot{u}_{ip}^{(1,l)} \right]; \quad \left[f_{ip}^{(2,l)} \right] = \left[M_{ip}^{(2,l)} \right]\left[\ddot{u}_{ip}^{(2,l)} \right]. \tag{7.139}$$

The rigid links' local frames are rotated with respect to the global frame $Oxyz$, as illustrated in Figure 7.14b. The following rotation equations relate the local- and global-frame inertia force and acceleration vectors:

$$\left[f_{ip}^{(1,l)} \right] = \left[R^{(r1)} \right]\left[f_{ip}^{(1)} \right]; \quad \left[f_{ip}^{(2,l)} \right] = \left[R^{(r2)} \right]\left[f_{ip}^{(2)} \right];$$
$$\left[\ddot{u}_{ip}^{(1,l)} \right] = \left[R^{(r1)} \right]\left[\ddot{u}_{ip}^{(1)} \right]; \quad \left[\ddot{u}_{ip,l}^{(2,l)} \right] = \left[R^{(r2)} \right]\left[\ddot{u}_{ip}^{(2)} \right], \tag{7.140}$$

where $\left[f_{ip}^{(1)} \right]$ and $\left[f_{ip}^{(2)} \right]$ are the global-frame inertia vectors, and the two rotation matrices are calculated using $\delta = \delta_{C_i}$ ($i = 1, 2$). Combining Eqs. (7.139) and (7.140) results in the following global-frame inertia submatrices:

$$\left[M_{ip}^{(1)} \right] = \left[R^{(r1)} \right]^T \left[M_{ip}^{(1,l)} \right]\left[R^{(r1)} \right];$$
$$\left[M_{ip}^{(2)} \right] = \left[R^{(r2)} \right]^T \left[M_{ip}^{(2,l)} \right]\left[R^{(r2)} \right]. \tag{7.141}$$

The in-plane inertia matrix of the full mechanism combines the inertia submatrices related to the in-plane motion of the two heavy rigid links as:

$$\left[M_{ip} \right]_{6\times6} = \begin{bmatrix} \left[M_{ip}^{(1)} \right]_{3\times3} & \left[0 \right]_{3\times3} \\ \left[0 \right]_{3\times3} & \left[M_{ip}^{(2)} \right]_{3\times3} \end{bmatrix}. \tag{7.142}$$

Note that the inertia matrix of Eq. (7.142) retains only the rigid-link contribution. However, as detailed in *Example 7.2*, the additional inertia matrix of Eq. (7.113) should be added to the rigid-link inertia matrix when the flexible hinges are considered heavy.

The in-plane compliance matrix of the mechanism is evaluated by expressing the displacements at points C_1 and C_2 resulting from the planar deformation of the two flexible hinges under the action of the two external loads $[f_{1,ip}]$ and $[f_{2,ip}]$ that are applied at C_1 and C_2. These global-frame displacements are calculated as:

$$\left[u_{ip}^{(1)} \right] = \left(\left[C_{C_1C_1,ip}^{(1)} \right] + \left[C_{C_1C_1,ip}^{(2)} \right] \right)\left[f_{1,ip} \right] + \left[C_{C_1C_2,ip}^{(2)} \right]\left[f_{2,ip} \right]$$
$$\left[u_{ip}^{(2)} \right] = \left[C_{C_2C_1,ip}^{(2)} \right]\left[f_{1,ip} \right] + \left[C_{C_2C_2,ip}^{(2)} \right]\left[f_{2,ip} \right] \tag{7.143}$$
$$\text{or} \quad \left[u_{ip} \right] = \left[C_{ip} \right]\left[f_{ip} \right]$$

The superscripts (1) and (2) in the right-hand sides of Eqs. (7.143) identify the two flexible hinges whose positions and local frames are shown in Figure 7.14c. The compliance matrix of the full mechanism corresponding to the DOF of Eq. (7.136) and the two displacement vectors of Eq. (7.143) is:

$$\left[C_{ip}\right]_{6\times6} = \begin{bmatrix} \left[C_{C_1C_1,ip}^{(1)}\right]_{3\times3} + \left[C_{C_1C_1,ip}^{(2)}\right]_{3\times3} & \left[C_{C_1C_2,ip}^{(2)}\right]_{3\times3} \\ \left[C_{C_2C_1,ip}^{(2)}\right]_{3\times3} & \left[C_{C_2C_2,ip}^{(2)}\right]_{3\times3} \end{bmatrix};$$

$$\left[u_{ip}\right] = \begin{bmatrix} \left[u_{ip}^{(1)}\right] & \left[u_{ip}^{(2)}\right] \end{bmatrix}^{T}$$

$$\left[f_{ip}\right] = \begin{bmatrix} \left[f_{1,ip}\right] & \left[f_{2,ip}\right] \end{bmatrix}^{T}$$

(7.144)

The submatrices of Eq. (7.144) are calculated as:

$$\begin{cases} \left[C_{C_1C_1,ip}^{(1)}\right]_{3\times3} = \left[T_{C_1H_1,ip}\right]^{T}\left[R^{(1)}\right]^{T}\left[C_{ip}^{(1)}\right]\left[R^{(1)}\right]\left[T_{C_1H_1,ip}\right] \\ \left[C_{C_1C_1,ip}^{(2)}\right]_{3\times3} = \left[T_{C_1H_2,ip}\right]^{T}\left[R^{(2)}\right]^{T}\left[C_{ip}^{(2)}\right]\left[R^{(2)}\right]\left[T_{C_1H_2,ip}\right] \\ \left[C_{C_1C_2,ip}^{(2)}\right]_{3\times3} = \left[C_{C_2C_1,ip}^{(2)}\right]_{3\times3}^{T} = \left[T_{C_1H_2,ip}\right]^{T}\left[R^{(2)}\right]^{T}\left[C_{ip}^{(2)}\right]\left[R^{(2)}\right]\left[T_{C_2H_2,ip}\right] \\ \left[C_{C_2C_2,ip}^{(2)}\right]_{3\times3} = \left[T_{C_2H_2,ip}\right]^{T}\left[R^{(2)}\right]^{T}\left[C_{ip}^{(2)}\right]\left[R^{(2)}\right]\left[T_{C_2H_2,ip}\right] \end{cases}, \quad (7.145)$$

where $\left[C_{ip}^{(1)}\right]$ and $\left[C_{ip}^{(2)}\right]$ are the in-plane compliance matrices of the two hinges. The translation matrices of Eq. (7.145) are calculated with the offsets highlighted in Figure 7.14a, while the rotation matrices of the same Eq. (7.145) are evaluated based on the angles δ_1 and δ_2 identified in Figure 7.14c.

7.4.1.2.2 Out-of-Plane Motion

The global-frame DOF for the out-of-plane motion are:

$$\left[u_{op}\right] = \begin{bmatrix} \left[u_{op}^{(1)}\right] \\ \left[u_{op}^{(2)}\right] \end{bmatrix} \quad \text{with} \quad \begin{cases} \left[u_{op}^{(1)}\right] = \begin{bmatrix} \theta_{C_1x} & \theta_{C_1y} & u_{C_1z} \end{bmatrix}^{T} \\ \left[u_{op}^{(2)}\right] = \begin{bmatrix} \theta_{C_2x} & \theta_{C_2y} & u_{C_2z} \end{bmatrix}^{T} \end{cases} \quad (7.146)$$

The following local-frame inertia matrices result from the heavy rigid links (1) and (2), which are formed based on the DOF of Eq. (7.146):

$$\left[M_{op}^{(1,l)}\right] = \begin{bmatrix} J_{1x} & 0 & 0 \\ 0 & J_{1y} & 0 \\ 0 & 0 & m_1 \end{bmatrix}; \quad \left[M_{op}^{(2,l)}\right] = \begin{bmatrix} J_{2x} & 0 & 0 \\ 0 & J_{2y} & 0 \\ 0 & 0 & m_2 \end{bmatrix}. \quad (7.147)$$

Applying a reasoning similar to the one used to derive the global-frame, in-plane inertia submatrices, the out-of-plane inertia matrices are expressed in terms of the local frame ones of Eqs. (7.147) and the rotation matrices of Eqs. (7.141) as:

$$\left[M_{op}^{(1)} \right] = \left[R^{(r1)} \right]^T \left[M_{op}^{(1,l)} \right] \left[R^{(r1)} \right];$$

$$\left[M_{op}^{(2)} \right] = \left[R^{(r2)} \right]^T \left[M_{op}^{(2,l)} \right] \left[R^{(r2)} \right], \tag{7.148}$$

which are assembled into the full inertia matrix:

$$\left[M_{op} \right]_{6\times6} = \begin{bmatrix} \left[M_{op}^{(1)} \right]_{3\times3} & \left[0 \right]_{3\times3} \\ \left[0 \right]_{3\times3} & \left[M_{op}^{(2)} \right]_{3\times3} \end{bmatrix}. \tag{7.149}$$

For heavy flexible hinges, the out-of-plane inertia matrix of Eq. (7.115) of *Example 7.2* should be added to the inertia matrix of Eq. (7.149), which only accounts for the rigid-link inertia.

Applying a similar reasoning to the one used in the in-plane case, it can be shown that the out-of-plane compliance matrix is:

$$\left[C_{op} \right]_{6\times6} = \begin{bmatrix} \left[C_{C_1C_1,op}^{(1)} \right]_{3\times3} + \left[C_{C_1C_1,op}^{(2)} \right]_{3\times3} & \left[C_{C_1C_2,op}^{(2)} \right]_{3\times3} \\ \left[C_{C_2C_1,op}^{(2)} \right]_{3\times3} & \left[C_{C_2C_2,op}^{(2)} \right]_{3\times3} \end{bmatrix};$$

$$\left[u_{op} \right] = \left[C_{op} \right] \left[f_{op} \right] \quad \text{with} \quad \begin{cases} \left[u_{op} \right] = \begin{bmatrix} \left[u_{op}^{(1)} \right] & \left[u_{op}^{(2)} \right] \end{bmatrix}^T \\ \left[f_{op} \right] = \begin{bmatrix} \left[f_{1,op} \right] & \left[f_{2,op} \right] \end{bmatrix}^T \end{cases} \tag{7.150}$$

whose submatrices are:

$$\begin{cases} \left[C_{C_1C_1,op}^{(1)} \right]_{3\times3} = \left[T_{C_1H_1,op} \right]^T \left[R^{(1)} \right]^T \left[C_{op}^{(1)} \right] \left[R^{(1)} \right] \left[T_{C_1H_1,op} \right] \\ \left[C_{C_1C_1,op}^{(2)} \right]_{3\times3} = \left[T_{C_1H_2,op} \right]^T \left[R^{(2)} \right]^T \left[C_{op}^{(2)} \right] \left[R^{(2)} \right] \left[T_{C_1H_2,op} \right] \\ \left[C_{C_1C_2,op}^{(2)} \right]_{3\times3} = \left[C_{C_2C_1,op}^{(2)} \right]_{3\times3} = \left[T_{C_1H_2,op} \right]^T \left[R^{(2)} \right]^T \left[C_{op}^{(2)} \right] \left[R^{(2)} \right] \left[T_{C_2H_2,op} \right] \\ \left[C_{C_2C_2,op}^{(2)} \right]_{3\times3} = \left[T_{C_2H_2,op} \right]^T \left[R^{(2)} \right]^T \left[C_{op}^{(2)} \right] \left[R^{(2)} \right] \left[T_{C_2H_2,op} \right] \end{cases} \tag{7.151}$$

Example 7.5

Calculate the in-plane and out-of-plane natural frequencies of the planar compliant mechanism of Figure 7.15. Consider that the heavy rigid links have the same constant cross-section defined by an area $A_r = 4\cdot10^{-4}\,\text{m}^2$ and a torsional moment of area $I_{r,x} = 1.333\cdot10^{-8}\,\text{m}^4$, a mass density $\rho = 7,800\,\text{kg/m}^3$, and the lengths $l_1 = 0.03\,\text{m}$ and $l_2 = 0.04\,\text{m}$. The two identical flexible hinges are massless; they have a length $l = 0.012\,\text{m}$, a constant rectangular cross-section with in-plane thickness $t = 0.001\,\text{m}$, out-of-plane width $w = 0.006\,\text{m}$ and their Young's modulus is $E = 2\cdot10^{11}\,\text{N/m}^2$. Also known is the angle $\delta = 10°$.

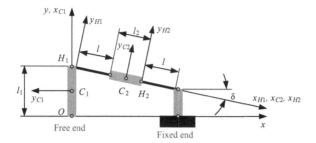

FIGURE 7.15 Fixed-free planar serial mechanism with two identical and collinear straight-axis flexible hinges and three rigid links.

Solution:

The nonzero components of the local-frame, in-plane inertia matrices of Eqs. (7.137) are:

$$m_i = \rho A_r l_i; \quad J_{iz}^{(i)} = \frac{m_i l_i^2}{12}, \quad i = 1,2. \tag{7.152}$$

The angles defining the rotation matrices of Eqs. (7.140) are: $\delta_{C_1} = \pi / 2, \delta_{C_2} = -\delta$.

The in-plane inertia matrix of the mechanism depicted in Figure 7.15 is calculated by means of Eqs. (7.137), (7.141) and (7.142).

The local-frame, in-plane compliance matrix of the two identical flexible hinges is:

$$\left[C_{ip}^{(1)}\right] = \left[C_{ip}^{(2)}\right] = \begin{bmatrix} C_{u_x-f_x}^{(h)} & 0 & 0 \\ 0 & C_{u_y-f_y}^{(h)} & C_{u_y-m_z}^{(h)} \\ 0 & C_{u_y-m_z}^{(h)} & C_{\theta_z-m_z}^{(h)} \end{bmatrix} \quad \text{with:} \tag{7.153}$$

$$\begin{cases} C_{u_x-f_x}^{(h)} = \dfrac{l}{EA} \\ C_{u_y-f_y}^{(h)} = \dfrac{l^3}{3EI_z}; \quad C_{u_y-m_z}^{(h)} = -\dfrac{l^2}{2EI_z}; \quad C_{\theta_z-m_z}^{(h)} = \dfrac{l}{EI_z} \end{cases} ; \quad A = wt, I_z = \dfrac{wt^3}{12}.$$

The offsets used for the compliance matrices of Eqs. (7.145) are:

$$\begin{cases} \Delta x_{C_1H_1} = 0; \quad \Delta y_{C_1H_1} = l_1 / 2; \\ \Delta x_{C_1H_2} = (l+l_2) \cdot \cos \delta; \quad \Delta y_{C_1H_2} = l_1 / 2 - (l+l_2) \cdot \sin \delta; \\ \Delta x_{C_2H_2} = (l_2 / 2) \cdot \cos \delta; \quad \Delta y_{C_2H_2} = -(l_2 / 2) \cdot \sin \delta, \end{cases} \tag{7.154}$$

while the angles of the two rotation matrices of Eqs. (7.145) are $\delta = \delta_1 = \delta_2$. The in-plane compliance matrix is numerically evaluated with Eqs. (7.153),

(7.144) and (7.145). The following values of the in-plane natural frequencies are calculated with the aid of Eq. (7.8): 71610, 25558, 7841, 3834, 594 and 140 rad/s.

The nonzero components of the local-frame, out-of-plane inertia matrices of Eqs. (7.147) are:

$$J_{ix} = \rho l_i I_{r,x}; \quad J_{iy} = J_{iz}, \quad i = 1, 2. \tag{7.155}$$

The out-of-plane inertia matrix of the entire mechanism is evaluated by means of Eqs. (7.147)–(7.149). The local-frame, out-of-plane compliance matrix of the two identical flexible hinges is:

$$\left[C_{op}^{(1)}\right] = \left[C_{op}^{(2)}\right] = \begin{bmatrix} C_{\theta_x - m_x}^{(h)} & 0 & 0 \\ 0 & C_{\theta_y - m_y}^{(h)} & C_{\theta_y - f_z}^{(h)} \\ 0 & C_{\theta_y - f_z}^{(h)} & C_{u_z - f_z}^{(h)} \end{bmatrix} \text{ with}$$

$$\begin{cases} C_{\theta_x - m_x}^{(h)} = \dfrac{l}{GI_t} \\ C_{\theta_y - m_y}^{(h)} = \dfrac{l}{EI_y}; \quad C_{\theta_y - f_z}^{(h)} = \dfrac{l^2}{2EI_y}; \quad C_{u_z - f_z}^{(h)} = \dfrac{l^3}{3EI_y} \end{cases}; \quad I_t = \dfrac{wh^3}{3}, I_y = \dfrac{w^3 h}{12}.$$

$$\tag{7.156}$$

The out-of-plane compliance matrix of the full mechanism is calculated with Eqs. (7.150) and (7.151). The out-of-plane dynamic matrix is evaluated as per Eq. (7.8) with the full-mechanism inertia and compliance matrices; the corresponding natural frequencies are 136410, 32549, 6824, 1917, 1640 and 232 rad/s.

Example 7.6

Consider the mechanism of *Example 7.5* and depicted in Figure 7.15. The force $f_2 = 100 \cdot e^{-2t}$ N acts perpendicularly on the rigid link at C_2, as illustrated in Figure 7.16. A force f_1 that is applied at C_1 at an angle $\alpha = 15°$ cancels the vertical displacement at C_1. Evaluate the magnitude of f_1 by plotting its time variation. Use all numerical values provided in *Example 7.5*.

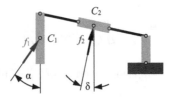

FIGURE 7.16 Mechanism of *Figure 7.15* with external forces.

Solution:

As per Eqs. (7.16) and (7.17), the in-plane displacement and forcing vectors are related in the Laplace domain as:

$$[U(s)] = [G(s)][F(s)]$$

with
$$
\begin{cases}
[G(s)] = \left(s^2[M] + [C]^{-1} \right)^{-1} \\
[U(s)] = \begin{bmatrix} U_{1x}(s) & U_{1y}(s) & \Theta_{1z}(s) & U_{2x}(s) & U_{2y}(s) & \Theta_{2z}(s) \end{bmatrix}^T. \\
[F(s)] = \begin{bmatrix} F_{1x}(s) & F_{1y}(s) & M_{1z}(s) & F_{2x}(s) & F_{2y}(s) & M_{2z}(s) \end{bmatrix}^T
\end{cases}
\tag{7.157}
$$

Note that there are no moments at C_1 and C_2, which means that $m_{1z}(t) = 0$, $m_{2z}(t) = 0$, or $M_{1z}(s) = 0$, $M_{2z}(s) = 0$. The subscript *"ip"* has been eliminated in Eq. (7.157). The example's displacement is $u_{1y}(t) = 0$, which means that $U_{1y}(s) = 0$. Explicitly, this condition is formulated from Eq. (7.157) as:

$$0 = G_{21}(s)F_{1x}(s) + G_{22}(s)F_{1y}(s) + G_{24}(s)F_{2x}(s) + G_{25}(s)F_{2y}(s), \tag{7.158}$$

where $G_{ij}(s)$ is the transfer function on the ith row and the jth column of the $[G(s)]$ transfer function matrix. Taking into account that:

$$
\begin{cases}
f_{1x}(t) = f_1(t) \cdot \sin\alpha; & f_{1y}(t) = f_1(t) \cdot \cos\alpha; \\
f_{2x}(t) = f_2(t) \cdot \sin\delta; & f_{2y}(t) = f_2(t) \cdot \cos\delta
\end{cases}
$$

or
$$
\begin{cases}
F_{1x}(s) = F_1(s) \cdot \sin\alpha; & F_{1y}(s) = F_1(s) \cdot \cos\alpha; \\
F_{2x}(s) = F_2(s) \cdot \sin\delta; & F_{2y}(s) = F_2(s) \cdot \cos\delta
\end{cases}
,
\tag{7.159}
$$

Eq. (7.158) results in:

$$F_1(s) = G(s) \cdot F_2(s) \quad \text{with} \quad G(s) = -\frac{G_{24}(s) \cdot \sin\delta + G_{25}(s) \cdot \cos\delta}{G_{21}(s) \cdot \sin\alpha + G_{22}(s) \cdot \cos\alpha}. \tag{7.160}$$

The transfer function $G(s)$ of Eq. (7.160) is obtained by means of the inertia matrix and the compliance matrix calculated in *Example 7.5*. Figure 7.17 is the plot of $f_1(t)$ that results from combining $G(s)$ with the particular $f_2(t)$.

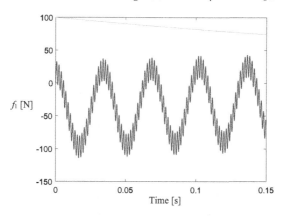

FIGURE 7.17 Time plot of force f_1 magnitude.

Example 7.7

Recalculate the in-plane natural frequencies of the planar mechanism of Figure 7.15 with the geometric and material properties given in *Example 7.5* when including the inertia of the flexible hinges with $\rho = 7,800\,kg/m^3$. Consider that $l_{11} = l_{12} = l_1/2$, $l_{21} = l_{22} = l_2/2$ with reference to Figures 7.10 and 7.15. Use both the lumped-parameter and the consistent-inertia properties of the two flexible hinges.

Solution:

The rotation matrices of Eqs. (7.141) and (7.145) are identical and evaluated for an angle equal to $-\delta$. The local-frame, in-plane inertia matrices of the two flexible hinges are calculated by means of Eqs. (7.91) for the consistent model, whereas Eqs. (7.56) and (7.94) are used to obtain the similar matrices by the lumped-parameter model. The 6×6 in-plane inertia matrix collecting contributions from both flexible hinges is calculated as per Eqs. (7.103) through (7.113). The total inertia matrix adds the inertia matrix due to rigid links $\left[M_{C,ip}^{(r)} \right]$ of Eq. (7.142) and the inertia matrix owing to flexible hinges $\left[M_{C,ip}^{(h)} \right]$ as:

$$\left[M_{C,ip} \right]_{6\times6} = \left[M_{C,ip}^{(r)} \right]_{6\times6} + \left[M_{C,ip}^{(h)} \right]_{6\times6}. \tag{7.161}$$

The in-plane compliance matrix of *Example 7.5* is combined with the inertia matrix of Eq. (7.161) to form dynamic matrices for both the consistent and the lumped-parameter models; the resulting natural frequencies are included in Table 7.1.

TABLE 7.1
In-Plane Natural Frequencies (in rad/s) According to: M_1 – Model with Heavy Rigid Links Only; M_2 – Model with Heavy Rigid Links and Consistent-Inertia Flexible Hinges; M_3 – Model with Heavy Rigid Links and Lumped-Parameter Inertia Flexible Hinges

	ω_{n6}	ω_{n5}	ω_{n4}	ω_{n3}	ω_{n2}	ω_{n1}
M_1	71,610	25,558	7,841	3,834	594	140
M_2	71,586	25,522	7,840	3,829	593	140
M_3	71,560	25,518	7,840	3,828	593	140

It can be seen that the differences between the three models' predictions are very small.

7.4.1.3 Overconstrained Chain with Multiple Flexible Hinges and Rigid Links

As discussed in Chapter 5, a common configuration of an overconstrained serial chain has one end fixed; in addition, there is another point (usually the opposite end) where zero-displacement boundary conditions are applied through external support. For such a chain, the reactions at the point with additional boundary conditions need to be first evaluated. With those reactions determined, the compliance matrices of the entire mechanism can be formed based on the mechanism DOF resulting from the motions of the mobile rigid links. Formation of the serial mechanism inertia matrices follows the procedure described in previous sections in this chapter. The following examples illustrate the procedure of formulating the in-plane and out-of-plane inertia and compliance matrices of a few overconstrained serial mechanisms.

Example 7.8

Consider the mechanism of *Example 5.8* and shown in Figure 5.20, which is redrawn in Figure 7.18. Calculate its in-plane natural frequencies by considering that the left rigid link of the mechanism is guided along the y axis. The two rigid links have the lengths $l_1 = 0.04$ m and $l_2 = 0.06$ m, their constant cross-section has an area $A_r = 0.001$ m², and their material mass density is $\rho = 7{,}800$ kg/m³. The two identical flexible hinges have a length $l = 0.012$ m and a constant rectangular cross-section of in-plane thickness $t = 0.001$ m and out-of-plane width $w = 0.008$ m. The hinge material is defined by Young's modulus $E = 2 \cdot 10^{11}$ N/m². The flexible hinges are assumed massless.

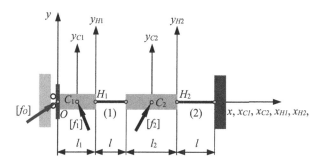

FIGURE 7.18   Fixed-guided serial flexible chain with collinear links.

Solution:

Because the left rigid link is guided, this link is only allowed translation along the y axis, and therefore, it has one DOF. The other rigid link has all three planar motions allowed; as a consequence, the in-plane motion of the mechanism is defined by four DOF, which are collected as:

$$\left[u_{ip}\right]=\begin{bmatrix}\left[u_{ip}^{(1)}\right]\\\left[u_{ip}^{(2)}\right]\end{bmatrix} \quad \text{with} \quad \begin{cases}\left[u_{ip}^{(1)}\right]=u_{C_1y}\\\left[u_{ip}^{(2)}\right]=\begin{bmatrix}u_{C_2x}&u_{C_2y}&\theta_{C_2z}\end{bmatrix}^T\end{cases}. \tag{7.162}$$

Corresponding to these DOF, the following global-frame inertia matrix is formed that accounts for the rigid-link inertia:

$$[M_{ip}]_{4\times4} = \begin{bmatrix} [M_{ip}^{(r1)}]_{1\times1} & [0]_{1\times3} \\ [0]_{3\times1} & [M_{ip}^{(r2)}]_{3\times3} \end{bmatrix} \quad \text{with}: \quad [M_{ip}^{(r1)}] = m_1;$$

$$[M_{ip}^{(r2)}] = \begin{bmatrix} m_2 & 0 & 0 \\ 0 & m_2 & 0 \\ 0 & 0 & J_{2z} \end{bmatrix}.$$

(7.163)

The dimension of the mechanism in-plane compliance matrix is consistent with the DOF of Eq. (7.162); therefore, it should be a 4 × 4 matrix. In order to derive it, assume that two load vectors are applied at C_1 and C_2, as illustrated in Figure 7.18 (where the load vectors are generic). The following equation expresses the in-plane displacements of the rigid-link midpoint C_1:

$$[u_{C_1,ip}] = \begin{bmatrix} u_{C_1 x} \\ u_{C_1 y} \\ \theta_{C_1 z} \end{bmatrix} = \begin{bmatrix} 0 \\ u_{C_1 y} \\ 0 \end{bmatrix} = [C_{C_1 O,ip}][f_{O,ip}] + [C_{C_1 C_1,ip}][f_{1,ip}] + [C_{C_1 C_2,ip}^{(2)}][f_{2,ip}]$$

$$\text{with } [C_{C_1 O,ip}] = [C_{C_1 O,ip}^{(1)}] + [C_{C_1 O,ip}^{(2)}]; [C_{C_1 C_1,ip}] = [C_{C_1 C_1,ip}^{(1)}] + [C_{C_1 C_1,ip}^{(2)}]; [f_{O,ip}] = \begin{bmatrix} f_{Ox} \\ 0 \\ m_{Oz} \end{bmatrix}.$$

(7.164)

The individual compliance matrices of Eq. (7.164) are calculated as:

$$[C_{C_1 C_2,ip}^{(2)}] = [T_{C_1 H_2,ip}]^T [C_{ip}^{(h)}][T_{C_2 H_2,ip}];$$

$$[C_{C_1 O,ip}^{(1)}] = [T_{C_1 H_1,ip}]^T [C_{ip}^{(h)}][T_{O H_1,ip}];$$

$$[C_{C_1 O,ip}^{(2)}] = [T_{C_1 H_2,ip}]^T [C_{ip}^{(h)}][T_{O H_2,ip}]; \quad (7.165)$$

$$[C_{C_1 C_1,ip}^{(1)}] = [T_{C_1 H_1,ip}]^T [C_{ip}^{(h)}][T_{C_1 H_1,ip}];$$

$$[C_{C_1 C_1,ip}^{(2)}] = [T_{C_1 H_2,ip}]^T [C_{ip}^{(h)}][T_{C_1 H_2,ip}].$$

The in-plane compliance matrix $[C_{ip}^{(h)}]$ of the flexible hinge is given in Eq. (7.153), and the translation matrices of Eqs. (7.165) are calculated with the following x-axis offsets (all the y-axis offsets are zero because the links are collinear):

$$\Delta x_{OH_1} = l_1; \quad \Delta x_{OH_2} = l_1 + l_2 + l; \quad \Delta x_{C_1 H_1} = l_1 / 2; \quad \Delta x_{C_1 H_2} = l_1 / 2 + l_2 + l; \quad (7.166)$$

Equation (7.164) is solved for the two nonzero reactions at O, which formally results in the reaction vector at O:

$$[f_{O,ip}] = [A_{ip}][f_{1,ip}] + [B_{ip}][f_{2,ip}], \quad (7.167)$$

where

$$
[A_{ip}] = \begin{bmatrix} A_{ip}(1,1) & A_{ip}(1,2) & A_{ip}(1,3) \\ 0 & 0 & 0 \\ A_{ip}(2,1) & A_{ip}(2,2) & A_{ip}(2,3) \end{bmatrix}; [B_{ip}] = \begin{bmatrix} B_{ip}(1,1) & B_{ip}(1,2) & B_{ip}(1,3) \\ 0 & 0 & 0 \\ B_{ip}(2,1) & B_{ip}(2,2) & B_{ip}(2,3) \end{bmatrix}.
$$

(7.168)

The nonzero components of the $[A_{ip}]$ matrix in Eq. (7.168) are the corresponding elements of the 2×3 matrix:

$$
\begin{bmatrix} A_{ip}(1,1) & A_{ip}(1,2) & A_{ip}(1,3) \\ A_{ip}(2,1) & A_{ip}(2,2) & A_{ip}(2,3) \end{bmatrix} = - \begin{bmatrix} C_{C_1O,ip}(1,1) & C_{C_1O,ip}(1,3) \\ C_{C_1O,ip}(3,1) & C_{C_1O,ip}(3,3) \end{bmatrix}^{-1}
$$

(7.169)

$$
\begin{bmatrix} C_{C_1C_1,ip}(1,1) & C_{C_1C_1,ip}(1,2) & C_{C_1C_1,ip}(1,3) \\ C_{C_1C_1,ip}(3,1) & C_{C_1C_1,ip}(3,2) & C_{C_1C_1,ip}(3,3) \end{bmatrix}.
$$

Similarly, the nonzero components of the $[B_{ip}]$ matrix in Eq. (7.168) are the corresponding elements of the 2×3 matrix:

$$
\begin{bmatrix} B_{ip}(1,1) & B_{ip}(1,2) & B_{ip}(1,3) \\ B_{ip}(2,1) & B_{ip}(2,2) & B_{ip}(2,3) \end{bmatrix} = - \begin{bmatrix} C_{C_1O,ip}(1,1) & C_{C_1O,ip}(1,3) \\ C_{C_1O,ip}(3,1) & C_{C_1O,ip}(3,3) \end{bmatrix}^{-1}
$$

(7.170)

$$
\begin{bmatrix} C_{C_1C_2,ip}^{(2)}(1,1) & C_{C_1C_2,ip}^{(2)}(1,2) & C_{C_1C_2,ip}^{(2)}(1,3) \\ C_{C_1C_2,ip}^{(2)}(3,1) & C_{C_1C_2,ip}^{(2)}(3,2) & C_{C_1C_2,ip}^{(2)}(3,3) \end{bmatrix}.
$$

The compliances in the right-hand sides of Eqs. (7.169) and (7.170) are components of the respective matrices calculated in Eqs. (7.164) and (7.165), whereas the pair (i, j) indicates the row–column position of those components in their matrices. Combining Eqs. (7.164) and (7.167) yields:

$$
[u_{C_1,ip}] = [C_{C_1C_1,ip}^*][f_{1,ip}] + [C_{C_1C_2,ip}^*][f_{2,ip}] \text{ with } \begin{cases} [C_{C_1C_1,ip}^*] = [C_{C_1O,ip}][A_{ip}] + [C_{C_1C_1,ip}] \\ [C_{C_1C_2,ip}^*] = [C_{C_1O,ip}][B_{ip}] + [C_{C_1C_2,ip}^{(2)}] \end{cases}
$$

(7.171)

Similarly, the in-plane displacements at C_2 are calculated as:

$$
[u_{C_2,ip}] = [C_{C_2O,ip}^{(2)}][f_{o,ip}] + [C_{C_2C_1,ip}^{(2)}][f_{1,ip}] + [C_{C_2C_2,ip}^{(2)}][f_{2,ip}]
$$

$$
= [C_{C_2C_1,ip}^*][f_{1,ip}] + [C_{C_2C_2,ip}^*][f_{2,ip}]
$$

(7.172)

$$
\text{with } \begin{cases} [C_{C_2C_1,ip}^*] = [C_{C_2O,ip}^{(2)}][A_{ip}] + [C_{C_2C_1,ip}^{(2)}] \\ [C_{C_1C_2,ip}^*] = [C_{C_2O,ip}^{(2)}][B_{ip}] + [C_{C_2C_2,ip}^{(2)}] \end{cases}.
$$

The matrices of Eqs. (7.172) are determined as:

$$\left[C^{(2)}_{C_2O,ip} \right] = \left[T_{C_2H_2,ip} \right]^T \left[C^{(h)}_{ip} \right] \left[T_{OH_2,ip} \right];$$

$$\left[C^{(2)}_{C_2C_1,ip} \right] = \left[C^{(2)}_{C_1C_2,ip} \right]^T = \left[T_{C_2H_2,ip} \right]^T \left[C^{(h)}_{ip} \right] \left[T_{C_1H_2,ip} \right]; \qquad (7.173)$$

$$\left[C^{(2)}_{C_2C_2,ip} \right] = \left[T_{C_2H_2,ip} \right]^T \left[C^{(h)}_{ip} \right] \left[T_{C_2H_2,ip} \right]$$

with the offset:

$$\Delta x_{C_2H_2} = l_2 / 2. \qquad (7.174)$$

The nonzero in-plane displacement and corresponding force vectors of the entire mechanism of Figure 7.18 are related as:

$$\left[u_{ip} \right] = \left[C_{ip} \right] \left[f_{ip} \right] \quad \text{with}$$

$$\left[u_{ip} \right] = \left[\begin{array}{cccc} u_{C_1y} & u_{C_2x} & u_{C_2y} & \theta_{C_2z} \end{array} \right]^T; \quad \left[f_{ip} \right] = \left[\begin{array}{cccc} f_{1y} & f_{2x} & f_{2y} & m_{2z} \end{array} \right]^T. \qquad (7.175)$$

The in-plane compliance matrix of Eq. (7.175) is:

$$\left[C_{ip} \right] = \left[\begin{array}{cccc} C^*_{C_1C_1,ip}(2,2) & C^*_{C_1C_2,ip}(2,1) & C^*_{C_1C_2,ip}(2,2) & C^*_{C_1C_2,ip}(2,3) \\ C^*_{C_2C_1,ip}(1,2) & C^*_{C_2C_2,ip}(1,1) & C^*_{C_2C_2,ip}(1,2) & C^*_{C_2C_2,ip}(1,3) \\ C^*_{C_2C_1,ip}(2,2) & C^*_{C_2C_2,ip}(2,1) & C^*_{C_2C_2,ip}(2,2) & C^*_{C_2C_2,ip}(2,3) \\ C^*_{C_2C_1,ip}(3,2) & C^*_{C_2C_2,ip}(3,1) & C^*_{C_2C_2,ip}(3,2) & C^*_{C_2C_2,ip}(3,3) \end{array} \right], \qquad (7.176)$$

with its components taken from the matrices of Eqs. (7.171) and (7.172).

The in-plane natural frequencies are calculated from the eigenvalues of the dynamic matrix $[D_{ip}] = [M_{ip}]^{-1}[C_{ip}]^{-1}$, and they are 23870, 3279, 364 and 69 rad/s.

Example 7.9

Calculate the out-of-plane natural frequencies of the mechanism pictured in Figure 7.18 of *Example 7.8*. Consider that the left rigid link is guided in translation along the z axis. Utilize all numerical values provided in *Example 7.8*.

Solution:

The out-of-plane motion of the mechanism is defined by five DOF, which are:

$$\left[u_{op} \right] = \left[\begin{array}{c} \left[u^{(1)}_{op} \right] \\ \left[u^{(2)}_{op} \right] \end{array} \right] \quad \text{with} \quad \begin{cases} \left[u^{(1)}_{op} \right] = \left[\begin{array}{cc} \theta_{C_1x} & u_{C_1z} \end{array} \right]^T \\ \left[u^{(2)}_{op} \right] = \left[\begin{array}{ccc} \theta_{C_2x} & \theta_{C_2y} & u_{C_2z} \end{array} \right]^T. \end{cases} \qquad (7.177)$$

Note that point C_1 is allowed rotation around the x axis, and therefore, there are two DOF at this point as indicated by $\left[u_{op}^{(1)}\right]$ in Eq. (7.177). Corresponding to these DOF, the following global-frame inertia matrix is formed:

$$\left[M_{op}\right]_{6\times6} = \begin{bmatrix} \left[M_{op}^{(r1)}\right]_{2\times2} & [0]_{2\times3} \\ [0]_{3\times2} & \left[M_{op}^{(r2)}\right]_{3\times3} \end{bmatrix} \text{ with:}$$

$$\left[M_{op}^{(r1)}\right] = \begin{bmatrix} J_{1x} & 0 \\ 0 & m_1 \end{bmatrix}; \quad \left[M_{op}^{(r2)}\right] = \begin{bmatrix} J_{2x} & 0 & 0 \\ 0 & J_{2y} & 0 \\ 0 & 0 & m_2 \end{bmatrix}.$$

(7.178)

The dimension of the full-mechanism out-of-plane compliance matrix is related to the DOF identified in Eq. (7.177); therefore, the compliance matrix has a 5×5 dimension. The matrix is formulated by formally applying two out-of-plane load vectors at C_1 and C_2 – see Figure 7.18, which identifies the generic loads – and evaluating the displacements at the two points. The out-of-plane displacements at point C_1 are:

$$\left[u_{C_1,op}\right] = \begin{bmatrix} \theta_{C_1x} \\ \theta_{C_1y} \\ u_{C_1z} \end{bmatrix} = \begin{bmatrix} \theta_{C_1x} \\ 0 \\ u_{C_1z} \end{bmatrix} = \left[C_{C_10,op}\right]\left[f_{0,op}\right] + \left[C_{C_1C_1,op}\right]\left[f_{1,op}\right] + \left[C_{C_1C_2,op}^{(2)}\right]\left[f_{2,op}\right]$$

with $\left[C_{C_10,op}\right] = \left[C_{C_10,op}^{(1)}\right] + \left[C_{C_10,op}^{(2)}\right]; \quad \left[C_{C_1C_1,op}\right]$

$$= \left[C_{C_1C_1,op}^{(1)}\right] + \left[C_{C_1C_1,op}^{(2)}\right]; \quad \left[f_{0,op}\right] = \begin{bmatrix} 0 \\ m_{Oy} \\ 0 \end{bmatrix}.$$

(7.179)

The new compliance matrices of Eq. (7.179) are:

$$\left[C_{C_1C_2,op}^{(2)}\right] = \left[T_{C_1H_2,op}\right]^T\left[C_{op}^{(h)}\right]\left[T_{C_2H_2,op}\right];$$

$$\left[C_{C_10,op}^{(1)}\right] = \left[T_{C_1H_1,op}\right]^T\left[C_{op}^{(h)}\right]\left[T_{OH_1,op}\right];$$

$$\left[C_{C_10,op}^{(2)}\right] = \left[T_{C_1H_2,op}\right]^T\left[C_{op}^{(h)}\right]\left[T_{OH_2,op}\right];$$

(7.180)

$$\left[C_{C_1C_1,op}^{(1)}\right] = \left[T_{C_1H_1,op}\right]^T\left[C_{op}^{(h)}\right]\left[T_{C_1H_1,op}\right];$$

$$\left[C_{C_1C_1,op}^{(2)}\right] = \left[T_{C_1H_2,op}\right]^T\left[C_{op}^{(h)}\right]\left[T_{C_1H_2,op}\right].$$

The out-of-plane compliance matrix $\left[C_{op}^{(h)}\right]$ of the flexible hinge is given in Eq. (7.156), and the translation matrices of Eqs. (7.180) are calculated with the x-offsets given in Eqs. (7.166), while the y-offsets are all zero. The reaction vector at O is formally expressed from Eq. (7.179) from the zero displacements as:

$$\left[f_{O,op}\right]=\left[A_{op}\right]\left[f_{1,op}\right]+\left[B_{op}\right]\left[f_{2,op}\right],\tag{7.181}$$

where

$$\left[A_{op}\right]=-\begin{bmatrix}0 & 0 & 0\\ C_{C_1C_1,op}(1,1) & C_{C_1C_1,op}(1,2) & C_{C_1C_1,op}(1,3)\\ 0 & 0 & 0\end{bmatrix}/C_{C_1O,op}(2,2);$$

$$\left[B_{op}\right]=-\begin{bmatrix}0 & 0 & 0\\ C^{(2)}_{C_1C_2,op}(1,1) & C^{(2)}_{C_1C_2,op}(1,2) & C^{(2)}_{C_1C_2,op}(1,3)\\ 0 & 0 & 0\end{bmatrix}/C_{C_1O,op}(2,2)$$

$$\tag{7.182}$$

The out-of-plane displacement vector at C_1 is expressed from Eqs. (7.179) and (7.181) as:

$$\left[u_{C_1,op}\right]=\left[C^*_{C_1C_1,op}\right]\left[f_{1,op}\right]+\left[C^*_{C_1C_2,op}\right]\left[f_{2,op}\right]\quad\text{with}$$

$$\begin{cases}\left[C^*_{C_1C_1,op}\right]=\left[C_{C_1O,op}\right]\left[A_{op}\right]+\left[C_{C_1C_1,op}\right]\\ \left[C^*_{C_1C_2,op}\right]=\left[C_{C_1O,op}\right]\left[B_{op}\right]+\left[C^{(2)}_{C_1C_2,op}\right].\end{cases}\tag{7.183}$$

The out-of-plane displacements at C_2 are calculated as:

$$\left[u_{C_2,op}\right]=\left[C^{(2)}_{C_2O,op}\right]\left[f_{O,op}\right]+\left[C^{(2)}_{C_2C_1,op}\right]\left[f_{1,op}\right]+\left[C^{(2)}_{C_2C_2,op}\right]\left[f_{2,op}\right]$$

$$=\left[C^*_{C_2C_1,op}\right]\left[f_{1,op}\right]+\left[C^*_{C_2C_2,op}\right]\left[f_{2,op}\right]\tag{7.184}$$

$$\text{with}\quad\begin{cases}\left[C^*_{C_2C_1,op}\right]=\left[C^{(2)}_{C_2O,op}\right]\left[A_{op}\right]+\left[C^{(2)}_{C_2C_1,op}\right]\\ \left[C^*_{C_2C_2,op}\right]=\left[C^{(2)}_{C_2O,op}\right]\left[B_{op}\right]+\left[C^{(2)}_{C_2C_2,op}\right].\end{cases}$$

The new matrices of Eq. (7.184) are:

$$\left[C^{(2)}_{C_2O,op}\right]=\left[T_{C_2H_2,op}\right]^T\left[C^{(h)}_{op}\right]\left[T_{OH_2,op}\right];$$

$$\left[C^{(2)}_{C_2C_1,op}\right]=\left[C^{(2)}_{C_1C_2,op}\right]^T=\left[T_{C_2H_2,op}\right]^T\left[C^{(h)}_{op}\right]\left[T_{C_1H_2,op}\right];\tag{7.185}$$

$$\left[C^{(2)}_{C_2C_2,op}\right]=\left[T_{C_2H_2,op}\right]^T\left[C^{(h)}_{op}\right]\left[T_{C_2H_2,op}\right]$$

The nonzero out-of-plane translation matrices of Eqs. (7.185) are calculated with the x-offsets of Eqs. (7.166) and zero y-axis offsets.

The out-of-plane force/displacement relationship for the mechanism is:

$$\left[u_{op} \right] = \left[C_{op} \right] \left[f_{op} \right] \quad \text{with}$$

$$\left[u_{op} \right] = \begin{bmatrix} \theta_{C_1 x} & u_{C_1 z} & \theta_{C_2 x} & \theta_{C_2 y} & u_{C_2 z} \end{bmatrix}^T ;$$

$$\left[f_{op} \right] = \begin{bmatrix} m_{C_1 x} & f_{C_1 z} & m_{C_2 x} & m_{C_2 y} & f_{C_2 z} \end{bmatrix}^T .$$

(7.186)

The compliance matrix of Eq. (7.186) is:

$$\left[C_{op} \right]$$

$$= \begin{bmatrix} C^*_{C_1 C_1, op}(1,1) & C^*_{C_1 C_1, op}(1,3) & C^*_{C_1 C_2, op}(1,1) & C^*_{C_1 C_2, op}(1,2) & C^*_{C_1 C_2, op}(1,3) \\ C^*_{C_1 C_1, op}(3,1) & C^*_{C_1 C_1, op}(3,3) & C^*_{C_1 C_2, op}(3,1) & C^*_{C_1 C_2, op}(3,2) & C^*_{C_1 C_2, op}(3,3) \\ C^*_{C_2 C_1, op}(1,1) & C^*_{C_2 C_1, op}(1,3) & C^*_{C_2 C_2, op}(1,1) & C^*_{C_2 C_2, op}(1,2) & C^*_{C_2 C_2, op}(1,3) \\ C^*_{C_2 C_1, op}(2,1) & C^*_{C_2 C_1, op}(2,3) & C^*_{C_2 C_2, op}(2,1) & C^*_{C_2 C_2, op}(2,2) & C^*_{C_2 C_2, op}(2,3) \\ C^*_{C_2 C_1, op}(3,1) & C^*_{C_2 C_1, op}(3,3) & C^*_{C_2 C_2, op}(3,1) & C^*_{C_2 C_2, op}(3,2) & C^*_{C_2 C_2, op}(3,3) \end{bmatrix}.$$

(7.187)

The components of $[C_{op}]$ are provided in Eqs. (7.183)–(7.185).

The out-of-plane eigenvalues of the dynamic matrix $[D_{op}] = [M_{op}]^{-1}[C_{op}]^{-1}$, with the inertia matrix of Eq. (7.178) and the compliance matrix of Eq. (7.187), result in the following natural frequencies: 202710, 82756, 26371, 4106 and 426 rad/s.

7.4.2 SPATIAL (3D) MECHANISMS

For a 3D series compliant mechanism that is free at one end and fixed at the opposite end, and which is formed of n heavy, moving rigid links, there are $6n$ independent coordinates that define the state of the mechanism, and so the mechanism has $6n$ DOF. Each rigid link i has six DOF associated with the three independent translations and three independent rotations at the link's centroid C_i; they are arranged into the displacement vectors:

$$\left[u_{C_i}^{(l)} \right] = \begin{bmatrix} u_{C_i x_i} & u_{C_i y_i} & \theta_{C_i z_i} & \theta_{C_i x_i} & \theta_{C_i y_i} & u_{C_i z_i} \end{bmatrix}^T$$

$$\left[u_{C_i} \right] = \begin{bmatrix} u_{C_i x} & u_{C_i y} & \theta_{C_i z} & \theta_{C_i x} & \theta_{C_i y} & u_{C_i z} \end{bmatrix}^T ,$$

(7.188)

where $\left[u_{C_i}^{(l)} \right]$ is the local-frame displacement vector and $\left[u_{C_i} \right]$ is the corresponding global-frame displacement vector.

The following local-frame, diagonal-form, inertia matrix can be formulated for the heavy link i:

$$\left[M^{(il)} \right] = \text{diag} \begin{pmatrix} m_i & m_i & J_{iz} & J_{ix} & J_{iy} & m_i \end{pmatrix},$$

(7.189)

where m_i is the mass and J_{ix}, J_{iy} and J_{iz} are the moments of inertia of the rigid link. The global-frame inertia matrix is obtained by means of a rotation as:

$$\left[M^{(i)} \right] = \left[R^{(i)} \right]^T \left[M^{(il)} \right] \left[R^{(i)} \right]. \tag{7.190}$$

The rotation matrix of Eq. (7.190) is calculated with Eq. (5.201). The full mechanism has a diagonal mass matrix, which combines the rigid-link matrices of Eq. (7.190):

$$\left[M \right]_{6n \times 6n} = \mathrm{diag}\left(\left[M^{(1)} \right]_{6 \times 6} \quad \left[M^{(2)} \right]_{6 \times 6} \quad \cdots \quad \left[M^{(i)} \right]_{6 \times 6} \quad \cdots \quad \left[M^{(n)} \right]_{6 \times 6} \right). \tag{7.191}$$

The compliance matrix of the 3D mechanism connects the displacements at the centers of the rigid links $\left[u_{C_i} \right]$ to the (assumed or actual) external loads $\left[f_{C_i} \right]$ at the same points in the form:

$$\begin{bmatrix} \left[u_{C_1} \right]_{6 \times 1} \\ \cdots \\ \left[u_{C_i} \right]_{6 \times 1} \\ \cdots \\ \left[u_{C_n} \right]_{6 \times 1} \end{bmatrix} = \begin{bmatrix} \left[C_{C_1 C_1} \right]_{6 \times 6} & \cdots & \left[C_{C_1 C_i} \right]_{6 \times 6} & \cdots & \left[C_{C_1 C_n} \right]_{6 \times 6} \\ \cdots & \cdots & \cdots & \cdots & \cdots \\ \left[C_{C_1 C_i} \right]_{6 \times 6}^T & \cdots & \left[C_{C_i C_i} \right]_{6 \times 6} & \cdots & \left[C_{C_i C_n} \right]_{6 \times 6} \\ \cdots & \cdots & \cdots & \cdots & \cdots \\ \left[C_{C_1 C_n} \right]_{6 \times 6}^T & \cdots & \cdots & \cdots & \left[C_{C_n C_n} \right]_{6 \times 6} \end{bmatrix} \begin{bmatrix} \left[f_{C_1} \right] \\ \cdots \\ \left[f_{C_i} \right] \\ \cdots \\ \left[f_{C_n} \right] \end{bmatrix}, \tag{7.192}$$

where the compliance submatrices are calculated following the rules applying to fixed-free mechanisms. The following *Example* studies a particular 3D flexible-hinge mechanism by providing the necessary calculation details.

Example 7.10

The fixed-free 3D Cartesian mechanism sketched in Figure 7.19 is formed of two heavy rigid links of lengths $l_1 = 0.05\,\mathrm{m}$, $l_2 = 0.04\,\mathrm{m}$, constant cross-sectional area defined by $W = 0.01\,\mathrm{m}$ (dimension in the xy-plane) and $H = 0.02\,\mathrm{m}$ (dimension parallel to the z axis). These links are connected by two identical, massless and straight-axis flexible hinges of length $l = 0.015\,\mathrm{m}$, diameter of constant circular cross-section $d = 0.002\,\mathrm{m}$, Young's modulus $E = 2 \cdot 10^{11}\,\mathrm{N/m^2}$ and Poisson's ratio $\mu = 0.3$. The rigid links have the mass density $\rho = 7{,}800\,\mathrm{kg/m^3}$. Calculate the natural frequencies of this mechanism.

Solution:

Because of its two rigid, mobile links, the mechanism has $6 \times 2 = 12$ DOF, which are the displacements of the two link centers:

$$\left[u_{C_i} \right] = \left[u_{C_i x} \quad u_{C_i y} \quad \theta_{C_i z} \quad \theta_{C_i x} \quad \theta_{C_i y} \quad u_{C_i z} \right]^T, \quad i = 1, 2. \tag{7.193}$$

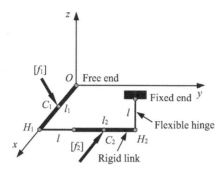

FIGURE 7.19 Fixed-free 3D Cartesian serial mechanism with two heavy rigid links and two massless flexible hinges.

The global-frame inertia matrix is:

$$[M]_{12\times12} = \begin{bmatrix} \left[M^{(r1)}\right]_{6\times6} & [0]_{6\times6} \\ [0]_{6\times6} & \left[M^{(r2)}\right]_{6\times6} \end{bmatrix}, \tag{7.194}$$

with the two inertia submatrices being expressed in Eqs. (7.189). The components of the local-frame, diagonal inertia submatrices of Eq. (7.194) are calculated as:

$$m_i = \rho WH l_i; \quad J_{iz} = J_{iy} = \frac{m_i l_i^2}{12}; \quad J_{ix} = \rho l_i \cdot \frac{W^2 + H^2}{12}, \quad i = 1,2. \tag{7.195}$$

The rotation matrices of Eq. (7.190), which are used to obtain the global-frame inertia matrices from the corresponding local-frame ones, are calculated based on Eq. (5.201) as:

$$\left[R^{(i)}\right] = \left[R^{(i)}_{\psi_i}\right]\left[R^{(i)}_{\theta_i}\right]\left[R^{(i)}_{\varphi_i}\right], \quad i = 1,2, \tag{7.196}$$

where

$$\left[R^{(1)}_{\psi_1}\right] = \left[R^{(1)}_{\theta_1}\right] = \left[R^{(1)}_{\varphi_1}\right] = \left[R^{(2)}_{\psi_2}\right] = \left[R^{(2)}_{\theta_2}\right] = [I]_{6\times6} \tag{7.197}$$

and $\left[R^{(2)}_{\varphi_2}\right]$ is determined from Eq. (5.200) with $\varphi_2 = \pi/2$.

As per the generic Eq. (7.192), the compliance matrix of the mechanism is:

$$[C]_{12\times12} = \begin{bmatrix} \left[C_{C_1C_1}\right]_{6\times6} & \left[C^{(2)}_{C_1C_2}\right]_{6\times6} \\ \left[C^{(2)}_{C_1C_2}\right]^T_{6\times6} & \left[C^{(2)}_{C_2C_2}\right]_{6\times6} \end{bmatrix}. \tag{7.198}$$

The matrix can formally be obtained by applying the two load vectors shown in Figure 7.19 and calculating the resulting displacements at C_1 and C_2. The compliance matrices of Eq. (7.198) are expressed as:

$$\left[C_{C_1 C_1} \right] = \left[C_{C_1 C_1}^{(1)} \right] + \left[C_{C_1 C_2}^{(2)} \right] \quad \text{with} \quad \begin{cases} \left[C_{C_1 C_1}^{(1)} \right] = \left[T_{C_1 H_1} \right]^T \left[R^{(2)} \right]^T \left[C^{(h)} \right] \left[R^{(2)} \right] \left[T_{C_1 H_1} \right] \\ \left[C_{C_1 C_1}^{(2)} \right] = \left[T_{C_1 H_2} \right]^T \left[R^{(h2)} \right]^T \left[C^{(h)} \right] \left[R^{(h2)} \right] \left[T_{C_1 H_2} \right] \end{cases}$$

$$\left[C_{C_1 C_2}^{(2)} \right] = \left[T_{C_1 H_2} \right]^T \left[R^{(h2)} \right]^T \left[C^{(h)} \right] \left[R^{(h2)} \right] \left[T_{C_2 H_2} \right]$$

$$\left[C_{C_2 C_2}^{(2)} \right] = \left[T_{C_2 H_2} \right]^T \left[R^{(h2)} \right]^T \left[C^{(h)} \right] \left[R^{(h2)} \right] \left[T_{C_2 H_2} \right]. \tag{7.199}$$

The translation matrices of Eqs. (7.199) are evaluated with Eq. (5.195) by means of the following corresponding offsets:

$$\begin{cases} \Delta x_{C_1 H_1} = l_1 / 2, \Delta y_{C_1 H_1} = 0, \Delta z_{C_1 H_1} = 0; \\ \Delta x_{C_1 H_2} = l_1 / 2, \Delta y_{C_1 H_2} = l + l_2 / 2, \Delta z_{C_1 H_2} = 0; \\ \Delta x_{C_2 H_2} = 0, \Delta y_{C_2 H_2} = l_2 / 2, \Delta z_{C_2 H_2} = 0; \end{cases} \tag{7.200}$$

The rotation matrix of Eq. (7.199) is calculated as:

$$\left[R^{(h2)} \right] = \left[R_\psi^{(h2)} \right] \left[R_\theta^{(h2)} \right] \left[R_\varphi^{(h2)} \right], \tag{7.201}$$

where

$$\left[R_\psi^{(h2)} \right] = \left[R_\varphi^{(h2)} \right] = [I]_{6 \times 6}, \tag{7.202}$$

and $\left[R_\theta^{(h2)} \right]$ is determined as in Eq. (5.199) with $\theta_2 = -\pi/2$.

The local-frame compliance matrix of the flexible hinge is calculated as:

$$\left[C^{(h)} \right] = \begin{bmatrix} C_{u_x - f_x} & 0 & 0 & 0 & 0 & 0 \\ 0 & C_{u_y - f_y} & C_{u_y - m_z} & 0 & 0 & 0 \\ 0 & C_{u_y - m_z} & C_{\theta_z - m_z} & 0 & 0 & 0 \\ 0 & 0 & 0 & C_{\theta_x - m_x} & 0 & 0 \\ 0 & 0 & 0 & 0 & C_{\theta_y - m_y} & C_{\theta_y - f_z} \\ 0 & 0 & 0 & 0 & C_{\theta_y - f_z} & C_{u_z - f_z} \end{bmatrix}, \tag{7.203}$$

where

$$C_{u_x - f_x} = \frac{l}{EA}, C_{u_y - f_y} = C_{u_z - f_z} = \frac{l^3}{3EI_z}, C_{u_y - m_z} = -C_{\theta_y - f_z}$$

$$= -\frac{l^2}{2EI_z}, C_{\theta_z - m_z} = C_{\theta_y - m_y} = \frac{l}{EI_z}, C_{\theta_x - m_x} = \frac{l}{GI_p} \tag{7.204}$$

$$\text{with}: \quad A = \frac{\pi d^2}{4}, I_z = \frac{\pi d^4}{64}, I_p = \frac{\pi d^4}{32}.$$

With the aid of Eqs. (7.194) and (7.198), the eigenvalues of the dynamic matrix $[D] = [M]^{-1}[C]^{-1}$ are calculated, which render the following natural frequencies: 53132, 51864, 7198, 7139, 4760, 2495, 2483, 1657, 687, 527, 193, 178 rad/s.

7.5 PARALLEL MECHANISMS

Several parallel mechanism configurations are analyzed in this section, including designs that are formed of one rigid link (platform) and either several hinges connected in parallel to the platform or several legs connected in parallel to the platform – the legs being serial combinations of multiple flexible hinges and rigid links. Discussed are mechanisms of planar structure with either in-plane or out-of-plane motion and mechanisms with spatial architecture and motion.

7.5.1 Planar (2D) Mechanisms

This section studies planar mechanisms comprising two or multiple flexible hinges/chains.

7.5.1.1 Straight-Axis Chain with Two Identical Flexible Hinges and One Rigid Link

The mechanism whose skeleton structure is shown in Figure 7.20 comprises two identical straight-axis flexible hinges connected to a middle rigid link. The three links are collinear and the mechanism is symmetric with respect to the y axis passing through its centroid. The free undamped mathematical model is formulated here for both the in-plane and out-of-plane vibrations by expressing the corresponding inertia and compliance matrices. The rigid link is assumed heavy, while the flexible hinges can be either massless or heavy. This problem is similar to the one of Section 7.4.1.1 that studies a mechanism with one rigid link and one flexible hinge. While there is no specific (numerical) example included here, the generic formulation can easily be applied to any straight-axis flexible-hinge mechanism configuration.

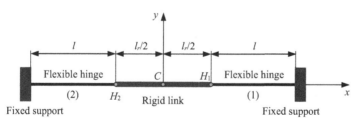

FIGURE 7.20 Skeleton representation of a basic parallel, collinear and symmetric mechanism formed of two identical straight-axis flexible hinges and one rigid link.

7.5.1.1.1 In-Plane Motion

The free undamped model of this system is expressed in Eq. (7.120). The inertia matrix $\left[M_{C,ip}^{(r)} \right]$ is expressed in Eq. (7.63) for the heavy rigid link. Contributions from the two flexible hinges can be added by means of either the consistent-inertia matrix of Eq. (7.53) or the lumped-inertia matrix of Eq. (7.56). These matrices are transferred to the frame $Cxyz$ from their local frames located at H_1 and H_2 as:

$$\begin{cases} \left[M_{C,ip}^{(1)} \right] = \left[T_{CH_1,ip} \right]^T \left[M_{ip}^{(h)} \right] \left[T_{CH_1,ip} \right] \\ \left[M_{C,ip}^{(2)} \right] = \left[T_{CH_2,ip} \right]^T \left[R^{(2)} \right]^T \left[M_{ip}^{(h)} \right] \left[R^{(2)} \right] \left[T_{CH_2,ip} \right], \end{cases} \tag{7.205}$$

where "1" denotes the hinge to the right of the rigid link in Figure 7.20 and "2" represents the other hinge. The translation matrices of Eq. (7.205) are calculated as:

$$\left[T_{CH_1,ip} \right] = \begin{bmatrix} 1 & 0 & 0 \\ 0 & 1 & 0 \\ 0 & -l_r/2 & 1 \end{bmatrix}; \quad \left[T_{CH_2,ip} \right] = \begin{bmatrix} 1 & 0 & 0 \\ 0 & 1 & 0 \\ 0 & l_r/2 & 1 \end{bmatrix}, \tag{7.206}$$

and the rotation matrix $[R^{(2)}]$ is determined using $\delta = \pi$. The mechanism inertia matrix adds the inertia matrix $\left[M_{C,ip}^{(r)} \right]$ of the rigid link to the two flexible-hinge inertia matrices of Eqs. (7.205), namely:

$$\left[M_{C,ip} \right] = \left[M_{C,ip}^{(r)} \right] + \left[M_{C,ip}^{(1)} \right] + \left[M_{C,ip}^{(2)} \right]. \tag{7.207}$$

The compliance matrix $[C_{C,ip}]$ results from the parallel combination of the two identical flexible hinges, which are also symmetric with respect to the y axis. The compliance of the half-mechanism situated to the right of y axis in Figure 7.20 is calculated as:

$$\left[C_{C,ip}^{(r)} \right] = \left[T_{CH_1,ip} \right]^T \left[C_{ip}^{(h)} \right] \left[T_{CH_1,ip} \right] \text{ with } \left[C_{ip}^{(h)} \right] = \begin{bmatrix} C_{u_x-f_x}^{(h)} & 0 & 0 \\ 0 & C_{u_y-f_y}^{(h)} & C_{u_y-m_z}^{(h)} \\ 0 & C_{u_y-m_z}^{(h)} & C_{\theta_z-m_z}^{(h)} \end{bmatrix}, \tag{7.208}$$

where the superscript "r" denotes right. The compliance matrix of the symmetric left half of the mechanism is formally obtained from the compliance matrix of the right half as:

$$\left[C_{C,ip}^{(l)} \right] = \begin{bmatrix} C_{C,ip}^{(r)}(1,1) & -C_{C,ip}^{(r)}(1,2) & C_{C,ip}^{(r)}(1,3) \\ -C_{C,ip}^{(r)}(1,2) & C_{C,ip}^{(r)}(2,2) & -C_{C,ip}^{(r)}(2,3) \\ C_{C,ip}^{(r)}(1,3) & -C_{C,ip}^{(r)}(2,3) & C_{C,ip}^{(r)}(3,3) \end{bmatrix}. \tag{7.209}$$

The full-mechanism compliance matrix at C is calculated as the parallel combination of the compliance matrices given in Eqs. (7.208) and (7.209):

$$\left[C_{C,ip}\right]=\left(\left[C_{C,ip}^{(r)}\right]^{-1}+\left[C_{C,ip}^{(l)}\right]^{-1}\right)^{-1}$$

$$=\begin{bmatrix} \dfrac{1}{2}\cdot C_{u_x-f_x}^{(h)} & 0 & 0 \\[4mm] 0 & \dfrac{1}{2}\cdot\left[C_{u_y-f_y}^{(h)}-\dfrac{\left(C_{u_y-m_z}^{(h)}\right)^2}{C_{\theta_z-m_z}^{(h)}}\right] & 0 \\[6mm] 0 & 0 & \dfrac{2\cdot\left(C_{\theta_z-m_z}^{(h)}\cdot C_{u_y-f_y}^{(h)}-\left(C_{u_y-m_z}^{(h)}\right)^2\right)}{4C_{u_y-f_y}^{(h)}-4l_r\cdot C_{u_y-m_z}^{(h)}+l_r^2\cdot C_{\theta_z-m_z}^{(h)}} \end{bmatrix}.$$

$$(7.210)$$

7.5.1.1.2 Out-of-Plane Motion

The mathematical model of the out-of-plane dynamics of this system is expressed in Eq. (7.124). The inertia matrix $\left[M_{C,op}^{(r)}\right]$ of the rigid link is provided in Eq. (7.63). Contributions of the two flexible hinges may be added by using either the consistent-inertia matrix of Eq. (7.54) or the lumped-inertia matrix of Eq. (7.56). These matrices need to be transferred from their local frames placed at H_1 and H_2 to the global frame at C as:

$$\begin{cases} \left[M_{C,op}^{(1)}\right]=\left[T_{CH_1,op}\right]^T\left[M_{op}^{(h)}\right]\left[T_{CH_1,op}\right] \\[3mm] \left[M_{C,op}^{(2)}\right]=\left[T_{CH_2,op}\right]^T\left[R^{(2)}\right]^T\left[M_{op}^{(h)}\right]\left[R^{(2)}\right]\left[T_{CH_2,op}\right], \end{cases} \quad (7.211)$$

where

$$\left[T_{CH_1,op}\right]=\begin{bmatrix} 1 & 0 & 0 \\ 0 & 1 & l_r/2 \\ 0 & 0 & 1 \end{bmatrix}; \quad \left[T_{CH_2,op}\right]=\begin{bmatrix} 1 & 0 & 0 \\ 0 & 1 & -l_r/2 \\ 0 & 0 & 1 \end{bmatrix}. \quad (7.212)$$

The full-mechanism inertia matrix is obtained by adding the rigid-link inertia matrix $\left[M_{C,op}^{(r)}\right]$ to the two hinges inertia matrices of Eqs. (7.211) as:

$$\left[M_{C,op}\right]=\left[M_{C,op}^{(r)}\right]+\left[M_{C,op}^{(1)}\right]+\left[M_{C,op}^{(2)}\right]. \quad (7.213)$$

The out-of-plane compliance matrix of the half-mechanism situated to the right of y axis in Figure 7.20 is formulated similarly to the in-plane compliance matrix as:

$$\left[C_{C,op}^{(r)} \right] = \left[T_{CH_1,op} \right]^T \left[C_{op}^{(h)} \right] \left[T_{CH_1,op} \right] \quad \text{with} \quad \left[C_{op}^{(h)} \right]$$

$$= \begin{bmatrix} C_{\theta_x-m_x}^{(h)} & 0 & 0 \\ 0 & C_{\theta_y-m_y}^{(h)} & C_{\theta_y-f_z}^{(h)} \\ 0 & C_{\theta_y-f_z}^{(h)} & C_{u_z-f_z}^{(h)} \end{bmatrix}. \tag{7.214}$$

The compliance matrix of the left half mechanism is calculated from the matrix defining the right half as:

$$\left[C_{C,op}^{(l)} \right] = \begin{bmatrix} C_{C,0p}^{(r)}(1,1) & -C_{C,op}^{(r)}(1,2) & C_{C,op}^{(r)}(1,3) \\ -C_{C,op}^{(r)}(1,2) & C_{C,op}^{(r)}(2,2) & -C_{C,op}^{(r)}(2,3) \\ C_{C,op}^{(r)}(1,3) & -C_{C,op}^{(r)}(2,3) & C_{C,op}^{(r)}(3,3) \end{bmatrix}. \tag{7.215}$$

The out-of-plane compliance matrix of the full mechanism is calculated as the parallel combination of the compliance matrices of the two halves – Eqs. (7.214) and (7.215):

$$\left[C_{C,op} \right] = \left(\left[C_{C,op}^{(r)} \right]^{-1} + \left[C_{C,op}^{(l)} \right]^{-1} \right)^{-1}$$

$$= \begin{bmatrix} \dfrac{1}{2} \cdot C_{\theta_x-m_x}^{(h)} & 0 & 0 \\ 0 & \dfrac{2 \cdot \left(C_{\theta_y-m_y}^{(h)} \cdot C_{u_z-f_z}^{(h)} - \left(C_{\theta_y-f_z}^{(h)} \right)^2 \right)}{4 C_{u_z-f_z}^{(h)} + 4 l_r \cdot C_{\theta_y-f_z}^{(h)} + l_r^2 \cdot C_{\theta_y-m_y}^{(h)}} & 0 \\ 0 & 0 & \dfrac{1}{2} \cdot \left[C_{u_z-f_z}^{(h)} - \dfrac{\left(C_{\theta_y-f_z}^{(h)} \right)^2}{C_{\theta_y-m_y}^{(h)}} \right] \end{bmatrix}. $$

$$\tag{7.216}$$

7.5.1.2 Mechanisms with Multiple Single-Hinge Chains

The prototype mechanism analyzed in this section is a planar device formed of a central circular rigid plate and several identical flexible hinges that are connected radially, symmetrically and in parallel to the plate. The central plate is heavy, whereas the flexible hinges are assumed massless. While a generic model can be developed to account for multiple flexible hinges, the following *Example* studies a configuration with three hinges, which is a common design.

Example 7.11

i. Calculate the out-of-plane natural frequencies of the compliant stage sketched in Figure 7.21 when the three identical and massless flexible hinges are of circular axis with a radius $R_1 = 0.03$ m, an opening angle $\alpha = 75°$, a constant rectangular cross-section with in-plane width $w = 0.005$ m, out-of-plane thickness $t = 0.001$ m and material properties: Young's modulus $E = 2 \cdot 10^{11}$ N/m^2 and Poisson's ratio $\mu = 0.33$. The rigid central plate is circular with a radius $R = 0.02$ m, a constant out-of-plane thickness $H = 0.012$ m and a mass density $\rho = 7,800$ kg/m^3;

ii. For an initial out-of-plane displacement at the center O of $[u_O] = [0\ 0\ 0.01]^T$ (where the nonzero displacement is measured in meters – m) and zero velocities at the same point, plot the time response of the z-axis displacement at O. Replot the time response of the same coordinate and for the same initial conditions when viscous damping acts on the rigid link, which is defined by the diagonal damping matrix: $[B_O] = \text{diag}(b, b, b_t)$ with $b_r = 10$ Nms and $b_t = 40$ Ns/m.

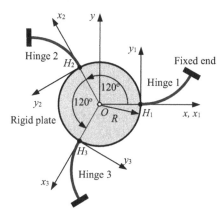

FIGURE 7.21 Top view of rigid heavy disk with three identical, massless, circular-axis flexible hinges with radial symmetry.

Solution:

i. The DOF of this mechanism describing the out-of-plane motion of the plate center of mass are θ_{Ox}, θ_{Oy} and u_{Oz}.

The stiffness matrix with respect to the rigid plate center O is calculated as:

$$[K_O] = [C_O^{(1)}]^{-1} + [C_O^{(2)}]^{-1} + [C_O^{(3)}]^{-1}, \qquad (7.217)$$

where the compliance matrices of the three flexible hinges are evaluated as:

$$\left[C_O^{(1)} \right] = \left[T_{OH_1} \right]^T \left[C_{H_1}^{(1)} \right] \left[T_{OH_1} \right]; \quad \left[C_O^{(2)} \right] = \left[R^{(2)} \right]^T \left[C_O^{(1)} \right] \left[R^{(2)} \right];$$

$$\left[C_O^{(3)} \right] = \left[R^{(3)} \right]^T \left[C_O^{(1)} \right] \left[R^{(3)} \right], \tag{7.218}$$

with the out-of-plane translation matrix being:

$$\left[T_{OH_1} \right] = \begin{bmatrix} 1 & 0 & -\Delta y_{OH_1} \\ 0 & 1 & \Delta x_{OH_1} \\ 0 & 0 & 1 \end{bmatrix}; \quad \Delta x_{OH_1} = R, \Delta y_{OH_1} = 0. \tag{7.219}$$

and the two rotation matrices calculated for $\delta_2 = 120°$ and $\delta_3 = 240°$. The out-of-plane compliance matrix of a circular-axis flexible hinge is expressed in Eqs. (2.109) and (2.110), and its components that correspond to the particular circular-axis design are provided in Eqs. (A2.27) through (A2.32).

The inertia matrix is related to the heavy center plate and for the three DOF is defined as:

$$[M_O] = \begin{bmatrix} J_x & 0 & 0 \\ 0 & J_y & 0 \\ 0 & 0 & m \end{bmatrix}; \quad m = \rho(\pi R^2) H, J_x = J_y = \frac{mR^2}{4}. \tag{7.220}$$

With the stiffness matrix of Eq. (7.217) and the inertia matrix of Eq. (7.220), the dynamic matrix $[D_O] = [M_O]^{-1}[K_O]$ is formed whose eigenvalues yield the following natural frequencies: 4,919 rad/s (double) and 1,502 rad/s. Note that both the stiffness matrix and inertia matrix are diagonal; moreover, the x- and y-components of both matrices are identical, and therefore, the natural frequencies corresponding to the x- and y-axis rotations are identical as well.

ii. For the given initial conditions, Eq. (7.9) becomes:

$$[U(s)] = s\left(s^2[M_O] + [K_O]\right)^{-1}[M_O][u(0)]. \tag{7.221}$$

Inverse Laplace transforming $[U(s)]$ of Eq. (7.221) yields the three time-domain displacements at the center O. Figure 7.22 plots the variation of u_{Oz} in terms of time.

When viscous damping is present and for the given initial conditions, Eq. (7.11) simplifies to:

$$[U(s)] = \left(s^2[M_O] + s[B_O] + [K_O]\right)^{-1}\left(s[M_O] + [B_O]\right)[u(0)]. \tag{7.222}$$

The disk center displacement vector $[u_O]$ is again found by inverse Laplace transforming $[U(s)]$ of Eq. (7.222), and Figure 7.23 is the plot of u_{Oz} as a function of time.

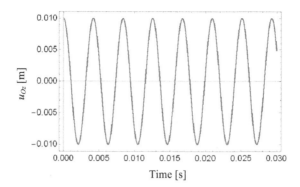

FIGURE 7.22 Time plot of rigid plate center z-axis displacement for the free undamped model.

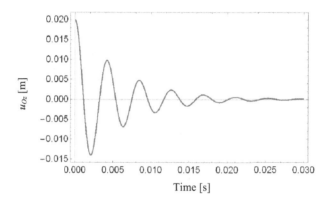

FIGURE 7.23 Time plot of rigid plate center z-axis displacement for the free damped model.

7.5.1.3 Mechanisms with Multiple Chains Formed of Serially Connected Rigid Links and Flexible Hinges

Figure 7.24a depicts a skeleton mechanism formed of a heavy rigid link (shown at the top of the figure) and m serial chains that are connected in parallel to the rigid link. Each serial chain is formed of several heavy rigid links and massless flexible hinges. The mechanism in this category can be planar or spatial.

The number of DOF is related to the number of rigid links; for the generic mechanism of Figure 7.24a, assuming that each of the m legs possesses a number n_j of rigid links, the total number of DOF is:

$$\#\mathrm{DOF} = 6 + 6\sum_{j=1}^{m} n_j = 6\left(1 + \sum_{j=1}^{m} n_j\right). \tag{7.223}$$

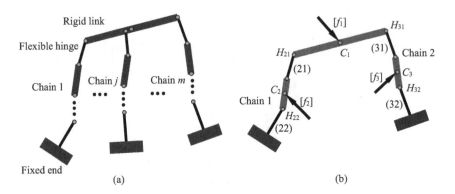

FIGURE 7.24 Skeleton representation of a mechanism formed of: (a) m serial chains connected in parallel to a rigid link; (b) two serial chains; each chain comprises two flexible hinges and one rigid link.

Equation (7.223) took into account that each rigid link (there are $1 + n$ such links) has six DOF associated with its center of mass three translations and three rotations in the 3D space.

Consider the particular configuration of Figure 7.24b, which has two serial chains and each chain has one rigid link; in this case, $n = 2$ and $n_1 = n_2 = 1$. As a consequence, Eq. (7.223) yields: #DOF = $6(1 + 1 + 1) = 18$. If the mechanism is planar, then nine DOF describe the in-plane motion and nine DOF characterize the out-of-plane motion. Let us formulate the compliance matrix and the inertia matrix of the mechanism shown in Figure 7.24b; while this configuration is particular, the resulting formulation can be extended/adapted to designs with a larger number of serial chains and chain rigid/flexible links. The derivation that follows is generic and so it can cover 3D, in-plane or out-of-plane loading/displacement.

7.5.1.3.1 Compliance Matrix

Assume the load vectors $[f_1]$, $[f_2]$ and $[f_3]$ are applied at the mass centers C_1, C_2 and C_3 of the three rigid links. The corresponding displacements at these points are the vectors $[u_1]$, $[u_2]$ and $[u_3]$. The aim is to derive the compliance matrix $[C]$ that connects the displacement vector $[u]$ to the force vector $[f]$ as:

$$[u] = [C][f] \quad \text{with}$$

$$[u] = \begin{bmatrix} [u_1] & [u_2] & [u_3] \end{bmatrix}^T ; \quad [f] = \begin{bmatrix} [f_1] & [f_2] & [f_3] \end{bmatrix}^T ;$$

$$[C] = \begin{bmatrix} [C_{11}] & [C_{12}] & [C_{13}] \\ [C_{21}] & [C_{22}] & [C_{23}] \\ [C_{31}] & [C_{32}] & [C_{33}] \end{bmatrix}$$

$$(7.224)$$

Assume that the force vector $[f_1]$ splits into two component vectors: $[f_{12}]$ acting on chain 1 and $[f_{13}]$ applied on chain 2 of the mechanism in Figure 7.24b such that:

$$[f_1] = [f_{12}] + [f_{13}]. \tag{7.225}$$

The two load vectors $[f_{12}]$ and $[f_{13}]$ are collocated at C_1. Formally, the displacement at point C_1 can be calculated in relation to either chain 1 or chain 2 as:

$$[u_1] = \left[C^{(2)}_{C_1C_1}\right][f_{12}] + \left[C^{(22)}_{C_1C_2}\right][f_2] = \left[C^{(3)}_{C_1C_1}\right][f_{13}] + \left[C^{(32)}_{C_1C_3}\right][f_3]$$

$$\text{with} \quad \begin{cases} \left[C^{(2)}_{C_1C_1}\right] = \left[C^{(21)}_{C_1C_1}\right] + \left[C^{(22)}_{C_1C_1}\right] \\[2mm] \left[C^{(3)}_{C_1C_1}\right] = \left[C^{(31)}_{C_1C_1}\right] + \left[C^{(32)}_{C_1C_1}\right] \end{cases}. \tag{7.226}$$

Combining Eqs. (7.225) and (7.226) allows solving for the two force vector components as:

$$\begin{bmatrix} [f_{12}] \\ [f_{13}] \end{bmatrix} = \begin{bmatrix} [A_{11}] & [A_{12}] & [A_{13}] \\ [A_{21}] & [A_{22}] & [A_{23}] \end{bmatrix} \begin{bmatrix} [f_1] \\ [f_2] \\ [f_3] \end{bmatrix}, \tag{7.227}$$

where

$$[A] = \begin{bmatrix} [A_{11}] & [A_{12}] & [A_{13}] \\ [A_{21}] & [A_{22}] & [A_{23}] \end{bmatrix}$$

$$= \begin{bmatrix} \left[C^{(2)}_{C_1C_1}\right] & -\left[C^{(3)}_{C_1C_1}\right] \\ [I] & [I] \end{bmatrix}^{-1} \begin{bmatrix} [0] & -\left[C^{(22)}_{C_1C_2}\right] & \left[C^{(32)}_{C_1C_3}\right] \\ [I] & [0] & [0] \end{bmatrix} \tag{7.228}$$

and $[I]$ is the identity matrix, while $[0]$ is the zero matrix – both of dimensions consistent with the other submatrices of Eq. (7.228). The displacement vector at C_1 can now be expressed by combining Eqs. (7.226) and (7.227):

$$[u_1] = [C_{11}][f_1] + [C_{12}][f_2] + [C_{13}][f_3] \quad \text{with}$$

$$[C_{11}] = \left[C^{(2)}_{C_1C_1}\right][A_{11}]; \quad [C_{12}] = \left[C^{(2)}_{C_1C_1}\right][A_{11}] + \left[C^{(22)}_{C_1C_1}\right]; \quad [C_{13}] = \left[C^{(2)}_{C_1C_1}\right][A_{13}]. \tag{7.229}$$

The displacement vector at C_2 is:

$$[u_2] = \left[C^{(22)}_{C_2C_1}\right][f_{12}] + \left[C^{(22)}_{C_2C_2}\right][f_2]. \tag{7.230}$$

With $[f_{12}]$ of Eq. (7.227), the displacement vector of Eq. (7.230) becomes:

$$[u_2] = [C_{21}][f_1] + [C_{22}][f_2] + [C_{23}][f_3] \quad \text{with}$$

$$[C_{21}] = \left[C_{C_2C_1}^{(22)}\right][A_{11}]; \quad [C_{22}] = \left[C_{C_2C_1}^{(22)}\right][A_{12}] + \left[C_{C_2C_2}^{(22)}\right]; \quad [C_{23}] = \left[C_{C_2C_1}^{(22)}\right][A_{13}].$$

(7.231)

Similarly, the displacement vector at C_3 is expressed as:

$$[u_3] = \left[C_{C_3C_1}^{(32)}\right][f_{13}] + \left[C_{C_3C_3}^{(32)}\right][f_3].$$

(7.232)

Substituting $[f_{13}]$ of Eq. (7.227) into Eq. (7.232), the displacement vector of this latter equation is:

$$[u_3] = [C_{31}][f_1] + [C_{32}][f_2] + [C_{33}][f_3] \quad \text{with}$$

$$[C_{31}] = \left[C_{C_3C_1}^{(32)}\right][A_{21}]; \quad [C_{32}] = \left[C_{C_3C_1}^{(32)}\right][A_{22}]; \quad [C_{33}] = \left[C_{C_3C_1}^{(32)}\right][A_{23}] + \left[C_{C_3C_3}^{(32)}\right].$$

(7.233)

Equations (7.229), (7.231) and (7.233) provide the nine compliance submatrices that form the compliance matrix of the full mechanism as per Eq. (7.224). Note that each compliance submatrix of Eq. (7.224) is square and of dimension 6×6 if the mechanism of Figure 7.24b is spatial; if the mechanism is planar, then the submatrices of Eq. (7.224) have a dimension of 3×3 for the in-plane formulation and an identical dimension for the out-of-plane model.

7.5.1.3.2   Inertia Matrix

The inertia matrix of the mechanism is diagonal and of the following form that is consistent with the displacement vector $[u]$ formulated in Eq. (7.224):

$$[M] = \begin{bmatrix} \left[M^{(r1)}\right] & [0] & [0] \\ [0] & \left[M^{(r2)}\right] & [0] \\ [0] & [0] & \left[M^{(r3)}\right] \end{bmatrix}.$$

(7.234)

The submatrices of Eq. (7.234) are related to the three rigid links; they are 6×6 diagonal matrices if the mechanism is spatial. For a planar mechanism, the submatrices are 3×3 diagonal for both the in-plane model and the out-of-plane model.

With the compliance matrix of Eq. (7.224) and the inertia matrix of Eq. (7.234) – both matrices are formulated in the global frame – the eigenvalues and the corresponding natural frequencies are calculated based on the dynamic matrix $[D] = [M]^{-1}[C]^{-1}$.

Example 7.12

Calculate the in-plane natural frequencies of the planar parallelogram mechanism shown in Figure 7.25 as a skeleton representation. Consider that the two serial chains are identical and parallel, being also perpendicular to the rigid link centered at C_1. The identical, massless flexible hinges are of right circular design with a radius $r = 0.008\,m$, an in-plane minimum thickness $t = 0.001\,m$, an out-of-plane constant width $w = 0.006\,m$ and a material Young's modulus $E = 2 \cdot 10^{11}\,N/m^2$. The rigid links are defined by the lengths $l_1 = 0.04\,m$ and $l_2 = 0.05\,m$, constant cross-sectional area $A_t = 0.0001\,m^2$ and mass density $\rho = 7,800\,kg/m^3$.

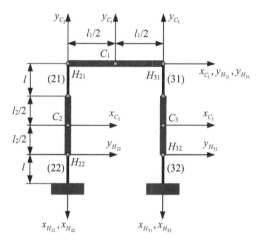

FIGURE 7.25   Skeleton representation of a planar parallel mechanism with two identical serial chains; each chain comprises two identical massless, straight-axis right circular flexible hinges and one heavy rigid link.

Solution:

The actual components of the three in-plane displacement vectors identified in Eq. (7.224) are:

$$[u_i] = \begin{bmatrix} u_{ix} & u_{iy} & \theta_{iz} \end{bmatrix}^T, \quad i = 1, 2, 3. \tag{7.235}$$

The four basic compliance matrices of Eqs. (7.226) are calculated as:

$$\left[C_{C_1C_1}^{(21)} \right] = \left[T_{C_1H_{21}} \right]^T [R]^T \left[C^{(h)} \right][R] \left[T_{C_1H_{21}} \right]; \quad \left[C_{C_1C_1}^{(22)} \right] = \left[T_{C_1H_{22}} \right]^T [R]^T \left[C^{(h)} \right][R] \left[T_{C_1H_{22}} \right];$$

$$\left[C_{C_1C_1}^{(31)} \right] = \left[T_{C_1H_{31}} \right]^T [R]^T \left[C^{(h)} \right][R] \left[T_{C_1H_{31}} \right]; \quad \left[C_{C_1C_1}^{(32)} \right] = \left[T_{C_1H_{32}} \right]^T [R]^T \left[C^{(h)} \right][R] \left[T_{C_1H_{32}} \right].$$

$$\tag{7.236}$$

The rotation matrix $[R]$ is calculated using an angle $\delta = -\pi/2$. The in-plane translation matrices of Eq. (7.236) are evaluated with the following offsets:

$$
\begin{cases}
\Delta x_{C_1 H_{21}} = -l_1 / 2; \quad \Delta y_{C_1 H_{21}} = 0; \\[4pt]
\Delta x_{C_1 H_{22}} = -l_1 / 2; \quad \Delta y_{C_1 H_{22}} = -(l + l_2); \\[4pt]
\Delta x_{C_1 H_{31}} = l_1 / 2; \quad \Delta y_{C_1 H_{31}} = 0; \\[4pt]
\Delta x_{C_1 H_{32}} = l_1 / 2; \quad \Delta y_{C_1 H_{32}} = -(l + l_2)
\end{cases}
\quad \text{with} \quad l = 2r. \tag{7.237}
$$

The other two compliance matrices of Eqs. (7.226) are formulated as:

$$
\begin{aligned}
\left[C_{C_1 C_2}^{(22)} \right] &= \left[T_{C_1 H_{22}} \right]^T [R]^T \left[C^{(h)} \right] [R] \left[T_{C_2 H_{22}} \right]; \\[4pt]
\left[C_{C_1 C_3}^{(32)} \right] &= \left[T_{C_1 H_{32}} \right]^T [R]^T \left[C^{(h)} \right] [R] \left[T_{C_3 H_{32}} \right].
\end{aligned} \tag{7.238}
$$

The new translation matrices of Eq. (7.238) have the following offsets:

$$
\Delta x_{C_2 H_{22}} = \Delta x_{C_3 H_{32}} = 0; \quad \Delta y_{C_2 H_{22}} = \Delta y_{C_3 H_{32}} = -l_2 / 2. \tag{7.239}
$$

The flexible-hinge, in-plane compliance matrix $[C^{(h)}]$ is calculated by means of Eqs. (A3.1). Also, note that:

$$
\left[C_{C_3 C_1}^{(32)} \right]^T = \left[C_{C_1 C_3}^{(32)} \right]; \quad \left[C_{C_3 C_3}^{(32)} \right] = \left[C_{C_2 C_2}^{(22)} \right]. \tag{7.240}
$$

Equations (7.236) through (7.240) enable to calculate the compliance matrix of Eq. (7.224) using all other necessary equations formulated in the theoretical section.

The inertia matrices of Eq. (7.234) are calculated as:

$$
\left[M^{(r1)} \right] = \begin{bmatrix} m_1 & 0 & 0 \\ 0 & m_1 & 0 \\ 0 & 0 & J_{1z} \end{bmatrix} \quad \text{with} \quad m_1 = \rho A_r l_1; \quad J_{1z} = \frac{m_1 l_1^2}{12} \tag{7.241}
$$

and

$$
\left[M^{(r2)} \right] = \left[M^{(r3)} \right] = [R]^T \begin{bmatrix} m_2 & 0 & 0 \\ 0 & m_2 & 0 \\ 0 & 0 & J_{2z} \end{bmatrix} [R] \quad \text{with} \quad m_2 = \rho A_r l_2; \quad J_{2z} = \frac{m_2 l_2^2}{12} \tag{7.242}
$$

The global-frame mass matrices of the identical rigid links 2 and 3 needed rotations since their x local directions (not shown in Figure 7.25) are along the vertical direction.

With the provided numerical values, the following natural frequencies are obtained: 199984, 132922, 66620, 60529, 59857, 54578, 32233, 26153 and 716 rad/s.

Example 7.13

The planar mechanism sketched in Figure 7.26 is formed of two heavy rigid links and two pairs of identical straight-axis massless flexible hinges. The mechanism is symmetric with respect to the y axis. A sinusoidal force $f = F \cdot \sin(\omega t)$ is applied at the center of mass C_1 of the bottom rigid link along the symmetry axis. Study the amplitude of the y displacement of C_1 by means of the frequency magnitude plot relating the input f to the output u_{1y}. Known are $l_1 = 0.02$ m, $l_2 = 0.012$ m, the constant rectangular cross-section of the four flexible hinges with in-plane thickness $t = 0.001$ m and out-of-plane width $w = 0.006$ m, Young's modulus $E = 2 \cdot 10^{11}$ N/m², the rigid link masses $m_1 = 0.2$ kg, $m_2 = 0.08$ kg, and the lengths $l_{r1} = 0.05$ m, $l_{r2} = 0.02$ m, $h = 0.03$ m. Also, consider viscous damping acting on the two rigid links with the coefficients: $b_{t1} = 20$ Ns/m, $b_{r1} = 8$ Nms, $b_{t2} = 15$ Ns/m and $b_{r2} = 6$ Nms.

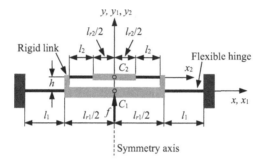

FIGURE 7.26 Skeleton representation of a planar symmetric mechanism with two rigid links and two pairs of identical hinges under sinusoidal input force.

Solution:

With respect to its in-plane motion, the mechanism has six DOF, which are the x, y translations and the z rotations at the rigid links' mass centers C_1 and C_2, namely:

$$[u] = \begin{bmatrix} [u_1] & [u_2] \end{bmatrix}^T; \quad \text{with} \quad \begin{cases} [u_1] = \begin{bmatrix} u_{1x} & u_{1y} & \theta_{1z} \end{bmatrix}^T \\ [u_2] = \begin{bmatrix} u_{2x} & u_{2y} & \theta_{2z} \end{bmatrix}^T \end{cases}. \tag{7.243}$$

To evaluate the transfer function matrix as per Eq. (7.16), the overall stiffness, inertia and damping matrices need to be determined.

The inertia matrix is formulated as:

$$[M] = \begin{bmatrix} \left[M^{(r1)} \right]_{3\times3} & [0]_{3\times3} \\ [0]_{3\times3} & \left[M^{(r2)} \right]_{3\times3} \end{bmatrix} \quad \text{with} \quad \left[M^{(ri)} \right] = \begin{bmatrix} m_i & 0 & 0 \\ 0 & m_i & 0 \\ 0 & 0 & J_{iz} \end{bmatrix}, \quad i = 1,2$$

$$\tag{7.244}$$

where $J_{iz} = m_i (l_{ri})^2 / 12$.

In order to formulate the mechanism compliance matrix $[C]$, which is the inverse of the stiffness matrix $[K]$, assume that two fictitious in-plane load vectors $[f_1]$ and $[f_2]$ – which are not shown in Figure 7.26 – are applied at C_1 and C_2, respectively. The displacements $[u_1]$ at C_1 and $[u_2]$ at C_2 are expressed as:

$$[u] = [C][f] \quad \text{or}$$

$$
\begin{bmatrix} [u_1] \\ [u_2] \end{bmatrix} = \begin{bmatrix} \left[C_{C_1}^{(1)}\right] + \left[T_{C_1C_2}\right]^T \left[C_{C_2}^{(2)}\right]\left[T_{C_1C_2}\right] & \left[T_{C_1C_2}\right]^T \left[C_{C_2}^{(2)}\right] \\ \left[C_{C_2}^{(2)}\right]\left[T_{C_1C_2}\right] & \left[C_{C_2}^{(2)}\right] \end{bmatrix} \begin{bmatrix} [f_1] \\ [f_2] \end{bmatrix}. \tag{7.245}
$$

Note that the displacement/load and the compliance matrix of Eq. (7.245) are expressed in the global frame xyz. Due to symmetry, the compliance matrix $\left[C_{C_1}^{(1)}\right]$ results from the parallel combinations of the two identical flexible hinges of length l_1, while $\left[C_{C_2}^{(2)}\right]$ results similarly as the parallel combinations of the two identical flexible hinges of length l_2. These compliance matrices are calculated based on Eq. (7.210) by appropriately using the dimensions of hinge 1 or hinge 2. The local-frame compliances of the two hinges are:

$$C_{u_x-f_x}^{(i)} = \frac{l_i}{EA}; \quad C_{u_y-f_y}^{(i)} = \frac{l_i^3}{3EI_z}; \quad C_{u_y-m_z}^{(i)} = -\frac{l_i^2}{2EI_z}; \quad C_{\theta_z-m_z}^{(i)} = \frac{l_i}{EI_z}, \quad i = 1,2$$

$$\tag{7.246}$$

$$\text{with} \quad A = wt; \quad I_z = \frac{wt^3}{12}.$$

The translation matrix of Eq. (7.245) is calculated with the offsets: $\Delta x_{C_1C_2} = 0$, $\Delta y_{C_1C_2} = h$.

The damping matrix is formulated similarly to the mass matrix of Eq. (7.244), namely:

$$[B] = \begin{bmatrix} [B_1]_{3\times3} & [0]_{3\times3} \\ [0]_{3\times3} & [B_2]_{3\times3} \end{bmatrix} \quad \text{with} \quad [B_i] = \begin{bmatrix} b_{ti} & 0 & 0 \\ 0 & b_{ti} & 0 \\ 0 & 0 & b_{ri} \end{bmatrix}, \quad i = 1,2. \tag{7.247}$$

The transfer function matrix $[G(s)]$ is obtained as per Eq. (7.16) by means of $[M]$ of Eq. (7.244), $[C]$ of Eqs. (7.245), (7.246) and $[B]$ of Eq. (7.247). Because the only force acting on the system is $f_{1y} = f$, the transfer function of interest relates the Laplace-domain displacement $U_{1y}(s)$ to the force $F_{1y}(s) = F(s)$. This transfer function is $G_{22}(s)$ – the component on the second row and second column in the matrix $[G(s)]$. The magnitude plot of the function $G_{22}(\omega j)$, which is the complex-form counterpart of $G_{22}(s)$, is graphed in Figure 7.27. Note that the natural frequencies of the mechanism of Figure 7.27 can also be determined from the

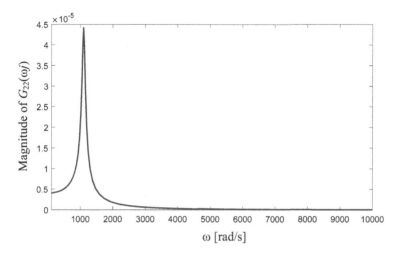

FIGURE 7.27   Frequency plot of the magnitude of $G_{22}(\omega j)$, $|G_{22}(\omega j)| = U_{2y}/F$.

eigenvalues of the dynamic matrix $[D] = ([C] \cdot [M])^{-1}$ – they are 207267, 49169, 4625, 3388, 1261 and 1103 rad/s. Figure 7.27 captures one prominent peak in the frequency response, which is relatively close to the smallest natural frequency.

7.5.2   SPATIAL (3D) MECHANISMS

The spatial configuration of a mechanism with a central rigid link and three identical flexible hinges that are connected in parallel to the central rigid link is studied next as an example illustrating the dynamic modeling approach to 3D parallel mechanisms.

Example 7.14

The 3D mechanism sketched in Figure 7.28a is formed of a heavy, disk-shaped rigid link of radius $R = 0.05$ m, out-of-plane thickness $h = 0.02$ m and mass density $\rho = 7,800 \, \text{kg/m}^3$. Three identical, massless, straight-axis, right circularly corner-filleted flexible hinges of circular cross-sections are connected in a radial symmetric manner to the disk periphery. The hinges are defined by the length $l = 0.03$ m, minimum diameter $d = 0.0015$ m, filet radius $r = 0.005$ m and material Young's modulus $E = 2 \cdot 10^{11} \, \text{N/m}^2$. The hinges, which are fixed at the opposite end, have a 3D inclination with respect to the central global frame $Oxyz$ defined by the Euler's angles $\varphi = 10°$, $\theta = 80°$ and $\psi = 0°$ – Figure 7.28b indicates the rotation angles φ and θ for the hinge 1. Calculate the natural frequencies of this mechanism.

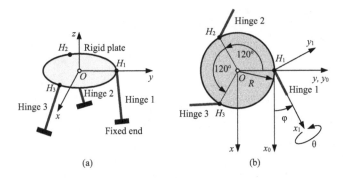

FIGURE 7.28 (a) Skeleton representation of 3D mechanism with central rigid heavy link and three identical straight-axis flexible hinges; (b) top view of mechanism with geometry and topology.

Solution:

The mechanism has six DOF, which are the disk center translations and rotations arranged in the vector:

$$[u_O] = \begin{bmatrix} u_{Ox} & u_{Oy} & \theta_{Oz} & \theta_{Ox} & \theta_{Oy} & u_{Oz} \end{bmatrix}^T. \tag{7.248}$$

The stiffness matrix is calculated with respect to the global frame $Oxyz$ as:

$$[K_O] = \begin{bmatrix} C_O^{(1)} \end{bmatrix}^{-1} + \begin{bmatrix} C_O^{(2)} \end{bmatrix}^{-1} + \begin{bmatrix} C_O^{(3)} \end{bmatrix}^{-1}, \tag{7.249}$$

where $\begin{bmatrix} C_O^{(1)} \end{bmatrix}, \begin{bmatrix} C_O^{(2)} \end{bmatrix}$ and $\begin{bmatrix} C_O^{(3)} \end{bmatrix}$ are the compliance matrices of the three hinges referenced to the global frame. The global-frame compliance matrix of the flexible hinge 1 is evaluated as:

$$\begin{bmatrix} C_O^{(1)} \end{bmatrix} = \begin{bmatrix} T_{OH_1} \end{bmatrix}^T \begin{bmatrix} C_{H_1}^{(1)} \end{bmatrix} \begin{bmatrix} T_{OH_1} \end{bmatrix}, \tag{7.250}$$

The translation matrix of Eq. (7.250) is:

$$[T_{OH_1}] = \begin{bmatrix} 1 & 0 & 0 & 0 & 0 & 0 \\ 0 & 1 & 0 & 0 & 0 & 0 \\ \Delta y_{OH_1} & -\Delta x_{OH_1} & 1 & 0 & 0 & 0 \\ 0 & \Delta z_{OH_1} & 0 & 1 & 0 & -\Delta y_{OH_1} \\ -\Delta z_{OH_1} & 0 & 0 & 0 & 1 & \Delta x_{OH_1} \\ 0 & 0 & 0 & 0 & 0 & 1 \end{bmatrix} \quad \text{with} \quad \begin{cases} \Delta x_{OH_1} = R; \\ \Delta y_{OH_1} = 0; \\ \Delta z_{OH_1} = 0. \end{cases} \tag{7.251}$$

The compliance matrix $\begin{bmatrix} C_{H_1}^{(1)} \end{bmatrix}$ is calculated in a local reference frame whose x_1 axis is along the longitudinal direction of the flexible hinge and which has been rotated by the three Euler angles. This compliance matrix is expressed as:

$$\begin{bmatrix} C_{H_1}^{(1)} \end{bmatrix} = [R]^T \begin{bmatrix} C^{(h)} \end{bmatrix} [R] \quad \text{with} \quad [R] = \begin{bmatrix} R_\psi \end{bmatrix} \begin{bmatrix} R_\theta \end{bmatrix} \begin{bmatrix} R_\varphi \end{bmatrix} = \begin{bmatrix} R_\theta \end{bmatrix} \begin{bmatrix} R_\varphi \end{bmatrix} \tag{7.252}$$

because $[R_\psi] = [I]$; the matrices $[R_\theta]$ and $[R_\phi]$ are provided in Eqs. (5.199) and (5.200). Figure 7.28b identifies the original longitudinal axis x_0 of the hinge 1. A rotation ϕ around the z_0 axis (which is placed at H_1 and is parallel to the global axis z) results in the frame $H_1 x_1 y_1 z_1$, also shown in Figure 7.28b – the longitudinal axis of the hinge 1 is now along the x_1 axis. Eventually, a rotation θ is used around the x_1 axis, which produces the final position of hinge 1. The matrix $[C^{(h)}]$ is the local-frame compliance matrix of the flexible hinge and is expressed as:

$$[C^{(h)}] = \begin{bmatrix} C_{u_x-f_x} & 0 & 0 & 0 & 0 & 0 \\ 0 & C_{u_y-f_y} & C_{u_y-m_z} & 0 & 0 & 0 \\ 0 & C_{u_y-m_z} & C_{\theta_z-m_z} & 0 & 0 & 0 \\ 0 & 0 & 0 & C_{\theta_x-m_x} & 0 & 0 \\ 0 & 0 & 0 & 0 & C_{\theta_z-m_z} & C_{u_y-m_z} \\ 0 & 0 & 0 & 0 & C_{u_y-m_z} & C_{u_z-f_z} \end{bmatrix}, \quad (7.253)$$

and its components are evaluated as in Section 3.2.2.3.2. The compliance matrices $[C_O^{(2)}]$ and $[C_O^{(3)}]$ are calculated by rotating the compliance matrix $[C_O^{(1)}]$ of Eq. (7.250) as:

$$[C_O^{(j)}] = [R^{(j)}]^T [C_O^{(1)}][R^{(j)}] \quad \text{with}$$

$$[R^{(j)}] = \begin{bmatrix} \cos\delta_j & \sin\delta_j & 0 & 0 & 0 & 0 \\ -\sin\delta_j & \cos\delta_j & 0 & 0 & 0 & 0 \\ 0 & 0 & 1 & 0 & 0 & 0 \\ 0 & 0 & 0 & \cos\delta_j & \sin\delta_j & 0 \\ 0 & 0 & 0 & -\sin\delta_j & \cos\delta_j & 0 \\ 0 & 0 & 0 & 0 & 0 & 1 \end{bmatrix}, \quad j = 2,3, \quad (7.254)$$

with $\delta_2 = 2\pi/3$ and $\delta_3 = 4\pi/3$.

The inertia matrix associated with the DOF of Eq. (7.248) is:

$$[M_O] = \begin{bmatrix} m & 0 & 0 & 0 & 0 & 0 \\ 0 & m & 0 & 0 & 0 & 0 \\ 0 & 0 & J_z & 0 & 0 & 0 \\ 0 & 0 & 0 & J_x & 0 & 0 \\ 0 & 0 & 0 & 0 & J_y & 0 \\ 0 & 0 & 0 & 0 & 0 & m \end{bmatrix} \quad \text{with} \quad \begin{cases} m = \rho(\pi R^2)h \\ J_x = J_y = \dfrac{mR^2}{4} \\ J_z = \dfrac{mR^2}{2}. \end{cases} \quad (7.255)$$

For the numerical values of this example, the following natural frequencies are obtained from the eigenvalues of the dynamic matrix $[D_O] = [M_O]^{-1}[K_O]$: 4,261 rad/s (double), 1,611 rad/s, 653 rad/s (double) and 353 rad/s.

REFERENCES

1. Shabana, A.A., *Dynamics of Multibody Systems*, Cambridge University Press, New York, 1998.
2. Boyer, F. and Coiffet, P., Symbolic modeling of a flexible manipulator via assembling of its generalized Newton-Euler model, *Mechanism and Machine Theory*, 31 (1), 45, 1996.
3. Vibet, C., Dynamics modeling of Lagrangian mechanisms from inertial matrix elements, *Computer Methods in Applied Mechanical Engineering*, 123, 317, 1995.
4. Surdilovic, D. and Vukobratovic, M., One method for efficient dynamic modeling of flexible manipulators, *Mechanism and Machine Theory*, 31 (3), 297, 1996.
5. Yu, W., Mathematical modeling of a class of flexible robot, *Applied Mathematical Modelling*, 19, 537, 1995.
6. Indri, M. and Tornambe, A., Lyapunov analysis of the approximate motion equations of flexible structures, *Systems & Control Letters*, 28, 31, 1996.
7. Boutaghou, Z.E. and Erdman, A.G., On various nonlinear rod theories for the dynamic analysis of multi-body systems, in: A.G. Erdman (ed.), *Modern Kinematics – Developments in the Last Forty Years*, John Wiley & Sons, Hoboken, NJ, 1993.
8. Simo, J.C. and Vu-Quoc, L., On the dynamics of flexible beams under large overall motions – the plane case: part 1 and part 2, *ASME Journal of Applied Mechanics*, 53, 849, 1986.
9. McPhee, J.J., Automatic generation of motion equations for planar mechanical systems using the new set of "branch coordinates", *Mechanism and Machine Theory*, 33 (6), 805, 1998.
10. Liew, K.M., Lee, S.E., and Liu, A.Q., Mixed-interface substructures for dynamic analysis of flexible multibody systems, *Engineering Structures*, 18 (7), 495, 1996.
11. Engstler, C. and Kaps, P., A comparison of one-step methods for multibody system dynamics in descriptor and state space form, *Applied Numerical Mathematics*, 24, 457, 1997.
12. Cui, K. and Haque, I., Symbolic equations of motion for hybrid multibody systems using a matrix-vector formulation, *Mechanism and Machine Theory*, 32 (6), 743, 1997.
13. Smith, S.T., *Flexures – Elements of Elastic Mechanisms*, Gordon and Breach Science Publishers, Amsterdam, 2000.
14. Rosner, M., Lammering, R., and Friedrich, R., Dynamic modeling and model order reduction of compliant mechanisms, *Precision Engineering*, 42, 85, 2015.
15. Gao, X., Solution methods for dynamic response of flexible mechanisms, in: A.G. Erdman (ed.), *Modern Kinematics – Developments in the Last Forty Years*, John Wiley & Sons, Hoboken, NJ, 1993.
16. Bagci, C. and Streit, D.A., Flexible manipulators, in: A.G. Erdman (ed.), *Modern Kinematics – Developments in the Last Forty Years*, John Wiley & Sons, Hoboken, NJ, 1993.
17. Tian, Y., Shirinzadeh, B., and Zhang, D., Design and dynamics of a 3-DOF flexure-based parallel mechanism for micro/nano manipulation, *Microelectronic Engineering*, 87 (2), 230, 2010.
18. Ling, M., A general two-port dynamic stiffness model and static/dynamic comparison for three bridge-type flexure displacement amplifiers, *Mechanical Systems and Signal Processing*, 119, 486, 2019.
19. Du, Y., Li, T., and Gao, G., Dynamic analysis of a flexure-based compliant stage, *Journal of Mechanical Sciences and Technology*, 32 (11), 5223, 2018.
20. Wang, F., Ma, Z., and Gao, W., Dynamic modeling and control of a novel XY positioning stage for semiconductor packaging, *Transactions of the Institute of Measurement and Control*, 37 (2), 177, 2015.

21. Polit, S. and Dong, J., Development of a high-bandwidth XY nanopositioning stage for high-rate micro-nanomanufacturing, *IEEE/ASME Transactions on Mechatronics*, 16 (4), 724, 2011.

22. Lin, J. et al., Design, analysis and testing of a new piezoelectric tool actuator for elliptical vibration turning, *Smart Materials and Structures*, 26 (8), 85008, 2017.

23. Meirovitch, L., *Principles and Techniques of Vibrations*, Prentice-Hall, Englewood Cliffs, NJ, 1997.

24. Den Hartog, J.P., *Mechanical Vibrations*, Dover, New York, 1985.

25. Thompson, W.T. and Dillon Dahleh, M., *Theory of Vibrations with Applications*, 5th ed., Prentice Hall, Englewood Cliffs, NJ, 1998.

26. Rao, S.S., *Mechanical Vibrations*, 5th ed., Prentice Hall, Englewood Cliffs, NJ, 2011.

27. Inman, D., *Engineering Vibration*, 4th ed., Pearson, Upper Saddle River, NJ, 2014.

28. Lobontiu, N., Goldfarb, M., and Garcia, M., Achieving maximum tip displacement during resonant excitation of piezoelectrically actuated beams, *Journal of Intelligent Material Systems and Structures*, 10, 900, 1999.

29. Lobontiu, N., Distributed-parameter dynamic model and optimized design of a four-link pendulum with flexure hinges, *Mechanism and Machine Theory*, 36, 653, 2001.

30. Lobontiu, N. and Garcia, E., Lumped-parameter inertia model for flexure hinges, *Mechanics Based Design of Structures and Machines*, 32 (1), 73, 2004.

31. Lobontiu, N., *Mechanical Microresonators: Modeling, Design, and Applications*, McGraw-Hill, New York, 2005.

32. Lobontiu, N., *Dynamics of Microelectromechanical Systems*, Springer, Boston, MA, 2007.

33. Lazan, B.J., *Damping of Materials and Members in Structural Mechanics*, Pergamon Press, Oxford, 1968.

34. Nashif, A.D., Jones, D.I.G., and Henderson, J.P., *Vibration Damping*, John Wiley & Sons, Hoboken, NJ, 1985.

35. Rivin, E.I., *Stiffness and Damping in Mechanical Design*, Marcel Dekker, New York, 1999.

8 Finite Element Analysis of Flexible-Hinge Mechanisms

This chapter utilizes the finite element method by means of the regular displacement approach to formulate element stiffness and inertia matrices of straight- and circular-axis flexible-hinge line elements. The elements possess axial, torsional and bending capabilities – the latter feature is modeled for both long configurations by means of the Euler–Bernoulli model and short designs that use the Timoshenko model. Flexible-hinge mechanisms are subsequently studied by means of their quasi-static finite element models. The free natural and undamped response of hinge mechanisms is also studied, as well as the forced response of compliant devices. Finite element flexible-hinge mechanism examples include serial, parallel, planar and three-dimensional configurations. The approach applied in this chapter is basic, and it follows the established linear finite element formulation, such as in Petyt [1], Reddy [2], Zienkiewicz et al. [3], Moaveni [4], Madenci and Guven [5] or Yang [6].

8.1 STRAIGHT-AXIS LINE ELEMENTS

Finite element models have been developed for straight-axis, non-prismatic flexible hinges, such as those studied by Gerardin and Cardona [7], Franciosi and Mecca [8], Jiang and Henshall [9], while works by Koster [10], Zhang and Fasse [11], Murin and Kutis [12] or Lobontiu and Garcia [13] placed an emphasis on the right circular design. Two-node or three-node line elements can be utilized to model a straight-axis flexible hinge and substitute the regular two- or three-dimensional meshing of finite element conventional software, as illustrated in Figure 8.1. Such a substitution sensibly reduces the problem dimension and complexity by decreasing the multiple degrees of freedom (DOF) of a two- or three-dimensional finite element mesh. This section formulates the stiffness and inertia matrices for two-/three-node, straight-axis line elements that are sensitive to axial, torsional and bending loads. Bending capabilities of both long beams (Euler–Bernoulli) and short beams (Timoshenko) are covered.

8.1.1 TWO-NODE ELEMENT

The stiffness and inertia matrices of a straight-line element with two end nodes are studied in this section; the element can deform axially, torsionally and in bending. The procedure of obtaining consistent and lumped elemental inertia

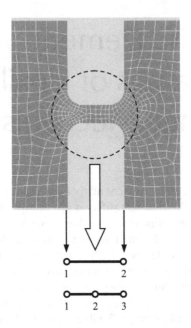

FIGURE 8.1 Straight-axis flexible hinge as a single-, two- or three-node, line element replacing the mesh formed of multiple 2D elements produced with conventional finite element software.

matrices is identical to the one utilized in Section 7.2.2.2 to derive inertia matrices of straight-axis flexible hinges by means of compliances; therefore, a detailed derivation of the inertia matrices is not reproduced here.

8.1.1.1 Axial Loading

The variable cross-section line element of Figure 8.2 has a length l and two end nodes 1 and 2, whose axial displacements are u_{1x} and u_{2x}.

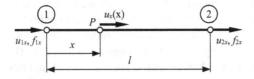

FIGURE 8.2 Two-node line element with axial deformation/motion capabilities.

The element nodal displacement vector and the corresponding nodal force vector are:

$$\left[u_x^{(e)}\right] = \begin{bmatrix} u_{1x} & u_{2x} \end{bmatrix}^T; \quad \left[f_x^{(e)}\right] = \begin{bmatrix} f_{1x} & f_{2x} \end{bmatrix}^T. \tag{8.1}$$

A generic point P placed at a distance x from the origin (identical to node 1) experiences an axial displacement $u_x(x)$ produced by the nodal loading. Because this element has two DOF (u_{1x} and u_{2x}), the displacement $u_x(x)$ can be expressed as polynomial in x with two constants (the number of constants is equal to the number of DOF) as:

$$u_x(x) = c_0 + c_1 \cdot x. \tag{8.2}$$

Using the boundary conditions:

$$u_x(0) = u_{1x}; \quad u_x(l) = u_{2x} \tag{8.3}$$

allows solving for the constants c_0 and c_1; the constants are substituted into Eq. (8.2), which becomes:

$$u_x(x,t) = \left(1 - \frac{x}{l}\right) \cdot u_{1x}(t) + \left(\frac{x}{l}\right) \cdot u_{2x}(t) \quad \text{or} \quad u_x(x,t) = [S_a(x)]^T \left[u_x^{(e)}(t)\right]. \tag{8.4}$$

where

$$[S_a(x)] = \begin{bmatrix} S_{a1}(x) & S_{a2}(x) \end{bmatrix}^T \quad \text{with} \quad \begin{cases} S_{a1}(x) = 1 - \dfrac{x}{l} \\[2mm] S_{a2}(x) = \dfrac{x}{l} \end{cases}. \tag{8.5}$$

$[S_a(x)]$ is the *shape-function vector*. Equation (8.4) is known as the *discretization* (or *interpolation*) equation; it expresses the continuous function $u_x(x, t)$ as a sum of products between the space-dependent shape functions and their corresponding time-dependent nodal displacements. The shape functions of Eq. (8.5) are also the ones provided in Eq. (7.68) in the context of axial vibrations of a free-free flexible hinge by the compliance model.

8.1.1.1.1 Element Stiffness Matrix

The strain energy that is stored in the line element of Figure 8.2 is as:

$$U_a^{(e)} = \frac{1}{2} \cdot E \int_0^l A(x) \left(\frac{\partial u_x(x,t)}{\partial x}\right)^2 dx. \tag{8.6}$$

The space derivative of Eq. (8.6) results from Eq. (8.4) as:

$$\frac{\partial u_x(x,t)}{\partial x} = \left[\frac{dS_a(x)}{dx}\right]^T \left[u_x^{(e)}(t)\right]. \tag{8.7}$$

By using the rules of matrix algebra, the following relationship can be formulated based on Eq. (8.7):

$$\left(\frac{\partial u_x(x,t)}{\partial x}\right)^2 = \left[u_x^{(e)}(t)\right]^T \left[\frac{dS_a(x)}{dx}\right]\left[\frac{dS_a(x)}{dx}\right]^T \left[u_x^{(e)}(t)\right]. \tag{8.8}$$

Substitution of Eq. (8.8) into Eq. (8.6) yields:

$$U_a^{(e)} = \frac{1}{2}\cdot\left[u_x^{(e)}(t)\right]^T\left(E\int_0^l A(x)\left[\frac{dS_a(x)}{dx}\right]\left[\frac{dS_a(x)}{dx}\right]^T dx\right)\left[u_x^{(e)}(t)\right]. \tag{8.9}$$

The strain (elastic) energy of a deformable system, which is defined by a generalized-coordinate vector $[u]$, is expressed as:

$$U = \frac{1}{2}[u]^T[K][u]. \tag{8.10}$$

where $[K]$ is the stiffness matrix. Comparing Eqs. (8.9) and (8.10) under the assumptions that the two systems (the finite element one and the general system) are identical and that the nodal displacement vector $\left[u_x^{(e)}(t)\right]$ is identical to the generalized-coordinate vector $[u]$, it follows that the 2×2 axially related element stiffness matrix is:

$$\left[K_a^{(e)}\right] = E\cdot\int_0^l A(x)\left[\frac{dS_a(x)}{dx}\right]\left[\frac{dS_a(x)}{dx}\right]^T dx. \tag{8.11}$$

For a constant cross-section bar element with $A(x) = A$, the elemental stiffness matrix becomes:

$$\left[K_a^{(e)}\right] = \begin{bmatrix} \dfrac{E\cdot A}{l} & -\dfrac{E\cdot A}{l} \\ -\dfrac{E\cdot A}{l} & \dfrac{E\cdot A}{l} \end{bmatrix} = \frac{E\cdot A}{l}\cdot\begin{bmatrix} 1 & -1 \\ -1 & 1 \end{bmatrix}. \tag{8.12}$$

after substituting the shape functions of Eq. (8.5) into Eq. (8.11).

8.1.1.1.2 Element Inertia Matrix

The element inertia matrix can be expressed in either consistent or lumped-parameter form, as detailed in Section 7.2.

The *consistent-inertia matrix* $\left[M_a^{(e)}\right]$ of a variable cross-section axial element is given in Eq. (7.67). For a constant cross-section member, the inertia matrix is provided in Eq. (7.69) where $m = \rho\cdot A\cdot l$ (the mass of the bar element) should be used instead of m_h, the mass of the flexible hinge.

Dividing the total element mass into two equal fractions and placing them at the two end nodes result in the following axial *lumped-inertia matrix*:

$$\left[M_a^{(e)} \right] = \frac{1}{2} \cdot \rho \left(\int_0^l A(x)\,dx \right) \cdot \begin{bmatrix} 1 & 0 \\ 0 & 1 \end{bmatrix}, \tag{8.13}$$

where the element cross-sectional area is variable. For a constant cross-section bar, with $A(x) = A$, Eq. (8.13) reduces to:

$$\left[M_a^{(e)} \right] = \begin{bmatrix} \dfrac{m}{2} & 0 \\ 0 & \dfrac{m}{2} \end{bmatrix} = \frac{m}{2} \cdot \begin{bmatrix} 1 & 0 \\ 0 & 1 \end{bmatrix}, \tag{8.14}$$

which is also captured in the in-plane inertia matrix of a flexible hinge in Eq. (7.94).

8.1.1.2 Torsional Loading

Figure 8.3 depicts a variable cross-section, two-node line element of length l, which is sensitive to torsional loading and deformation.

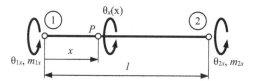

FIGURE 8.3 Two-node line element with torsional deformation and rotary motion capabilities.

The nodal displacement and load vectors are:

$$\left[\theta_x^{(e)} \right] = \begin{bmatrix} \theta_{1x} & \theta_{2x} \end{bmatrix}^T; \quad \left[m_x^{(e)} \right] = \begin{bmatrix} m_{1x} & m_{2x} \end{bmatrix}^T \tag{8.15}$$

where "m" denotes the moment. Because the element of Figure 8.3 has two DOF (θ_{1x} and θ_{2x}), the torsional deformation at an arbitrary point P is expressed as:

$$\theta_x(x,t) = \left[S_a(x) \right]^T \left[\theta_x^{(e)}(t) \right] \tag{8.16}$$

with the shape functions of Eq. (8.5).

8.1.1.2.1 Element Stiffness Matrix

The torsion strain energy of the element shown in Figure 8.3 is expressed as:

$$U_t^{(e)} = \frac{1}{2} \cdot G \int_0^l I_t(x) \left(\frac{\partial \theta_x(x,t)}{\partial x} \right)^2 dx, \tag{8.17}$$

where $I_t(x)$ is the torsional moment of area with respect to the centroidal axis. Following a derivation similar to the one used for the axial deformation of a line element, the 2×2 torsion-related element stiffness matrix is calculated as:

$$[K_t^{(e)}] = G \cdot \int_0^l I_t(x) \left[\frac{dS_a(x)}{dx} \right] \left[\frac{dS_a(x)}{dx} \right]^T dx. \tag{8.18}$$

When the element has a constant cross-section, which means $I_t(x) = I_t$, Eq. (8.18) becomes:

$$[K_t^{(e)}] = \begin{bmatrix} \dfrac{G \cdot I_t}{l} & -\dfrac{G \cdot I_t}{l} \\ -\dfrac{G \cdot I_t}{l} & \dfrac{G \cdot I_t}{l} \end{bmatrix} = \frac{G \cdot I_t}{l} \cdot \begin{bmatrix} 1 & -1 \\ -1 & 1 \end{bmatrix}. \tag{8.19}$$

8.1.1.2.2　Element Inertia Matrix

The element inertia matrix can be expressed as a consistent matrix or as a lumped matrix. Following the approach described in Section 7.2.2.2, the elemental consistent-inertia matrix $[M_t^{(e)}]$ is the one provided in Eq. (7.73). For constant cross-section with $I_t(x) = I_t$, the inertia matrix becomes the one given in Eq. (7.74), where $J_x = \rho\, I_t \cdot l$ is the inertia mass moment of the bar element with respect to the centroidal axis and should be used instead of $J_{h,x}$ – the axial moment of inertia of the flexible hinge.

The lumped-inertia element matrix results by dividing the total mass moment of inertia into two equal fractions at the two nodes:

$$[M_t^{(e)}] = \frac{1}{2} \cdot \rho \left(\int_0^l I_t(x)\,dx \right) \cdot \begin{bmatrix} 1 & 0 \\ 0 & 1 \end{bmatrix}. \tag{8.20}$$

For a constant cross-section bar, with $I_t(x) = I_t$, Eq. (8.20) becomes:

$$[M_t^{(e)}] = \begin{bmatrix} \dfrac{J_x}{2} & 0 \\ 0 & \dfrac{J_x}{2} \end{bmatrix} = \frac{J_x}{2} \cdot \begin{bmatrix} 1 & 0 \\ 0 & 1 \end{bmatrix}, \tag{8.21}$$

which is also expressed in Eq. (7.95) as part of the out-of-plane, flexible-hinge inertia matrix.

8.1.1.3　Bending

Two different models are regularly employed to derive element stiffness and inertia matrices for beam elements: the Euler–Bernoulli model applies to long beams (with the length approximately three to five times larger than the largest cross-sectional dimension) and the Timoshenko model relates to short beams. The Timoshenko model considers the additional effects of shear deformation and rotary inertia.

8.1.1.3.1 Long Beams and the Euler–Bernoulli Model

Bending around the z axis is studied first based on the two-node line element of Figure 8.4a. The element has a variable cross-section and a length l.

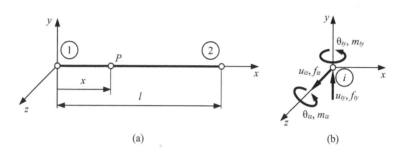

(a)   (b)

FIGURE 8.4   (a) Two-node beam element; (b) bending-related nodal displacements and forces/moments at node i ($i = 1, 2$).

At an arbitrary point P, the relevant deformation parameters in the xy plane (which correspond to bending around the z axis) are the deflection $u_y(x)$ along the y axis and the slope (rotation) $\theta_z(x)$ around the z axis. Correspondingly, the DOF at nodes 1 and 2 are the deflections and slopes that are collected in the element nodal displacement vector $\left[u_{bz}^{(e)} \right]$; similarly, the nodal load vector $\left[f_{bz}^{(e)} \right]$ is also formed. The two vectors (whose components are identified in Figure 8.4b) are:

$$\left[u_{bz}^{(e)} \right] = \left[\begin{array}{cccc} u_{1y} & \theta_{1z} & u_{2y} & \theta_{2z} \end{array} \right]^T; \quad \left[f_{bz}^{(e)} \right] = \left[\begin{array}{cccc} f_{1y} & m_{1z} & f_{2y} & m_{2z} \end{array} \right]^T.$$

$$(8.22)$$

The subscript "bz" indicates bending around the z axis. Because of the four nodal DOF u_{1y}, θ_{1z}, u_{2y} and θ_{2z}, the deflection can be interpolated as a polynomial with four coefficients. The deflection and slope (which is the deflection derivative with respect to x) are expressed as:

$$\begin{cases} u_y(x) = c_0 + c_1 \cdot x + c_2 \cdot x^2 + c_3 \cdot x^3; \\ \theta_z(x) = \dfrac{du_y(x)}{dx} = c_1 + 2c_2 \cdot x + 3c_3 \cdot x^2 \end{cases}$$

$$(8.23)$$

The following boundary conditions are used in combination with Eqs. (8.23):

$$u_y(0) = u_{1y}; \quad \theta_z(0) = \theta_{1z}; \quad u_y(l) = u_{2y}; \quad \theta_z(l) = \theta_{2z} \qquad (8.24)$$

to solve for the constants c_0 through c_3; the constants are then substituted back into Eq. (8.23). As a result, the deflection $u_y(x)$ is expressed as in Eqs. (7.86) and (7.87), namely:

$$u_y(x,t) = [S_b(x)]^T [u_{bz}^{(e)}(t)]; \quad [S_b(x)] = \begin{bmatrix} S_{b1}(x) & S_{b2}(x) & S_{b3}(x) & S_{b4}(x) \end{bmatrix}^T,$$
(8.25)

with the shape functions of Eqs. (7.44) and (7.85), which are:

$$S_{b1}(x) = 1 - 3 \cdot \frac{x^2}{l^2} + 2 \cdot \frac{x^3}{l^3}; \quad S_{b2}(x) = x - 2 \cdot \frac{x^2}{l} + \frac{x^3}{l^2};$$

$$S_{b3}(x) = 3 \cdot \frac{x^2}{l^2} - 2 \cdot \frac{x^3}{l^3}; \quad S_{b4}(x) = -\frac{x^2}{l} + \frac{x^3}{l^2}.$$
(8.26)

The elastic strain energy stored in the beam element of Figure 8.4 is:

$$U_{bz}^{(e)} = \frac{1}{2} \cdot E \int_0^l I_z(x) \left(\frac{\partial \theta_z(x,t)}{\partial x} \right)^2 dx = \frac{1}{2} \cdot E \int_0^l I_z(x) \left(\frac{\partial^2 u_y(x,t)}{\partial x^2} \right)^2 dx. \quad (8.27)$$

The discretization Eq. (8.25) results in:

$$\left(\frac{\partial^2 u_y(x,t)}{\partial x^2} \right)^2 = [u_{bz}^{(e)}(t)]^T \left[\frac{d^2 S_b(x)}{dx^2} \right] \left[\frac{d^2 S_b(x)}{dx^2} \right]^T [u_{bz}^{(e)}(t)]. \quad (8.28)$$

which enables expressing Eq. (8.27) as:

$$U_{bz}^{(e)} = \frac{1}{2} \cdot [u_{bz}^{(e)}(t)]^T \left(E \int_0^l I_z(x) \left[\frac{d^2 S_b(x)}{dx^2} \right] \left[\frac{d^2 S_b(x)}{dx^2} \right]^T dx \right) [u_{bz}^{(e)}(t)]. \quad (8.29)$$

By comparing Eq. (8.10), which expresses the strain energy of a generic deformable system defined by the generalized-coordinate vector $[u]$, to Eq. (8.29), it follows that the 4×4 bending-related elemental stiffness matrix is:

$$[K_{bz}^{(e)}] = E \int_0^l I_z(x) \left[\frac{d^2 S_b(x)}{dx^2} \right] \left[\frac{d^2 S_b(x)}{dx^2} \right]^T dx. \quad (8.30)$$

For a constant cross-section beam element, with $I_z(x) = I_z$, Eq. (8.30) becomes:

$$[K_{bz}^{(e)}] = \frac{2E \cdot I_z}{l^3} \begin{bmatrix} 6 & 3l & -6 & 3l \\ 3l & 2l^2 & -3l & l^2 \\ -6 & -3l & 6 & -3l \\ 3l & l^2 & -3l & 2l^2 \end{bmatrix}. \quad (8.31)$$

The z-axis, bending-related inertia matrices can, again, be formulated in consistent or lumped forms. The consistent element inertia matrix of the line element sketched in Figure 8.4a is calculated similarly to the procedure described in Section 7.2.2.2 and is expressed in Eq. (7.88). For a constant cross-section beam (with $A(x) = A$), the inertia matrix becomes the one of Eq. (7.89), where $m = \rho \cdot A \cdot l$ is the total mass of the element that should be used instead of the flexible-hinge mass m_h.

The total mass of the element can be divided between the two nodal translation DOF, and the resulting lumped-inertia matrix is:

$$\left[M_{bz}^{(e)} \right] = \frac{1}{2} \cdot \rho \left(\int_0^l A(x)\,dx \right) \begin{bmatrix} 1 & 0 & 0 & 0 \\ 0 & 0 & 0 & 0 \\ 0 & 0 & 1 & 0 \\ 0 & 0 & 0 & 0 \end{bmatrix}. \tag{8.32}$$

For a constant cross-section beam, Eq. (8.32) becomes:

$$\left[M_{bz}^{(e)} \right] = \begin{bmatrix} \dfrac{m}{2} & 0 & 0 & 0 \\ 0 & 0 & 0 & 0 \\ 0 & 0 & \dfrac{m}{2} & 0 \\ 0 & 0 & 0 & 0 \end{bmatrix}, \tag{8.33}$$

which is also included in the compliance-based, in-plane inertia matrix of a flexible hinge in Eq. (7.95).

Similar results and element matrices are obtained when bending around the y axis is assessed due to forces aligned with the z axis and moments oriented along the y axis – see Figure 8.4b. The out-of-plane nodal displacement vector and the corresponding load vector are:

$$\left[u_{by}^{(e)} \right] = \begin{bmatrix} u_{1z} & \theta_{1y} & u_{2z} & \theta_{2y} \end{bmatrix}^T ; \left[f_{by}^{(e)} \right] = \begin{bmatrix} f_{1z} & m_{1y} & f_{2z} & m_{2y} \end{bmatrix}^T. \tag{8.34}$$

The resulting 4×4 element stiffness matrix is similar to the one of Eq. (8.30), namely:

$$\left[K_{by}^{(e)} \right] = E \int_0^l I_y(x) \left[\frac{d^2 S_b(x)}{dx^2} \right] \left[\frac{d^2 S_b(x)}{dx^2} \right]^T dx. \tag{8.35}$$

For a constant cross-section beam element, the stiffness matrix of Eq. (8.31) should be used with I_y instead of I_z. The inertia matrices resulting from bending around the y axis are identical to those modeling the bending around the z axis.

8.1.1.3.2   Short Beams and the Timoshenko Model

Short beams where the length is less than three to five times the maximum cross-sectional dimension need to include in their model shear deformations and rotary inertia. For bending around the z axis and a constant cross-section beam the deflection and slope are connected by means of the following equilibrium equation – see Richards [14]:

$$\alpha_s GA\left(\theta_z - \frac{du_y(x)}{dx}\right) - EI_z \cdot \frac{d^2\theta_z(x)}{dx^2} = 0, \tag{8.36}$$

where α_s is the shear coefficient, which depends on the cross-section shape and is also studied/utilized in Chapters 2–4. Note that some finite element texts are using the symbol κ for the shear coefficient. Because of the connection Eq. (8.36), the slope cannot be obtained directly as the space derivative of the deflection; the deflection and slope are expressed separately as in Reddy [15]:

$$\begin{cases} u_y(x) = c_0 + c_1 x + c_2 x^2 + c_3 x^3 \\ \theta_z(x) = c_4 + c_5 x + c_6 x^2 \end{cases}. \tag{8.37}$$

By calculating the derivatives of $u_y(x)$ and $\theta_z(x)$ from Eqs. (8.37) and substituting them into Eq. (8.36), a second-degree algebraic equation in x results. Because the polynomial associated with it has to be zero for all acceptable values of x, the only possibility is that the three coefficients be zero. This leads to:

$$\begin{cases} c_4 = c_1 + 6\beta_z c_3 \\ c_5 = 2c_2 \\ c_6 = 3c_3 \end{cases}, \tag{8.38}$$

where

$$\beta_z = \frac{EI_z}{\alpha_s GA}. \tag{8.39}$$

By enforcing the four boundary conditions associated with the deflections and slopes at the element's nodes 1 and 2 – Eqs. (8.24), two sets of shape functions are obtained, which result in the discretized deflection and slope:

$$\begin{cases} u_y(x,t) = \left[S_{u_y}(x) \right]^T \left[u_{bz}^{(e)}(t) \right] \\ \theta_z(x,t) = \left[S_{\theta_z}(x) \right]^T \left[u_{bz}^{(e)}(t) \right] \end{cases}. \tag{8.40}$$

where $\left[u_{bz}^{(e)}(t) \right]$ is defined in Eq. (8.22) and the two shape-function vectors are:

$$
\begin{cases}
\left[S_{u_y}(x) \right] = \begin{bmatrix} S_{u_y 1}(x) & S_{u_y 2}(x) & S_{u_y 3}(x) & S_{u_y 4}(x) \end{bmatrix}^T \\[2mm]
\left[S_{\theta_z}(x) \right] = \begin{bmatrix} S_{\theta_z 1}(x) & S_{\theta_z 2}(x) & S_{\theta_z 3}(x) & S_{\theta_z 4}(x) \end{bmatrix}^T
\end{cases}
\tag{8.41}
$$

The shape functions associated with u_y in Eqs. (8.40) and (8.41) are:

$$
\begin{cases}
S_{u_y 1}(x) = 1 - \dfrac{12\beta_z l}{l^2\left(12\beta_z + l^2\right)} \cdot x - \dfrac{3l^2}{l^2\left(12\beta_z + l^2\right)} \cdot x^2 + \dfrac{2l}{l^2\left(12\beta_z + l^2\right)} \cdot x^3 \\[4mm]
S_{u_y 2}(x) = \dfrac{6\beta_z + l^2}{12\beta_z + l^2} \cdot x - \dfrac{2\left(3\beta_z + l^2\right)}{l\left(12\beta_z + l^2\right)} \cdot x^2 + \dfrac{1}{12\beta_z + l^2} \cdot x^3 \\[4mm]
S_{u_y 3}(x) = \dfrac{12\beta_z}{l\left(12\beta_z + l^2\right)} \cdot x + \dfrac{3}{12\beta_z + l^2} \cdot x^2 - \dfrac{2}{l\left(12\beta_z + l^2\right)} \cdot x^3 \\[4mm]
S_{u_y 4}(x) = -\dfrac{6\beta_z}{12\beta_z + l^2} \cdot x + \dfrac{6\beta_z - l^2}{l\left(12\beta_z + l^2\right)} \cdot x^2 + \dfrac{1}{12\beta_z + l^2} \cdot x^3,
\end{cases}
\tag{8.42}
$$

and the ones forming the shape-function vector corresponding to θ_z are:

$$
\begin{cases}
S_{\theta_z 1}(x) = -\dfrac{6}{12\beta_z + l^2} \cdot x + \dfrac{6}{l\left(12\beta_z + l^2\right)} \cdot x^2 \\[4mm]
S_{\theta_z 2}(x) = 1 - \dfrac{4\left(3\beta_z + l^2\right)}{l\left(12\beta_z + l^2\right)} \cdot x + \dfrac{3}{12\beta_z + l^2} \cdot x^2 \\[4mm]
S_{\theta_z 3}(x) = \dfrac{6}{12\beta_z + l^2} \cdot x - \dfrac{6}{l\left(12\beta_z + l^2\right)} \cdot x^2 \\[4mm]
S_{\theta_z 4}(x) = \dfrac{2\left(6\beta_z - l^2\right)}{l\left(12\beta_z + l^2\right)} \cdot x + \dfrac{3}{12\beta_z + l^2} \cdot x^2.
\end{cases}
\tag{8.43}
$$

The strain energy stored in the short element includes contributions from both bending and shear in the form – see Petyt [1] and Richards [14], for instance:

$$
U_{bz}^{(e)} = \frac{1}{2} \cdot E \int_0^l I_z(x)\left(\frac{\partial \theta_z(x,t)}{\partial x} \right)^2 dx + \frac{\alpha_s G}{2} \int_0^l A(x)\left(\theta_z^{sh}(x,t) \right)^2 dx
\tag{8.44}
$$

with $\quad \theta_z^{sh}(x,t) = \dfrac{\partial u_y(x,t)}{\partial x} - \theta_z(x,t).$

The angle $\theta_z^{sh}(x)$ represents the additional rotation due to shear.

An elegant method allowing to formulate the element stiffness matrix is based on *Castigliano's first* (or *force*) *theorem*, according to which the force applied to a linearly elastic member is equal to the partial derivative of the strain energy in terms of the displacement at that point along the force direction – as discussed in Yang [6]. A similar definition allows us to calculate a moment in terms of the corresponding rotation angle. Based on this theorem, the four bending-related nodal loads (forces and moments) of Eqs. (8.22) are expressed as:

$$f_{jy} = \frac{\partial U_{bz}^{(e)}}{\partial u_{jy}}; \quad m_{jz} = \frac{\partial U_{bz}^{(e)}}{\partial \theta_{jz}}, \quad j = 1,2. \tag{8.45}$$

Combining Eqs. (8.40)–(8.45) results in an equation that connects the nodal displacement and load vectors of Eq. (8.22) as:

$$\left[f_{bz}^{(e)} \right] = \left[K_{bz}^{(e)} \right]\left[u_{bz}^{(e)} \right], \tag{8.46}$$

where $\left[K_{bz}^{(e)} \right]$ is the element stiffness matrix.

For a constant cross-section member, the components of the 4×4 symmetric stiffness matrix of Eq. (8.46) are:

$$\begin{cases} K_{bz}^{(e)}(1,1) = -K_{bz}^{(e)}(1,3) = K_{bz}^{(e)}(3,3) = \dfrac{12\left(EI_z l^2 + 12\beta_z^2 \alpha_s GA\right)}{l\left(12\beta_z + l^2\right)^2}; \\[4mm] K_{bz}^{(e)}(1,2) = K_{bz}^{(e)}(1,4) = -K_{bz}^{(e)}(2,3) = -K_{bz}^{(e)}(3,4) = \dfrac{6\left(EI_z l^2 + 12\beta_z^2 \alpha_s GA\right)}{\left(12\beta_z + l^2\right)^2}; \\[4mm] K_{bz}^{(e)}(2,2) = K_{bz}^{(e)}(4,4) = \dfrac{4\left(EI_z\left(36\beta_z^2 + 6\beta_z l^2 + l^4\right) + 9\beta_z^2 \alpha_s GAl^2\right)}{l\left(12\beta_z + l^2\right)^2}; \\[4mm] K_{bz}^{(e)}(2,4) = \dfrac{2EI_z\left(-72\beta_z^2 - 12\beta_z l^2 + l^4\right) + 36\beta_z^2 \alpha_s GAl^2}{l\left(12\beta_z + l^2\right)^2}. \end{cases} \tag{8.47}$$

The kinetic energy related to the z-axis bending vibrations of the beam element of Figure 8.4 is:

$$T_{bz}^{(e)} = \frac{1}{2} \cdot \rho \left[\int_0^l A(x)\left(\frac{\partial u_y(x,t)}{\partial t}\right)^2 dx + \int_0^l I_z(x)\left(\frac{\partial \theta_z(x,t)}{\partial t}\right)^2 dx \right]. \tag{8.48}$$

By using the discretization Eqs. (8.40), it can be shown that:

$$\begin{cases} \left(\dfrac{\partial u_y(x,t)}{\partial t}\right)^2 = \left[\dot{u}_{bz}^{(e)}(t)\right]^T \left[S_{u_y}(x)\right]\left[S_{u_y}(x)\right]^T\left[\dot{u}_{bz}^{(e)}(t)\right] \\[2em] \left(\dfrac{\partial \theta_z(x,t)}{\partial t}\right)^2 = \left[\dot{u}_{bz}^{(e)}(t)\right]^T \left[S_{\theta_z}(x)\right]\left[S_{\theta_z}(x)\right]^T\left[\dot{u}_{bz}^{(e)}(t)\right] \end{cases} \tag{8.49}$$

Equations (8.49) are substituted into Eq. (8.48), and the resulting equation is compared to Eq. (7.22), which expresses the kinetic energy T of a generic system defined by the generalized-coordinate vector $[u]$ as: $T = (1/2)[\dot{u}]^T[M][\dot{u}]$. It follows that the consistent-model inertia matrix of the short beam undergoing z-axis bending vibrations is:

$$\left[M_{bz}^{(e)}\right] = \left[M_{by}^{(e)}\right] = \rho\left(\int_0^l A(x)\left[S_{u_y}(x)\right]\left[S_{u_y}(x)\right]^T dx + \int_0^l I_z(x)\left[S_{\theta_z}(x)\right]\left[S_{\theta_z}(x)\right]^T dx\right). \tag{8.50}$$

Equation (8.50) states that the element inertia matrices modeling bending around the z and y axes are identical, which can readily be demonstrated. When the cross-section is constant, the 4×4 symmetric inertia matrix of Eq. (8.50) has the following components:

$$\begin{cases} M_{bz}^{(e)}(1,1) = M_{bz}^{(e)}(3,3) = \dfrac{\rho l\left(13Al^4 + 42(7\beta_z A + I_z)l^2 + 1680\beta_z^2 A\right)}{35\left(12\beta_z + l^2\right)^2}; \\[2em] M_{bz}^{(e)}(1,2) = -M_{bz}^{(e)}(3,4) = \dfrac{\rho l^2\left(11Al^4 + 21(11\beta_z A + I_z)l^2 + 1260\beta_z(A\beta_z - I_z)\right)}{210\left(12\beta_z + l^2\right)^2}; \\[2em] M_{bz}^{(e)}(1,3) = \dfrac{3\rho l\left(3Al^4 + 28(3A\beta_z - I_z)l^2 + 560\beta_z^2 A\right)}{70\left(12\beta_z + l^2\right)^2}; \\[2em] M_{bz}^{(e)}(1,4) = -M_{bz}^{(e)}(2,3) = -\dfrac{\rho l^2\left(13Al^4 + 42(9A\beta_z - I_z)l^2 + 2520\beta_z(I_z + A\beta_z)\right)}{420\left(12\beta_z + l^2\right)^2}; \\[2em] M_{bz}^{(e)}(2,2) = M_{bz}^{(e)}(4,4) = \dfrac{\rho l\left(Al^6 + 7(3\beta_z A + 2I_z)l^4 + 42\beta_z(5I_z + 3A\beta_z)l^2 + 5040\beta_z^2 I_z\right)}{105\left(12\beta_z + l^2\right)^2}; \\[2em] M_{bz}^{(e)}(2,4) = -\dfrac{\rho l\left(3Al^6 + 14(6\beta_z A + I_z)l^4 + 168\beta_z(5I_z + 3A\beta_z)l^2 - 10080\beta_z^2 I_z\right)}{140\left(12\beta_z + l^2\right)^2}. \end{cases} \tag{8.51}$$

When rotary effects are considered, the mechanical moments of inertia of the two beam halves can be lumped at the element nodes. Each of them is:

$$J_z = \frac{1}{3} \cdot \frac{1}{2} \cdot \rho \left(\int_0^l A(x)\,dx \right) \cdot \left(\frac{l}{2} \right)^2 = \frac{\rho \cdot l^2}{24} \cdot \left(\int_0^l A(x)\,dx \right). \qquad (8.52)$$

As a consequence, the 4×4 lumped-parameter inertia matrix combines the translation components of Eq. (8.32) with the rotation ones of Eq. (8.52) and is of the diagonal form:

$$\left[M_{bz}^{(e)} \right] = \left[M_{by}^{(e)} \right] = \frac{\rho}{2} \cdot \left(\int_0^l A(x)\,dx \right) \mathrm{diag} \left(1 \quad \frac{l^2}{12} \quad 1 \quad \frac{l^2}{12} \right). \qquad (8.53)$$

When the cross-section is constant, Eq. (8.53) becomes:

$$\left[M_{bz}^{(e)} \right] = \frac{m}{2} \cdot \begin{bmatrix} 1 & 0 & 0 & 0 \\ 0 & \dfrac{l^2}{12} & 0 & 0 \\ 0 & 0 & 1 & 0 \\ 0 & 0 & 0 & \dfrac{l^2}{12} \end{bmatrix} \qquad (8.54)$$

8.1.1.4 In-Plane and Out-of-Plane Element Matrices

For a two-node line element and when considering axial effects and bending around the z axis, the following nodal displacement vector is assembled:

$$\left[u_{ip}^{(e)} \right] = \begin{bmatrix} u_{1x} & u_{1y} & \theta_{1z} & u_{2x} & u_{2y} & \theta_{2z} \end{bmatrix}^T. \qquad (8.55)$$

The in-plane element stiffness matrix corresponding to this displacement vector collects its components from Eqs. (8.11) and (8.30) or (8.46) in the form:

$$\left[K_{ip}^{(e)} \right]$$

$$= \begin{bmatrix} K_a^{(e)}(1,1) & 0 & 0 & K_a^{(e)}(1,2) & 0 & 0 \\ 0 & K_{bz}^{(e)}(1,1) & K_{bz}^{(e)}(1,2) & 0 & K_{bz}^{(e)}(1,3) & K_{bz}^{(e)}(1,4) \\ 0 & K_{bz}^{(e)}(1,2) & K_{bz}^{(e)}(2,2) & 0 & K_{bz}^{(e)}(2,3) & K_{bz}^{(e)}(2,4) \\ K_a^{(e)}(1,1) & 0 & 0 & K_a^{(e)}(2,2) & 0 & 0 \\ 0 & K_{bz}^{(e)}(1,3) & K_{bz}^{(e)}(2,3) & 0 & K_{bz}^{(e)}(3,3) & K_{bz}^{(e)}(3,4) \\ 0 & K_{bz}^{(e)}(1,4) & K_{bz}^{(e)}(2,4) & 0 & K_{bz}^{(e)}(3,4) & K_{bz}^{(e)}(4,4) \end{bmatrix}.$$

$$(8.56)$$

The in-plane element inertia matrix has a similar structure to the stiffness matrix of Eq. (8.56) – just use M instead of K. The axial components are provided in Eq. (7.67), while the z-bending ones are given in Eq. (7.88) for long beams and in Eq. (8.50) for short beams based on the consistent approach. Equations (8.13) and (8.53) should be utilized with the lumped-parameter approach.

The out-of-plane nodal displacement vector describes the rotation about the x axis and the bending around the y axis, and is formed as:

$$\left[u_{op}^{(e)} \right] = \left[\begin{array}{cccccc} \theta_{1x} & \theta_{1y} & u_{1z} & \theta_{2x} & \theta_{2y} & u_{2z} \end{array} \right]^T. \tag{8.57}$$

Corresponding to the displacement vector of Eq. (8.57) and using components from Eqs. (8.18) and (8.35), the out-of-plane element stiffness matrix is assembled as:

$$\left[K_{op}^{(e)} \right]$$

$$= \begin{bmatrix} K_t^{(e)}(1,1) & 0 & 0 & K_t^{(e)}(1,2) & 0 & 0 \\ 0 & K_{by}^{(e)}(2,2) & K_{by}^{(e)}(1,2) & 0 & K_{by}^{(e)}(2,4) & K_{by}^{(e)}(2,3) \\ 0 & K_{by}^{(e)}(1,2) & K_{by}^{(e)}(1,1) & 0 & K_{by}^{(e)}(1,4) & K_{by}^{(e)}(1,3) \\ K_t^{(e)}(1,2) & 0 & 0 & K_t^{(e)}(2,2) & 0 & 0 \\ 0 & K_{by}^{(e)}(2,4) & K_{by}^{(e)}(1,4) & 0 & K_{by}^{(e)}(4,4) & K_{by}^{(e)}(3,4) \\ 0 & K_{by}^{(e)}(2,3) & K_{by}^{(e)}(1,3) & 0 & K_{by}^{(e)}(3,4) & K_{by}^{(e)}(3,3) \end{bmatrix}. \tag{8.58}$$

The out-of-plane element inertia matrix has the same structure with the stiffness matrix of Eq. (8.58) – one just needs to use M instead of K. The axial components are given in Eq. (7.73) and the y-bending ones are expressed in Eq. (7.88) for Euler–Bernoulli beams and in Eq. (8.50) for Timoshenko beams by means of the consistent approach. The lumped-parameter out-of-plane element inertia matrix is formed from Eqs. (8.20) and (8.53).

Example 8.1

The fixed-free right elliptical flexible hinge illustrated in Figure 8.5 has a rectangular cross-section with a constant out-of-plane width $w = 0.007$ m and a minimum in-plane thickness $t = 0.001$ m. Known are the material Young's modulus $E = 2 \cdot 10^{11} \, N/m^2$, Poisson's ratio $\mu = 0.3$ and the mass density $\rho = 7,800 \, kg/m^3$ as well as the shear coefficient $\alpha_s = 1.176$. The ellipse large semi-axis length a varies from $a = b$ to $a = 10b$ (where b is the length of the small ellipse semi-axis). Using a two-node Euler–Bernoulli beam element and a two-node Timoshenko beam element, compare the two models' predictions by evaluating the deflection at the free end produced by a force $f_{1y} = 1 \, N$ applied at that point, as well as the first in-plane natural frequency.

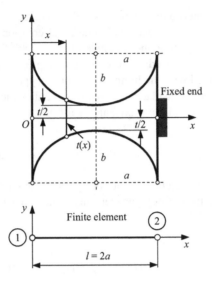

FIGURE 8.5 Front view with geometric parameters of fixed-free right elliptical flexible hinge and a two-node line element.

Solution:

Figure 8.5a identifies the variable thickness $t(x)$, which is calculated as:

$$t(x) = t + 2b\left(1 - \sqrt{1 - \left(1 - \frac{x}{a}\right)^2}\right). \tag{8.59}$$

This parameter is needed to calculate $I_z(x)$ and $A(x)$ in the stiffness matrices of Eqs. (8.30) and (8.46), as well as in the inertia matrices of Eqs. (7.88) and (8.50). Because node 2 of the finite element shown in Figure 8.5 is fixed, the deflection and rotation at that node are zero, namely: $u_{2y} = 0$ and $\theta_{2z} = 0$. As a direct consequence, the following reduced static model results from the dimension-4 equation defining the static response of the full element:

$$\begin{bmatrix} K_{bz}^{(e)}(1,1) & K_{bz}^{(e)}(1,2) \\ K_{bz}^{(e)}(1,2) & K_{bz}^{(e)}(2,2) \end{bmatrix} \begin{bmatrix} u_{1y} \\ \theta_{1z} \end{bmatrix} = \begin{bmatrix} f_{1y} \\ m_{1z} \end{bmatrix} \rightarrow \begin{bmatrix} u_{1y} \\ \theta_{1z} \end{bmatrix}$$

$$= \begin{bmatrix} K_{bz}^{(e)}(1,1) & K_{bz}^{(e)}(1,2) \\ K_{bz}^{(e)}(1,2) & K_{bz}^{(e)}(2,2) \end{bmatrix}^{-1} \begin{bmatrix} f_{1y} \\ m_{1z} \end{bmatrix}. \tag{8.60}$$

Section 8.3 presents a systematic approach to reducing dimensions and matrices of multiple-element finite element models due to zero boundary conditions resulting from external supports.

Because the forcing vector of Eq. (8.60) consists of $f_{1y} = 1$ N and $m_{1z} = 0$, the deflection u_{1y} can be determined by both the Euler–Bernoulli model and the Timoshenko model, and the following relative deflection is calculated:

$$r_u = \frac{u_{1y}^T - u_{1y}^{EB}}{u_{1y}^T} \cdot 100, \qquad (8.61)$$

where the superscripts "*EB*" and "*T*" denote the Euler–Bernoulli and Timoshenko models. Figure 8.6 plots this ratio in terms of the ellipse semi-axis length ratio a/b. It can be seen that the relative difference between the free-end deflection by the two beam models is less than 2% for semi-axis length ratios larger than 3.

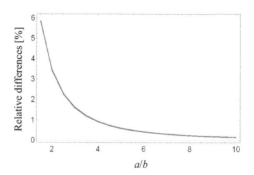

FIGURE 8.6 Plots of relative differences between Euler–Bernoulli and Timoshenko model predictions of the free-end deflection and first natural frequency in terms of the semi-axes length ratio for a fixed-free right elliptical flexible hinge.

In a similar manner, the following reduced dynamic matrix is formed:

$$[D] = \begin{bmatrix} M_{bz}^{(e)}(1,1) & M_{bz}^{(e)}(1,2) \\ M_{bz}^{(e)}(1,2) & M_{bz}^{(e)}(2,2) \end{bmatrix}^{-1} \begin{bmatrix} K_{bz}^{(e)}(1,1) & K_{bz}^{(e)}(1,2) \\ K_{bz}^{(e)}(1,2) & K_{bz}^{(e)}(2,2) \end{bmatrix}, \qquad (8.62)$$

whose eigenvalues yield the two nonzero natural frequencies. The Euler–Bernoulli model and the Timoshenko model predictions of the first natural frequency result in the following relative deviation:

$$r_\omega = \frac{\omega_{n1}^T - \omega_{n1}^{EB}}{\omega_{n1}^T} \cdot 100, \qquad (8.63)$$

which is also plotted in Figure 8.6. As it can be seen, the relative natural frequency difference is very similar in profile and values to the deflection relative difference (it is actually superimposed over the relative deflection curve).

8.1.1.5   Inertia Matrices of Rigid Elements

Finite element inertia matrices are formulated here for two-node, straight-line elements that are rigid. Axial, torsional and bending inertia matrices are analyzed separately.

8.1.1.5.1 Bar Element with Axial Effects and Torsional Effects

Consider the two-node bar element of Figure 8.2. Because the element does not deform, its two nodal displacements are equal, $u_{1x} = u_{2x}$. This constraint indicates that the element has only one DOF. As a consequence, the displacement field of Eq. (8.2) can be expressed in terms of a single constant as: $u_x(x) = u_{1x} = u_{2x} = c_0$. This relationship can formally be written as:

$$u_x(x,t) = \left(\frac{1}{2}\right) \cdot u_{1x}(t) + \left(\frac{1}{2}\right) \cdot u_{2x}(t) \quad \text{or} \quad u_x(x,t) = \left[S_a(x)\right]^T \left[u_x^{(e)}(t)\right], \quad (8.64)$$

where the shape-function vector is:

$$\left[S_a(x)\right] = \begin{bmatrix} S_{a1}(x) & S_{a2}(x) \end{bmatrix}^T \quad \text{with} \quad S_{a1}(x) = S_{a2}(x) = \frac{1}{2}. \quad (8.65)$$

This shape-function vector is combined with Eq. (7.67) to obtain the element inertia matrix of a rigid element with variable cross-section; for a constant cross-section bar element of area A and length l, Eqs. (7.67) and (8.65) yield:

$$\left[M_a^{(e)}\right] = \begin{bmatrix} \dfrac{m}{4} & \dfrac{m}{4} \\ \dfrac{m}{4} & \dfrac{m}{4} \end{bmatrix}, \quad (8.66)$$

where m is the member mass.

For a two-node rigid line element with x-axis rotation DOF, like the one shown in Figure 8.3, the inertia matrix is found using a similar approach; the shape functions of Eq. (8.65) are used in conjunction with Eq. (7.73) to calculate the inertia matrix of a variable cross-section, straight-axis rigid line element. For a constant cross-section element of torsional moment of area I_t, the inertia matrix is:

$$\left[M_t^{(e)}\right] = \begin{bmatrix} \dfrac{J_t}{4} & \dfrac{J_t}{4} \\ \dfrac{J_t}{4} & \dfrac{J_t}{4} \end{bmatrix}, \quad (8.67)$$

where the torsional moment of inertia is $J_t = \rho \cdot I_t \cdot l$.

8.1.1.5.2 Beam Element with Bending Effects

The two-node beam element of Figure 8.4 is considered rigid. As such, the rotation of this rigid body is identical for all points, including the end nodes. For z-axis bending, this means $\theta_z(x) = \theta_{1z} = \theta_{2z}$. Another connection between the nodal DOF is: $\tan\theta_z(x) = \theta_z(x) = (u_{2y} - u_{1y})/l$, which corresponds to small deformations/displacements. Because of these two constraints, there are only two nodal parameters that are independent; as a consequence, the rigid element has two

DOF, and the deflection distribution can be expressed in terms of two constants in polynomial form as:

$$u_y(x) = c_0 + c_1 \cdot x; \quad \theta_z(x) = \frac{du_y(x)}{dx} = c_1. \tag{8.68}$$

The boundary conditions are:

$$u_y(0) = u_{1y}; \quad u_y(l) = u_{2y}. \tag{8.69}$$

Equations (8.69) are used in conjunction with the deflection distribution of Eq. (8.68) to find the constants c_0 and c_1. These constants are substituted back in $u_y(x)$ of Eq. (8.68), which can be written in the discretized form of Eq. (8.25). The four shape functions are:

$$S_{b1}(x) = 1 - \frac{x}{l}; \quad S_{b2}(x) = S_{b4}(x) = 0; \quad S_{b3}(x) = \frac{x}{l}. \tag{8.70}$$

These shape functions are substituted into Eq. (7.88) to calculate the inertia matrix of a rigid, beam-type element. For a constant cross-section, the inertia matrices for z-axis bending and y-axis bending are:

$$\left[M_{bz}^{(e)} \right] = \left[M_{by}^{(e)} \right] = \begin{bmatrix} \dfrac{m}{3} & 0 & \dfrac{m}{6} & 0 \\ 0 & 0 & 0 & 0 \\ \dfrac{m}{6} & 0 & \dfrac{m}{3} & 0 \\ 0 & 0 & 0 & 0 \end{bmatrix}. \tag{8.71}$$

8.1.1.5.3 In-Plane and Out-of-Plane Inertia Matrices for Rigid Members

It is now possible to assemble the in-plane and out-of-plane inertia matrices for a two-node line element with six DOF per node. The in-plane matrix corresponds to u_x, u_y and θ_z, while the out-of-plane matrix is connected to θ_x, θ_y and u_z. The in-plane inertia matrix of a constant cross-section member combines Eqs. (8.66) and (8.71) as:

$$\left[M_{ip}^{(e)} \right] = \begin{bmatrix} m/4 & 0 & 0 & m/4 & 0 & 0 \\ 0 & m/3 & 0 & 0 & m/6 & 0 \\ 0 & 0 & 0 & 0 & 0 & 0 \\ m/4 & 0 & 0 & m/4 & 0 & 0 \\ 0 & m/6 & 0 & 0 & m/3 & 0 \\ 0 & 0 & 0 & 0 & 0 & 0 \end{bmatrix}. \tag{8.72}$$

Similarly, the following out-of-plane inertia matrix is assembled from Eqs. (8.67) and (8.71) for a constant cross-section:

$$
\left[M_{op}^{(e)} \right] = \begin{bmatrix}
J_t/4 & 0 & 0 & J_t/4 & 0 & 0 \\
0 & 0 & 0 & 0 & 0 & 0 \\
0 & 0 & m/3 & 0 & 0 & m/6 \\
J_t/4 & 0 & 0 & J_t/4 & 0 & 0 \\
0 & 0 & 0 & 0 & 0 & 0 \\
0 & 0 & m/6 & 0 & 0 & m/3
\end{bmatrix}. \tag{8.73}
$$

8.1.2 THREE-NODE ELEMENT

The stiffness and inertia matrices of a straight-line element with three nodes (two at the ends and one at the midpoint, as illustrated in Figure 8.7) are briefly presented here. This element can deform and vibrate axially, torsionally and in bending. Compared to the two-node element of Section 8.1.1, the three-node element produces more accurate predictions due to the increased number of DOF.

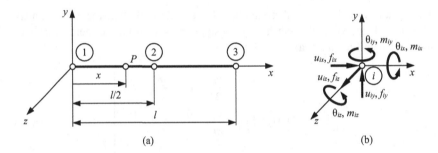

FIGURE 8.7 (a) Three-node straight-line element; (b) nodal DOF and forces/moments ($i = 1, 2, 3$).

The generic elemental matrices of the two-node line element that were formulated in Section 8.1.1 are also valid for the three-node element. Due to the extra node of this latter element, the shape functions are different, as discussed next.

8.1.2.1 Axial Effects and Torsion

The nodal displacement and load vectors for axial effects and torsion are:

$$
\left[u_x^{(e)} \right] = \begin{bmatrix} u_{1x} & u_{2x} & u_{3x} \end{bmatrix}^T; \quad \left[\theta_x^{(e)} \right] = \begin{bmatrix} \theta_{1x} & \theta_{2x} & \theta_{3x} \end{bmatrix}^T
$$

$$
\left[f_x^{(e)} \right] = \begin{bmatrix} f_{1x} & f_{2x} & f_{3x} \end{bmatrix}^T; \quad \left[m_x^{(e)} \right] = \begin{bmatrix} m_{1x} & m_{2x} & m_{3x} \end{bmatrix}^T. \tag{8.74}
$$

Because of its three DOF, the axial displacement $u_x(x)$ of the arbitrary point P can be represented in polynomial form as:

$$
u_x(x) = c_0 + c_1 \cdot x + c_2 \cdot x^2. \tag{8.75}
$$

Using the boundary conditions:

$$u_x(0) = u_{1x}; \quad u_x(l/2) = u_{2x}; \quad u_x(l) = u_{3x} \tag{8.76}$$

in conjunction with Eqs. (8.75) allows solving for the constants c_0, c_1 and c_2. These constants are substituted back into Eq. (8.75), which becomes of the discretized form given in Eq. (8.4) where the shape functions are:

$$[S_a(x)] = \begin{bmatrix} S_{a1}(x) & S_{a2}(x) & S_{a2}(x) \end{bmatrix}^T \text{ with } \begin{cases} S_{a1}(x) = 1 - 3 \cdot \dfrac{x}{l} + 2 \cdot \dfrac{x^2}{l^2} \\[2mm] S_{a2}(x) = 4 \cdot \dfrac{x}{l} - 4 \cdot \dfrac{x^2}{l^2} \\[2mm] S_{a3}(x) = -\dfrac{x}{l} + 2 \cdot \dfrac{x^2}{l^2}. \end{cases} \tag{8.77}$$

The 3×3 element stiffness matrix related to the x-axis translation elasticity of the three-node line element of Figure 8.7a is calculated based on the definition Eq. (8.11) with the shape functions of Eq. (8.77). For a constant cross-section member with $A(x) = A$, the stiffness matrix is:

$$\begin{bmatrix} K_a^{(e)} \end{bmatrix} = \frac{E \cdot A}{3l} \begin{bmatrix} 7 & -8 & 1 \\ -8 & 16 & -8 \\ 1 & -8 & 7 \end{bmatrix} \tag{8.78}$$

The consistent-model elemental inertia matrix characterizing the translation vibrations along the x axis is determined with the general Eq. (7.67) and the shape functions of Eq. (8.77). The following matrix results for a constant cross-section member:

$$\begin{bmatrix} M_a^{(e)} \end{bmatrix} = \frac{m}{15} \begin{bmatrix} 2 & 1 & -\dfrac{1}{2} \\[2mm] 1 & 8 & 1 \\[2mm] -\dfrac{1}{2} & 1 & 2 \end{bmatrix}. \tag{8.79}$$

The lumped-model inertia matrix is calculated as:

$$\begin{bmatrix} M_a^{(e)} \end{bmatrix} = \frac{1}{3} \cdot \rho \cdot \left(\int_0^l A(x)\,dx \right) \cdot \begin{bmatrix} 1 & 0 & 0 \\ 0 & 1 & 0 \\ 0 & 0 & 1 \end{bmatrix}, \tag{8.80}$$

which, for a constant cross-section member, becomes:

$$\left[M_a^{(e)} \right] = \frac{m}{3} \cdot \begin{bmatrix} 1 & 0 & 0 \\ 0 & 1 & 0 \\ 0 & 0 & 1 \end{bmatrix}. \tag{8.81}$$

The 3×3 torsion-related element stiffness matrix is generically expressed in Eq. (8.18) and is calculated with the same shape functions of Eq. (8.77). For a member with constant cross-section with $I_t(x) = I_t$, the stiffness matrix is expressed as:

$$\left[K_t^{(e)} \right] = \frac{G \cdot I_t}{3l} \cdot \begin{bmatrix} 7 & -8 & 1 \\ -8 & 16 & -8 \\ 1 & -8 & 7 \end{bmatrix}. \tag{8.82}$$

Equations (7.73) are combined with the shape functions of Eq. (8.77) to formulate the consistent-model inertia matrix of a variable cross-section line element. For a constant cross-section element, the consistent torsion inertia matrix is:

$$\left[M_t^{(e)} \right] = \frac{J_t}{15} \cdot \begin{bmatrix} 2 & 1 & -\dfrac{1}{2} \\ 1 & 8 & 1 \\ -\dfrac{1}{2} & 1 & 2 \end{bmatrix}, \tag{8.83}$$

where J_t is the mass moment of inertia about the x axis. The lumped-parameter inertia matrix related to torsion is calculated similarly to the one corresponding to axial translation in Eq. (8.80) as:

$$\left[M_t^{(e)} \right] = \frac{1}{3} \cdot \rho \cdot \left(\int_0^l I_t(x) dx \right) \cdot \begin{bmatrix} 1 & 0 & 0 \\ 0 & 1 & 0 \\ 0 & 0 & 1 \end{bmatrix}. \tag{8.84}$$

Equation (8.84) simplifies to the following form for a constant cross-section member:

$$\left[M_t^{(e)} \right] = \frac{J_t}{3} \cdot \begin{bmatrix} 1 & 0 & 0 \\ 0 & 1 & 0 \\ 0 & 0 & 1 \end{bmatrix}. \tag{8.85}$$

8.1.2.2 Bending

Euler–Bernoulli and Timoshenko models are utilized in this section to derive element stiffness and inertia matrices resulting from bending.

8.1.2.2.1 Long Beams and the Euler–Bernoulli Model

The nodal displacement vector of a beam element with three nodes and two DOF per node (which are specified deflection u_y and rotation θ_z) is:

$$\left[u_{bz}^{(e)} \right] = \left[\begin{array}{cccccc} u_{1y} & \theta_{1z} & u_{2y} & \theta_{2z} & u_{3y} & \theta_{3z} \end{array} \right]^T. \tag{8.86}$$

The six nodal DOF of Eq. (8.86) allow using a polynomial distribution function of the deflection at the arbitrary point P in Figure 8.7 with six constants, namely:

$$\begin{cases} u_y(x) = c_0 + c_1 \cdot x + c_2 \cdot x^2 + c_3 \cdot x^3 + c_4 \cdot x^4 + c_5 \cdot x^5 \\ \theta_z(x) = \dfrac{du_y(x)}{dx} = c_1 + 2c_2 \cdot x + 3c_3 \cdot x^2 + 4c_4 \cdot x^3 + 5c_5 \cdot x^4 \end{cases} \tag{8.87}$$

The boundary conditions at the three nodes are:

$$u_y(0) = u_{1y}; \quad \theta_z(0) = \theta_{1z}; \quad u_y(l/2) = u_{2y}; \quad \theta_z(l/2) = \theta_{2z};$$
$$u_y(l) = u_{3y}; \quad \theta_z(l) = \theta_{3z}. \tag{8.88}$$

They are used in conjunction with Eqs. (8.87) to solve for the seven constants c_0 through c_6. These constants are substituted back into $u_y(x)$ of Eq. (8.87), which result in the discretized relationship of Eq. (8.25) with the shape-function vector:

$$\left[S_b(x) \right] = \left[\begin{array}{cccccc} S_{b1}(x) & S_{b2}(x) & S_{b3}(x) & S_{b4}(x) & S_{b5}(x) & S_{b6}(x) \end{array} \right]^T. \tag{8.89}$$

The individual shape functions of Eq. (8.89) are:

$$\begin{cases} S_{b1}(x) = 1 - 23 \cdot \dfrac{x^2}{l^2} + 66 \cdot \dfrac{x^3}{l^3} - 68 \cdot \dfrac{x^4}{l^4} + 25 \cdot \dfrac{x^5}{l^5}; \\[2mm] S_{b2}(x) = x - 6 \cdot \dfrac{x^2}{l} + 13 \cdot \dfrac{x^3}{l^2} - 12 \cdot \dfrac{x^4}{l^3} + 4 \cdot \dfrac{x^5}{l^4}; \\[2mm] S_{b3}(x) = 16 \cdot \dfrac{x^2}{l^2} - 32 \cdot \dfrac{x^3}{l^3} + 16 \cdot \dfrac{x^4}{l^4}; \\[2mm] S_{b4}(x) = -8 \cdot \dfrac{x^2}{l} + 32 \cdot \dfrac{x^3}{l^2} - 40 \cdot \dfrac{x^4}{l^3} + 16 \cdot \dfrac{x^5}{l^4}; \\[2mm] S_{b5}(x) = 7 \cdot \dfrac{x^2}{l^2} - 34 \cdot \dfrac{x^3}{l^3} + 52 \cdot \dfrac{x^4}{l^4} - 24 \cdot \dfrac{x^5}{l^5}; \\[2mm] S_{b6}(x) = -\dfrac{x^2}{l} + 5 \cdot \dfrac{x^3}{l^2} - 8 \cdot \dfrac{x^4}{l^3} + 4 \cdot \dfrac{x^5}{l^4}. \end{cases} \tag{8.90}$$

With the shape functions of Eq. (8.90), Eq. (8.30) is used to calculate the 6×6 elemental stiffness matrix of a beam element with three nodes. For a constant cross-section beam with $I_z(x) = I_z$, the stiffness matrix is:

$$\left[K_{bz}^{(e)} \right] = \frac{E \cdot I_z}{l^3}
\begin{bmatrix}
\dfrac{5092}{35} & \dfrac{1138}{35} \cdot l & -\dfrac{512}{5} & \dfrac{384}{7} \cdot l & -\dfrac{1508}{35} & \dfrac{242}{35} \cdot l \\[8pt]
\dfrac{1138}{35} \cdot l & \dfrac{332}{35} \cdot l^2 & -\dfrac{128}{5} \cdot l & \dfrac{64}{7} \cdot l^2 & -\dfrac{242}{35} \cdot l & \dfrac{38}{35} \cdot l^2 \\[8pt]
-\dfrac{512}{5} & -\dfrac{128}{5} \cdot l & \dfrac{1024}{5} & 0 & -\dfrac{512}{5} & \dfrac{128}{5} \cdot l \\[8pt]
\dfrac{384}{7} \cdot l & \dfrac{64}{7} \cdot l^2 & 0 & \dfrac{256}{7} \cdot l^2 & -\dfrac{384}{7} \cdot l & \dfrac{64}{7} \cdot l^2 \\[8pt]
-\dfrac{1508}{35} & -\dfrac{242}{35} \cdot l & -\dfrac{512}{5} & -\dfrac{384}{7} \cdot l & \dfrac{5092}{35} & -\dfrac{1138}{35} \cdot l \\[8pt]
\dfrac{242}{35} \cdot l & \dfrac{38}{35} \cdot l^2 & \dfrac{128}{5} \cdot l & \dfrac{64}{7} \cdot l^2 & -\dfrac{1138}{35} \cdot l & \dfrac{332}{35} \cdot l^2
\end{bmatrix}.$$

$$(8.91)$$

The consistent-inertia matrix is calculated as per Eq. (7.88) using the shape functions of Eqs. (8.90). When the cross-section is constant, the following inertia matrix is obtained:

$$\left[M_{bz}^{(e)} \right] = m
\begin{bmatrix}
\dfrac{523}{3465} & \dfrac{19}{2310} \cdot l & \dfrac{4}{63} & -\dfrac{8}{693} \cdot l & \dfrac{131}{6930} & -\dfrac{29}{13860} \cdot l \\[8pt]
\dfrac{19}{2310} \cdot l & \dfrac{2}{3465} \cdot l^2 & \dfrac{2}{315} \cdot l & -\dfrac{1}{1155} \cdot l^2 & \dfrac{29}{13860} \cdot l & -\dfrac{1}{4620} \cdot l^2 \\[8pt]
\dfrac{4}{63} & \dfrac{2}{315} \cdot l & \dfrac{128}{315} & 0 & \dfrac{4}{63} & -\dfrac{2}{315} \cdot l \\[8pt]
-\dfrac{8}{693} \cdot l & -\dfrac{1}{1155} \cdot l^2 & 0 & \dfrac{32}{3465} \cdot l^2 & \dfrac{8}{693} \cdot l & -\dfrac{1}{1155} \cdot l^2 \\[8pt]
\dfrac{131}{6930} & \dfrac{29}{13860} \cdot l & \dfrac{4}{63} & \dfrac{8}{693} \cdot l & \dfrac{523}{3465} & -\dfrac{19}{2310} \cdot l \\[8pt]
-\dfrac{29}{13860} \cdot l & -\dfrac{1}{4620} \cdot l^2 & -\dfrac{2}{315} \cdot l & -\dfrac{1}{1155} \cdot l^2 & -\dfrac{19}{2310} \cdot l & \dfrac{2}{3465} \cdot l^2
\end{bmatrix}.$$

$$(8.92)$$

For a variable cross-section beam, the lumped-inertia matrix is calculated as:

$$\left[M_{bz}^{(e)} \right] = \frac{1}{3} \cdot \rho \cdot \left(\int_0^l A(x)\,dx \right) \cdot \begin{bmatrix} 1 & 0 & 0 & 0 & 0 & 0 \\ 0 & 0 & 0 & 0 & 0 & 0 \\ 0 & 0 & 1 & 0 & 0 & 0 \\ 0 & 0 & 0 & 0 & 0 & 0 \\ 0 & 0 & 0 & 0 & 1 & 0 \\ 0 & 0 & 0 & 0 & 0 & 0 \end{bmatrix}. \tag{8.93}$$

When the cross-section is constant, the matrix of Eq. (8.93) becomes:

$$\left[M_{bz}^{(e)} \right] = \begin{bmatrix} \dfrac{m}{3} & 0 & 0 & 0 & 0 & 0 \\ 0 & 0 & 0 & 0 & 0 & 0 \\ 0 & 0 & \dfrac{m}{3} & 0 & 0 & 0 \\ 0 & 0 & 0 & 0 & 0 & 0 \\ 0 & 0 & 0 & 0 & \dfrac{m}{3} & 0 \\ 0 & 0 & 0 & 0 & 0 & 0 \end{bmatrix}. \tag{8.94}$$

8.1.2.2.2 Short Beams and the Timoshenko Model

The following separate distribution functions can be used for deflection and slope when short beams with shear deformations and rotary inertia are considered:

$$\begin{cases} u_y(x) = c_0 + c_1 x + c_2 x^2 + c_3 x^3 + c_4 x^4 + c_5 x^5 \\ \theta_z(x) = c_6 + c_7 x + c_8 x^2 + c_9 x^3 + c_{10} x^4 \end{cases} \tag{8.95}$$

Equation (8.36) is used to relate the slope to the deflection, which results in the following equations connecting the polynomial coefficients of Eqs. (8.95):

$$\begin{cases} c_6 = c_1 + 6\beta_z \left(c_3 + 20\beta_z c_5 \right) \\ c_7 = 2\left(c_2 + 12\beta_z c_4 \right) \\ c_8 = 3\left(c_3 + 20\beta_z c_5 \right) \\ c_9 = 4c_4 \\ c_{10} = 5c_5 \end{cases} \tag{8.96}$$

with β_z of Eq. (8.39). The boundary conditions of Eqs. (8.88) are used to obtain the discretization relationships for the deflection and slope of Eqs. (8.40); they allow identifying the shape-function vectors $\left[S_{u_y}(x)\right], \left[S_{\theta_z}(x)\right]$. The shape-function components are provided explicitly in Lobontiu and Garcia [13]. The element stiffness matrix can now be calculated as in Eq. (8.46), while the element consistent-inertia matrix is determined as in Eq. (8.50).

Similar to the two-node short-beam element, the lumped-model inertia matrix of a three-node, Timoshenko-beam element of variable cross-section includes the rotary inertia fraction expressed in Eq. (8.52) and is of the form:

$$\left[M_{bz}^{(e)}\right] = \left[M_{by}^{(e)}\right] = \frac{\rho}{2} \cdot \left(\int_0^l A(x)\,dx\right) \cdot \text{diag}\left(1 \quad \frac{l^2}{12} \quad 1 \quad \frac{l^2}{12} \quad 1 \quad \frac{l^2}{12}\right). \quad (8.97)$$

For a constant cross-section member, Eq. (8.97) becomes:

$$\left[M_{bz}^{(e)}\right] = \frac{m}{2} \cdot \text{diag}\left(1 \quad \frac{l^2}{12} \quad 1 \quad \frac{l^2}{12} \quad 1 \quad \frac{l^2}{12}\right). \quad (8.98)$$

8.2 CIRCULAR-AXIS LINE ELEMENTS

Finite element formulation of curvilinear-axis and circular-axis elements has been utilized to model and solve static and dynamic problems related to a wide variety of solid mechanics applications – see Ganapathi et al. [16], Kang and Riedel [17], Dayyani et al. [18], Cazzani et al. [19], Tufekci et al. [20], Borkovic et al. [21], Connor and Brebbia [22], Petyt and Fleischer [23], Rao [24], Yang and Kim [25], Lebeck and Knowlton [26], Akhtar [27], Benedetti and Tralli [28], Choi and Lim [29], as well as Khrishnan and Suresh [30], Wu and Chiang [31,32], Kim et al. [33], Saffari et al. [34], Kim et al. [35], Shahbah et al. [36], for instance. This section offers a systematic study of planar circular-axis finite elements with two end nodes that can be used to model flexible hinges. It formulates in-plane and out-of-plane element stiffness and inertia matrices for relatively long members based on the Euler–Bernoulli theory and for short members using Timoshenko's model.

8.2.1 LONG ELEMENTS AND THE EULER–BERNOULLI MODEL

8.2.1.1 In-Plane Matrices

There are several models of two-node, planar, circular-axis finite elements that define the in-plane stiffness and inertia matrices – see Yang [6] for a comprehensive review of those models. This section studies a three-DOF per node model, which is similar to the Euler–Bernoulli model developed in Section 8.1.1.3.1. Its performance compares well with that of a more complex model that uses four DOF per node, as also discussed in this section.

8.2.1.1.1 Linear-Cubic Model

Consider the two-node planar, circular-axis element of Figure 8.8, which is under planar load and whose corresponding displacements/deformations are also planar.

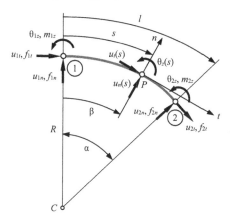

FIGURE 8.8 Planar, circular-axis flexible-hinge element with two nodes and three in-plane DOF per node.

The in-plane deformations at the arbitrary point P, which is situated at a circumferential distance s from end node 1, are the tangential component $u_t(s)$, the normal (radial) deflection $u_n(s)$ and the z-axis rotation $\theta_z(s)$. The corresponding in-plane nodal displacement and forcing vectors are:

$$\begin{cases} \left[u_{ip}^{(e)} \right] = \begin{bmatrix} u_{1t} & u_{1n} & \theta_{1z} & u_{2t} & u_{2n} & \theta_{2z} \end{bmatrix}^T \\ \left[f_{ip}^{(e)} \right] = \begin{bmatrix} f_{1t} & f_{1n} & m_{1z} & f_{2t} & f_{2n} & m_{2z} \end{bmatrix}^T \end{cases}. \tag{8.99}$$

Corresponding to the six DOF of Eq. (8.99), the following interpolation polynomials of $u_t(s)$ and $u_n(s)$ can be used, which form the simplest model, also known as the *linear-cubic model* – see Yang [6], Connor and Brebbia [22] and Petyt and Fleischer [23]:

$$\begin{cases} u_t(s) = c_0 + c_1 \cdot s \\ u_n(s) = c_2 + c_3 \cdot s + c_4 \cdot s^2 + c_5 \cdot s^3 \end{cases}. \tag{8.100}$$

In Eq. (8.100), the tangential displacement is assumed to vary linearly with the coordinate s, whereas the normal displacement varies as a cubic polynomial in s – this is why the model is called linear-cubic. As per Yang [6] and Kardestuncer [37], the z-axis rotation coordinate is expressed as:

$$\theta_z(s) = \frac{du_n(s)}{ds} - \frac{1}{R} \cdot u_t(s) = c_3 - \frac{1}{R} \cdot c_0 + \left(2c_4 - \frac{1}{R} \cdot c_1 \right) \cdot s + 3c_5 \cdot s^2 \tag{8.101}$$

The following boundary conditions are enforced in conjunction with Eqs. (8.100) and (8.101):

$$u_t(0) = u_{1t}; \quad u_n(0) = u_{1n}; \quad \theta_z(0) = \theta_{1z}; \quad u_t(l) = u_{2t}; \quad u_n(l) = u_{2n}; \quad \theta_z(l) = \theta_{2z}$$

(8.102)

where $l = \alpha \cdot R$ is the element length, as illustrated in Figure 8.8. The six Eqs. (8.102) allow solving for the constants c_0 through c_5; they are substituted back into Eqs. (8.100), which can be written in the following discretized form:

$$u_t(s,t) = [S_t(s)]^T \left[u_{ip}^{(e)}(t) \right] \text{ with } [S_t(s)] = \begin{bmatrix} S_{t1}(s) & 0 & 0 & S_{t2}(s) & 0 & 0 \end{bmatrix}^T$$

$$u_n(s,t) = [S_n(s)]^T \left[u_{ip}^{(e)}(t) \right] \text{ with}$$

(8.103)

$$[S_n(s)] = \begin{bmatrix} S_{n1}(s) & S_{n2}(s) & S_{n3}(s) & S_{n4}(s) & S_{n5}(s) & S_{n6}(s) \end{bmatrix}^T.$$

where the component shape functions are:

$$S_{t1} = 1 - \frac{s}{l}; \quad S_{t2} = \frac{s}{l};$$

$$\begin{cases} S_{n1}(s) = \dfrac{s}{R} - 2 \cdot \dfrac{s^2}{R \cdot l} + \dfrac{s^3}{R \cdot l^2}; \ S_{n2}(s) = 1 - 3 \cdot \dfrac{s^2}{l^2} + 2 \cdot \dfrac{s^3}{l^3}; \ S_{n3}(x) = s - 2 \cdot \dfrac{s^2}{l} + \dfrac{s^3}{l^2}; \\[2mm] S_{n4}(s) = -\dfrac{s^2}{R \cdot l} + \dfrac{s^3}{R \cdot l^2}; \ S_{n5}(x) = 3 \cdot \dfrac{s^2}{l^2} - 2 \cdot \dfrac{s^3}{l^3}; \ S_{n6}(x) = -\dfrac{s^2}{l} + \dfrac{s^3}{l^2}. \end{cases}$$

(8.104)

The elastic strain energy stored in the element of Figure 8.8 is expressed as in Yang [6] and Kardestuncer [37], namely:

$$U_{ip}^{(e)} = \frac{E}{2} \cdot \int_0^l \left(A(s) \cdot \varepsilon^2 + I_z(s) \cdot \kappa_z^2 \right) ds,$$

(8.105)

where $A(s)$ and $I_z(s)$ are the cross-sectional area and z-axis moment of area. The circumferential strain ε and the curvature κ_z (produced through planar deformation) are:

$$\varepsilon = \frac{du_t(s)}{ds} + \frac{1}{R} \cdot u_n(s); \quad \kappa_z = \frac{d^2 u_n(s)}{ds^2} - \frac{1}{R} \cdot \frac{du_t(s)}{ds}.$$

(8.106)

Substituting the strain and curvature of Eqs. (8.106) into Eq. (8.105) results in the following strain energy expression:

$$U_{ip}^{(e)} = \frac{E}{2} \cdot \left[\int_0^l A(s) \left(\frac{\partial u_t(s,t)}{\partial s} + \frac{1}{R} \cdot u_n(s,t) \right)^2 ds + \int_0^l I_z(s) \left(\frac{\partial^2 u_n(s,t)}{\partial s^2} - \frac{1}{R} \cdot \frac{\partial u_t(s,t)}{\partial s} \right)^2 ds \right]$$

$$(8.107)$$

The six nodal loads of Eq. (8.100) are expressed by means of Castigliano's first (force) theorem as:

$$f_{jt} = \frac{\partial U_{ip}^{(e)}}{\partial u_{jt}}; \quad f_{jn} = \frac{\partial U_{ip}^{(e)}}{\partial u_{jn}}; \quad m_{jz} = \frac{\partial U_{ip}^{(e)}}{\partial \theta_{jz}}, \quad j = 1, 2. \tag{8.108}$$

After performing the derivations required in Eq. (8.108) by means of the strain energy given in Eq. (8.107), the resulting six equations are combined into the vector-matrix form:

$$\left[f_{ip}^{(e)} \right] = \left[K_{ip}^{(e)} \right] \left[u_{ip}^{(e)} \right], \tag{8.109}$$

where $\left[K_{ip}^{(e)} \right]$ is the 6×6 element in-plane stiffness matrix.

The kinetic energy of the finite element of Figure 8.8 is calculated based on the elementary kinetic energy of the mass $dm = \rho \cdot A(s)ds$ in translation along the tangent and normal axes and also using the discretization Eqs. (8.103), namely:

$$T_{ip}^{(e)} = \frac{1}{2} \cdot \rho \int_0^l \left[\left(\frac{\partial u_t(s,t)}{\partial t} \right)^2 + \left(\frac{\partial u_n(s,t)}{\partial t} \right)^2 \right] A(s) ds$$

$$= \frac{1}{2} \cdot \left[\dot{u}_{ip}^{(e)}(t) \right] \left(\rho \left(\int_0^l A(s) \left[S_t(s) \right] \left[S_t(s) \right]^T ds + \int_0^l A(s) \left[S_n(s) \right] \left[S_n(s) \right]^T ds \right) \right) \left[\dot{u}_{ip}^{(e)}(t) \right]^T.$$

$$(8.110)$$

Comparing Eq. (8.110) with Eq. (7.22), which gives the kinetic energy of a lumped-parameter system, it follows that the 6×6 consistent element inertia matrix of the circular-axis element undergoing in-plane vibrations is:

$$\left[M_{ip}^{(e)} \right] = \rho \left(\int_0^l A(s) \left[S_t(s) \right] \left[S_t(s) \right]^T ds + \int_0^l A(s) \left[S_n(s) \right] \left[S_n(s) \right]^T ds \right). \tag{8.111}$$

8.2.1.1.2   Cubic-Cubic Model

As discussed in Yang [6], Petyt and Fleischer [23] and Cowper et al. [38], the linear-cubic model does not represent exactly the rigid-body displacement of an element. Other models are available that use distribution functions with more terms in the tangential displacement $u_t(s)$, as mentioned in the same references. One such model, known as the *cubic-cubic model*, is defined by the following functions:

$$\begin{cases} u_t(s) = c_0 + c_1 \cdot s + c_2 \cdot s^2 + c_3 \cdot s^3 \\ u_n(s) = c_4 + c_5 \cdot s + c_6 \cdot s^2 + c_7 \cdot s^3 \end{cases} \quad (8.112)$$

and its name reflects the distribution of $u_t(s)$ and $u_n(s)$, which are both cubic polynomials in the variable s. The model uses the following nodal displacement and forcing vectors:

$$\begin{aligned} & \left[u_{ip}^{(e)} \right] \\ & = \left[\begin{array}{cccccccc} u_{1t} & \left(\dfrac{du_t}{ds}\right)_1 = u'_{1t} & u_{1n} & \left(\dfrac{du_n}{ds}\right)_1 = u'_{1n} & u_{2t} & \left(\dfrac{du_t}{ds}\right)_2 = u'_{2t} & u_{2n} & \left(\dfrac{du_n}{ds}\right)_2 = u'_{2n} \end{array} \right]^T \end{aligned}$$

$$\left[f_{ip}^{(e)} \right] = \left[\begin{array}{cccccccc} f_{1t} & f'_{1t} & f_{1n} & f'_{1n} & f_{2t} & f'_{2t} & f_{2n} & f'_{2n} \end{array} \right]^T. \quad (8.113)$$

Note that the only nodal DOF with physical relevance (in the sense that they are actual translational displacements) in $\left[u_{ip}^{(e)} \right]$ are u_t and u_n. Similarly, only the nodal forces f_t and f_n are physically significant in the nodal force vector $\left[f_{ip}^{(e)} \right]$ of Eq. (8.113). Utilizing the eight nodal coordinates of Eq. (8.113) as boundary conditions in the polynomials of Eqs. (8.112) enables to express the displacements of Eq. (8.112) as in Eq. (8.103) where:

$$\begin{cases} [S_t(s)] = \left[\begin{array}{cccccccc} S_1(s) & S_2(s) & 0 & 0 & S_3(s) & S_4(s) & 0 & 0 \end{array} \right]^T \\ [S_n(s)] = \left[\begin{array}{cccccccc} 0 & 0 & S_1(s) & S_2(s) & 0 & 0 & S_3(s) & S_4(s) \end{array} \right]^T \end{cases}, \quad (8.114)$$

and the individual shape functions are:

$$S_1(s) = 1 - 3 \cdot \frac{s^2}{l^2} + 2 \cdot \frac{s^3}{l^3}; \quad S_2(s) = s - 2 \cdot \frac{s^2}{l} + \frac{s^3}{l^2};$$

$$S_3(s) = 3 \cdot \frac{s^2}{l^2} - 2 \cdot \frac{s^3}{l^3}; \quad S_4(s) = -\frac{s^2}{l} + \frac{s^3}{l^2}. \quad (8.115)$$

Equations (8.103), (8.114) and (8.115), together with the necessary derivatives, are substituted in the strain energy of Eq. (8.107). Using Castigliano's first theorem, the nodal forcing components f_{1t}, f_{1n}, f_{2t} and f_{2n} are calculated as in Eq. (8.108). The other four forcing components identified in Eq. (8.113) are determined as:

$$f'_{jt} = \frac{\partial U_{ip}^{(e)}}{\partial u'_{jt}}; \quad f'_{jn} = \frac{\partial U_{ip}^{(e)}}{\partial u'_{jn}}, \quad j = 1,2. \tag{8.116}$$

After performing all necessary operations, Eqs. (8.108) and (8.116) are rendered into the generic form of Eq. (8.109), which identifies the 9×9 element stiffness matrix $\left[K_{ip}^{(e)} \right]$.

The 9×9 element inertia matrix is calculated as in Eq. (8.111) by means of the shape functions given in Eqs. (8.114) and (8.115).

Example 8.2

The finite element of Figure 8.8 is fixed at node 2. The member has a constant, rectangular cross-section defined by in-plane thickness $t = 0.001$ m and out-of-plane width $w = 0.008$ m. The member's material is homogeneous with Young's modulus $E = 2 \cdot 10^{11}$ N/m^2 and mass density $\rho = 7{,}800$ kg/m^3. Compare the linear-cubic model to the cubic-cubic model by graphically studying the relative differences between the two model predictions in the radial displacement u_{1n} at node 1 when a unit-force vector is applied at node 1. Make a similar comparison by studying the first (smallest) natural frequency ω_{n1}. Assume the opening angle α spans the [4°; 30°] interval and the radius R varies within the [0.03 m; 0.06 m] range.

Solution:

Because node 2 is fixed, the stiffness and inertia matrices to be used in static and dynamic calculations are of reduced dimensions. Similar to the procedure outlined in *Example 8.1*, the in-plane displacement vectors at node 1 are expressed in terms of the unit load as:

$$\begin{bmatrix} u_{1t} \\ u_{1n} \\ \theta_{1z} \end{bmatrix} = \begin{bmatrix} K_{ip}^{(e)}(1,1) & K_{ip}^{(e)}(1,2) & K_{ip}^{(e)}(1,3) \\ K_{ip}^{(e)}(1,2) & K_{ip}^{(e)}(2,2) & K_{ip}^{(e)}(2,3) \\ K_{ip}^{(e)}(1,3) & K_{ip}^{(e)}(2,3) & K_{ip}^{(e)}(3,3) \end{bmatrix} \begin{bmatrix} 1 \\ 1 \\ 1 \end{bmatrix} \tag{8.117}$$

and

$$\begin{bmatrix} u_{1t} \\ u'_{1t} \\ u_{1n} \\ u'_{1n} \end{bmatrix} = \begin{bmatrix} K_{ip}^{(e)}(1,1) & K_{ip}^{(e)}(1,2) & K_{ip}^{(e)}(1,3) & K_{ip}^{(e)}(1,4) \\ K_{ip}^{(e)}(1,2) & K_{ip}^{(e)}(2,2) & K_{ip}^{(e)}(2,3) & K_{ip}^{(e)}(2,4) \\ K_{ip}^{(e)}(1,3) & K_{ip}^{(e)}(2,3) & K_{ip}^{(e)}(3,3) & K_{ip}^{(e)}(3,4) \\ K_{ip}^{(e)}(1,4) & K_{ip,24}^{(e)}(2,4) & K_{ip}^{(e)}(3,4) & K_{ip}^{(e)}(4,4) \end{bmatrix} \begin{bmatrix} 1 \\ 1 \\ 1 \\ 1 \end{bmatrix}. \tag{8.118}$$

Equation (8.117) is based on the linear-cubic model, whereas Eq. (8.118) corresponds to the cubic-cubic model. The displacements u_{1n} are determined from Eqs. (8.117) and (8.118); they enable to calculate the ratio of the difference between the two values of u_{1n} to the smaller of the two values.

Similarly, reduced matrices are extracted from the full element inertia matrices resulting from the two models. The following dynamic matrices can subsequently be formed with the reduced stiffness matrices of Eqs. (8.117) and (8.118):

$$[D]_{3\times3} = \begin{bmatrix} M_{ip}^{(e)}(1,1) & M_{ip}^{(e)}(1,2) & M_{ip}^{(e)}(1,3) \\ M_{ip}^{(e)}(1,2) & M_{ip}^{(e)}(2,2) & M_{ip}^{(e)}(2,3) \\ M_{ip}^{(e)}(1,3) & M_{ip}^{(e)}(2,3) & M_{ip}^{(e)}(3,3) \end{bmatrix}^{-1} \begin{bmatrix} K_{ip}^{(e)}(1,1) & K_{ip}^{(e)}(1,2) & K_{ip}^{(e)}(1,3) \\ K_{ip}^{(e)}(1,2) & K_{ip}^{(e)}(2,2) & K_{ip}^{(e)}(2,3) \\ K_{ip}^{(e)}(1,3) & K_{ip}^{(e)}(2,3) & K_{ip}^{(e)}(3,3) \end{bmatrix}$$

$$[D]_{4\times4} = \begin{bmatrix} M_{ip}^{(e)}(1,1) & M_{ip}^{(e)}(1,2) & M_{ip}^{(e)}(1,3) & M_{ip}^{(e)}(1,4) \\ M_{ip}^{(e)}(1,2) & M_{ip}^{(e)}(2,2) & M_{ip}^{(e)}(2,3) & M_{ip}^{(e)}(2,4) \\ M_{ip}^{(e)}(1,3) & M_{ip}^{(e)}(2,3) & M_{ip}^{(e)}(3,3) & M_{ip}^{(e)}(3,4) \\ M_{ip}^{(e)}(1,4) & M_{ip,24}^{(e)}(2,4) & M_{ip}^{(e)}(3,4) & M_{ip}^{(e)}(4,4) \end{bmatrix}^{-1}$$

$$\begin{bmatrix} K_{ip}^{(e)}(1,1) & K_{ip}^{(e)}(1,2) & K_{ip}^{(e)}(1,3) & K_{ip}^{(e)}(1,4) \\ K_{ip}^{(e)}(1,2) & K_{ip}^{(e)}(2,2) & K_{ip}^{(e)}(2,3) & K_{ip}^{(e)}(2,4) \\ K_{ip}^{(e)}(1,3) & K_{ip}^{(e)}(2,3) & K_{ip}^{(e)}(3,3) & K_{ip}^{(e)}(3,4) \\ K_{ip}^{(e)}(1,4) & K_{ip,24}^{(e)}(2,4) & K_{ip}^{(e)}(3,4) & K_{ip}^{(e)}(4,4) \end{bmatrix}.$$

$$(8.119)$$

where the first dynamic matrix results from the linear-cubic model, while the second one is produced by the cubic-cubic model. The smallest natural frequencies are found from the eigenvalues of the two matrices of Eq. (8.119); the ratio of the difference between these two frequencies to the smaller of the two frequencies is subsequently calculated.

Figure 8.9 plots the relative deflection ratio and the relative first natural frequency ratio in terms of the circular element opening angle α, whereas Figure 8.10 graphs the same ratios as functions of the radius R. It can be seen that the ratios increase with both parameters increasing, and the percentage deviations between the two models' predictions are very small. This conclusion enables the utilization of the simpler linear-cubic model in further finite element analysis simulation.

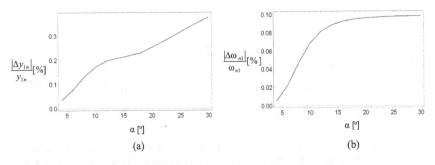

FIGURE 8.9 Plots in terms of the element opening angle α of the relative variation in: (a) radial displacement at node 1; (b) first natural frequency.

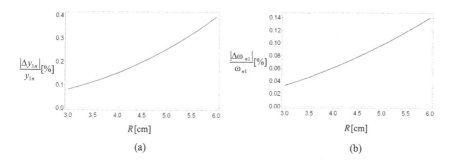

(a)   (b)

FIGURE 8.10 Plots in terms of the element radius R of the relative variation in: (a) radial displacement at node 1; (b) first natural frequency.

8.2.1.2 Out-of-Plane Matrices

The planar, circular-axis element of Figure 8.8 is sketched in Figure 8.11 by highlighting its out-of-plane nodal DOF and loads.

The out-of-plane deformations at the arbitrary point P, which is situated at a distance s measured along the circumference from node 1 (s is not shown in Figure 8.11), are the tangential-axis rotation $\theta_t(s)$, the normal (radial)-axis rotation $\theta_n(s)$ and the z-axis (out-of-plane) deflection $u_z(s)$. The out-of-plane nodal displacement vector collects these displacements at the nodes 1 and 2; similarly, the nodal load vector is formed of moments about the t- and n-axes, as well as of z-axis forces at the two nodes:

$$
\begin{cases}
\left[u_{op}^{(e)} \right] = \begin{bmatrix} \theta_{1t} & \theta_{1n} & u_{1z} & \theta_{2t} & \theta_{2n} & u_{2z} \end{bmatrix}^T \\
\left[f_{op}^{(e)} \right] = \begin{bmatrix} m_{1t} & m_{1n} & f_{1z} & m_{2t} & m_{2n} & f_{2z} \end{bmatrix}^T
\end{cases}
\tag{8.120}
$$

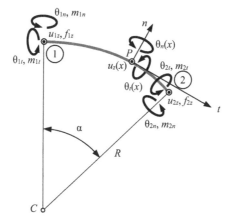

FIGURE 8.11 Planar, circular-axis flexible-hinge element with two nodes and three out-of-plane DOF per node.

The simplest model to describe the distribution of the three displacements at point P is the linear-cubic model, according to which:

$$
\begin{cases}
\theta_t(s) = c_0 + c_1 \cdot s \\[2mm]
u_z(s) = c_2 + c_3 \cdot s + c_4 \cdot s^2 + c_5 \cdot s^3 \\[2mm]
\theta_n(s) = \dfrac{du_z(s)}{ds} = c_3 + 2c_4 \cdot s + 3c_5 \cdot s^2
\end{cases}
\tag{8.121}
$$

The third Eq. (8.121) states that, similarly to a straight-axis beam, the slope/rotation $\theta_n(s)$ is the s-derivative of the corresponding deflection $u_z(s)$.

The following boundary conditions express the three nodal displacements at the two nodes:

$$
\theta_t(0) = \theta_{1t}; \quad \theta_n(0) = \theta_{1n}; \quad u_z(0) = u_{1z}; \quad \theta_t(l) = \theta_{2t}; \quad \theta_n(l) = \theta_{2n}; \quad u_z(l) = u_{2z}.
\tag{8.122}
$$

where $l = \alpha \cdot R$. These conditions are utilized in conjunction with Eqs. (8.121), which enables to find the constants c_0 through c_5; they are substituted back into the first two Eqs. (8.121), which are expressed in discretized form as:

$$
\theta_t(s,t) = \left[S_{\theta_t}(s) \right]^T \left[u_{op}^{(e)}(t) \right] \quad \text{with}
$$

$$
\left[S_{\theta_t}(s) \right] = \left[\begin{array}{cccccc} S_{\theta_t 1}(s) & 0 & 0 & S_{\theta_t 2}(s) & 0 & 0 \end{array} \right]^T
\tag{8.123}
$$

$$
u_z(s,t) = \left[S_z(s) \right]^T \left[u_{op}^{(e)}(t) \right] \quad \text{with}
$$

$$
\left[S_z(s) \right] = \left[\begin{array}{cccccc} 0 & S_{z1}(s) & S_{z2}(s) & 0 & S_{z3}(s) & S_{z4}(s) \end{array} \right]^T.
$$

The individual shape functions of the vectors given in Eqs (8.123) are:

$$
S_{\theta_t 1} = 1 - \frac{s}{l}; \quad S_{\theta_t 2} = \frac{s}{l};
$$

$$
\begin{cases}
S_{z1}(s) = s - 2 \cdot \dfrac{s^2}{l} + \dfrac{s^3}{l^2}; & S_{z2}(s) = 1 - 3 \cdot \dfrac{s^2}{l^2} + 2 \cdot \dfrac{s^3}{l^3}; \\[3mm]
S_{z3}(s) = -\dfrac{s^2}{l} + \dfrac{s^3}{l^2}; & S_{z4}(x) = 3 \cdot \dfrac{s^2}{l^2} - 2 \cdot \dfrac{s^3}{l^3}.
\end{cases}
\tag{8.124}
$$

The elastic strain energy of the element sketched in Figure 8.11 is expressed as in Howson and Jemah [39]:

$$U_{ip}^{(e)} = \frac{E}{2} \cdot \int_0^l I_n(s) \cdot \kappa_n^2 \, ds + \frac{G}{2} \cdot \int_0^l I_t(s) \cdot \gamma^2 \, ds, \tag{8.125}$$

where $I_n(s)$ and $I_t(s)$ are the cross-sectional moments of area about the n and t axes, respectively. The shear strain γ produced through torsion and the normal curvature κ_n are expressed as in Howson and Jemah [39]:

$$\gamma = \frac{d\theta_t(s)}{ds} - \frac{1}{R} \cdot \frac{du_z(s)}{ds}; \quad \kappa_n = \frac{d^2 u_z(s)}{ds^2} + \frac{1}{R} \cdot \theta_t(s). \tag{8.126}$$

The strain and curvature of Eqs. (8.126) are substituted into the strain energy of Eq. (8.125), which becomes:

$$U_{op}^{(e)} = \frac{E}{2} \cdot \int_0^l I_n(s) \left(\frac{\partial^2 u_z(s,t)}{\partial s^2} + \frac{1}{R} \cdot \theta_t(s,t) \right)^2 ds + \frac{G}{2} \cdot \int_0^l I_t(s) \left(\frac{\partial \theta_t(s,t)}{\partial s} - \frac{1}{R} \cdot \frac{\partial u_z(s,t)}{\partial s} \right)^2 ds \tag{8.127}$$

The element stiffness matrix is formulated by means of Castigliano's force theorem, according to which the nodal load components are calculated as:

$$m_{jt} = \frac{\partial U_{op}^{(e)}}{\partial \theta_{jt}}; \quad m_{jn} = \frac{\partial U_{op}^{(e)}}{\partial \theta_{jn}}; \quad f_{jz} = \frac{\partial U_{op}^{(e)}}{\partial u_{jz}}, \quad j = 1,2 \tag{8.128}$$

The shape functions of Eqs. (8.123) are used in conjunction with the strain energy of Eq. (8.127) to obtain the loads of Eqs. (8.128), which can ultimately be expressed in vector-matrix form as:

$$\left[f_{op}^{(e)} \right] = \left[K_{op}^{(e)} \right] \left[u_{op}^{(e)} \right]. \tag{8.129}$$

where $\left[K_{op}^{(e)} \right]$ is the element out-of-plane stiffness matrix.

The total kinetic energy of the finite element is the sum of all elementary kinetic energy fractions resulting from translation along the z axis and rotation around the t axis, namely:

$$T_{op}^{(e)} = \frac{1}{2} \cdot \rho \cdot \left(\int_0^l A(s) \left(\frac{\partial u_z(s,t)}{\partial t} \right)^2 ds + \int_0^l I_t(s) \cdot \left(\frac{\partial \theta_t(s,t)}{\partial t} \right)^2 ds \right)$$

$$= \frac{1}{2} \cdot \left[\dot{u}_{op}^{(e)}(t) \right] \left(\rho \left(\int_0^l A(s) [S_z(s)][S_z(s)]^T ds + \int_0^l I_t(s) [S_{\theta_t}(s)][S_{\theta_t}(s)]^T ds \right) \right)$$

$$\left[\dot{u}_{op}^{(e)}(t) \right]^T. \tag{8.130}$$

Comparing Eq. (8.130) to Eq. (7.22), the out-of-plane consistent element inertia matrix of the circular-axis element is:

$$\left[M_{op}^{(e)} \right] = \rho \left(\int_0^l A(s) \left[S_z(s) \right] \left[S_z(s) \right]^T ds + \int_0^l I_t(s) \left[S_{\theta_t}(s) \right] \left[S_{\theta_t}(s) \right]^T ds \right). \quad (8.131)$$

8.2.2 SHORT ELEMENTS AND THE TIMOSHENKO MODEL

For circular-axis flexible hinges that have small ratios of the length to the maximum cross-sectional dimension and/or a small angular span α, the added effects of shear deformation and rotary inertia have to be considered – this is achieved through the Timoshenko model, as discussed in the straight-axis element section in this chapter. A similar approach is applied here in order to formulate the element stiffness and inertia matrices separately for in-plane and out-of-plane motions.

8.2.2.1 In-Plane Matrices

The nodal displacement and force vectors of Eq. (8.99) used for a relatively long and thin member to derive elemental matrices with the Euler–Bernoulli model are also utilized here – see Figure 8.8. While the expressions of the strain ε and curvature κ_z given in Eq. (8.106) are still valid, there is an extra rotation angle produced through shear; this extra angle θ_z^{sh}, as discussed in Khrishnan and Suresh [30] and Shahbah et al. [36] among others, is expressed as:

$$\theta_z^{sh}(s) = \theta_z(s) - \frac{du_n(s)}{ds}, \quad (8.132)$$

Because of this additional rotation angle, the z-axis rotation $\theta_z(s)$ cannot be expressed in terms of $u_t(s)$ and $u_n(s)$ as in Eq. (8.101) and needs to be discretized separately; however, the tangential and radial displacements $u_t(s)$ and $u_n(s)$ are expressed as in Eq. (8.100) based on the linear-cubic model. The resulting distribution equations are:

$$\begin{cases} u_t(s) = c_0 + c_1 \cdot s \\ u_n(s) = c_2 + c_3 \cdot s + c_4 \cdot s^2 + c_5 \cdot s^3 . \\ \theta_z(s) = c_6 + c_7 \cdot s + c_8 \cdot s^2 \end{cases} \quad (8.133)$$

The radial displacement $u_n(s)$ and rotation $\theta_z(s)$ are not independent; as discussed in Khrishnan and Suresh [30] and Shahbah et al. [36], the following equilibrium equation (which results from the differential relationship between bending moment and shear force) can be used to relate these two parameters:

$$\beta_z \left(\frac{\partial^3 u_n(s)}{\partial s^3} - \frac{1}{R} \cdot \frac{\partial^2 u_t(s)}{\partial s^2} \right) = \theta_z(s) - \frac{\partial u_n(s)}{\partial s} \quad \text{with} \quad \beta_z = \frac{EI_z}{\alpha_s GA}. \quad (8.134)$$

The six boundary conditions of Eqs. (8.102) are still valid with this model; they are used in conjunction with Eqs. (8.133) and (8.134) to find the constants c_0 through c_8; the expressions of these constants are then substituted back into Eqs. (8.133), which are written in discretized form as:

$$u_t(s,t) = [S_t(s)]^T [u_{ip}^{(e)}(t)] \text{ with } [S_t(s)] = \begin{bmatrix} S_{t1}(s) & 0 & 0 & S_{t2}(s) & 0 & 0 \end{bmatrix}^T$$

$$u_n(s,t) = [S_n(s)]^T [u_{ip}^{(e)}(t)] \text{ with}$$

$$[S_n(s)] = \begin{bmatrix} 0 & S_{n1}(s) & S_{n2}(s) & 0 & S_{n3}(s) & S_{n4}(s) \end{bmatrix}^T \quad (8.135)$$

$$\theta_z(s,t) = [S_{\theta_z}(s)]^T [u_{ip}^{(e)}(t)] \text{ with}$$

$$[S_n(s)] = \begin{bmatrix} 0 & S_{\theta_z 1}(s) & S_{\theta_z 2}(s) & 0 & S_{\theta_z 3}(s) & S_{\theta_z 4}(s) \end{bmatrix}^T.$$

The individual shape functions of Eqs. (8.135) are too complex and are not included here.

The strain energy of the finite element of Figure 8.8 during its planar elastic deformation is similar to the one of the Euler–Bernoulli element – see Eq. (8.105); it includes an extra term related to shear deformation, namely:

$$U_{ip}^{(e)} = \frac{E}{2} \cdot \int_0^l \left(A(s) \cdot \varepsilon^2 + I_z(s) \cdot \kappa_z^2 \right) ds + \frac{G}{2} \cdot \int_0^l \alpha_s A(s) \cdot \left(\theta_z^{sh}(s) \right)^2 ds. \quad (8.136)$$

The strain and curvature of Eqs. (8.106), as well as the shear angle of Eq. (8.132), are substituted into the strain energy of Eq. (8.136), which becomes:

$$U_{ip}^{(e)} = \frac{E}{2} \cdot \left(\int_0^l A(s) \left(\frac{\partial u_t(s,t)}{\partial s} + \frac{1}{R} \cdot u_n(s,t) \right)^2 ds + \int_0^l I_z(s) \left(\frac{\partial^2 u_n(s,t)}{\partial s^2} - \frac{1}{R} \cdot u_t(s,t) \right)^2 ds \right)$$

$$+ \frac{G\alpha_s}{2} \cdot \int_0^l A(s) \left(\theta_z(s,t) - \frac{\partial u_n(s,t)}{\partial s} \right)^2 ds \quad (8.137)$$

The discretization relationships of Eqs. (8.135) are substituted into the strain energy of Eq. (8.137). Castigliano's force theorem is subsequently applied in the form of Eq. (8.108), which results in a nodal load/displacement connection as in Eq. (8.109), where $\left[K_{ip}^{(e)} \right]$ is the element in-plane stiffness matrix of a circular-axis Timoshenko element.

The element kinetic energy adds a rotary component to the two translational fractions corresponding to the Euler–Bernoulli model of Eq. (8.110); this energy is:

$$T_{ip}^{(e)} = \frac{1}{2} \cdot \rho \left(\int_0^l A(s) \cdot \left(\frac{\partial u_t(s,t)}{\partial t} \right)^2 ds + \int_0^l A(s) \cdot \left(\frac{\partial u_n(s,t)}{\partial t} \right)^2 ds + \int_0^l I_z(s) \cdot \left(\frac{\partial \theta_z(s,t)}{\partial t} \right) \right)$$

$$(8.138)$$

Using the discretization Eqs. (8.135) in conjunction with Eq. (8.138) results in the element kinetic energy being expressed as in the generic Eq. (7.22), which identifies the element inertia matrix of the finite element of Figure 8.8 as:

$$\left[M_{ip}^{(e)} \right] = \rho \left(\int_0^l A(s) \left[S_t(s) \right] \left[S_t(s) \right]^T ds + \int_0^l A(s) \left[S_n(s) \right] \left[S_n(s) \right]^T ds \right.$$

$$\left. + \int_0^l I_z(s) \left[S_{\theta_z}(s) \right] \left[S_{\theta_z}(s) \right]^T dsw \right).$$

$$(8.139)$$

8.2.2.2 Out-of-Plane Matrices

The nodal displacement and load vectors of the two-node element sketched in Figure 8.11 are the ones of Eq. (8.120). Because of shear effects, the rotation around the normal (n) axis cannot be expressed simply as in Eq. (8.121); instead, the total rotation angle comprises the rotation angle θ_n^{sh} due to shear, namely:

$$\theta_n(s) = \theta_n^{sh}(s) + \frac{du_z(s)}{ds}. \tag{8.140}$$

As a result, the following distribution functions are proposed for the simplest model:

$$\begin{cases} \theta_t(s) = c_0 + c_1 \cdot s \\ u_z(s) = c_2 + c_3 \cdot s + c_4 \cdot s^2 + c_5 \cdot s^3 \\ \theta_n(s) = c_6 + c_7 \cdot s + c_8 \cdot s^2 \end{cases} \tag{8.141}$$

As detailed in Howson and Jemah [39] and Davis et al. [40], an equilibrium equation can be used, which connects the bending moment (around the normal axis), the torque (around the tangent axis) and the z-axis force; this equation can be written in terms of displacements as:

$$\beta_n \cdot \left(\frac{\partial^3 u_z(s,t)}{\partial s^3} + \frac{1}{R} \cdot \frac{\partial \theta_t(s,t)}{\partial s} \right) + \gamma_t \cdot \left(\frac{\partial \theta_t(s,t)}{\partial s} - \frac{1}{R} \cdot \theta_n(s,t) \right) = \theta_n(s,t) - \frac{\partial u_z(s,t)}{\partial s}$$

$$\text{with} \quad \beta_n = \frac{EI_n}{\alpha_s GA}; \ \gamma_t = \frac{I_t}{\alpha_s RA} \tag{8.142}$$

Equations (8.141) and (8.142) are combined with the six boundary conditions of Eqs. (8.122); they allow solving for the constants of Eqs. (8.141), which enable to reformulate the latter equations in the discretized form:

$$\theta_t(s,t) = \left[S_{\theta_t}(s) \right]^T \left[u_{op}^{(e)}(t) \right] \quad \text{with}$$

$$\left[S_{\theta_t}(s) \right] = \left[\begin{array}{cccccc} S_{\theta_t 1}(s) & 0 & 0 & S_{\theta_t 2}(s) & 0 & 0 \end{array} \right]^T$$

$$\theta_n(s,t) = \left[S_{\theta_n}(s) \right]^T \left[u_{op}^{(e)}(t) \right] \quad \text{with}$$

$$\left[S_n(s) \right] = \left[\begin{array}{cccccc} S_{\theta_n 1}(s) & S_{\theta_n 2}(s) & S_{\theta_n 3}(s) & S_{\theta_n 4}(s) & S_{\theta_n 5}(s) & S_{\theta_n 6}(s) \end{array} \right]^T$$

$$u_z(s,t) = \left[S_z(s) \right]^T \left[u_{op}^{(e)}(t) \right] \quad \text{with}$$

$$\left[S_z(s) \right] = \left[\begin{array}{cccccc} S_{z1}(s) & S_{z2}(s) & S_{z3}(s) & S_{z4}(s) & S_{z5}(s) & S_{z6}(s) \end{array} \right]^T. \tag{8.143}$$

The shape functions of Eqs. (8.143) are not included here being too complex.

The strain energy is formed based on Eqs. (8.125)–(8.127), which consider the bending and torsion effects related to a regular Euler–Bernoulli member; in addition, the shear elastic energy is added based on the Timoshenko model; as indicated in Howson and Jemah [39] and Davis et al. [40], this strain energy is ultimately expressed as:

$$U_{op}^{(e)} = \frac{E}{2} \cdot \int_0^l I_n(s) \cdot \left(\frac{\partial^2 u_z(s,t)}{\partial s^2} + \frac{1}{R} \cdot \theta_t(s,t) \right)^2 ds + \frac{G}{2} \cdot \int_0^l I_t(s) \cdot$$

$$\times \left(\frac{\partial \theta_t(s,t)}{\partial s} - \frac{1}{R} \cdot \theta_n(s,t) \right)^2 ds + \frac{\alpha_s G}{2} \cdot \int_0^l A(s) \cdot \left(\theta_n(s,t) - \frac{\partial u_z(s,t)}{\partial s} \right)^2 ds \tag{8.144}$$

Castigliano's force theorem is applied to derive the element stiffness matrix. Combining Eqs. (8.143) and (8.144), the six nodal load components are calculated as in Eqs. (8.128), which are rendered into the load-displacement Eq. (8.129) – this relationship provides the out-of-plane stiffness matrix $\left[K_{op}^{(e)} \right]$ of a circular-axis Timoshenko element.

The element kinetic energy sums contributions from translation along the z axis and rotations around the tangent and normal axes, namely:

$$T_{op}^{(e)}$$

$$= \frac{1}{2} \cdot \rho \left(\int_0^l A(s) \cdot \left(\frac{\partial u_z(s,t)}{\partial t} \right)^2 ds + \int_0^l I_t(s) \cdot \left(\frac{\partial \theta_t(s,t)}{\partial t} \right)^2 ds + \int_0^l I_n(s) \cdot \left(\frac{\partial \theta_n(s,t)}{\partial t} \right)^2 ds \right).$$

(8.145)

The discretization relationships of Eqs. (8.143) are substituted into Eqs. (8.145); the resulting energy has the form of the generic Eq. (7.22), and the consistent element inertia matrix corresponding to out-of-plane motion of the element of Figure 8.11 is:

$$\left[M_{op}^{(e)} \right] = \rho \cdot \left(\int_0^l A(s) [S_z(s)][S_z(s)]^T ds + \int_0^l I_t(s) [S_{\theta_t}(s)][S_{\theta_t}(s)]^T ds \right.$$

(8.146)

$$\left. + \int_0^l I_n(s) [S_{\theta_n}(s)][S_{\theta_n}(s)]^T ds \right).$$

Example 8.3

A planar, circular-axis flexible hinge is shown in Figure 8.12; it has a rectangular cross-section whose in-plane thickness varies linearly from a minimum value t_1 at one end to a maximum value t_2 at the opposite end. Considering that the end of thickness t_2 is fixed, study the variations of the in-plane natural frequencies when calculated with the Euler–Bernoulli model vs. the Timoshenko model in terms of the thickness variation $\Delta t = t_2 - t_1$. Known are $t_1 = 0.001$ m, $w = 0.008$ m (constant out-of-plane width), $R = 0.018$ m, $\alpha = 10°$, $E = 2.1 \cdot 10^{11}$ N/m², $\mu = 0.3$, $\alpha_s = 1.176$ and $\rho = 7,800$ kg/m³. Use the two-node finite element of Figure 8.8.

Solution:

The variable thickness $t(s)$ at a variable distance s from the origin 1 is calculated as:

$$t(s) = \frac{t_2 - t_1}{l} \cdot s + t_1, \quad l = \alpha \cdot R. \tag{8.147}$$

With it, the variable cross-sectional area and z-axis moment of area are: $A(s) = w \cdot t(s)$, $I_z(s) = w \cdot t(s)^3/12$.

The full, 6×6 element stiffness matrix corresponding to the Euler–Bernoulli model is calculated by using the shape functions of Eqs. (8.104) based on the

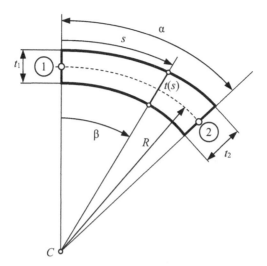

FIGURE 8.12 Planar, circular-axis flexible-hinge element with linearly variable in-plane thickness $t(s)$.

definition Eqs. (8.107)–(8.109). The similar Timoshenko-model element stiffness matrix is determined by the shape functions of Eqs. (8.135) that are combined with the definition Eqs. (8.107)–(8.109). The Euler–Bernoulli element inertia matrix is evaluated as in Eq. (8.111) with the shape functions of Eqs. (8.104), while the similar matrix related to the Timoshenko model is determined as in Eq. (8.139) with the shape functions of Eqs. (8.135).

The nonzero, in-plane displacement vector is formed of the three displacements at the free node 1, namely, u_{1t}, u_{1n} and θ_{1z}. The resulting reduced 3×3 stiffness and inertia matrices are of the form provided in Eqs. (8.119) of *Example 8.2*, which also expresses the resulting dynamic matrix – this is valid for both models. The latter matrix provides the three natural frequencies when the parameter Δt varies from 0 to 1 mm. To study the variation of the natural frequencies, the following ratios are formed:

$$\frac{\left|\Delta\omega_{ni}\right|}{\omega_{ni}} = \frac{\left|\omega_{ni}^{EB} - \omega_{ni}^{T}\right|}{\min\left(\omega_{ni}^{EB}, \omega_{ni}^{T}\right)}, \quad i = 1,2,3, \tag{8.148}$$

where the superscripts "*EB*" and "*T*" stand for Euler–Bernoulli and Timoshenko, respectively. The relative differences between the two models' predictions of the first (smallest) natural frequencies grow almost linearly from approximately 5% to 18%, as shown in Figure 8.13. The relative differences corresponding to the other two natural frequencies are not plotted, but the second natural frequency variation is almost constant and equal to 0.09%, while the relative variation in the third natural frequencies slightly increases from 24% to 28% with an increase in Δt.

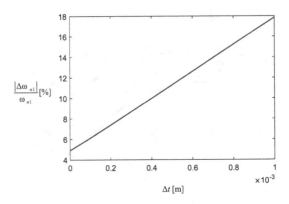

FIGURE 8.13 Plot of the relative first natural frequency variation in terms of the thickness parameter Δt.

8.3   FLEXIBLE-HINGE MECHANISMS

The finite element free and forced responses of mechanisms with flexible links/ hinges have received considerable coverage by means of a variety of modeling/solution methods and applications – see Thompson and Sung [41], Wang and Wang [42], Saxena and Ananthasuresh [43], Yu and Smith [44], Sriram and Mruthyunjaya [45], Xianmin et al. [46], Wang [47], Chen [48], Besseling and Gong [49], Gao [50], Turcic and Midha [51], Chu and Pan [52], Song and Haug [53], Gao et al. [54], Sekulovic and Salatic [55], Hac and Osinski [56] or Hac [57]. Similar research was reported by Piras [58], Albanesi [59], Dubay [60], Muhammad et al. [61], Beloti et al. [62], Tian et al. [63], Kermanian [64] and My [65].

Several examples of mechanisms, both serial and parallel, are modeled and analyzed in this section by using the line elements formulated in the previous sections. Planar and spatial flexible mechanism configurations are studied by formulating both the quasi-static and the dynamic responses. The mechanisms are modeled as frames, i.e., devices that are structurally formed by rigidly connecting flexible hinges and other rigid members.

8.3.1   GLOBAL-FRAME ELEMENT STIFFNESS AND INERTIA MATRICES

Planar and spatial mechanisms are most often formed of flexible hinges, which have various orientations/directions. Because the nodal displacements and loads need to be expressed in a unique reference frame (the *global frame*), the elemental stiffness and inertia matrices have to be correspondingly transformed into the global frame from their element *local frames*. This section addresses transforming elemental matrices of straight-axis and circular-axis planar flexible hinges by means of rotation.

8.3.1.1   Straight-Axis Flexible Hinges

Figure 8.14 shows a two-node, straight-axis line element – while only the in-plane DOF are symbolized in the figure, the element is assumed to have six DOF per node, and therefore, it also possesses out-of-plane capabilities.

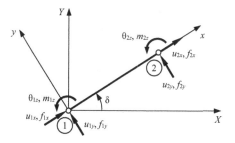

FIGURE 8.14   Straight-axis line element with local reference frame xyz and global reference frame XYZ.

A rotation matrix $[R^{(e)}]$ connects the local-frame displacement vector $\left[u_{i,ip}^{(e,l)}\right] = \begin{bmatrix} u_{ix} & u_{iy} & \theta_{iz} \end{bmatrix}^T$ to the global-frame vector $\left[u_{i,ip}^{(e,g)}\right] = \begin{bmatrix} u_{iX} & u_{iY} & \theta_{iZ} \end{bmatrix}^T$; similarly, the local-frame nodal force vector $\left[f_{i,ip}^{(e,l)}\right] = \begin{bmatrix} f_{ix} & f_{iy} & m_{iz} \end{bmatrix}^T$ is connected to its global-frame counterpart $\left[f_{i,ip}^{(e,g)}\right] = \begin{bmatrix} f_{iX} & f_{iY} & m_{iZ} \end{bmatrix}^T$ – where $i = 1, 2$ indicates any of the two nodes; these relationships are:

$$\begin{aligned} \left[u_{ip}^{(e,l)}\right] &= [R]\left[u_{ip}^{(e,g)}\right] \\ \left[f_{ip}^{(e,l)}\right] &= [R]\left[f_{ip}^{(e,g)}\right] \end{aligned} \quad \text{with } [R] = \begin{bmatrix} \left[R^{(e)}\right] & [0] \\ [0] & \left[R^{(e)}\right] \end{bmatrix};$$

$$\left[R^{(e)}\right] = \begin{bmatrix} \cos\delta & \sin\delta & 0 \\ -\sin\delta & \cos\delta & 0 \\ 0 & 0 & 1 \end{bmatrix} \tag{8.149}$$

In the local frame xyz, the in-plane nodal force vector is related to the corresponding nodal displacement vector by means of the stiffness matrix:

$$\left[f_{ip}^{(e)}\right] = \left[K_{ip}^{(e)}\right]\left[u_{ip}^{(e)}\right]. \tag{8.150}$$

Combining Eqs. (8.149) and (8.150) results in the global-frame (transformed) load-displacement relationship by means of a corresponding stiffness matrix:

$$\left[f_{ip}\right] = \left[K_{ip}\right]\left[u_{ip}\right] \quad \text{with} \quad \left[K_{ip}\right] = [R]^T\left[K_{ip}^{(e)}\right][R]. \tag{8.151}$$

A similar formulation that would use accelerations instead of displacements and inertia loads instead of static nodal loads enables transforming the local-frame element inertia matrix $\left[M_{ip}^{(e)}\right]$ into the global one $[M_{ip}]$ as:

$$\left[M_{ip}\right] = [R]^T\left[M_{ip}^{(e)}\right][R]. \tag{8.152}$$

It can be shown that the out-of-plane elemental stiffness and inertia matrices can be transformed from the original, local frame into the global frame by means of the same $[R]$ matrix as:

$$\left[K_{op}\right]=[R]^{T}\left[K_{op}^{(e)}\right][R]; \quad \left[M_{op}\right]=[R]^{T}\left[M_{op}^{(e)}\right][R]. \tag{8.153}$$

The out-of-plane, local-frame and global-frame displacement vectors are $\left[u_{op}^{(e)}\right]=\left[\begin{array}{ccc} \theta_x & \theta_y & u_z \end{array}\right]^{T}$ and $[u_{op}]=[\theta_X\ \theta_Y\ u_Z]^{T}$, while the corresponding nodal-force vectors are $\left[f_{op}^{(e)}\right]=\left[\begin{array}{ccc} m_x & m_y & f_z \end{array}\right]^{T}$ and $[f_{op}]=[m_X\ m_Y\ f_Z]^{T}$.

8.3.1.2 Circular-Axis Flexible Hinges

Similar to the straight-axis element depicted in Figure 8.14, only the in-plane nodal loads and displacements are indicated for the two-node, planar, circular-axis element of Figure 8.15. The translation displacements are along the tangent and normal directions at the two nodes, while the rotations at the two nodes are around the z axis, which is perpendicular to the plane of the element. Assume that the local frame at node 1 has its x_1 axis along the tangent and the y_1 axis along the normal at that node; similarly, the frame located at node 2 has its x_2 axis along the tangent and the y_2 axis directed along the normal at that node.

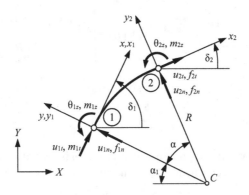

FIGURE 8.15 Circular-axis line element with local and global reference frames.

The nodal displacements and loads, as well as the related element stiffness and inertia matrix components, are formulated in terms of the end-point frames $x_1y_1z_1$ and $x_2y_2z_2$, which are rotated by an angle α in the global plane XY. It is therefore useful to express these matrices in a single reference frame, like xyz, which is identical to $x_1y_1z_1$, as illustrated in Figure 8.15. The subsequent derivation applies to both in-plane and out-of-plane formulations, and therefore, the subscripts normally used to indicate the respective cases ("*ip*" or "*op*") are not utilized. The displacements and loads at node 2 (which are defined in the $x_2y_2z_2$ reference frame) are expressed in terms of the corresponding vectors formulated in the xyz frame as:

$$\begin{cases} \left[u_2^{(e)} \right] = \left[R_2^{(e)} \right] \left[u_2^{(e,l)} \right] \\ \left[f_2^{(e)} \right] = \left[R_2^{(e)} \right] \left[f_2^{(e,l)} \right] \end{cases} \text{with} \quad \left[R_2^{(e)} \right] = \begin{bmatrix} \cos(-\alpha) & \sin(-\alpha) & 0 \\ -\sin(-\alpha) & \cos(-\alpha) & 0 \\ 0 & 0 & 1 \end{bmatrix}. \quad (8.154)$$

where the superscript "e" denotes the vectors in the $x_2 y_2 z_2$ frame, while "e,l" identifies the vectors in the local frame xyz. The original static element model is:

$$\begin{bmatrix} \left[f_1^{(e)} \right] \\ \left[f_2^{(e)} \right] \end{bmatrix} = \begin{bmatrix} \left[K^{(e)}(1,1) \right] & \left[K^{(e)}(1,2) \right] \\ \left[K^{(e)}(1,2) \right]^T & \left[K^{(e)}(2,2) \right] \end{bmatrix} \begin{bmatrix} \left[u_1^{(e)} \right] \\ \left[u_2^{(e)} \right] \end{bmatrix}. \quad (8.155)$$

Combining Eqs. (8.154) and (8.155) results in the following equation:

$$\begin{bmatrix} \left[f_1^{(e,l)} \right] \\ \left[f_2^{(e,l)} \right] \end{bmatrix} = \left[K^{(e,l)} \right] \begin{bmatrix} \left[u_1^{(e,l)} \right] \\ \left[u_2^{(e,l)} \right] \end{bmatrix} \text{with}$$

$$\left[K^{(e,l)} \right] = \begin{bmatrix} \left[K^{(e)}(1,1) \right] & \left[K^{(e)}(1,2) \right] \left[R_2^{(e)} \right] \\ \left(\left[K^{(e)}(1,2) \right] \left[R_2^{(e)} \right] \right)^T & \left[R_2^{(e)} \right]^T \left[K^{(e,l)}(2,2) \right] \left[R_2^{(e)} \right] \end{bmatrix}. \quad (8.156)$$

where $[K^{(e,l)}]$ is the element stiffness matrix expressed in the local frame xyz located at node 1.

The local-frame inertia matrix is formulated similarly as:

$$\left[M^{(e,l)} \right] = \begin{bmatrix} \left[M^{(e)}(1,1) \right] & \left[M^{(e)}(1,2) \right] \left[R_2^{(e)} \right] \\ \left(\left[M^{(e)}(1,2) \right] \left[R_2^{(e)} \right] \right)^T & \left[R_2^{(e)} \right]^T \left[M^{(e,l)}(2,2) \right] \left[R_2^{(e)} \right] \end{bmatrix}$$

$$\text{where} \quad \left[M^{(e)} \right] = \begin{bmatrix} \left[M^{(e)}(1,1) \right] & \left[M^{(e)}(1,2) \right] \\ \left[M^{(e)}(1,2) \right]^T & \left[M^{(e)}(2,2) \right] \end{bmatrix}. \quad (8.157)$$

The local-frame stiffness and inertia matrices of Eqs. (8.156) and (8.157) can be transferred further to the global frame XYZ of Figure 8.15 as:

$$\left[K^{(e,g)} \right] = [R]^T \left[K^{(e,l)} \right] [R]; \quad \left[M^{(e,g)} \right] = [R]^T \left[M^{(e,l)} \right] [R]. \quad (8.158)$$

where the superscript "g" means global, and the rotation matrix is:

$$[R] = \begin{bmatrix} [R_1] & [0] \\ [0] & [R_1] \end{bmatrix} \quad \text{with} \quad [R_1] = \begin{bmatrix} \cos\delta_1 & \sin\delta_1 & 0 \\ -\sin\delta_1 & \cos\delta_1 & 0 \\ 0 & 0 & 1 \end{bmatrix}. \quad (8.159)$$

The rotation angle of Eq. (8.159), which is shown in Figure 8.15, is $\delta_1 = \pi/2 - \alpha_1$.

8.3.2 THE ASSEMBLING PROCESS

The strain energy of a finite element model comprising n elements can be written by summing the strain energies of individual elements, namely:

$$U = \frac{1}{2} \cdot \sum_{i=1}^{n} \left[u^{(i)} \right]^T \left[K^{(i)} \right] \left[u^{(i)} \right]. \quad (8.160)$$

The summation of Eq. (8.160) is written in vector form as:

$$U = \frac{1}{2} \cdot [\tilde{u}]^T [\tilde{K}][\tilde{u}], \quad (8.161)$$

where

$$[\tilde{u}] = \begin{bmatrix} \left[u^{(1)} \right] & \left[u^{(2)} \right] & \cdots & \left[u^{(n)} \right] \end{bmatrix}^T \quad (8.162)$$

and

$$[\tilde{K}] = \begin{bmatrix} \left[K^{(1)} \right] & 0 & 0 & \cdots & 0 \\ 0 & \left[K^{(2)} \right] & 0 & \cdots & 0 \\ 0 & 0 & \left[K^{(3)} \right] & \cdots & 0 \\ \cdots & \cdots & \cdots & \cdots & \cdots \\ 0 & 0 & 0 & \cdots & \left[K^{(n)} \right] \end{bmatrix}. \quad (8.163)$$

Compatibility should exist between the vector collecting all elemental displacements of Eq. (8.162) and the vector comprising all nodal displacements $[u]$ of the generic form:

$$[\tilde{u}] = [Bo][u], \quad (8.164)$$

where $[Bo]$ is a Boolean matrix containing only 1 and 0 terms in it. By means of this equation, the total strain energy of Eq. (8.161) becomes:

$$U = \frac{1}{2} \cdot [u]^T [Bo]^T \left[\tilde{K} \right] [Bo][u]. \tag{8.165}$$

The strain energy of a system can be expressed in terms of its generic coordinates as in Eq. (8.10), where $[K]$ is the assembled stiffness matrix of the entire finite element model formed of n elements. By comparing Eqs. (8.165) and (8.10), it follows that:

$$[K] = [Bo]^T \left[\tilde{K} \right] [Bo]. \tag{8.166}$$

A similar procedure can be applied in terms of the kinetic energy of a system, and this provides the following global-frame assembled inertia matrix:

$$[M] = [Bo]^T \left[\tilde{M} \right] [Bo], \tag{8.167}$$

with

$$[\tilde{M}] = \begin{bmatrix} \left[M^{(1)} \right] & 0 & 0 & \cdots & 0 \\ 0 & \left[M^{(2)} \right] & 0 & \cdots & 0 \\ 0 & 0 & \left[M^{(3)} \right] & \cdots & 0 \\ \cdots & \cdots & \cdots & \cdots & \cdots \\ 0 & 0 & 0 & \cdots & \left[M^{(n)} \right] \end{bmatrix} \tag{8.168}$$

being the inertia matrix that collects all elements' inertia matrices on its main diagonal. It can also be shown that the nodal load vector corresponding to the system is:

$$[F] = [Bo]^T \left[\tilde{F} \right], \tag{8.169}$$

where

$$\left[\tilde{F} \right] = \left[\ \left[F^{(1)} \right] \ \ \left[F^{(2)} \right] \ \ \cdots \ \ \left[F^{(n)} \right] \ \right]^T \tag{8.170}$$

is formed of all element loads.

Example 8.4

Obtain the assembled stiffness and inertia matrices for the finite element model of Figure 8.16 that is formed of two collinear, straight-axis, two-node line elements with axial DOF only.

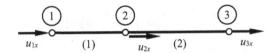

FIGURE 8.16   Two-element bar.

Solution:

The nodal displacement vector $[u]$ of the two-element bar and the vector $[\tilde{u}]$ collecting all the element nodal displacements are:

$$[u] = \begin{bmatrix} u_{1x} & u_{2x} & u_{3x} \end{bmatrix}^T ;$$

$$[\tilde{u}] = \begin{bmatrix} [u^{(1)}] & [u^{(2)}] \end{bmatrix}^T = \begin{bmatrix} u_{1x} & u_{2x} & u_{2x} & u_{3x} \end{bmatrix}^T . \tag{8.171}$$

It is straightforward to establish that the connection between these two vectors is realized as in Eq. (8.164) by the following Boolean matrix:

$$[Bo] = \begin{bmatrix} 1 & 0 & 0 \\ 0 & 1 & 0 \\ 0 & 1 & 0 \\ 0 & 0 & 1 \end{bmatrix}. \tag{8.172}$$

The stiffness and inertia matrices of the two elements of Figure 8.16 are of the form:

$$\left[K^{(i)}\right] = \begin{bmatrix} K^{(i)}(1,1) & K^{(i)}(1,2) \\ K^{(i)}(1,2) & K^{(i)}(2,2) \end{bmatrix}; \quad \left[M^{(i)}\right] = \begin{bmatrix} M^{(i)}(1,1) & M^{(i)}(1,2) \\ M^{(i)}(1,2) & M^{(i)}(2,2) \end{bmatrix}, \quad i = 1, 2. \tag{8.173}$$

The diagonal stiffness and the mass matrices of Eqs. (8.163) and (8.168) are:

$$\left[\tilde{K}\right] = \begin{bmatrix} \left[K^{(1)}\right] & [0] \\ [0] & \left[K^{(2)}\right] \end{bmatrix} = \begin{bmatrix} K^{(1)}(1,1) & K^{(1)}(1,2) & 0 & 0 \\ K^{(1)}(1,2) & K^{(1)}(2,2) & 0 & 0 \\ 0 & 0 & K^{(2)}(1,1) & K^{(2)}(1,2) \\ 0 & 0 & K^{(2)}(1,2) & K^{(2)}(2,2) \end{bmatrix};$$

$$[\tilde{M}] = \begin{bmatrix} [M^{(1)}] & [0] \\ [0] & [M^{(2)}] \end{bmatrix} = \begin{bmatrix} M^{(1)}(1,1) & M^{(1)}(1,2) & 0 & 0 \\ M^{(1)}(1,2) & M^{(1)}(2,2) & 0 & 0 \\ 0 & 0 & M^{(2)}(1,1) & M^{(2)}(1,2) \\ 0 & 0 & M^{(2)}(1,2) & M^{(2)}(2,2) \end{bmatrix}$$

(8.174)

By combining now Eqs. (8.172)–(8.174), the assembled stiffness and inertia matrices are obtained as per Eqs. (8.166) and (8.167) as:

$$\begin{cases} [K] = \begin{bmatrix} K^{(1)}(1,1) & K^{(1)}(1,2) & 0 \\ K^{(1)}(1,2) & K^{(1)}(2,2)+K^{(2)}(1,1) & K^{(2)}(1,2) \\ 0 & K^{(2)}(1,2) & K^{(2)}(2,2) \end{bmatrix} \\ [M] = \begin{bmatrix} M^{(1)}(1,1) & M^{(1)}(1,2) & 0 \\ M^{(1)}(1,2) & M^{(1)}(2,2)+M^{(2)}(1,1) & M^{(2)}(1,2) \\ 0 & M^{(2)}(1,2) & M^{(2)}(2,2) \end{bmatrix} \end{cases}$$

(8.175)

Equations (8.175) suggest a superposition of the matrix components resulting from the two adjacent elements at the common node (and common DOF). Figure 8.17 illustrates this concept by highlighting the blocks pertaining to the element matrices in the assembled matrices $[K]$ and $[M]$ by using generic element matrices $[A^{(1)}]$ and $[A^{(2)}]$ to form the generic assembled matrix $[A]$. The thicker-line box emphasizes the two-element matrix components that are added to result in the element of the assembled matrix.

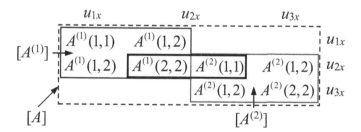

FIGURE 8.17 Graphical representation of combining two generic element matrices $[A^{(1)}]$ and $[A^{(2)}]$ of two adjacent line bar elements into the assembled matrix $[A]$.

There are other methods of assembling elementary matrices, such as the nodal force balance that gathers elastic or inertia forces/moments from neighboring elements at common nodes, but we use in this chapter the superposition method illustrated in Figure 8.17.

8.3.3 QUASI-STATIC RESPONSE

The equation governing the quasi-static response of a flexible-hinge mechanism is of the form:

$$[K][u] = [f] \qquad (8.176)$$

For a flexible-hinge mechanism which is sufficiently supported, the nodal displacement vector $[u]$ comprises zero components corresponding to blocked motions (DOF) through external support. The zero-displacement components in $[u]$ result in unknown reaction loads in $[f]$. Therefore, Eq. (8.176) can be rearranged by assembling a zero-displacement vector in $[u]$, which corresponds to a vector of reactions $[f_r]$ in $[f]$; the rearranged Eq. (8.176) is:

$$\begin{bmatrix} [K_{red}] & [K^*] \\ [K^{**}] & [K^{***}] \end{bmatrix} \begin{Bmatrix} [u_{red}] \\ [0] \end{Bmatrix} = \begin{Bmatrix} [f_{ext}] \\ [f_r] \end{Bmatrix}, \qquad (8.177)$$

where $[f_{ext}]$ denotes the external load vector. Two separate matrix equations are comprised in Eq. (8.177) that can be solved for the nonzero-displacement vector $[u_{red}]$, where "*red*" denotes reduced, and the reaction vector $[f_r]$ as:

$$[u_{red}] = [K_{red}]^{-1} [f_{ext}];$$
$$[f_r] = [K^{**}][u_{red}] = [K^{**}][K_{red}]^{-1} [f_{ext}] \qquad (8.178)$$

It should be noted that the reduced stiffness matrix $[K_{red}]$ is obtained from the full assembled stiffness matrix $[K]$ by eliminating the rows and columns that correspond to zero displacements. In the symbolic 2×2 matrix $[K]$ of Eqs. (8.177), the $[K_{red}]$ matrix is obtained by eliminating the second row and second column of $[K]$ – the eliminated portions correspond to the $[0]$ displacement vector.

Example 8.5

The mechanism of Figure 8.18a is formed of two identical right circularly corner-filleted flexible hinges of rectangular cross-section and a connecting link. A force is applied along the symmetry axis of the mechanism. Based on Figure 8.18b, use a line element with three nodes for the flexible hinge and another line element with two nodes for the connecting link in order to evaluate the deflection u_y along the symmetry axis in terms of the ratio of the fillet radius to the hinge length r/l. The elements have bending deformation capabilities and are considered long. The geometric parameters are: $l = 0.02$ m, $t = 0.001$ m (the hinge minimum thickness), $w = 0.007$ m (out-of-plane constant width of the hinge), $l_1 = 0.04$ m, and $h_r = w_r = 0.015$ m (the constant in-plane height and out-of-plane width of the connecting link). The device is built from a material with $E = 2 \cdot 10^{11}$ N/m² and the applied force is $f_y = 500$ N.

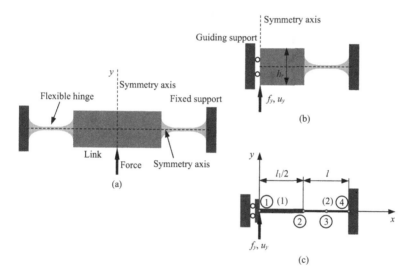

FIGURE 8.18 Symmetric mechanism with two identical right circularly corner-filleted flexible hinges, one connecting link and symmetric load: (a) schematic of full mechanism; (b) schematic of half-mechanism; (c) finite element model of half-mechanism.

Solution:

The finite element model is sketched in Figure 8.18c. The two elements (1) and (2) result in the overall nodal displacement vector:

$$[u] = \begin{bmatrix} u_{1y} & \theta_{1z} & u_{2y} & \theta_{2z} & u_{3y} & \theta_{3z} & u_{4y} & \theta_{4z} \end{bmatrix}^T \quad (8.179)$$

The stiffness matrix of the connecting-link element (1) is given in Eq. (8.31). The z-axis moment of area of this element is calculated as $I_z = w_r \cdot h_r^3/12$. The front-view geometry of the right circularly corner-filleted flexible hinge is sketched in Figure 8.19.

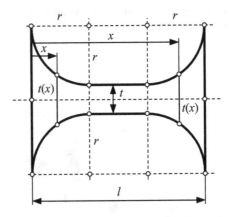

FIGURE 8.19 Front-view geometry of right circularly corner-filleted flexible hinge.

The in-plane thickness of the three segments composing the hinge is calculated as:

$$t(x) = \begin{cases} t + 2r\left[1 - \sqrt{1 - \left(1 - \dfrac{x}{r}\right)^2}\right], & 0 \leq x \leq r \\[3mm] t, & r \leq x \leq l - r \\[3mm] t + 2r\left[1 - \sqrt{1 - \left(1 - \dfrac{l-x}{r}\right)^2}\right], & l - r \leq x \leq l \end{cases} \quad . \quad (8.180)$$

The thickness of Eq. (8.180) is used to calculate the moment of area $I_z(x) = w \cdot t(x)^3/12$, which is needed to evaluate the stiffness matrix defined in Eq. (8.30) by means of the shape functions of Eq. (8.90). Note that the integral over the $[0, l]$ interval (which yields the flexible-hinge stiffness matrix in Eq. (8.30)) is expressed as a sum of three integrals over the subintervals defined in Eq. (8.180).

The assembled static equation for the two-element model of Figure 8.18c is:

$$
\begin{bmatrix}
K^{(1)}(1,1) & K^{(1)}(1,2) & K^{(1)}(1,3) & K^{(1)}(1,4) & 0 & 0 & 0 & 0 \\
K^{(1)}(1,2) & K^{(1)}(2,2) & K^{(1)}(2,3) & K^{(1)}(2,4) & 0 & 0 & 0 & 0 \\
K^{(1)}(1,3) & K^{(1)}(2,3) & K^{(1)}(3,3)+K^{(2)}(1,1) & K^{(1)}(3,4)+K^{(2)}(1,2) & K^{(2)}(1,3) & K^{(2)}(1,4) & K^{(2)}(1,5) & K^{(2)}(1,6) \\
K^{(1)}(1,4) & K^{(1)}(2,4) & K^{(1)}(3,4)+K^{(2)}(1,2) & K^{(1)}(4,4)+K^{(2)}(2,2) & K^{(2)}(2,3) & K^{(2)}(2,4) & K^{(2)}(2,5) & K^{(2)}(2,6) \\
0 & 0 & K^{(2)}(1,3) & K^{(2)}(2,3) & K^{(2)}(3,3) & K^{(2)}(3,4) & K^{(2)}(3,5) & K^{(2)}(3,6) \\
0 & 0 & K^{(2)}(1,4) & K^{(2)}(2,4) & K^{(2)}(3,4) & K^{(2)}(4,4) & K^{(2)}(4,5) & K^{(2)}(4,6) \\
0 & 0 & K^{(2)}(1,5) & K^{(2)}(2,5) & K^{(2)}(3,5) & K^{(2)}(4,5) & K^{(2)}(5,5) & K^{(2)}(5,6) \\
0 & 0 & K^{(2)}(1,6) & K^{(2)}(2,6) & K^{(2)}(3,6) & K^{(2)}(4,6) & K^{(2)}(5,6) & K^{(2)}(6,6)
\end{bmatrix}
\begin{bmatrix}
u_{1y} \\ \theta_{1z} \\ u_{2y} \\ \theta_{2z} \\ u_{3y} \\ \theta_{3z} \\ u_{4y} \\ \theta_{4z}
\end{bmatrix}
=
\begin{bmatrix}
f_y \\ M_{1z} \\ 0 \\ 0 \\ 0 \\ 0 \\ R_{4y} \\ M_{4z}
\end{bmatrix}
$$

(8.181)

The following nodal displacements are zero due to boundary conditions: θ_{1z}, u_{4y} and θ_{4z}. These zero boundary conditions formally eliminate rows and columns 2, 7 and 8 of the assembled stiffness matrix, as well as the same components of the forcing vector in the right-hand side of Eq. (8.181) – these forcing components are actually the (unknown) reactions at nodes 1 and 4. The reduced system resulting after eliminating the components related to zero displacements in Eq. (8.181) is:

$$\begin{bmatrix}
K^{(1)}(1,1) & K^{(1)}(1,3) & K^{(1)}(1,4) & 0 & 0 \\
K^{(1)}(1,3) & K^{(1)}(3,3)+K^{(2)}(1,1) & K^{(1)}(3,4)+K^{(2)}(1,2) & K^{(2)}(1,3) & K^{(2)}(1,4) \\
K^{(1)}(1,4) & K^{(1)}(3,4)+K^{(2)}(1,2) & K^{(1)}(4,4)+K^{(2)}(2,2) & K^{(2)}(2,3) & K^{(2)}(2,4) \\
0 & K^{(2)}(1,3) & K^{(2)}(2,3) & K^{(2)}(3,3) & K^{(2)}(3,4) \\
0 & K^{(2)}(1,4) & K^{(2)}(2,4) & K^{(2)}(3,4) & K^{(2)}(4,4)
\end{bmatrix}$$

$$\cdot \begin{bmatrix} u_{1y} \\ u_{2y} \\ \theta_{2z} \\ u_{3y} \\ \theta_{3z} \end{bmatrix} = \begin{bmatrix} f_y \\ 0 \\ 0 \\ 0 \\ 0 \end{bmatrix} \quad \text{or} \quad [K_{red}][u_{red}] = [f_{red}] \tag{8.182}$$

Equation (8.182) results in:

$$[u_{red}] = [K_{red}]^{-1}[f_{red}], \tag{8.183}$$

and the first component of the reduced displacement vector $[u_{red}]$ is u_{1y}. With the numerical values of this example, the deflection at node 1 is plotted in Figure 8.20 – it can be seen that as the fillet radius increases from 1/10 to 1/2 of the hinge length, the deflection decreases 54 times, from $u_{1y} = 1.783$ mm to $u_{1y} = 0.033$ mm.

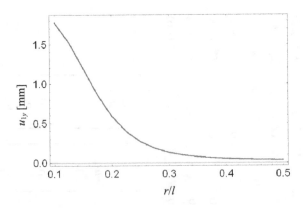

FIGURE 8.20 Plot of deflection u_{1y} in terms of the r/l ratio.

Example 8.6

The fixed-free serial planar mechanism of Figure 8.21a comprises two identical right circular flexible hinges and two constant cross-section links, and is acted upon by two forces f_1 and f_4. The planar dimensions of the flexible hinges and the links are shown in Figure 8.21b; all members have rectangular cross-sections. Use two-node Timoshenko beam elements to model the flexible hinges and two-node Euler–Bernoulli beam elements for the other two links by also including axial deformation capabilities in both element types. Graphically study the variation of the deflection u_{1Y} at node 1 in terms of the ratio r/t. Known are $t = 0.001$ m, $w = 0.006$ m (flexible-hinge constant out-of-plane width), $l_1 = 0.08$ m, $l_2 = 0.05$ m, $t_r = w_r = 0.015$ m (cross-sectional dimensions of the stiffer links), $\delta = 45°$, $E = 2 \cdot 10^{11}$ N/m², $\mu = 0.3$, $\alpha_s = 1.176$, $f_1 = 50$ N and $f_4 = 200$ N. The forces are perpendicular to the links they act on.

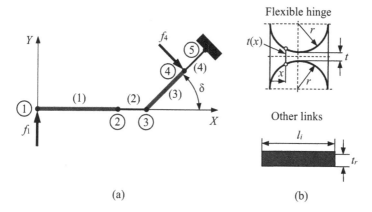

(a)   (b)

FIGURE 8.21   Serial planar flexible-hinge mechanism: (a) skeleton representation; (b) planar geometry of members.

Solution:

The nonzero, planar-motion nodal displacement vector of the mechanism in the global frame XY is:

$$[u_{red}]$$

$$= \begin{bmatrix} u_{1X} & u_{1Y} & \theta_{1Z} & u_{2X} & u_{2Y} & \theta_{2Z} & u_{3X} & u_{3Y} & \theta_{3Z} & u_{4X} & u_{4Y} & \theta_{4Z} \end{bmatrix}^T$$

$$(8.184)$$

The local-frame stiffness matrix of the flexible hinges is of the form given in Eq. (8.56) with the axial components determined based on Eqs. (8.5), (8.11) and the bending components calculated as in Eqs. (8.41)–(8.43), (8.45) and (8.46).

The variable thickness of the right circular hinge is the one in the first row of Eq. (8.180). It allows calculation of the variable cross-section area $A(x) = w{\cdot}t(x)$ and moment of area $I_z(x) = w{\cdot}t(x)^3/12$. The local-frame stiffness matrices of the two constant cross-section links are calculated as in Eqs. (8.12), (8.31) and (8.56). While the global-frame stiffness matrices of elements (1) and (2) are identical to their local ones, the following rotation matrix:

$$[R]_{6\times 6} = \begin{bmatrix} [R_1] & [0]_{3\times 3} \\ [0]_{3\times 3} & [R_1] \end{bmatrix}; \quad [R_1] = \begin{bmatrix} \cos\delta & \sin\delta & 0 \\ -\sin\delta & \cos\delta & 0 \\ 0 & 0 & 1 \end{bmatrix} \quad (8.185)$$

is used as in Eq. (8.151) to obtain the global-frame stiffness matrices of elements (3) and (4) from their corresponding local-frame matrices. The reduced, assembled stiffness matrix is a 12×12 matrix, which is not included here being too large, but its formation follows the rules of Section 8.3.2. The external global-frame load vector is:

$$[f_{ext}] = \begin{bmatrix} 0 & f_1 & 0 & 0 & 0 & 0 & 0 & 0 & 0 & f_4 \cdot \cos\delta & -f_4 \cdot \sin\delta & 0 \end{bmatrix}^T.$$

$$(8.186)$$

The shear modulus is: $G = E/[2(1+\mu)]$. The vector $[u_{red}]$ of Eq. (8.184) is calculated as per Eq. (8.178), and the plot of Figure 8.22 shows the variation of the y-axis displacement at node 1 in terms of the ratio r/t. When this ratio increases from 2 to 10, the deflection decreases approximately by a factor of 6.

Note: A right circular hinge of rectangular cross-section was modeled in Lobontiu and Garcia [13] as a three-node, six-DOF per node line element, that uses the Euler–Bernoulli model for smaller cross-section portions and the Timoshenko model for thicker section segments.

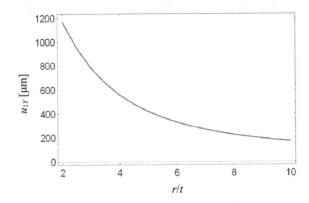

FIGURE 8.22   Plot of deflection u_{1Y} in terms of the r/t ratio.

Example 8.7

Figure 8.23a is the skeleton representation of a doubly-symmetric planar, displacement-amplification mechanism, which comprises four identical circular-axis flexible hinges of a constant rectangular cross-section and four connecting straight-axis links. The mechanism is subjected to symmetric loading by two identical forces along the horizontal axis. Use long finite elements with three DOF per node to model the in-plane quasi-static response of the mechanism. Study the variation of the displacement amplification (d.a.), which is the ratio of the vertical displacement at point B to the horizontal displacement at point A, in terms of the circular-axis flexible-hinge radius R ranging in the [0.02 m, 0.08 m] interval and the opening angle α varying in the [30°, 45°] interval – see Figure 8.8. Known are $l_1 = 0.01$ m, $l_2 = 0.015$ m, $t_1 = t_2 = 0.008$ m (in-plane thicknesses of the straight-axis links), $t = 0.001$ m (in-plane thickness of the flexible hinges), $w = 0.008$ m (out-of-plane width of all links) and $\alpha_1 = 60°$ – see Figure 8.15. The material Young's modulus is $E = 2.1 \cdot 10^{11}$ N/m², and the shear coefficient is $\alpha_s = 1.176$.

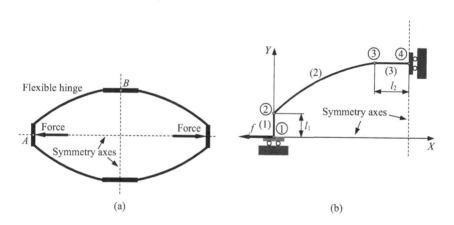

(a)   (b)

FIGURE 8.23 Planar symmetric displacement-amplification mechanism with circular-axis flexible hinges and straight-axis connecting links: (a) skeleton configuration; (b) quarter finite element model.

Solution:

Due to structural and loading symmetry, the full-mechanism static response is equivalent to the behavior of the quarter mechanism pictured in Figure 8.23b, which has guided ends along the two symmetry axes.

The local-frame, in-plane stiffness matrices of the straight-axis elements (1) and (3) are calculated with the aid of Eqs. (8.12), (8.31) and (8.56). The similar matrix of the circular-axis Euler–Bernoulli element (2) is determined based on Eqs. (8.108) and (8.109) with the shape functions defined in Eqs. (8.104). The global-frame stiffness matrix of element 1 is calculated as in Eq. (8.151) with the rotation matrix:

$$\left[R^{(1)}\right]=\begin{bmatrix} \left[R_1^{(1)}\right] & \left[0\right]_{3\times3} \\ \left[0\right]_{3\times3} & \left[R_1^{(1)}\right] \end{bmatrix}; \quad \left[R_1^{(1)}\right]=\begin{bmatrix} \cos(\pi/2) & \sin(\pi/2) & 0 \\ -\sin(\pi/2) & \cos(\pi/2) & 0 \\ 0 & 0 & 1 \end{bmatrix}$$

$$=\begin{bmatrix} 0 & 1 & 0 \\ -1 & 0 & 0 \\ 0 & 0 & 1 \end{bmatrix}. \tag{8.187}$$

The global-frame stiffness matrix of element (2) is evaluated as in Eqs. (8.156), (8.158) and (8.159) with $\delta_1 = \pi/2 - \alpha_1$. Element (3) has its local axes parallel to the corresponding global-frame axes, and therefore, its global-frame stiffness matrix is identical to the local-frame one.

The nonzero in-plane nodal displacement vector and its corresponding external load vector are:

$$\begin{cases} \left[u_{red}\right]=\begin{bmatrix} u_{1X} & u_{2X} & u_{2Y} & \theta_{2Z} & u_{3X} & u_{3Y} & \theta_{3Z} & u_{4Y} \end{bmatrix}^T \\ \left[f_{red}\right]=\begin{bmatrix} -f & 0 & 0 & 0 & 0 & 0 & 0 & 0 \end{bmatrix}^T. \end{cases} \tag{8.188}$$

The reduced stiffness matrix corresponds to the nonzero nodal displacements listed in $[u_{red}]$ of Eq. (8.188); it is obtained after eliminating the rows and columns corresponding to zero displacements from the full, 12×12 assembled stiffness matrix, namely:

$$
[K_{red}] =
\begin{bmatrix}
K^{(1)}(1,1) & K^{(1)}(1,4) & K^{(1)}(1,5) & K^{(1)}(1,6) & 0 & 0 & 0 & 0 \\[4pt]
K^{(1)}(1,4) & K^{(1)}(4,4)+K^{(2)}(1,1) & K^{(1)}(4,5)+K^{(2)}(1,2) & K^{(1)}(4,6)+K^{(2)}(1,3) & K^{(2)}(1,4) & K^{(2)}(1,5) & K^{(2)}(1,6) & 0 \\[4pt]
K^{(1)}(1,5) & K^{(1)}(4,5)+K^{(2)}(1,2) & K^{(1)}(5,5)+K^{(2)}(2,2) & K^{(1)}(5,6)+K^{(2)}(2,3) & K^{(2)}(2,4) & K^{(2)}(2,5) & K^{(2)}(2,6) & 0 \\[4pt]
K^{(1)}(1,6) & K^{(1)}(4,6)+K^{(2)}(1,3) & K^{(1)}(5,6)+K^{(2)}(2,3) & K^{(1)}(6,6)+K^{(2)}(3,3) & K^{(2)}(3,4) & K^{(2)}(3,5) & K^{(2)}(3,6) & 0 \\[4pt]
0 & K^{(2)}(1,4) & K^{(2)}(2,4) & K^{(2)}(3,4) & K^{(2)}(4,4)+K^{(3)}(1,1) & K^{(2)}(1,5) & K^{(2)}(1,6) & K^{(3)}(1,5) \\[4pt]
0 & K^{(2)}(1,5) & K^{(2)}(2,5) & K^{(2)}(3,5) & K^{(2)}(4,5)+K^{(3)}(1,2) & K^{(2)}(5,5)+K^{(3)}(2,2) & K^{(2)}(2,3)+K^{(3)}(2,3) & K^{(3)}(2,5) \\[4pt]
0 & K^{(2)}(1,6) & K^{(2)}(2,6) & K^{(2)}(3,6) & K^{(2)}(4,6)+K^{(3)}(1,3) & K^{(2)}(5,6)+K^{(3)}(2,3) & K^{(2)}(6,6)+K^{(3)}(3,3) & K^{(3)}(3,5) \\[4pt]
0 & 0 & 0 & 0 & K^{(3)}(1,5) & K^{(3)}(2,5) & K^{(3)}(3,5) & K^{(3)}(5,5)
\end{bmatrix}
$$

$$(8.189)$$

The reduced displacement vector is calculated as in Eq. (8.183) by means of the inverse of the stiffness matrix given in Eq. (8.189) and the load vector of Eq. (8.188). From it, the displacement amplification is calculated as $d.a. = |u_{4Y}| / |u_{1X}|$. The graphs of Figure 8.24 plot the variation of $d.a.$ with R and α. While the displacement amplification increases monotonically with R, it also increases with α up to approximately 40°, followed by a range of small decrease.

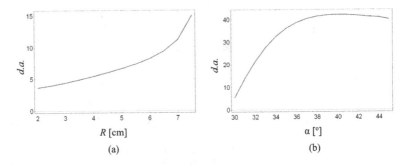

FIGURE 8.24 Plots of displacement amplification in terms of the circular-axis element: (a) radius R; (b) center angle α.

Example 8.8

Three identical, planar circular-axis flexible hinges of a constant circular cross-section are connected spatially, as illustrated in Figure 8.25a and b. The hinges are radially symmetric with respect to their connection point O, as shown in Figure 8.25a, and are fixed at their opposite ends. Known are the hinge cross-sectional diameter $d = 0.002$ m and the material properties $E = 2.1 \cdot 10^{11}$ N/m² and $\mu = 0.3$. Consider that a load is applied at the center O consisting of the forces $f_{OX} = f_{OZ} = 10$ N and $f_{OY} = 120$ N. Graphically study the variation of the displacement ratio u_{OY}/u_{OX} (u_{OX} and u_{OY} are the displacements at O along the global-frame axes X and Y) when the hinge center angle α varies in the [5°, 20°] domain and the radius R is within the [0.01 m, 0.04 m] interval. Use Euler–Bernoulli two-node elements.

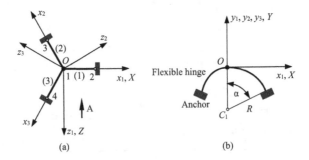

FIGURE 8.25 Three-dimensional skeleton mechanism with three identical and planar circular-axis flexible hinges: (a) top view showing the radial symmetry and identifying the nodes and elements; (b) side view from A.

Solution:

The three-dimensional stiffness matrix of element (1) is first formulated in its local frame $Ox_1y_1z_1$ based on the following displacement and load vectors:

$$\left[u^{(e,l)} \right]_{12\times1} = \left[\begin{array}{cc} \left[u_1^{(e,l)} \right]_{6\times1} & \left[u_2^{(e,l)} \right]_{6\times1} \end{array} \right]^T \quad \text{with}$$

$$\left[u_i^{(e,l)} \right]_{6\times1} = \left[\begin{array}{cc} \left[u_{i,ip}^{(e,l)} \right]_{3\times1} & \left[u_{i,op}^{(e,l)} \right]_{3\times1} \end{array} \right]^T$$

$$\left[f^{(e,l)} \right]_{12\times1} = \left[\begin{array}{cc} \left[f_1^{(e,l)} \right]_{6\times1} & \left[f_2^{(e,l)} \right]_{6\times1} \end{array} \right]^T \quad \text{with}$$

$$\left[f_i^{(e,l)} \right]_{6\times1} = \left[\begin{array}{cc} \left[f_{i,ip}^{(e,l)} \right]_{3\times1} & \left[f_{i,op}^{(e,l)} \right]_{3\times1} \end{array} \right]^T , \tag{8.190}$$

where $i = 1, 2$ denotes nodes 1 and 2. The vectors of Eqs. (8.190) are valid for any of the three elements in terms of their local frames. The displacements and loads of Eqs. (8.190) are used in conjunction with the local-frame element stiffness matrix of Eq. (8.156) – which is valid for both the in-plane and out-of-plane cases – to formulate the following 12×12, local-frame element stiffness matrix:

$$\left[K^{(i,l)} \right]_{12\times12}$$

$$= \begin{bmatrix} \left[K_{ip}^{(i,l)}(1,1) \right]_{3\times3} & [0] & \left[K_{ip}^{(i,l)}(1,2) \right]_{3\times3} & [0] \\ [0] & \left[K_{op}^{(i,l)}(1,1) \right]_{3\times3} & [0] & \left[K_{op}^{(i,l)}(1,2) \right]_{3\times3} \\ \left[K_{ip}^{(i,l)}(1,2) \right]_{3\times3}^T & [0] & \left[K_{ip}^{(i,l)}(2,2) \right]_{3\times3} & [0] \\ [0] & \left[K_{op}^{(i,l)}(1,2) \right]_{3\times3}^T & [0] & \left[K_{op}^{(i,l)}(2,2) \right]_{3\times3} \end{bmatrix}$$

$$\tag{8.191}$$

and $i = 1, 2, 3$ – which indicates, again, that the matrix of Eq. (8.191) is valid for any of the three flexible hinges in its local frame. As per Eq. (8.156), the submatrices of Eq. (8.191) are:

$$\begin{cases} \left[K_{ip}^{(i,l)}(1,1) \right] = \left[K_{ip}^{(e)}(1,1) \right]; \quad \left[K_{op}^{(i,l)}(1,1) \right] = \left[K_{op}^{(e)}(1,1) \right] \\ \left[K_{ip}^{(i,l)}(1,2) \right] = \left[K_{ip}^{(e)}(1,2) \right] \left[R_2^{(e)} \right]; \quad \left[K_{op}^{(i,l)}(1,2) \right] = \left[K_{op}^{(e)}(1,2) \right] \left[R_2^{(e)} \right] \\ \left[K_{ip}^{(i,l)}(2,2) \right] = \left[R_2^{(e)} \right]^T \left[K_{ip}^{(e)}(2,2) \right] \left[R_2^{(e)} \right]; \quad \left[K_{op}^{(i,l)}(2,2) \right] = \left[R_2^{(e)} \right]^T \left[K_{op}^{(e)}(2,2) \right] \left[R_2^{(e)} \right], \end{cases}$$

$$\tag{8.192}$$

with $\left[R_2^{(e)} \right]$ provided in Eq. (8.154). The 3×3 in-plane submatrices in the right-hand sides of Eqs. (8.192) result from the 6×6 in-plane, local-frame, element stiffness matrix expressed by means of Eqs. (8.108) and (8.109) via the linear-cubic shape functions of Eqs. (8.103) and (8.104). In a similar manner, the out-of-plane submatrices of Eqs. (8.192) are taken from the out-of-plane, local-frame, element stiffness matrix, which is evaluated based on Eqs. (8.123), (8.124), (8.128) and (8.129).

As illustrated in Figure 8.25, the local frame of element (1) is identical to the global frame $OXYZ$, but the local frames of elements (2) and (3) are rotated with respect to the global frame. Therefore, the resulting stiffness matrices of elements (2) and (3) need three-dimensional rotations to express them in the global frame and are expressed as:

$$\left[K^{(i)} \right] = \begin{bmatrix} \left[K^{(i)}(1,1) \right] & \left[K^{(i)}(1,2) \right] \\ \left[K^{(i)}(1,2) \right]^T & \left[K^{(i)}(2,2) \right] \end{bmatrix} = \left[R^{(i)} \right]^T \left[K^{(i,l)} \right] \left[R^{(i)} \right]$$

$$\text{(8.193)}$$

$$\text{with } \left[R^{(i)} \right] = \begin{bmatrix} \left[R_b^{(i)} \right]_{6 \times 6} & \left[0 \right]_{6 \times 6} \\ \left[0 \right]_{6 \times 6} & \left[R_b^{(i)} \right]_{6 \times 6} \end{bmatrix}, \quad i = 2,3$$

and $[K^{(i,l)}]$ given in Eq. (8.191). The basic rotation matrix $\left[R_b^{(i)} \right]$ of Eq. (8.193) is the product of three separate rotation matrices, as detailed in Chapter 5, in Eqs. (5.197), (5.198), (5.200) and (5.201). The rotation angles are: $\varphi_2 = \varphi_3 = 90°$, $\theta_2 = 120°$, $\theta_3 = 240°$, $\psi_2 = \psi_3 = -90°$.

The three elements are linked in parallel with respect to point O; as a consequence, the stiffness matrix of the mechanism is calculated by summing the global-frame stiffness matrices of these elements. Moreover, because nodes 2, 3 and 4 are fixed, the following reduced global-frame, load/displacement equation results:

$$[f_1] = [K(1,1)][u_1] \quad \text{or} \quad [u_1] = [K(1,1)]^{-1}[f_1] \quad \text{with}$$

$$[K_{red}] = [K(1,1)] = \left[K^{(1)}(1,1) \right] + \left[K^{(2)}(1,1) \right] + \left[K^{(3)}(1,1) \right].$$

$$\text{(8.194)}$$

The loads and displacements of Eq. (8.194) are:

$$[f_1] = \begin{bmatrix} f_{1x} & f_{1y} & m_{1z} & m_{1x} & m_{1y} & f_{1z} \end{bmatrix}^T$$

$$[u_1] = \begin{bmatrix} u_{1x} & u_{1y} & \theta_{1z} & \theta_{1x} & \theta_{1y} & u_{1z} \end{bmatrix}^T.$$

$$\text{(8.195)}$$

Figure 8.26a and b plots the variation of u_{OY}/u_{OX} in terms of R and α, respectively. As the plots show, the ratio of the out-of-plane displacement to the in-plane displacement at O increases with an increase in both the radius R and the opening angle α.

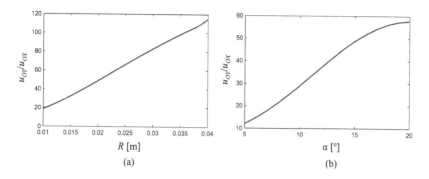

FIGURE 8.26 Plots of the center point O displacement ratio u_{OY}/u_{OX} in terms of the hinge: (a) radius R; (b) center angle α.

8.3.4 Natural Response and Frequencies

The free, undamped behavior of a flexible-hinge mechanism is characterized by the dynamic equation:

$$[M][\ddot{u}] + [K][u] = [0],$$ (8.196)

where $[M]$ and $[K]$ are the assembled (overall) matrices of the mechanism. Similar to the quasi-static modeling, Eq. (8.196) can be partitioned as:

$$\begin{bmatrix} [M_{red}] & [M^*] \\ [M^{**}] & [M^{***}] \end{bmatrix} \begin{bmatrix} [\ddot{u}_{red}] \\ [0] \end{bmatrix} + \begin{bmatrix} [K_{red}] & [K^*] \\ [K^{**}] & [K^{***}] \end{bmatrix} \begin{bmatrix} [u_{red}] \\ [0] \end{bmatrix} = \begin{bmatrix} [0] \\ [f_r] \end{bmatrix}$$ (8.197)

in order to outline the zero displacements produced by fixed boundary conditions. Equation (8.197) identifies the zero displacements corresponding to external supports that block the respective displacements. The zero displacements are directly related to reaction loads that are gathered in a $[f_r]$ vector. Because the nonzero displacements (collected in the $[u_{red}]$ vector) are of interest, the relevant portion of Eq. (8.197) is:

$$[M_{red}][\ddot{u}_{red}] + [K_{red}][u_{red}] = [0].$$ (8.198)

Harmonic solution is sought for Eq. (8.198) of the form:

$$[u_{red}] = [U_{red}]\sin(\omega_n t + \varphi).$$ (8.199)

which, together with its second derivative, is substituted in Eq. (8.198) – the equation becomes:

$$\left(-\omega_n^2 [M_{red}] + [K_{red}]\right)[U_{red}]\sin(\omega_n t + \varphi) = [0].\qquad (8.200)$$

Equation (8.200) can also be written in eigenvalue form as:

$$[M_{red}]^{-1}[K_{red}][U_{red}] = \omega_n^2[U_{red}] \quad \text{or} \quad [D_{red}][U_{red}] = \lambda[U_{red}],\quad (8.201)$$

where the eigenvalues λ (the squares of the natural frequencies ω_n) are those of the reduced dynamic matrix:

$$[D_{red}] = [M_{red}]^{-1}[K_{red}].\qquad (8.202)$$

Example 8.9

Study the variation of the first natural frequency of the mechanism shown in Figure 8.18b of *Example 8.5* in terms of the ratio r/l. Use the finite element model of Figure 8.18c together with the numerical values of *Example 8.5*. Consider that element (1) is defined according to the rigid-element formulation. The material density is $\rho = 7,800\,\text{kg/m}^3$.

Solution:

As discussed in *Example 8.5*, the following nodal displacements are nonzero: $u_{1y}, u_{2y}, \theta_{2z}, u_{3y}$ and θ_{3z}. This resulted in a reduced stiffness matrix $[K_{red}]$ that is expressed in Eq. (8.182). Similarly, the reduced inertia matrix is:

$[M_{red}]$

$$
= \begin{bmatrix}
M^{(1)}(1,1) & M^{(1)}(1,3) & M^{(1)}(1,4) & 0 & 0 \\
M^{(1)}(1,3) & M^{(1)}(3,3)+M^{(2)}(1,1) & M^{(1)}(3,4)+M^{(2)}(1,2) & M^{(2)}(1,3) & M^{(2)}(1,4) \\
M^{(1)}(1,4) & M^{(1)}(3,4)+M^{(2)}(1,2) & M^{(1)}(4,4)+M^{(2)}(2,2) & M^{(2)}(2,3) & M^{(2)}(2,4) \\
0 & M^{(2)}(1,3) & M^{(2)}(2,3) & M^{(2)}(3,3) & M^{(2)}(3,4) \\
0 & M^{(2)}(1,4) & M^{(2)}(2,4) & M^{(2)}(3,4) & M^{(2)}(4,4)
\end{bmatrix}
$$

$$(8.203)$$

The inertia matrix of the flexible-hinge mechanism is calculated based on the definition Eq. (7.88) with the shape functions of Eq. (8.90), while the inertia matrix of the rigid element is determined with Eq. (8.71). The corresponding reduced dynamic matrix is calculated as in Eq. (8.202). With the numerical values of *Example 8.5* and this *Example*, Figure 8.27 is the plot of the first natural frequency in terms of the ratio r/l. It can be seen that as the fillet radius r relatively increases with respect to the flexible-hinge length l, the first natural frequency also increases.

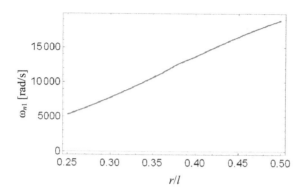

FIGURE 8.27   Plot of first natural frequency ω_{n1} in terms of the r/l ratio.

Example 8.10

Consider the planar mechanism of *Example 8.6* and shown in Figure 8.21a. Using the finite element model of *Example 8.6* and the parameters provided there, study the variation of the ratio of the largest natural frequency to the smallest natural frequency in terms of the ratio r/t. Use the consistent for-mulation for the inertia matrix of elements 1 and 3. The material density is $\rho = 7,800\,\text{kg/m}^3$.

Solution:

A procedure similar to the one detailed for the elemental stiffness matrices of *Example 8.6* is applied to determine the reduced inertia matrix $[M_{red}]$ of this mechanism. The local-frame inertia matrices of the flexible hinges and the other links have the form of Eq. (8.56). The axial-vibration components of the flexible-hinge inertia matrix in the local frame are determined as in Eqs. (8.13), (7.67), while the ones corresponding to bending are calculated as in Eqs. (8.41)–(8.43) and (8.50). For the constant cross-section links, the axial-related inertia components are given in Eq. (7.69) and those that depend on bending are expressed in Eq. (7.89). Similar to the stiffness matrices, the local-frame inertia matrices of elements (1) and (2) are also valid in the global frame, while for elements (3) and (4), the local-frame inertia matrices need to be transformed as in Eq. (8.152) by means of the rotation matrix of Eq. (8.185). The assembled, reduced inertia matrix of the full mechanism follows the nodal displacements of Eq. (8.184). The natural frequencies are found based on the eigenvalues of the reduced dynamic matrix of Eq. (8.202). The plot of Figure 8.28 shows the variation of the ratio of the largest natural frequency to the smallest one, $\omega_{n,max}/\omega_{n,min}$, in terms of the r/t ratio. As r/t increases from 2 to 10, the frequency ratio decreases almost three times.

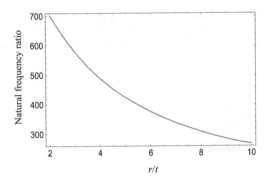

FIGURE 8.28   Plot of natural frequency ratio $\omega_{n,\max}/\omega_{n,\min}$ in terms of the r/t ratio.

8.3.5   TIME-DOMAIN RESPONSE

A typical time-domain finite element mathematical model is expressed by the assembled equation:

$$[M][\ddot{u}]+[B][\dot{u}]+[K][u]=[f],\qquad(8.204)$$

where $[M]$, $[B]$ and $[K]$ are the inertia, damping and stiffness matrices, $[f]$ is the time-domain load/forcing vector, and $[u]$ is the unknown nodal displacement vector. In order to directly find the nodal displacement vector, time discretization is usually employed in addition to the spatial one. Generally, time discretization and space discretization are applied separately but they can also be operated simultaneously, like in convection-dominated problems – see Zienkiewicz et al. [3].

For time discretization, the infinite time domain is partitioned into a finite number of time stations, separated by a time step Δt as $t_{n+1}=t_n+\Delta t$. Based on this discretization, recurrence relationships enable to determine the system variables at the moment t_{n+1} in terms of the known values of the system variables at the moment t_n and the forcing factors. There are two major avenues of formulating a finite element model in the time domain. One modality uses the weighted residual approach, whereas the other utilizes the truncated Taylor series expansion of the unknown time-dependent nodal vector. The latter method results in a *time-stepping scheme* whereby a polynomial is selected to approximate the nodal vector of the following form:

$$[u]\approx[u]_n+\left[\frac{du}{dt}\right]_n\cdot(t-t_n)+\frac{1}{2!}\left[\frac{d^2u}{dt^2}\right]_n\cdot(t-t_n)^2+\cdots+\frac{1}{m!}\left[\frac{d^mu}{dt^m}\right]_n\cdot(t-t_n)^m.$$
$$(8.205)$$

When the values of the polynomial and its derivatives up to the order m are known at the moment n, similar values of the polynomial and its derivatives corresponding to the moment $n+1$ can be computed by means of the equation above.

A time-stepping scheme needs to be *consistent* and *stable*. Consistency requires that the degree of the approximating polynomial be at least equal to the order of the differential equation modeling the system. For the dynamics of a structural microsystem, for instance, which is typically governed by a second-order differential equation, the approximating polynomial needs to be at least a second-degree one. In order to be stable, a time-stepping scheme cannot produce a solution, which increases indefinitely with time (diverges) or a solution which is oscillatory. Certain algorithms comply with the stability requirement when the time step is less than a critical value: $\Delta t < \Delta t_{cr}$. Such algorithms are named *conditionally stable*, as contrasted to the *unconditionally stable* ones, which provide a stable solution independently of the time step magnitude.

Time-stepping schemes, as the one described by Eq. (8.205), are named *single-step*, when the unknowns at time $n+1$, which are $d^2[u]_{n+1}/dt^2$, $d[u]_{n+1}/dt$ and $[u]_{n+1}$, are expressed in terms of the same amounts at the previous time station, $d^2[u]_n/dt$, $d[u]_n/dt$ and $[u]_n$. These algorithms are largely implemented in finite element software, and they can be designed based on a constant or a variable time step to solve linear first- and second-order problems. Another class of solvers for time-dependent finite element problems comprises *multiple-step algorithms* where the unknown $[u]_{n+1}$ is determined as a function of $[u]_n$, $[u]_{n-1}$, ..., $[u]_p$, and therefore, the nodal vector derivatives are eliminated from the recurrence relationships. For *nonlinear problems*, special algorithms do exist, and more details can be found in specialized finite element texts, such as in Zienkiewicz et al. [3].

Another criterion of classifying the time-stepping schemes regards the algebraic equations system which has to be solved at each time step. *Implicit algorithms* solve coupled equations, and this involves more computational resources, such as the case is with consistent formulation of the system matrices. *Explicit algorithms* result in equations describing the system response at any given time moment that are decoupled, such as the situation is where lumped matrix formulation is utilized.

Popular time-stepping schemes that are implemented in finite element time-domain modeling include difference (central, backward, forward) methods, Runge–Kutta algorithms, the Houbolt method or the Newmark scheme – see Newmark [66]. However, for the relatively simple finite element models formulated in this chapter, time integration can be performed by means of any of the procedures mentioned in Chapter 7, such the Laplace transform or the transfer function approach – the latter method is illustrated in the following *Example*.

Example 8.11

A force $f_{OY} = 100 \cdot e^{-t} \cdot \sin(10t)$ N acts at point O on the mechanism of Figure 8.25 in *Example 8.8*. Plot the time variation of the displacement u_{OY} when known are the hinge parameters: diameter $d = 0.002$ m, radius $R = 0.02$ m, angle $\alpha = 15°$, as well as the material properties $E = 2.1 \cdot 10^{11}$ N/m^2, $\mu = 0.3$ and $\rho = 7,800$ kg/m^3. Consider that proportional damping acts on the three hinges, which is defined by $\alpha_h = 0.2$ and $\beta = 0.3$.

Solution:

The local-frame inertia matrix of any of the three elements is expressed similar to the stiffness matrix of Eq. (8.191) as:

$$\left[M^{(i,l)} \right]_{12\times12}$$

$$= \begin{bmatrix} \left[M_{ip}^{(i,l)}(1,1) \right]_{3\times3} & [0] & \left[M_{ip}^{(i,l)}(1,2) \right]_{3\times3} & [0] \\ [0] & \left[M_{op}^{(i,l)}(1,1) \right]_{3\times3} & [0] & \left[M_{op}^{(i,l)}(1,2) \right]_{3\times3} \\ \left[M_{ip}^{(i,l)}(1,2) \right]_{3\times3}^{T} & [0] & \left[M_{ip}^{(i,l)}(2,2) \right]_{3\times3} & [0] \\ [0] & \left[M_{op}^{(i,l)}(1,2) \right]_{3\times3}^{T} & [0] & \left[M_{op}^{(i,l)}(2,2) \right]_{3\times3} \end{bmatrix}.$$

$$(8.206)$$

Based on Eq. (8.157), the submatrices of Eq. (8.206) are:

$$\begin{cases} \left[M_{ip}^{(i,l)}(1,1) \right] = \left[M_{ip}^{(e)}(1,1) \right]; \quad \left[M_{op}^{(i,l)}(1,1) \right] = \left[M_{op}^{(e)}(1,1) \right] \\ \left[M_{ip}^{(i,l)}(1,2) \right] = \left[M_{ip}^{(e)}(1,2) \right]\left[R_{2}^{(e)} \right]; \quad \left[M_{op}^{(i,l)}(1,2) \right] = \left[M_{op}^{(e)}(1,2) \right]\left[R_{2}^{(e)} \right] \\ \left[M_{ip}^{(i,l)}(2,2) \right] = \left[R_{2}^{(e)} \right]^{T}\left[M_{ip}^{(e)}(2,2) \right]\left[R_{2}^{(e)} \right]; \\ \left[M_{op}^{(i,l)}(2,2) \right] = \left[R_{2}^{(e)} \right]^{T}\left[M_{op}^{(e)}(2,2) \right]\left[R_{2}^{(e)} \right], \end{cases}$$

$$(8.207)$$

where $i = 1, 2, 3$. The submatrices in the right-hand sides of Eqs. (8.207) result from Eqs. (8.103), (8.104), (8.111), (8.123), (8.124) and (8.131).

The inertia matrices of elements (2) and (3) need to be rotated in order to transform them from their local frames; the global-frame inertia matrices of these two elements are calculated as:

$$\left[M^{(i)} \right] = \begin{bmatrix} \left[M^{(i)}(1,1) \right] & \left[M^{(i)}(1,2) \right] \\ \left[M^{(i)}(1,2) \right]^{T} & \left[M^{(i)}(2,2) \right] \end{bmatrix} = \left[R^{(i)} \right]^{T}\left[M^{(i,l)} \right]\left[R^{(i)} \right], \quad i = 2,3,$$

$$(8.208)$$

with the rotation matrix of Eq. (8.193). The mechanism reduced inertia matrix is determined similarly to the reduced stiffness matrix of Eq. (8.194):

$$[M_{red}] = [M(1,1)] = \left[M^{(1)}(1,1) \right] + \left[M^{(2)}(1,1) \right] + \left[M^{(3)}(1,1) \right] \qquad (8.209)$$

The reduced damping matrix is calculated based on an equation similar to Eq. (8.209), namely:

$$[B_{red}] = [B(1,1)] = \left[B^{(1)}(1,1) \right] + \left[B^{(2)}(1,1) \right] + \left[B^{(3)}(1,1) \right], \qquad (8.210)$$

where

$$\left[B^{(i)}(1,1) \right] = \alpha_{h}\left[M^{(i)}(1,1) \right] + \beta\left[K^{(i)}(1,1) \right], \quad i = 1,2,3. \qquad (8.211)$$

The reduced dynamic equation is:

$$[M_{red}][\ddot{u}_{red}] + [B_{red}][\dot{u}_{red}] + [K_{red}][u_{red}] = [f_{red}], \tag{8.212}$$

with

$$[u_{red}] = \begin{bmatrix} u_{1x} & u_{1y} & \theta_{1z} & \theta_{1x} & \theta_{1y} & u_{1z} \end{bmatrix}^T$$

$$[f_{red}] = \begin{bmatrix} f_{1x} & f_{1y} & m_{1z} & m_{1x} & m_{1y} & f_{1z} \end{bmatrix}^T \tag{8.213}$$

$$= \begin{bmatrix} 0 & f_{Oy} & 0 & 0 & 0 & 0 \end{bmatrix}^T.$$

Node 1 coincides with the central point O. It should also be noted that a combination of Eqs. (8.212) and (8.213) demonstrates that:

$$[B_{red}] = \alpha_h [M_{red}] + \beta [K_{red}]. \tag{8.214}$$

Using the transfer function approach to plot the time response, the following transfer function matrix is formed:

$$[G(s)] = \left(s^2 \left[M_{red}^{(e)} \right] + s \left[B_{red}^{(e)} \right] + \left[K_{red}^{(e)} \right] \right)^{-1}$$

$$= \left[\left(s^2 + \alpha_h s \right) \left[M_{red}^{(e)} \right] + \left(\beta s + 1 \right) \left[K_{red}^{(e)} \right] \right]^{-1}. \tag{8.215}$$

The nonzero input in the forcing vector of Eq. (8.213) is combined with the scalar transfer function $G_{22}(s)$ of the transfer function matrix $[G(s)]$ calculated in Eq. (8.215), and the resulting plot of point O along the y axis is shown in Figure 8.29.

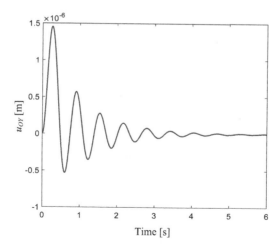

FIGURE 8.29 Time plot of the out-of-plane displacement of mechanism center point.

REFERENCES

1. Petyt, M., *Introduction to Finite Element Vibration Analysis*, Cambridge University Press, Cambridge, 1990.
2. Reddy, J.N., *Introduction to the Finite Element Method*, 4th ed., McGraw-Hill, New York, 2019.
3. Zienkiewicz, O.C., Taylor, R.L., and Zhu, J.Z., *The Finite Element Method: Its Basis and Fundamentals*, 7th ed., Butterworth-Heinemann, Oxford, 2013.
4. Moaveni, S., *Finite Element Analysis: Theory and Application with ANSYS*, 2nd ed., Pearson Education, Upper Saddle River, NJ, 2003.
5. Madenci, E. and Guven, I., *The Finite Element and Applications in Engineering Using ANSYS*, 2nd ed., Springer, New York, 2015.
6. Yang, T.Y., *Finite Element Structural Analysis*, Prentice Hall, Englewood Cliffs, NJ, 1986.
7. Gerardin, M. and Cardona, A., *Flexible Multibody Dynamics: A Finite Element Approach*, John Wiley and Sons, Chichester, NY, 2001.
8. Franciosi, C. and Mecca, M., Some finite elements for the static analysis of beams with varying cross section, *Computers and Structures*, 69, 191, 1998.
9. Jiang, W.G. and Henshall, J.L., A coupling cross-section finite element model for torsion analysis of prismatic beams, *European Journal of Mechanics, A/Solids*, 21, 513, 2002.
10. Koster, M., *Constructieprincipies voor het Nauwkeurig Bewegen en Positioneren*, Twente University Press, Twente, Holland, 1998.
11. Zhang, S. and Fasse, E.D., A finite-element-based method to determine the spatial stiffness properties of a notch hinge, *ASME Journal of Mechanical Design*, 123 (1), 141, 2001.
12. Murin, J. and Kutis, V., 3D-beam element with continuous variation of the cross-sectional area, *Computers and Structures*, 80 (3–4), 329, 2002.
13. Lobontiu, N. and Garcia, E., Circular-hinge line element for finite element analysis of compliant mechanisms, *ASME Journal of Mechanical Design*, 127, 766, 2005.
14. Richards, T.H., *Energy Methods in Stress Analysis*, Ellis Horwood, Chichester, NY, 1977.
15. Reddy, J.N., On locking-free shear deformable beam finite elements, *Computational Methods in Applied Mechanics*, 149, 113, 1997.
16. Ganapathi, M. et al., Shear flexible curved spline element for static analysis, *Finite Element in Analysis and Design*, 32, 181, 1999.
17. Kang, B. and Riedel, C.H., On the validity of planar, thick curved beam models derived with respect to centroidal and neutral axes, *Wave Motion*, 49, 1, 2012.
18. Dayyani, I., Friswell, M.I., and Saavedra Flores, E.I., A general super element for curved beams, *International Journal of Solids and Structures*, 51, 2931, 2014.
19. Cazzani, A., Malagu, M., and Turco, E., Isogeometric analysis of plane-curved beams, *Mathematics and Mechanics of Solids*, 21 (5), 562, 2016.
20. Tufekci, E., Eroglu, U., and Aya, S.A., A new two-noded curved beam finite element formulation based on exact solution, *Engineering with Computers*, 33, 261, 2017.
21. Borkovic, A. et al., Rotation-free isogeometric analysis of an arbitrarily curved plane Bernoulli-Euler beam, *Computer Methods in Applied Mechanics and Engineering*, 334, 238, 2018.
22. Connor, J.J. and Brebbia, C.A., *Finite Element Techniques in Structural Mechanics*, University of Southampton Publications, Southampton, 1970.
23. Petyt, M. and Fleischer, C.C., Free vibration of a curved beam, *Journal of Sound and Vibration*, 18 (1), 17, 1971.

24. Rao, S.S., Effects of transverse shear and rotatory inertia on the coupled twist-bending vibrations of circular rings, *Journal of Sound and Vibration*, 16 (4), 551, 1971.

25. Yang, T.Y. and Kim, H.W., Vibration and bucking of shells under initial stresses, *AIAA Journal*, 11 (11), 1525, 1973.

26. Lebeck, A.O. and Knowlton, J.S., A finite element for the three-dimensional deformation of a circular ring, *International Journal of Numerical Methods in Engineering*, 21 (3), 421, 1985.

27. Akhtar, M., Element stiffness of circular member, *Journal of Structural Engineering*, 113, 867, 1987.

28. Benedetti, A. and Tralli, A., A new hybrid F.E. model for arbitrary curved beam – I. Linear analysis, *Computers and Structures*, 33 (6), 1437, 1989.

29. Choi, J.-K. and Lim, J.K., General curved beam elements based on the assumed strain fields, *Computers and Structures*, 55 (3), 379, 1995.

30. Khrishnan, A. and Suresh, Y.J., A simple cubic linear element for static and free vibration analysis of curved beams, *Computers & Structures*, 68, 473, 1998.

31. Wu, J.-S. and Chiang, L.K., Free vibration of a circularly curved Timoshenko beam normal to its initial plane using finite curved beam elements, *Computers & Structures*, 82, 2525, 2004.

32. Wu, J.-S. and Chiang, L.K., A new approach for displacement functions of a curved Timoshenko beam element in motions normal to its plane, *International Journal for Numerical Methods in Engineering*, 64, 1375, 2005.

33. Kim, C.B. et al., A finite thin circular beam element for in-plane vibration analysis of curved beams, *Journal of Mechanical Science and Technology*, 19 (12), 2187, 2005.

34. Saffari, H., Fadaee, M.J., and Tatabaei, R., Developing a formulation based upon curvature for analysis of nonprismatic curved beams, *Mathematical Problems in Engineering*, Article ID 46215, 1, 2007.

35. Kim, B.Y. et al., A finite thin circular beam element for out-of-plane vibration analysis of curved beams, *Journal of Mechanical Science and Technology*, 23, 1396, 2009.

36. Shahbah, A. et al., New shape functions for non-uniform Timoshenko beams with arbitrary varying curvature using displacements functions, *Mecanica*, 48, 159, 2013.

37. Kardestuncer, H. (editor), *Finite Element Handbook*, McGraw-Hill, New York, 1987.

38. Cowper, G.R., Lindberg, G.M., and Olson, M.D., A shallow shell finite element of triangular shape, *International Journal of Solids and Structures*, 6, 1133, 1970.

39. Howson, W.P. and Jemah, A.K., Exact out-of-plane natural frequencies of curved Timoshenko beams, *Journal of Engineering Mechanics*, 125 (2), 19, 1999.

40. Davis, R., Henshell, R.D., and Warburton, G.B., Curved beam finite elements for coupled bending and torsional vibration, *Earthquake Engineering and Structural Dynamics*, 1, 165, 1972.

41. Thompson, B.S. and Sung, C.K., A survey of finite element techniques for mechanism design, *Mechanism and Machine Theory*, 21 (4), 351, 1986.

42. Wang, Y. and Wang, Z., A time finite element method for dynamic analysis of elastic mechanisms in link coordinate systems, *Computers and Structures*, 79 (2), 223, 2001.

43. Saxena, A. and Ananthasuresh, G.K., Topology synthesis of compliant mechanisms for non-linear forced-deflection and curved path specifications, *ASME Journal of Mechanical Design*, 123 (1), 33, 2001.

44. Yu, Y.-Q. and Smith, M. R., The effect of cross-sectional parameters on the dynamics of elastic mechanisms, *Mechanism and Machine Theory*, 31 (7), 947, 1996.
45. Sriram, B.R. and Mruthyunjaya, T.S., Dynamics of flexible-link mechanisms, *Computers and Structures*, 56 (6), 1029, 1995.
46. Xianmin, Z., Jike, L., and Yunwen, S., A high frequency analysis method for closed flexible mechanism systems, *Mechanism and Machine Theory*, 33 (8), 1117, 1998.
47. Wang, Y.X., Multifrequency resonances of flexible linkages, *Mechanism and Machine Theory*, 33 (3), 255, 1998.
48. Chen, W., Dynamic modeling of multi-link flexible robotic manipulators, *Computers and Structures*, 79 (2), 183, 2001.
49. Besseling, J.F. and Gong, D.G., Numerical simulation of spatial mechanisms and manipulators with flexible links, *Finite Element in Analysis and Design*, 18, 121, 1994.
50. Gao, X., Solution methods for dynamic response of flexible mechanisms, in: A.G. Erdman (ed.), *Modern Kinematics: Developments in the Last Forty Years*, John Wiley and Sons, New York, 1993.
51. Turcic, D.A. and Midha, A., Dynamic analysis of elastic mechanism systems-Part 1: applications, *ASME Journal of Dynamic Systems, Measurement and Control*, 106, 249, 1984.
52. Chu, S.C. and Pan, K.C., Dynamic response of high-speed slider-crank mechanism with an elastic connecting rod, *ASME Journal of Engineering for Industry*, 97 (2), 542, 1975.
53. Song, J.O. and Haug, E.J., Dynamic analysis of planar flexible mechanisms, *Computer Methods in Applied Mechanics and Engineering*, 24, 358, 1980.
54. Gao, X.C., King, Z.Y., and Zhang, Q.X., A closed-form linear multi-step algorithm for the steady-state response of high-speed flexible mechanisms, *Mechanism and Machine Theory*, 23 (5), 361, 1988.
55. Sekulovic, M. and Salatic, R., Nonlinear analysis of frames with flexible connections, *Computers and Structures*, 79 (11), 1097, 2001.
56. Hac, M. and Osinski, J., Finite element formulation of rigid body motion in dynamic analysis of mechanisms, *Computers & Structures*, 57 (2), 213, 1995.
57. Hac, M., Dynamics of flexible mechanisms with mutual dependence between rigid body motion and longitudinal deformation of links, *Mechanism and Machine Theory*, 30 (6), 837, 1995.
58. Piras, G., Dynamic finite-element analysis of a planar high-speed, high-precision parallel manipulator with flexible links, *Mechanism and Machine Theory*, 40 (7), 849, 2005.
59. Albanesi, A.E., A new method to design compliant mechanisms based on the inverse beam finite element model, *Mechanism and Machine Theory*, 65, 14, 2013.
60. Dubay, R., Finite element based model predictive control for active vibration suppression of one-link flexible manipulator, *ISA Transactions*, 53 (5), 1609, 2014.
61. Muhammad, A.K., Okamoto, S., and Lee, J.H., Comparison of proportional-derivative and active control of vibration of a flexible single-link manipulator using finite-element method, *Artificial Life and Robotics*, 19 (4), 375, 2014.
62. Belotti, R. et al., An updating method for finite element models of flexible-link mechanisms based on an equivalent rigid-link system, *Shock and Vibration*, 1, 2018.

63. Tian, Y. et al., A unified element stiffness matrix model for variable cross-section flexure hinges in compliant mechanisms for micro/nano positioning, *Microsystem Technologies*, 25 (11), 4257, 2019.
64. Kermanian, A., Dynamic analysis of flexible parallel robots via enhanced co-rotational and rigid finite element formulations, *Mechanism and Machine Theory*, 139, 144, 2019.
65. My, C.A., An efficient finite element formulation of dynamics for flexible robot with different type of joints, *Mechanism and Machine Theory*, 134, 267, 2019.
66. Newmark, N.M., A method for computation of structural dynamics, *Journal of ASCE*, 85, 67, 1959.

9 Miscellaneous Topics

This final chapter presents a few more topics on flexible hinges and flexible-hinge mechanisms, such as stress concentration, yielding and related maximum load, precision of rotation of straight- and circular-axis flexible hinges, temperature effects on hinge compliances, layered flexible hinges, and piezoelectric actuation and sensing. Illustrative examples from each topic are included, which are solved by either the compliance-based method or the finite element procedure.

9.1 STRESS CONCENTRATION IN FLEXIBLE HINGES AND FLEXIBLE-HINGE MECHANISMS

The stress concentration of flexible hinges and flexible-hinge mechanisms is briefly discussed in this section. The focus is on straight-axis flexible hinges and planar hinge mechanisms that are subjected to in-plane or out-of-plane quasi-static load. Stress concentration factors that are available in the literature and compliance matrices are combined into a formulation enabling to predict the maximum loads and displacements of flexible hinges and their mechanisms.

9.1.1 STRESS CONCENTRATORS

Stress concentrators or raisers are geometric discontinuities or irregularities in a mechanical part. Their presence alters the elementary stress formulas that are valid for members having a constant cross-section or a section with only gradual change in its defining contour. As a consequence, highly localized stresses are set at points where stress raisers are present. The stress concentrators can act at microscopic scale in the form of cracks, inclusions or voids, but are also macroscopic in configurations such as holes, notches, grooves, threads, fillets, stepped sections or sharp reentrant corners. The net effect of stress raisers is shortening the life of both ductile and brittle materials. The stress concentrators are particularly important in flexible hinges that are actually discontinuities, particularly in the areas of connection to adjacent rigid links. The sharp corners of a flexible hinge with either constant or variable cross-section are generally avoided by using fillets at the respective areas. The stress raisers are mathematically characterized by a *stress concentration factor*, $K_t > 1$, which is defined as the ratio of the maximum or peak stress to the nominal stress. For normal stresses σ that are produced through bending or axial loading and for shear stresses τ generated through shear or torsion, the respective stress concentration factors are expressed as:

$$K_{tn} = \frac{\sigma_{max}}{\sigma_{nom}}; \quad K_{ts} = \frac{\tau_{max}}{\tau_{nom}}. \tag{9.1}$$

The subscript "t" in the stress concentration factor notation indicates that its value is theoretical, whereas the subscripts "n" and "s" denote the normal and shear nature of the stress. Consider, for instance, the straight-axis, right circular flexible hinge of Figure 9.1, which is acted upon by symmetric axial load. As the figure shows it, the stress value at the bottom of the circular groove is maximum and it decreases when moving away from the bottom toward the symmetry axis. Assume the hinge has rectangular cross-section with a constant out-of-plane width w. The nominal stress is calculated by using the net cross-sectional area at the minimum-thickness neck region in Figure 9.1, namely:

$$\sigma_{nom} = \frac{f_x}{w \cdot t}.\qquad(9.2)$$

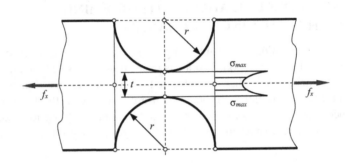

FIGURE 9.1   Straight-axis right circular flexible hinge with geometry, symmetrical axial loading and stress distribution across the minimum-thickness zone.

A *notch sensitivity factor q*, which ranges from 0 to 1 – see Peterson [1] or Pilkey and Pilkey [2], is usually introduced in order to define the *effective stress concentration factor K_e* and to express the maximum normal and shear stresses of Eq. (9.1) as:

$$K_e = q \cdot (K_t - 1) + 1 \rightarrow \begin{cases} \sigma_{\max} = [q \cdot (K_{tn} - 1) + 1] \cdot \sigma_{nom} \\ \tau_{\max} = [q \cdot (K_{ts} - 1) + 1] \cdot \tau_{nom} \end{cases}.\qquad(9.3)$$

The notch sensitivity factor is usually determined experimentally. As per Eq. (9.3), when $q = 0$, the stress concentration has no effect on the strength of a member; conversely, $K_e = K_t$ when $q = 1$, and as a consequence, the theoretical stress concentration factor acts at its maximum weight. A value of 1 can be used for the notch sensitivity factor q in conservative designs that require higher safety factors, and this is the approach used here.

The stress concentration factor, also termed *form factor* in order to emphasize that it is mainly determined by the geometry of a particular raiser, can be evaluated analytically by means of the theory of elasticity. Finite element techniques permit the direct evaluation of the peak stresses at regions with stress concentrators, and therefore enable the indirect calculation of the stress concentration factor. Experimental laboratory methods for stress analysis such as photoelasticity, other optomechanical techniques or strain gauges can be employed in static, fatigue or impact tests as alternative means of assessing the stress concentration factor.

Empirical equations and related charts for notches that can be used as flexible hinges in compliant mechanisms have been studied by Peterson [1], Pilkey and Pilkey [2], Young and Budynas [3], Budynas and Nisbett [4], Isida [5], Ling [6], Noda et al. [7], Gray et al. [8], Durelli [9], Zappalorto et al. [10], Zheng and Niemi [11], Panasyuk et al. [12], Taylor et al. [13], Radaj [14], Gomez et al. [15] and Neuber [16] for various geometries such as right circular, circularly corner-filleted or V-shaped under axial loading, bending or torsion and with rectangular or circular cross-sections. Hyperbolic and parabolic notches are investigated in Zappalorto et al. [17], while Guimin et al. [18] propose stress concentration equations that are obtained through finite element analysis for conic-section flexure hinges.

9.1.2 IN-PLANE STRESS CONCENTRATION

Flexible hinges under the action of in-plane forces/moments are subjected to axial and bending deformations. When ignoring the shear effects, the resulting normal stresses are perpendicular to the hinge cross-section. The axial and bending stresses can therefore be added algebraically to result in a total normal stress, whose effect is magnified by the stress concentrators associated with a particular hinge geometry.

Consider the generic flexible hinge of Figure 9.2. The hinge has filleted areas at its end points that contribute to reduce the stress concentration and to adequately design/tailor the elastic behavior of the hinge. The hinge has a length l, rectangular cross-section of minimum thickness t and constant out-of-plane width w. The in-plane loads f_x, f_y and m_z act at the free end O, while the opposite end is assumed fixed.

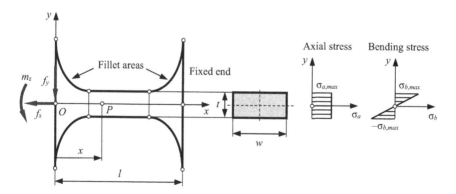

FIGURE 9.2 Generic fixed-free, straight-axis, rectangular cross-section flexible hinge with filleted end areas, in-plane load and resulting axial/bending normal stresses.

This particular choice of end loads results in the following internal reactions (stress resultants) at an arbitrary point P whose axial position is defined by an abscissa x:

$$N = f_x; \quad M_b = f_y \cdot x + m_z, \tag{9.4}$$

where N is the normal/axial force and M_b is the bending moment at P. The force N produces normal axial stresses σ_a, while the moment M_b results in normal bending stresses σ_b whose nominal values are:

$$\sigma_a = \frac{N}{A_{net}} = \frac{f_x}{w \cdot t}; \quad \sigma_b = \frac{M_b \cdot y}{I_{z,net}} = \frac{12(f_y \cdot x + m_z) \cdot y}{w \cdot t^3}. \tag{9.5}$$

The variation of the normal axial and bending stresses is also illustrated in Figure 9.2. While the axial normal strain is positive and constant across the rectangular section height, the bending stress varies linearly being zero at the section centroid and maximum on the outer fibers where $y = \pm t/2$.

Superimposing the axial and bending effects, the maximum normal stress will occur on the upper outer fiber for $x = l$ (which renders the bending moment of Eq. (9.4) maximum), according to the equation:

$$\sigma_{max} = \sigma_{a,max} + \sigma_{b,max}. \tag{9.6}$$

The constant stresses generated through axial loading are expressed based on Eqs. (9.1) and (9.5) as:

$$\sigma_{a,max} = K_{ta} \cdot \frac{f_x}{w \cdot t}, \tag{9.7}$$

where K_{ta} is the theoretical stress concentration factor in axial loading (in tension for the particular f_x of Figure 9.2, but it can also cover compression). The normal bending stresses are maximum on the upper outer fiber at the fixed end ($y = +t/2$ and $x = l$) and are obtained from Eqs. (9.1) and (9.5) as:

$$\sigma_{b,max} = K_{tb} \cdot \frac{6 \cdot (f_y \cdot l + m_z)}{w \cdot t^2}, \tag{9.8}$$

where K_{tb} is the theoretical stress concentration factor in bending. Both K_{ta} and K_{tb} are provided in the literature for various hinge geometries. By substituting Eqs. (9.7) and (9.8) into Eq. (9.6) results in:

$$\sigma_{max} = \frac{K_{ta}}{w \cdot t} \cdot f_x + \frac{6 \cdot K_{tb} \cdot l}{w \cdot t^2} \cdot f_y + \frac{6 \cdot K_{tb}}{w \cdot t^2} \cdot m_z = [G_K]^T [f_{ip}], \tag{9.9}$$

where $[f_{ip}] = [f_x f_y\, m_z]^T$ is the in-plane load vector and:

$$[G_K]^T = \begin{bmatrix} G_{K,x} & G_{K,y} & G_{K,z} \end{bmatrix} = \begin{bmatrix} \dfrac{K_{ta}}{w \cdot t} & \dfrac{6 \cdot K_{tb} \cdot l}{w \cdot t^2} & \dfrac{6 \cdot K_{tb}}{w \cdot t^2} \end{bmatrix} \qquad (9.10)$$

is the *geometric raiser vector*, which comprises geometric parameters and stress concentration factors.

The maximum normal stress σ_{max} is usually the allowable limit σ_{allow}, which, for ductile materials, is determined by means of a *factor of safety* $n \geq 1$ from the yield strength σ_Y as:

$$n = \frac{\sigma_Y}{\sigma_{allow}} \quad \text{or} \quad \sigma_{allow} = \frac{\sigma_Y}{n}. \qquad (9.11)$$

As a result, Eq. (9.9) can be used to establish the maximum levels of the in-plane load vector acting on a flexure hinge. The equation has an infinite number of solutions as it consists of three unknown load components. It is possible, for instance, to select a solution from the following ones:

$$[f_{ip,max}] = \begin{bmatrix} f_{x,max} \\ f_{y,max} \\ m_{z,max} \end{bmatrix} = \begin{cases} \begin{bmatrix} \sigma_{allow}/G_{K,x} & 1/G_{K,y} & -1/G_{K,z} \end{bmatrix}^T \\[2mm] \begin{bmatrix} \sigma_{allow}/G_{K,x} & -1/G_{K,y} & 1/G_{K,z} \end{bmatrix}^T \\[2mm] \begin{bmatrix} 1/G_{K,x} & \sigma_{allow}/G_{K,y} & -1/G_{K,z} \end{bmatrix}^T \\[2mm] \begin{bmatrix} -1/G_{K,x} & \sigma_{allow}/G_{K,y} & 1/G_{K,z} \end{bmatrix}^T \\[2mm] \begin{bmatrix} 1/G_{K,x} & -1/G_{K,y} & \sigma_{allow}/G_{K,z} \end{bmatrix}^T \\[2mm] \begin{bmatrix} -1/G_{K,x} & 1/G_{K,y} & \sigma_{allow}/G_{K,z} \end{bmatrix}^T \end{cases} . \qquad (9.12)$$

Another choice is to allocate fractions of the allowable normal stress to each of the three normal stresses and to formulate them in conjunction with Eq. (9.10) as:

$$\begin{cases} \sigma_{a,max} = G_{K,x} \cdot f_{x,max} = c_a \cdot \sigma_{allow} \\[2mm] \sigma_{f,max} = G_{K,y} \cdot f_{y,max} = c_f \cdot \sigma_{allow} \;, \\[2mm] \sigma_{m,max} = G_{K,z} \cdot m_{z,max} = c_m \cdot \sigma_{allow} \end{cases} \qquad (9.13)$$

where c_a, c_f and c_m are fractions of the axial, bending (through force) and bending (through moment) stresses. The resulting maximum load vector is therefore:

$$\left[f_{ip,\max}\right] = \begin{bmatrix} f_{x,\max} \\ f_{y,\max} \\ m_{z,\max} \end{bmatrix} = \sigma_{allow} \cdot \begin{bmatrix} \dfrac{c_a}{G_{K,x}} \\[2mm] \dfrac{c_f}{G_{K,y}} \\[2mm] \dfrac{c_m}{G_{K,z}} \end{bmatrix} \quad \text{with} \quad c_a + c_f + c_m = 1. \qquad (9.14)$$

The variant of Eq. (9.14) is particularly useful in designing loads with specified levels of participation of the axial, bending force and bending moment load components that are applied at the free end of a flexible hinge.

Equations (9.9), (9.10), (9.12) and (9.14) relate the allowable stress to the maximum load for a given flexible-hinge geometry. It is useful to establish a similar connection between the allowable stress levels and the deformations of a flexible hinge, which are needed in order to evaluate the workspace covered by the moving end of the hinge with respect to the opposite end. The compliance-based deformation/load equation is expressed as:

$$\left[u_{ip,\max}\right] = \left[C_{ip}\right]\left[f_{ip,\max}\right] = \left[K_{ip}\right]^{-1}\left[f_{ip,\max}\right], \qquad (9.15)$$

with $[u_{ip,\max}]$ being the maximum-value evaluation of the in-plane displacement vector $[u_{ip}] = [u_x \; u_y \; \theta_z]^T$ at the moving end O of the hinge, while $[C_{ip}]$ and $[K_{ip}]$ are the in-plane compliance and stiffness matrices. The maximum load vector of Eq. (9.15) can be determined as in Eqs. (9.12) or (9.14).

Load and geometric raiser vector variants:

The previous formulation was based on a particular choice of the load applied at the free end of a generic straight-axis, rectangular cross-section flexible hinge, and it resulted in a maximum (positive) stress located at the hinge's fixed end on the upper fiber. However, for an arbitrary loading of a flexible hinge in a planar compliant mechanism, the in-plane free-end loads may be different from those of Figure 9.2 and may result in maximum stresses that are located at points other than the upper-fiber, fixed-end point and of values different from the one of Eq. (9.9). It is therefore worthwhile to identify all possible load combinations and to calculate the maximum stress and its location for each variant, which will result in distinct geometry raiser vectors for each separate case enabling to express the maximum normal stress in a manner similar to the one of Eq. (9.9). Figure 9.3a illustrates a generic straight-axis flexure hinge with positive in-plane loads at one end and fixed at the opposite end. Figure 9.3b and c shows the nominal (minimum) rectangular cross-sections at the two ends with relevant points. It is assumed that the nominal cross-section is defined by the dimensions t and w with $t < w$ for in-plane bending and that the hinge is filleted at its ends.

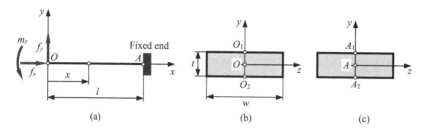

FIGURE 9.3 Generic, fixed-free straight-axis flexible hinge: (a) skeleton representation; (b) enlarged cross-section at the free end; (c) enlarged cross-section at the fixed end.

Figure 9.4 graphs the bending moment diagrams for the possible combinations of f_y and m_z.

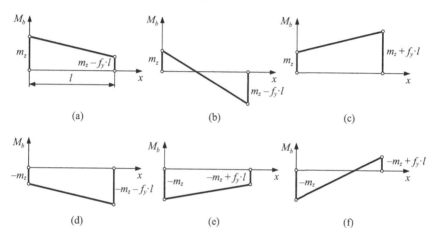

FIGURE 9.4 Bending moment diagrams corresponding to: (a) $m_z > 0, f_y > 0$ and $m_z > f_y \cdot l$; (b) $m_z > 0, f_y > 0$ and $m_z < f_y \cdot l$; (c) $m_z > 0$ and $f_y < 0$; (d) $m_z < 0$ and $f_y > 0$; (e) $m_z < 0, f_y < 0$ and $|m_z| > f_y \cdot l$; (f) $m_z < 0, f_y < 0$ and $|m_z| < f_y \cdot l$.

Depending on the sign of the axial load f_x, two conditions are possible, which are analyzed next.

When $f_x > 0$ (as in Figure 9.3a), the internal axial reaction produces compressive (negative) axial normal stress, which is the one of Eqs. (9.5) but negative. In order to obtain a total normal stress that is maximum (in absolute value), it is necessary to identify the positions along the hinge axis x and those along the cross-section y-direction that correspond to the maximum negative bending stress. The following situations can arise that need to be studied based on the relationship between f_y and m_z, as well as on their signs:

Case 1: $m_z > 0, f_y > 0$, and $m_z > f_y \cdot l$

The bending moment is expressed as:

$$M_b = m_z - f_y \cdot x \tag{9.16}$$

and is plotted in Figure 9.4a. The maximum moment occurs for $x = 0$ (at the cross-section passing through the free end O) and at point O_2 on that cross-section of Figure 9.3b where the stress is negative due to $y = -t/2$. The total maximum stress and the corresponding geometric stress raiser of Eq. (9.9) are:

$$\sigma_{max} = -\frac{K_{ta}}{w \cdot t} \cdot f_x - \frac{6 \cdot K_{tb}}{w \cdot t^2} \cdot m_z; \quad [G_K]^T = \left[\begin{array}{ccc} -\dfrac{K_{ta}}{w \cdot t} & 0 & -\dfrac{6 \cdot K_{tb}}{w \cdot t^2} \end{array} \right]. \quad (9.17)$$

Case 2: $m_z > 0, f_y > 0$, and $m_z < f_y \cdot l$

The bending moment of Eq. (9.16) is plotted in Figure 9.4b. Two situations are possible: when $m_z > f_y \cdot l - m_z$, the maximum bending moment is at O_2 and Case 1 covers this situation; when $m_z < f_y \cdot l - m_z$, the absolute maximum bending moment is negative and corresponds to the fixed end of the beam. In order to keep the bending stress negative, point A_1 of Figure 9.3c has to be selected to maximize the bending stress. The total maximum normal stress and the corresponding geometric raiser vector are therefore:

$$\sigma_{max} = -\frac{K_{ta}}{w \cdot t} \cdot f_x - \frac{6 \cdot K_{tb} \cdot l}{w \cdot t^2} \cdot f_y + \frac{6 \cdot K_{tb}}{w \cdot t^2} \cdot m_z;$$

$$[G_K]^T = \left[\begin{array}{ccc} -\dfrac{K_{ta}}{w \cdot t} & -\dfrac{6 \cdot K_{tb} \cdot l}{w \cdot t^2} & \dfrac{6 \cdot K_{tb}}{w \cdot t^2} \end{array} \right]. \quad (9.18)$$

Case 3: $m_z > 0$ and $f_y < 0$

The bending moment is plotted in Figure 9.4c based on the equation:

$$M_b = m_z + f_y \cdot x \quad (9.19)$$

and is maximum for $x = l$ at the cross-section passing through the fixed end A. The negative stress is maximum in absolute value at point A_2 of Figure 9.3c where the total normal stress and the geometric raiser vector are:

$$\sigma_{max} = -\frac{K_{ta}}{w \cdot t} \cdot f_x - \frac{6 \cdot K_{tb} \cdot l}{w \cdot t^2} \cdot f_y - \frac{6 \cdot K_{tb}}{w \cdot t^2} \cdot m_z;$$

$$[G_K]^T = \left[\begin{array}{ccc} -\dfrac{K_{ta}}{w \cdot t} & -\dfrac{6 \cdot K_{tb} \cdot l}{w \cdot t^2} & -\dfrac{6 \cdot K_{tb}}{w \cdot t^2} \end{array} \right]. \quad (9.20)$$

Case 4: $m_z < 0$ and $f_y > 0$

Figure 9.4d plots the following bending moment:

$$M_b = -m_z - f_y \cdot x. \quad (9.21)$$

In this case, the absolute-value maximum bending moment occurs at the fixed end A and the negative bending stress has a maximum absolute value at A_1 in

Figure 9.3c. As a consequence, the total maximum normal stress and the resulting raiser vector are the ones given in Eq. (9.20).

Case 5: $m_z < 0, f_y < 0$, and $|m_z| > |f_y \cdot l|$
The bending moment is:

$$M_b = -m_z + f_y \cdot x \qquad (9.22)$$

and is plotted in Figure 9.4e. Its absolute maximum value occurs at the cross-section located at O and the maximum negative bending stress corresponds to the upper fiber at point O_1 in Figure 9.3b. As a consequence, the total maximum normal negative stress and the related geometric raiser vector are:

$$\sigma_{max} = -\frac{K_{ta}}{w \cdot t} \cdot f_x - \frac{6 \cdot K_{tb}}{w \cdot t^2} \cdot m_z; \quad [G_K] = \left[\begin{array}{ccc} -\dfrac{K_{ta}}{w \cdot t} & 0 & -\dfrac{6 \cdot K_{tb}}{w \cdot t^2} \end{array} \right]. \qquad (9.23)$$

which is identical to the raiser vector of Case 1 in Eq. (9.17). It can be verified that this model is also valid for loads where $m_z < 0$ and $|m_z| > |-m_z + f_y \cdot l|$.

Case 6: $m_z < 0, f_y < 0$ and $|m_z| < |f_y \cdot l|$
The bending moment is plotted in Figure 9.4f being defined as:

$$M_b = -m_z + f_y \cdot x. \qquad (9.24)$$

which is identical to the M_b of Eq. (.22). When $|m_z| > |-m_z + f_y \cdot l|$, the absolute-value maximum bending moment is at section O and this scenario is analyzed in Case 5. For loads where $|m_z| < |-m_z + f_y \cdot l|$, the bending moment is maximum in absolute value at the fixed-end cross-section and corresponds to the points on the lower fiber, which is at A_2. It can be shown that the total maximum normal stress and the corresponding raiser vector are those of Eq. (9.18).

When $f_x < 0$, the normal axial stress is positive, which requires to identify the location and value of the maximum positive normal stresses that are produced through bending. The six cases analyzed previously for $f_x > 0$ can be studied considering that $f_x < 0$, and similar conclusions are derived by pairing the corresponding load cases in the two categories. For each load case with $f_x > 0$, the corresponding case with $f_x < 0$ (which is of interest here) has the maximum stresses located at the same cross-section but on the opposite outer fibers; in addition, when $f_x < 0$, the maximum stress has a sign opposite to the one of the related case with $f_x > 0$.

Example 9.1

A symmetric, straight-axis right circular flexible hinge defined by a radius $r = 0.01$ m, a minimum thickness $t = 0.002$ m and a constant, out-of-plane width $w = 0.008$ m is manufactured of steel whose Young's modulus is $E = 2.1 \cdot 10^{11}$ N/m^2 and allowable normal stress is $\sigma_{allow} = 150 \cdot 10^6$ N/m^2. The hinge is acted upon by in-plane loading consisting of f_x, f_y and m_z, as illustrated in

Figure 9.5. Graphically study the variation of the displacement ratio $u_{y,max}/u_{x,max}$ at the free end O in terms of the axial-stress fraction c_a spanning the [0.05, 0.9] range – utilize equal bending stress fractions $c_f = c_m$.

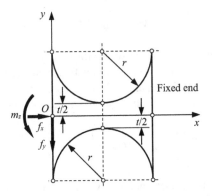

FIGURE 9.5 Fixed-free, straight-axis right circular flexible hinge with geometry and in-plane loading.

Solution:

The load case of Figure 9.5 is identical to the one described by the bending moment of Eq. (9.19) and the maximum normal stress and corresponding geometric raiser vector expressed in Eq. (9.20) namely:

$$\left[G_K \right]^T = \left[\quad -\frac{K_{ta}}{w \cdot t} \quad -\frac{12 \cdot K_{tb} \cdot r}{w \cdot t^2} \quad -\frac{6 \cdot K_{tb}}{w \cdot t^2} \quad \right], \tag{9.25}$$

which considered that the length of this particular hinge is $l = 2r$.

The right circular flexible hinge of Figure 9.5 can be considered a semi-circular notch; as such, Pilkey and Pilkey [2], Young and Budynas [3], Isida [5] and Ling [6] recommend the following stress concentration equations for normal stresses produced by axial loading and bending:

$$\begin{cases} K_{ta} = 3.0365 - 3.472 \cdot \left(\dfrac{2r}{t+2r} \right) + 1.009 \cdot \left(\dfrac{2r}{t+2r} \right)^2 + 0.405 \cdot \left(\dfrac{2r}{t+2r} \right)^3 \\[3mm] K_{tb} = 3.0365 - 6.637 \cdot \left(\dfrac{2r}{t+2r} \right) + 8.229 \cdot \left(\dfrac{2r}{t+2r} \right)^2 - 3.636 \cdot \left(\dfrac{2r}{t+2r} \right)^3 \end{cases} \tag{9.26}$$

With the numerical values of this example, Eq. (9.26) yields $K_{ta} = 1.0183$ and $K_{tb} = 1.07191$. The compliances of the right circular flexible hinge are provided in Eqs. (A3.1), which are assembled in the in-plane compliance matrix $[C_{ip}]$ as in Eq. (3.35). The maximum displacements at the free end O can now be calculated with Eqs. (9.14) and (9.15) considering that $c_f = c_m = (1 - c_a)/2$. Figure 9.6 plots the displacement ratio in terms of the axial-stress fraction c_a – it can be seen that as the weight of the axial load/stress increases, the displacement ratio decreases from a value of 22.2 for $c_a = 0.05$ to a value of 0.13 when $c_a = 0.9$, which indicates that the free-end axial displacement relatively increases with respect to the y deflection of the free end when c_a increases.

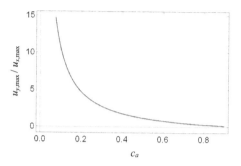

FIGURE 9.6 Plot of free-end displacement ratio $u_{y,max}/u_{x,max}$ in terms of the axial-stress fraction c_a for a right circular flexible hinge under planar load.

Example 9.2

Consider the planar mechanism of *Example 5.5* and sketched in Figure 5.14, which is redrawn in Figure 9.7. The device is formed of two rigid links AB and CD and two identical straight-axis right circular hinges of rectangular cross-section, like the one of Figure 3.11a. Evaluate the maximum planar loads that can be applied at points A and D, which will result in maximum deformations of the two hinges. For that load, calculate the deformations of the flexible hinges, as well as the displacements at A and D. Known are the material Young's modulus $E = 2.1 \cdot 10^{11}$ N/m², allowable normal stress $\sigma_{allow} = 150 \cdot 10^6$ N/m², the hinge geometric parameters: semi-angle $\alpha = 30°$, minimum in-plane thickness $t = 0.002$ m, constant out-of-plane width $w = 0.007$ m, radius $r = 0.02$ m and the rigid-link parameters: $l_{AB} = 0.03$ m, $l_{CD} = 0.02$ m, $\delta = 45°$. Use the following stress fractions: $c_a^{(1)} = 0.1, c_f^{(1)} = 0.5, c_m^{(1)} = 0.4$ for hinge (1) and $c_a^{(2)} = 0.2, c_f^{(2)} = 0.3, c_m^{(2)} = 0.5$ for hinge (2). Assume that the external loads generate local-frame loads on the two hinges that consist of positive (compressive) axial loads, negative y-axis forces and positive z-axis moments.

Solution:

The maximum, local-frame, in-plane loads acting at the end B of hinge (1) and end D of hinge (2) can be calculated as in Eq. (9.14):

$$\left[f_{max}^{(i)} \right] = \begin{bmatrix} f_{x,max}^{(i)} \\ f_{y,max}^{(i)} \\ m_{z,max}^{(i)} \end{bmatrix} = \sigma_{allow} \cdot \begin{bmatrix} \dfrac{c_a^{(i)}}{G_{K,x}} \\ \dfrac{c_f^{(i)}}{G_{K,y}} \\ \dfrac{c_m^{(i)}}{G_{K,z}} \end{bmatrix} \quad \text{with} \quad i = 1, 2. \qquad (9.27)$$

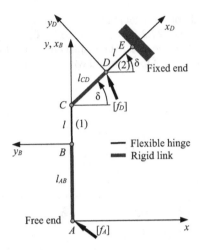

FIGURE 9.7   Free-fixed planar serial chain with two identical, straight-axis circular flexible hinges, two rigid links and two external loads.

For the required direction of the loads at B and D, the stress intensity components of Eq. (9.27) are those given in Eq. (9.20):

$$G_{K,x} = -\frac{K_{ta}}{w \cdot t}; \quad G_{K,y} = -\frac{6 \cdot K_{tb} \cdot l}{w \cdot t^2}; \quad G_{K,z} = -\frac{6 \cdot K_{tb}}{w \cdot t^2}, \qquad (9.28)$$

with $l = 2r \cdot \sin\alpha$ being the length of the flexible hinge. The stress intensity factors are provided graphically in Pilkey and Pilkey [2]. For an infinite ratio of the in-plane thickness of the rigid portion of the circular hinge link (outside the flexure hinge) to the minimum in-plane thickness t, as well as for $r/t = 10$, the two stress intensity factors resulting from the plots provided in Pilkey and Pilkey [2], particularly from Chart 2.27, are $K_{ta} = K_{tb} = 1.03$. The loads of Eqs. (9.27) are connected to the actual loads to be externally applied at B and D as:

$$\begin{cases} \left[f_{\max}^{(1)} \right] = \left[Tr_{AB} \right] \left[f_A \right] \\ \left[f_{\max}^{(2)} \right] = \left[Tr_{AD} \right] \left[f_A \right] + \left[R^{(2)} \right] \left[f_D \right] \end{cases} \rightarrow \begin{cases} \left[f_A \right] = \left[Tr_{AB} \right]^{-1} \left[f_{\max}^{(1)} \right] \\ \left[f_D \right] = \left[R^{(2)} \right]^{T} \left(\left[f_{\max}^{(2)} \right] - \left[Tr_{AD} \right] \left[Tr_{AB} \right]^{-1} \left[f_{\max}^{(1)} \right] \right), \end{cases}$$
$$(9.29)$$

where

$$\left[Tr_{AB} \right] = \left[R^{(1)} \right] \left[T_{AB} \right]; \quad \left[Tr_{AD} \right] = \left[R^{(2)} \right] \left[T_{AD} \right]. \qquad (9.30)$$

The rotation matrices of Eqs. (9.29) and (9.30) are calculated by means of Eq. (3.21) with $\delta_1 = 90°$ and $\delta_2 = \delta$. The translation matrices of Eqs. (9.30) are evaluated as in Eq. (3.15) with the following offsets:

$$\begin{cases} \Delta x_{AB} = 0 \\ \Delta y_{AB} = l_{AB} \end{cases}; \quad \begin{cases} \Delta x_{AD} = l_{CD} \cos\delta \\ \Delta y_{AD} = l_{AB} + l + l_{CD} \sin\delta \end{cases}. \qquad (9.31)$$

The maximum in-plane deformation of each flexure hinge can be calculated as per Eq. (9.15):

$$\left[u_{max}^{(i)}\right] = \left[C_{ip}\right]\left[f_{max}^{(i)}\right] \quad \text{with} \quad i = 1,2 \tag{9.32}$$

in terms of the maximum hinge loads of Eqs. (9.27).
  The displacements at points A and D are calculated as:

$$\begin{cases} [u_A] = \left([Tr_{AB}]^T [C_{ip}][Tr_{AB}] + [Tr_{AD}]^T [C_{ip}][Tr_{AD}]\right)[f_A] + [Tr_{AD}]^T [C_{ip}][R^{(2)}][f_D] \\ [u_D] = [R^{(2)}]^T [C_{ip}]\left([Tr_{AD}][f_A] + [R^{(2)}][f_D]\right) \end{cases} \tag{9.33}$$

The in-plane compliance matrix $[C_{ip}]$ is evaluated by means of Eqs. (2.53) and (3.35).
  The following numerical loads are obtained: $f_{Ax} = 16.99\,N$, $f_{Ay} = -203.88\,N$, $m_{Az} = -0.78$ Nm, $f_{Dx} = -408.22\,N$, $f_{Dy} = 88.5\,N$ and $m_{Dz} = -4.43$ Nm. The deformation components of the two hinges are calculated as in Eqs. (9.32): $u_x^{(1)} = -8.76 \cdot 10^{-7}\,m$, $u_y^{(1)} = -1.13 \cdot 10^{-5}\,m$, $\theta_z^{(1)} = 0.0005$ rad, $u_x^{(2)} = -1.75 \cdot 10^{-6}\,m$, $u_y^{(2)} = 1.94 \cdot 10^{-5}\,m$ and $\theta_z^{(2)} = -0.001$ rad. The displacements at A and D result from Eqs. (9.33) as: $u_{Ax} = -0.0318\,m$, $u_{Ay} = 0.0205\,m$, $\theta_{Az} = -0.526$ rad, $u_{Dx} = -0.0028\,m$, $u_{Dy} = 0.0103$ m and $\theta_{Dz} = -0.527$ rad.

9.1.3 OUT-OF-PLANE STRESS CONCENTRATION

Out-of-plane deformations of flexible hinges are usually produced through bending and torsion. While bending generates normal stresses (perpendicular to the hinge cross-section), torsion results in shear stresses (in the plane of the hinge cross-section). Because of the different nature and directions of the two stresses, they cannot simply be added algebraically as in the case of in-plane deformation. Instead, theories of material yielding need to be utilized to obtain an equivalent stress. The analysis in this section is limited to hinge configurations with circular cross-section, but it can be extended to designs of rectangular cross-sections.

Consider the generic flexible hinge with end filleted regions of Figure 9.8. The hinge has a length l and circular cross-section of minimum diameter d. The out-of-plane loads m_x, m_y and f_z are applied at the free-end point O – the opposite end is assumed fixed. While not included here, a more thorough analysis of the various possible combinations of the force and moment producing bending can be performed to identify the hinge ends where the bending moment is maximum.

For the particular loading of Figure 9.8, the internal reactions (stress resultants) at an arbitrary point P located at a distance x from O are:

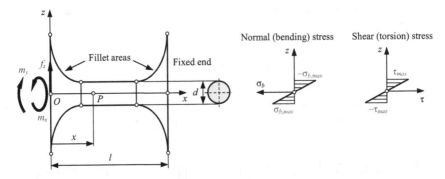

FIGURE 9.8 Generic fixed-free, straight-axis, circular cross-section flexible hinge with filleted end areas and out-of-plane loading.

$$M_t = m_x; \quad M_b = f_z \cdot x + m_y, \tag{9.34}$$

where M_t is the x-axis torsional moment and M_b is the bending moment about the y axis. The resulting shear stresses τ and bending stresses σ are:

$$\tau = \frac{M_t \cdot r}{I_{p,net}} = \frac{32 \cdot m_x \cdot r}{\pi \cdot d^4}; \quad \sigma = \frac{M_b \cdot r}{I_{y,net}} = \frac{64 \left(f_z \cdot x + m_y \right) \cdot r}{\pi \cdot d^4}, \tag{9.35}$$

where r is a variable radius of the circular cross-section. The variation of the shear and normal stresses is also shown in Figure 9.8. Both stresses vary linearly with the radial distance r and reach the maximum values on the outer surface where $r = d/2$ for the net (minimum) section.

Like in in-plane loading, the maximum stresses at the fixed end include the effects of stress concentration and are expressed as:

$$\tau_{max} = K_{tt} \cdot \frac{16}{\pi \cdot d^3} \cdot m_x; \quad \sigma_{max} = K_{tb} \cdot \frac{32 \cdot (f_z \cdot l + m_y)}{\pi \cdot d^3} = \frac{32 K_{tb}}{\pi \cdot d^3} \cdot m_y + \frac{32 K_{tb} \cdot l}{\pi \cdot d^3} \cdot f_z$$

or

$$\begin{cases} \tau_{max} = G_{K,x} \cdot m_{x,max} \\ \sigma_{m,max} = G_{K,y} \cdot m_{y,max} \quad \text{with}: \ G_{K,x} = \frac{16 \cdot K_{tt}}{\pi \cdot d^3}; \ G_{K,y} = \frac{32 \cdot K_{tb}}{\pi \cdot d^3}; \ G_{K,z} = \frac{32 \cdot K_{tb} \cdot l}{\pi \cdot d^3} \\ \sigma_{f,max} = G_{K,z} \cdot f_{z,max}, \end{cases}$$

$$\tag{9.36}$$

where K_{tt} is the theoretical stress concentration factor in torsion and K_{tb} is the theoretical stress concentration due to bending. The factors multiplying the maximum loads in Eqs. (9.36) are geometric raisers.

Material yield failure theories/criteria:

As mentioned earlier, the shear and normal stresses directions are perpendicular. Superimposing the shear and normal stress is achieved by means of theories (or criteria) of material failure, as briefly discussed in this section.

Mechanical components usually fail due to one or more of the following principal mechanisms: yield failure, fracture failure or fatigue failure. The failure phenomenon is currently seen as an irreversible process that develops locally and through which the macrostructure and operational performance of the component are negatively altered. Ductile materials (especially metals) fail by yielding, whereby the deformations exceed the proportionality limit and reach the non-recoverable plastic domain. Ductile materials, such as typical metals including mild steel, aluminum, titanium, copper, magnesium, and some of their alloys, as well as nonmetals like Teflon, are capable to undergo large plastic deformations before fracture. Brittle materials, such as cast iron, concrete, glass or ceramic compounds, silicon and silicon-based compounds, fail by fracture as they incur relatively small deformations before breaking apart, compared to ductile materials. Fatigue failure applies to both ductile and brittle materials and is caused by stresses less than the ultimate strength. Such stresses might occur statically on a less frequent basis, when the loads are not applied repeatedly but failure manifests itself in time because cracks, for instance, develop to critical sizes. More frequently though, the fatigue failure is generated by cyclic loading, whereby loads and deformations are applied to the mechanical component more than once in a repetitive manner.

Other mechanisms of failure, in addition to the ones mentioned above, are excessive elastic (recoverable) deformation, low stiffness and operation at resonant frequencies, time deformation under unchanged loading conditions (creep), and unstable response at critical loads (buckling). The vast majority of flexible hinges and mechanisms thereof are fabricated out of metallic/ductile materials, and the brief exposure that follows will focus on yielding failure theories only.

In a complex state of stress, as frequently encountered in real-life applications, the failure can be predicted by means of theories of failures. Fundamentally, any theory of failure attempts to establish an equivalence between the real complex state of stress and a simple, uniaxial state of stress (tension/compression) condition by applying a specific prediction criterion – see Ugural and Fenster [19], Timoshenko [20] or Solecki and Conant [21].

For ductile and isotropic materials, there are three main failure theories that can be utilized to predict the critical condition in a member under a complex state of stress. They are the maximum energy of deformation theory (von Mises), the maximum shear stress theory (Tresca) and the maximum total strain energy theory (Beltrami–Haigh). Other theories/criteria better apply to brittle materials. They include the maximum principal stress theory (Rankine), the maximum principal strain theory (Saint Venant) and the maximum internal friction theory (Coulomb–Mohr), as indicated in Ugural and Fenster [19] or Volterra and Gaines [22]. The failure theories for orthotropic materials, such as the Norris criterion (see Cook and Young [23] for more details) or the criteria dedicated to composite materials, such as those of Halpin–Tsai or Greszczuk (as shown in Kobayashi [24]), are not studied here.

Experimental results indicated, according to Ugural and Fenster [19] and Solecki and Conant [21], for instance, that the maximum energy of deformation theory (von Mises) and the maximum shear stress theory (Tresca) provide best predictions for ductile-material members under a complex state of stress. They are addressed in the following with emphasis being placed on the von Mises failure criterion since this theory, in addition to being slightly more accurate than the Tresca theory, has an elegant mathematical formulation that is extensively employed in tools of engineering calculation, like the finite element technique.

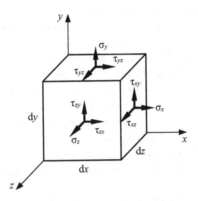

FIGURE 9.9   Stresses on an elementary cube.

Figure 9.9 illustrates the normal and shear stresses on the positive faces of an elementary/infinitesimal cube that has been removed from an elastic body that is subjected to a complex state of stress.

Mechanics of materials theory (see Solecki and Conant [21], for instance) shows that the principal stresses σ_1, σ_2 and σ_3 (one of them being the maximum stress) are the roots of the third-degree equation in σ:

$$\sigma^3 - I_1 \cdot \sigma^2 + I_2 \cdot \sigma - I_3 = 0, \tag{9.37}$$

I_1, I_2 and I_3 are the stress invariants (they do not depend on the way the reference frame is directed), which are defined in terms of the normal and shear stresses acting on the elementary cube's faces of Figure 9.9 as:

$$\begin{cases} I_1 = \sigma_x + \sigma_y + \sigma_z \\ I_2 = \sigma_x \cdot \sigma_y + \sigma_y \cdot \sigma_z + \sigma_z \cdot \sigma_x - \tau_{xy}^2 - \tau_{yz}^2 - \tau_{zx}^2 \\ I_3 = \sigma_x \cdot \sigma_y \cdot \sigma_z + 2\tau_{xy} \cdot \tau_{yz} \cdot \tau_{zx} - \sigma_x \cdot \tau_{yz}^2 - \sigma_y \cdot \tau_{zx}^2 - \sigma_z \cdot \tau_{xy}^2 \end{cases} \tag{9.38}$$

The Tresca theory (first mentioned by Coulomb) specifies that a ductile-material member under complex loading will yield when the maximum shear stress in the body reaches the critical shear stress in the simple tension/compression test. According to this theory, the *equivalent yield stress* is given by:

$$\sigma_{eq} = \max\left(|\sigma_1 - \sigma_2|, |\sigma_2 - \sigma_3|, |\sigma_3 - \sigma_1|\right). \tag{9.39}$$

The von Mises criterion (also known as the Huber–von Mises–Hencky criterion, owing to the researchers who have formulated it) states that the failure of a member under a complex state of stress occurs when its energy of distortion equals the critical energy of distortion in the uniaxial tension/compression test. Generally, the strain energy of an elastic body can be separated into two parts: one part produces the volume deformation without modifying the shape of the loaded body and is generated by a so-called hydrostatic load (actually an average stress). The other energy component does not affect the volume but alters the shape. The latter one is known as distortion or deviatoric energy and is generated by deviatoric stresses that represent the difference between the normal stresses and the average stress. Numerous experiments have been carried out (see Ugural and Fenster [19], Solecki and Conant [21] or Cook and Young [23]), which have proven that elastic bodies usually do not fail under a hydrostatic state of stress alone. Therefore, a criterion that would predict failure based on the strain energy should be centered upon the deviatoric portion of the strain energy. The von Mises criterion follows this direction and uses the following equivalent stress:

$$\sigma_{eq} = \sqrt{\frac{(\sigma_1 - \sigma_2)^2 + (\sigma_2 - \sigma_3)^2 + (\sigma_3 - \sigma_1)^2}{2}} = \sqrt{I_1^2 - 3I_2^2}. \tag{9.40}$$

By substituting I_1 and I_2 of Eq. (9.38) into Eq. (9.40) results in:

$$\sigma_{eq} = \sqrt{\sigma_x^2 + \sigma_y^2 + \sigma_z^2 - (\sigma_x \cdot \sigma_y + \sigma_y \cdot \sigma_z + \sigma_z \cdot \sigma_x) + 3(\tau_{xy}^2 + \tau_{yz}^2 + \tau_{zx}^2)}. \tag{9.41}$$

A condition that is often encountered in engineering applications is that of *plane stress* where all the nonzero stresses are located in one plane, such as *xy*. As a result, the other stresses are zero, namely:

$$\sigma_z = 0; \quad \tau_{yz} = 0; \quad \tau_{zx} = 0. \tag{9.42}$$

In a plane stress condition, the generic von Mises criterion of Eq. (9.41) becomes:

$$\sigma_{eq} = \sqrt{\sigma_x^2 + \sigma_y^2 - \sigma_x \cdot \sigma_y + 3\tau_{xy}^2}. \tag{9.43}$$

Moreover, for a state of plane stress where σ_y is zero, Eq. (9.43) simplifies to:

$$\sigma_{eq} = \sqrt{\sigma_x^2 + 3\tau_{xy}^2} \quad \text{or} \quad \sigma_{eq} = \sqrt{\sigma^2 + 3\tau^2}, \tag{9.44}$$

which is a well-known equation that applies in a variety of cases, including the bending/torsion of shafts and flexible hinges.

The maximum shear and normal stresses of Eq. (9.36) are now combined by means of von Mises criterion – Eq. (9.44) – as:

$$\sigma_{eq,max} = \sqrt{\sigma_{max}^2 + 3\tau_{max}^2} = \sqrt{\left(\sigma_{m,max} + \sigma_{f,max}\right)^2 + 3\tau_{max}^2}$$

$$= \sqrt{\left(G_{K,y} \cdot m_{y,max} + G_{K,z} \cdot f_{z,max}\right)^2 + 3\left(G_{K,x} \cdot m_{x,max}\right)^2}. \qquad (9.45)$$

When the external loads are known and the flexible hinge is designed/ specified (which means that its stress concentration factors are determined), Eq. (9.45) can be used to verify whether the maximum equivalent stress is smaller than the allowable normal stress, which would validate the design and load selection.

When the external load needs to be evaluated, a procedure similar to the one utilized for the in-plane case can be applied. The maximum stresses of Eq. (9.36) can be assumed to be fractions of the allowable normal stress σ_{allow}, namely:

$$\tau_{max} = c_t \cdot \sigma_{allow}, \sigma_{m,max} = c_m \cdot \sigma_{allow}, \sigma_{f,max} = c_f \cdot \sigma_{allow}, \qquad (9.46)$$

where c_t, c_m and c_f are the fractions defining the participation of the torsion, bending-moment and bending-force stresses. These fractions can be chosen adequately in order to allow various portions of the torsion and bending loads to form the total hinge load. Comparing Eqs. (9.45) and (9.46) indicates that the three fractions are related as:

$$\left(c_m + c_f\right)^2 + 3c_t^2 = 1. \qquad (9.47)$$

Combining Eqs. (9.46) with the geometric raiser components of Eqs. (9.36) allows expressing the three maximum loads as:

$$m_{x,max} = \frac{c_t}{G_{K,x}} \cdot \sigma_{allow}, m_{y,max} = \frac{c_m}{G_{K,y}} \cdot \sigma_{allow}, f_{z,max} = \frac{c_f}{G_{K,z}} \cdot \sigma_{allow}. \qquad (9.48)$$

Once the maximum end-point loads have been calculated with Eq. (9.48), the maximum displacements at the same point can be calculated by means of the flexible-hinge out-of-plane compliance matrix $[C_{op}]$ as:

$$\left[u_{op,max}\right] = \left[C_{op}\right]\left[f_{op,max}\right] \quad \text{with:} \quad \begin{cases} \left[u_{op,max}\right] = \begin{bmatrix} \theta_{x,max} & \theta_{y,max} & u_{z,max} \end{bmatrix}^T \\ \left[u_{op,max}\right] = \begin{bmatrix} m_{x,max} & m_{y,max} & f_{z,max} \end{bmatrix}^T \end{cases}.$$

$$(9.49)$$

Example 9.3

The right circular flexible hinge of Figure 9.10 has a circular cross-section and is acted upon at its free end by the loads shown in the same figure. Known are the minimum diameter $d = 0.001$ m, the radius $r = 0.01$ m, the dimension $t = 3r = 0.03$ m, as well as the material properties $E = 2.1 \cdot 10^{11}$ N/m^2 and $G = 8.08 \cdot 10^{10}$ N/m^2. Consider that the torsion fraction is $c_t = 0.2$ and the two bending fractions are equal. Calculate the maximum values of the three end loads, as well as the corresponding free-end displacements.

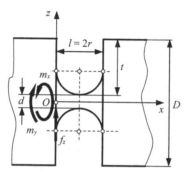

FIGURE 9.10   Straight-axis, right circular flexible hinge of circular cross-section with out-of-plane loads.

Solution:

The bending stress concentration K_{tb} and torsion stress concentration K_{tt} for this flexible-hinge configuration are expressed in Pilkey and Pilkey [2] as:

$$K_t = C_1 + C_2 \cdot \left(\frac{2t}{D}\right) + C_3 \cdot \left(\frac{2t}{D}\right)^2 + C_4 \cdot \left(\frac{2t}{D}\right)^3 \quad \text{with} \quad D = d + 2t. \quad (9.50)$$

Two particular sets of constants are defined separately for bending and torsion, which are calculated in terms of the t/r ratio – those particular equations are provided in Pilkey and Pilkey [2] at pp. 122 and 128. The numerical values are $K_{tb} = 1.81$ and $K_{tt} = 1.01$. From Eq. (9.47), it follows that the equal bending fractions are:

$$c_m = c_f = \frac{\sqrt{1 - 3c_t^2}}{2} = 0.469. \quad (9.51)$$

Equations (9.36) and (9.48) are combined to obtain the following maximum values of the three end loads shown in Figure 9.10: $m_{x,max} = 0.00714$ Nm, $m_{y,max} = 0.0046$ Nm and $f_{z,max} = 0.229$ N. The out-of-plane compliance matrix of the right circular hinge is evaluated by means of Eqs. (2.58), (3.61) and (3.62). With it and the maximum load vector, the following maximum end-point displacements are obtained as per Eq. (9.49): $\theta_{x,max} = 0.0028$ rad, $\theta_{y,max} = 0.002$ rad and $u_{z,max} = 2.1 \cdot 10^{-5}$ m.

9.2   ROTATION PRECISION OF FLEXIBLE HINGES

Planar flexible hinges, especially those that deform/operate in their plane and are designed to enable relative rotation of adjacent rigid links, do not generate a pure rotary motion due to coupled bending, axial and shear effects. The precision of rotation can be assessed by analyzing the displacement of the symmetry center for transversely symmetric flexible hinges, as discussed in Lobontiu et al. [25], for instance. This section studies a new quantifier that evaluates the rotary motion of a generic flexure hinge, either of straight axis or of circular axis, by comparing the position of the end point of a flexible hinge with the position of a point that lies at the end of a typical rotation link and undergoes the same rotation as the hinge end point.

9.2.1   STRAIGHT-AXIS FLEXIBLE HINGES

Consider a regular rotation joint like the one of Figure 9.11a, whose mobile, rigid link of length l undergoes an angular rotation θ_z with respect to a fixed link; as a result, the end point A of the mobile link moves circularly to A_r. Assume that the arm of the rigid link is replaced by a flexible hinge of length l, as illustrated in Figure 9.11b. Under the combined action of the planar load formed of f_x, f_y and m_z, the end point A of the flexure hinge displaces to a position A_f that differs from A_r when the rigid link at the end of the hinge undergoes the same rotation θ_z.

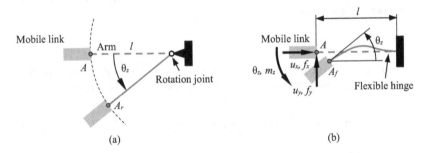

(a)   (b)

FIGURE 9.11   Relative planar motion between rigid links as produced by a: (a) rotation joint; (b) flexible hinge.

More details of the two connectors' original and final positions with the relevant geometric parameters are shown in Figure 9.12. A direct measure of the difference between the final positions of the two moving links is the length of the segment A_rA_f, which is:

$$\left| A_r A_f \right| = d = \sqrt{\left(x_{A_r} - x_{A_f} \right)^2 + \left(y_{A_r} - y_{A_f} \right)^2} \tag{9.52}$$

and is expressed in terms of the planar coordinates of the points A_r and A_f in Figure 9.12. The x and y coordinates of point A_r are:

$$x_{A_r} = l - l \cdot \cos\theta_z; \quad y_{A_r} = -l \cdot \sin\theta_z. \tag{9.53}$$

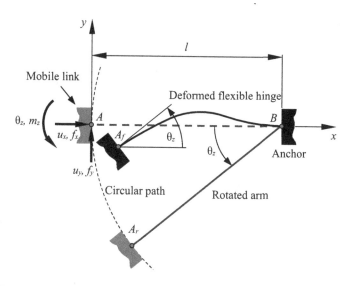

FIGURE 9.12 Skeleton representation of generic straight-axis flexible hinge in original and deformed conditions, as well as a rotation joint in original and displaced positions.

The planar coordinates of A_f are actually the displacements of the flexible-hinge free-end point, which, under the assumption of small deformations, are expressed as:

$$x_{A_f} = C_{u_x-f_x} \cdot f_x; \quad y_{A_f} = C_{u_y-f_y} \cdot f_y + C_{u_y-m_z} \cdot m_z; \quad \theta_z = C_{u_y-m_z} \cdot f_y + C_{\theta_z-m_z} \cdot m_z,$$

$$(9.54)$$

where $C_{u_x-f_x}$ is the axial compliance and $C_{u_y-f_y}$, $C_{u_y-m_z}$ and $C_{\theta_z-m_z}$ are the bending compliances. Substituting the slope θ_z from the third Eq. (9.54) into Eq. (9.53), and then combining Eqs. (9.52)–(9.54) result in:

$$d = \left(\left\{ l \cdot \left[1 - \cos\left(C_{\theta_z-m_z} \cdot m_z + C_{u_y-m_z} \cdot f_y \right) \right] - C_{u_x-f_x} \cdot f_x \right\}^2 \right.$$

$$\left. + \left[l \cdot \sin\left(C_{\theta_z-m_z} \cdot m_z + C_{u_y-m_z} \cdot f_y \right) + \left(C_{u_y-f_y} \cdot f_y + C_{u_y-m_z} \cdot m_z \right) \right]^2 \right)^{\frac{1}{2}}. \quad (9.55)$$

For unit loads applied at A, namely, $f_x = 1, f_y = 1, m_z = 1$, Eq. (9.55) simplifies to:

$$d = \left(\left\{ l \cdot \left[1 - \cos\left(C_{\theta_z-m_z} + C_{u_y-m_z} \right) \right] - C_{u_x-f_x} \right\}^2 \right.$$

$$\left. + \left[l \cdot \sin\left(C_{\theta_z-m_z} + C_{u_y-m_z} \right) + \left(C_{u_y-f_y} + C_{u_y-m_z} \right) \right]^2 \right)^{\frac{1}{2}} \quad (9.56)$$

Example 9.4

Compare the rotation precision of a constant cross-section flexible hinge with that of a right circular flexible hinge in terms of the radius r. Both configurations have rectangular cross-sections with an out-of-plane width $w = 0.006\,m$ and a minimum in-plane thickness $t = 0.001\,m$. The hinges are designed from the same isotropic material. Consider that unit loads act on the hinge free end.

Solution:

The two hinge designs have the same length l; therefore, $l = 2r$. The deviation distance of Eq. (9.56) is utilized to evaluate the following percentage ratio:

$$\Delta d_r = \frac{d_1 - d_2}{d_1} \cdot 100[\%], \qquad (9.57)$$

where the subscript 1 denotes the constant cross-section hinge and 2 denotes the right circular configuration. The compliances of Eq. (9.56) are given in Eqs. (2.26) for hinge 1 and in Eqs. (A3.1) for hinge 2. The plot of Figure 9.13 graphs the difference of Eq. (9.57), which indicates that the deviation of the right circular hinge is smaller than the similar deviation of the constant cross-section hinge; as a result, the right circular design is more precise than its constant cross-section counterpart in terms of approximating the pure circular motion of a rotary joint. Figure 9.13 also shows that as the length l of the hinges increases (since $l = 2r$), the right circular design generates a better-quality rotation than the constant cross-section hinge.

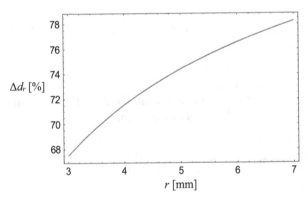

FIGURE 9.13 Plot of percentage relative difference Δd_r in terms of the radius r.

9.2.2 CIRCULAR-AXIS FLEXIBLE HINGES

Figure 9.14 shows the original positions of a rotary joint (its rotation center is at B, and the radius of rotation is l) and a circular-axis flexible hinge – see Lobontiu [26], Lobontiu and Cullin [27] – whose original length is also l.

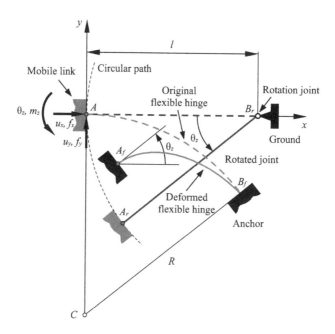

FIGURE 9.14 Skeleton representation of generic circular-axis flexible hinge in original and deformed conditions, as well as a rotation joint in original and displaced positions.

The direction of the tangent at the circular hinge free end A is identical to the direction of the rotary joint. It is again assumed that the rotation θ_z of the rotary joint is identical to the slope of the flexure's tip in its deformed position.

While the final planar coordinates of the rotary joint tip remain those of Eqs. (9.53), the deformed hinge's tip (A_f) coordinates (together with the tip slope) are expressed in terms of in-plane compliances and loads as:

$$\begin{cases} x_{A_f} = C_{u_x-f_x} \cdot f_x - C_{u_x-f_y} \cdot f_y - C_{u_x-m_z} \cdot m_z \\ y_{A_f} = -C_{u_x-f_y} \cdot f_x + C_{u_y-f_y} \cdot f_y + C_{u_y-m_z} \cdot m_z. \\ \theta_z = -C_{u_x-m_z} \cdot f_x + C_{u_y-m_z} \cdot f_y + C_{\theta_z-m_z} \cdot m_z \end{cases} \qquad (9.58)$$

Equations (9.58) were obtained based on the in-plane compliance matrix of Eq. (2.107) – that matrix is expressed in the reference frame $Axyz$ of Figure 9.14; note that Figure 2.24b, which was used to define the compliance matrix of Eq. (2.107), has an axis denoted by y^* instead of the y axis of Figure 9.14 here. The distance between the ends A_r and A_f is calculated as:

$$|A_r A_f| = d = \left(\left\{ l \cdot \left[1 - \cos\left(-C_{u_x-m_z} \cdot f_x + C_{u_y-m_z} \cdot f_y + C_{\theta_z-m_z} \cdot m_z \right) \right] \right. \right.$$

$$\left. - \left(C_{u_x-f_x} \cdot f_x - C_{u_x-f_y} \cdot f_y - C_{u_x-m_z} \cdot m_z \right) \right\}^2$$

$$+ \left[l \cdot \sin\left(-C_{u_x-m_z} \cdot f_x + C_{u_y-m_z} \cdot f_y + C_{\theta_z-m_z} \cdot m_z \right) \right.$$

$$\left. \left. - C_{u_x-f_y} \cdot f_x + C_{u_y-f_y} \cdot f_y + C_{u_y-m_z} \cdot m_z \right]^2 \right)^{\frac{1}{2}} \qquad (9.59)$$

For unit loads, Eq. (9.59) becomes:

$$d = \left(\left\{ l \cdot \left[1 - \cos\left(-C_{u_x-m_z} + C_{u_y-m_z} + C_{\theta_z-m_z} \right) \right] - \left(C_{u_x-f_x} - C_{u_x-f_y} - C_{u_x-m_z} \right) \right\}^2 \right.$$

$$\left. + \left[l \cdot \sin\left(-C_{u_x-m_z} + C_{u_y-m_z} + C_{\theta_z-m_z} \right) - C_{u_x-f_y} + C_{u_y-f_y} + C_{u_y-m_z} \right]^2 \right)^{\frac{1}{2}} \qquad (9.60)$$

Example 9.5

Compare the rotation precision of a straight-axis flexible hinge with that of a circular-axis flexible hinge of radius R in terms of the opening angle α of the circular-axis hinge. The two configurations have identical constant rectangular cross-sections of out-of-plane width $w = 0.006$ m and in-plane thickness $t = 0.001$ m; they also have the same length l and are designed from the same isotropic material. Consider that unit loads act at the free ends of hinges.

Solution:

Because the two hinges have the same length l, the radius of the circular-axis design is calculated as $R = l/\alpha$. The following percentage relative difference is used to compare the precision of rotation of the two hinge designs:

$$\Delta d_r = \frac{d_c - d_s}{d_c} \cdot 100[\%], \qquad (9.61)$$

where d_c represents the deviation of the circular-axis hinge and is calculated with Eq. (9.60), and d_s is the deviation of the straight-axis hinge and is given in Eq. (9.56). The compliances of the straight-axis hinge are determined with Eqs. (2.26), whereas the compliances needed for the circular-axis hinge are calculated based on Eqs. (A2.22) through (A2.26).

Figure 9.15 is the plot of Δd_r of Eq. (9.61) in terms of the angle α. For the angle range used here, the variation in Δd_r is very small and around the approximate value of 67%; this indicates that the straight-axis hinge is more precise in rotation than a circular-axis flexure hinge of identical length. Larger changes in the opening angle (curvature of the circular design) do not substantially affect the rotation precision difference between the two configurations (for an opening angle of 45°, for instance, the resulting Δd_r was 66.59%).

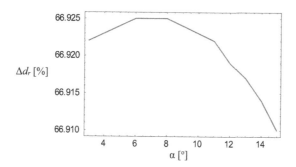

FIGURE 9.15 Plot of percentage relative difference Δd_r in terms of the center angle α.

9.3 FLEXIBLE-HINGE COMPLIANCE THERMAL SENSITIVITY

Flexible-hinge dimensions are affected by temperature variation, which, in turn, alters the elastic properties of these connectors and their mechanisms. This section studies the thermal sensitivity of straight- and circular-axis flexible-hinge compliances. Thermal stresses and deformations of elastic line members can also be evaluated by the finite element method, as discussed in Nicholson [28] and Hetnarski and Eslami [29], for instance. Static and dynamic modeling of hinge-based compliant mechanisms under combined mechanical and thermal load has been studied by Li and Zhou [30], Manoach and Ribeiro [31], Li and Xi [32], Hou and Zhang [33], Zhang and Hou [34] and Burgreen [35] by means of conventional finite element software or analytically.

9.3.1 STRAIGHT-AXIS FLEXIBLE HINGES

A linear dimension d of an isotropic-material mechanical member experiences an expansion Δd when exposed to a temperature increase ΔT and becomes d_{th}:

$$d_{th} = d + \Delta d = d + \alpha_{th} \cdot d \cdot \Delta T = (1 + \alpha_{th} \cdot \Delta T) \cdot d, \qquad (9.62)$$

where α_{th} is the coefficient of linear thermal expansion. Equation (9.62) enables to study the variation of flexible-hinge compliances with temperature. The compliance thermal sensitivity of straight-axis flexible hinges of rectangular and circular cross-section is expressed here.

9.3.1.1 Rectangular Cross-Section

A prismatic flexible hinge of rectangular cross-section has an initial length l and its cross-section initial dimensions $w(x)$ and $t(x)$ are considered variable with the abscissa x. The final values of the following relevant cross-sectional properties (that are needed for compliance calculations) are expressed by means of Eqs. (9.62) as:

$$
\begin{cases}
A(x)_{th} = w(x)_{th} \cdot t(x)_{th} = \left(1 + \alpha_{th} \cdot \Delta T\right)^2 \cdot A(x) \\[2mm]
I_z(x)_{th} = \dfrac{w(x)_{th} \cdot t(x)_{th}^3}{12} = \left(1 + \alpha_{th} \cdot \Delta T\right)^4 \cdot I_z(x) \\[2mm]
I_y(x)_{th} = \dfrac{t(x)_{th} \cdot w(x)_{th}^3}{12} = \left(1 + \alpha_{th} \cdot \Delta T\right)^4 \cdot I_y(x) \\[2mm]
I_t(x)_{th} = \dfrac{w(x)_{th} \cdot t(x)_{th}^3}{3} = \left(1 + \alpha_{th} \cdot \Delta T\right)^4 \cdot I_t(x)
\end{cases}
\tag{9.63}
$$

where

$$
A(x) = w(x) \cdot t(x); \;\; I_z(x) = \frac{w(x) \cdot t(x)^3}{12}; \;\; I_y(x) = \frac{t(x) \cdot w(x)^3}{12}; \;\; I_t(x) = \frac{w(x) \cdot t(x)^3}{3}
\tag{9.64}
$$

are the initial-temperature cross-sectional properties. Note that while the torsional moment of area $I_t(x)$ is formulated for the particular designs with $t(x) \ll w(x)$, the other two possibilities discussed in Chapter 2 result in the same expression of Eq. (9.63).

9.3.1.1.1 In-Plane Compliances

The temperature-dependent axial compliance is calculated as:

$$
\begin{aligned}
C_{u_x - f_x, th} &= \frac{1}{E} \cdot \int_0^{l_{th}} \frac{dx}{A(x)_{th}} = \frac{1}{E \cdot \left(1 + \alpha_{th} \cdot \Delta T\right)^2} \cdot \left(\int_0^l \frac{dx}{A(x)} + \int_l^{l_{th}} \frac{dx}{A(x)} \right) \\[3mm]
&= \frac{1}{\left(1 + \alpha_{th} \cdot \Delta T\right)^2} \cdot \left(C_{u_x - f_x} + \frac{1}{E} \cdot \int_l^{l_{th}} \frac{dx}{A(x)} \right) \\[3mm]
&= \frac{1}{\left(1 + \alpha_{th} \cdot \Delta T\right)^2} \cdot \left(C_{u_x - f_x} + \Delta C_{u_x - f_x} \right), \Delta C_{u_x - f_x} = \frac{1}{E} \cdot \int_l^{l_{th}} \frac{dx}{A(x)}.
\end{aligned}
\tag{9.65}
$$

Equation (9.65) took into consideration that the temperature-dependent length is $l_{th} = (1 + \alpha_{th} \cdot \Delta T) \cdot l$ as per Eq. (9.62). In a similar manner, the following temperature-dependent, bending-related compliances result after the temperature increase:

$$\left\{ \begin{array}{ll} C_{u_y-f_y,th} = \dfrac{1}{\left(1+\alpha_{th}\cdot\Delta T\right)^4}\cdot\left(C_{u_y-f_y}+\Delta C_{u_y-f_y}\right), & \Delta C_{u_y-f_y} = \dfrac{1}{E}\cdot\displaystyle\int_l^{l_{th}}\dfrac{x^2 dx}{I_z(x)} \\[4mm] C_{u_y-m_z,th} = \dfrac{1}{\left(1+\alpha_{th}\cdot\Delta T\right)^4}\cdot\left(C_{u_y-m_z}+\Delta C_{u_y-m_z}\right), & \Delta C_{u_y-m_z} = -\dfrac{1}{E}\cdot\displaystyle\int_l^{l_{th}}\dfrac{x dx}{I_z(x)} \\[4mm] C_{\theta_z-m_z,th} = \dfrac{1}{\left(1+\alpha_{th}\cdot\Delta T\right)^4}\cdot\left(C_{\theta_z-m_z}+\Delta C_{\theta_z-m_z}\right), & \Delta C_{\theta_z-m_z} = \dfrac{1}{E}\cdot\displaystyle\int_l^{l_{th}}\dfrac{dx}{I_z(x)} \end{array} \right. .$$

$$(9.66)$$

The compliance $C_{u_y-f_y,th}$ of Eq. (9.66) is based on the long-beam Euler–Bernoulli model. Equations (9.65) and (9.66) can be written in matrix form as:

$$\left[C_{ip,th}\right]=[A]\cdot\left(\left[C_{ip}\right]+\left[\Delta C_{ip}\right]\right),\qquad(9.67)$$

with

$$\left[C_{ip,th}\right]=\begin{bmatrix} C_{u_x-f_x,th} & 0 & 0 \\ 0 & C_{u_y-f_y,th} & C_{u_y-m_z,th} \\ 0 & C_{u_y-m_z,th} & C_{\theta_z-m_z,th} \end{bmatrix};$$

$$\left[\Delta C_{ip}\right]=\begin{bmatrix} \Delta C_{u_x-f_x} & 0 & 0 \\ 0 & \Delta C_{u_y-f_y} & \Delta C_{u_y-m_z} \\ 0 & \Delta C_{u_y-m_z} & \Delta C_{\theta_z-m_z} \end{bmatrix};\qquad(9.68)$$

$$[A]=\dfrac{1}{\left(1+\alpha_{th}\cdot\Delta T\right)^2}\cdot\begin{bmatrix} 1 & 0 & 0 \\ 0 & \dfrac{1}{\left(1+\alpha_{th}\cdot\Delta T\right)^2} & 0 \\ 0 & 0 & \dfrac{1}{\left(1+\alpha\cdot\Delta T\right)^2} \end{bmatrix},$$

$[C_{ip}]$ is the original in-plane compliance matrix of the flexible hinge in Eq. (9.68).

For relatively short flexible hinges and according to the Timoshenko model, the direct-bending compliance is expressed as in Eq. (2.10), namely:

$$C_{u_y-f_y,s,th} = C_{u_y-f_y,th} + 2\alpha_s\cdot\left(1+\mu\right)\cdot C_{u_x-f_x,th},\qquad(9.69)$$

where $C_{u_y-f_y,th}$ and $C_{u_x-f_x,th}$ are given in Eqs. (9.65) and (9.66).

Equation (9.67) allows us to calculate the difference (error) between the temperature-affected in-plane compliance matrix and the similar original compliance matrix as:

$$\left[C_{ip,th}\right]-\left[C_{ip}\right]=\left(\left[A\right]-\left[I\right]\right)\cdot\left[C_{ip}\right]+\left[A\right]\cdot\left[\Delta C_{ip}\right], \qquad (9.70)$$

where $[I]$ is the 3×3 identity matrix.

Example 9.6

A right circular micro flexible hinge of polysilicon is exposed to temperature increase. Graphically study the relative variation of the in-plane bending compliance $C_{u_y-f_y}$ in terms of the temperature variation ΔT. Known are the radius $r = 5$ μm, minimum thickness $t = 0.6$ μm, out-of-plane width $w = 3$ μm and Young's modulus $E = 1.5\cdot10^{11}\,\text{N/m}^2$. Consider that ΔT varies in the [40°; 120°] interval.

Solution:

Based on Eq. (9.66), the following relative error function is formed:

$$\text{error}\left(C_{u_y-f_y}\right)=\frac{\left|C_{u_y-f_y,th}-C_{u_y-f_y}\right|}{C_{u_y-f_y}}$$

$$=\left|\frac{1}{\left(1+\alpha_{th}\cdot\Delta T\right)^4}-1+\frac{12}{E\cdot w\cdot\left(1+\alpha_{th}\cdot\Delta T\right)^4\cdot C_{u_y-f_y}}\cdot\int_{l}^{l_{th}}\frac{x^2}{t(x)^3}\cdot dx\right|. \qquad (9.71)$$

with $\quad t(x)=t+2\left(r-\sqrt{2r\cdot x-x^2}\right)$

Figure 9.16, which plots this error function in terms of the temperature variation, indicates that the percentage changes due to temperature increase are very small for the selected temperature range.

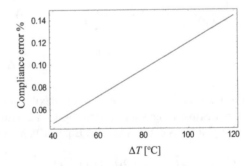

FIGURE 9.16 Plot of percentage relative error of the $C_{u_y-f_y}$ compliance in terms of ΔT.

9.3.1.1.2 *Out-of-Plane Compliances*

After a temperature increase ΔT, the out-of-plane compliances of a straight-axis flexible hinge become:

$$
\begin{cases}
C_{\theta_x-m_x,th} = \dfrac{1}{\left(1+\alpha_{th}\cdot\Delta T\right)^4}\cdot\left(C_{\theta_x-m_x}+\Delta C_{\theta_x-m_x}\right), & \Delta C_{\theta_x-m_x} = \dfrac{1}{G}\cdot\int\limits_{l}^{l_{th}}\dfrac{dx}{I_t(x)} \\[4mm]
C_{\theta_y-m_y,th} = \dfrac{1}{\left(1+\alpha_{th}\cdot\Delta T\right)^4}\cdot\left(C_{\theta_y-m_y}+\Delta C_{\theta_y-m_y}\right), & \Delta C_{\theta_y-m_y} = \dfrac{1}{E}\cdot\int\limits_{l}^{l_{th}}\dfrac{dx}{I_y(x)} \\[4mm]
C_{\theta_y-f_z,th} = \dfrac{1}{\left(1+\alpha_{th}\cdot\Delta T\right)^4}\cdot\left(C_{\theta_y-f_z}+\Delta C_{\theta_y-f_z}\right), & \Delta C_{\theta_y-f_z} = \dfrac{1}{E}\cdot\int\limits_{l}^{l_{th}}\dfrac{x\,dx}{I_y(x)} \\[4mm]
C_{u_z-f_z,th} = \dfrac{1}{\left(1+\alpha_{th}\cdot\Delta T\right)^4}\cdot\left(C_{u_z-f_z}+\Delta C_{u_z-f_z}\right), & \Delta C_{u_z-f_z} = \dfrac{1}{E}\cdot\int\limits_{l}^{l_{th}}\dfrac{x^2\,dx}{I_y(x)}
\end{cases}
$$

$$(9.72)$$

Equations (9.72) are written in matrix form as:

$$
\left[C_{op,th}\right]=[B]\cdot\left(\left[C_{op}\right]+\left[\Delta C_{op}\right]\right), \tag{9.73}
$$

with

$$
\left[C_{op,th}\right]=
\begin{bmatrix}
C_{\theta_x-m_x,th} & 0 & 0 \\
0 & C_{\theta_y-m_y,th} & C_{\theta_y-f_z,th} \\
0 & C_{\theta_y-f_z,th} & C_{u_z-f_z,th}
\end{bmatrix};
$$

$$
\left[\Delta C_{op}\right]=
\begin{bmatrix}
\Delta C_{\theta_x-m_x} & 0 & 0 \\
0 & \Delta C_{\theta_y-m_y} & \Delta C_{\theta_y-f_z} \\
0 & \Delta C_{\theta_y-f_z} & \Delta C_{u_z-f_z}
\end{bmatrix}; \tag{9.74}
$$

$$
[B]=\dfrac{1}{\left(1+\alpha_{th}\cdot\Delta T\right)^4}\cdot
\begin{bmatrix}
1 & 0 & 0 \\
0 & 1 & 0 \\
0 & 0 & 1
\end{bmatrix}.
$$

$[C_{op}]$ of Eq. (9.73) is the original out-of-plane compliance matrix of the flexible hinge. The error between the temperature-dependent out-of-plane compliance matrix and the similar original out-of-plane compliance matrix is evaluated by means of Eq. (9.73) as:

$$
\left[C_{op,th}\right]-\left[C_{op}\right]=\left([B]-[I]\right)\cdot\left[C_{op}\right]+[B]\cdot\left[\Delta C_{op}\right]. \tag{9.75}
$$

The direct-bending compliance of Eq. (9.72) is valid for long flexible hinges; for short hinges, this compliance is expressed similarly to Eq. (9.69), namely:

$$C_{u_z-f_z,s,th} = C_{u_z-f_z,th} + 2\alpha_s \cdot (1+\mu) \cdot C_{u_x-f_x,th}. \tag{9.76}$$

9.3.1.2   Circular Cross-Section

As per Eq. (9.62), the variable diameter $d(x)$ becomes:

$$d(x)_{th} = (1+\alpha_{th} \cdot \Delta T) \cdot d(x) \tag{9.77}$$

after the member experiences a temperature increase ΔT. The corresponding final cross-sectional properties are:

$$\begin{cases} A(x)_{th} = \dfrac{\pi \cdot d(x)_{th}^2}{4} = (1+\alpha_{th} \cdot \Delta T)^2 \cdot A(x) \\[3mm] I_z(x)_{th} = I_y(x)_{th} = \dfrac{\pi \cdot d(x)_{th}^2}{64} = (1+\alpha_{th} \cdot \Delta T)^4 \cdot I_z(x). \\[3mm] I_p(x)_{th} = \dfrac{\pi \cdot d(x)_{th}^2}{32} = (1+\alpha_{th} \cdot \Delta T)^4 \cdot I_p(x) \end{cases} \tag{9.78}$$

The following are initial cross-sectional properties:

$$A(x) = \frac{\pi \cdot d(x)^2}{4}; \quad I_z(x) = I_y(x) = \frac{\pi \cdot d(x)^4}{64}; \quad I_p(x) = \frac{\pi \cdot d(x)^4}{32}. \tag{9.79}$$

All compliance equations developed for the rectangular cross-section hinges in the previous Section 9.3.1.1 are also valid for the circular cross-section hinge configuration when used in conjunction with Eqs. (9.78) and (9.79).

9.3.2   Circular-Axis Flexible Hinges

The temperature-generated cross-sectional dimension variation of circular-axis flexible hinges is identical to that of straight-axis hinges. Assuming there is no stress induced in the flexible hinge by a temperature increase ΔT, it follows that the radius R remains unchanged, while the center angle varies from its original value α to α_h. Expressing the final arc length in two different ways results in the final center angle α_h being:

$$R \cdot \alpha_h = (1+\alpha_{th} \cdot \Delta T) \cdot R \cdot \alpha \rightarrow \alpha_h = (1+\alpha_{th} \cdot \Delta T) \cdot \alpha, \tag{9.80}$$

where the subscript "h" denotes the final value of the hinge opening angle.

The temperature-related in-plane and out-of-plane compliances of circular-axis flexible hinges of rectangular and circular cross-section are derived in this section.

9.3.2.1 Rectangular Cross-Section

9.3.2.1.1 In-Plane Compliances

Following an approach similar to the one used for straight-axis flexible hinges, the in-plane compliances of a generic circular-axis flexible hinges after exposure to a temperature increase of ΔT can be expressed. Equations (2.121), for instance, yield:

$$
\begin{cases}
C^{(b)}_{u_x-f_x,th} = \dfrac{1}{(1+\alpha_{th}\cdot\Delta T)^4}\cdot\left(C^{(b)}_{u_x-f_x}+\Delta C^{(b)}_{u_x-f_x}\right), & \Delta C^{(b)}_{u_x-f_x} = \dfrac{R^3}{E}\cdot\displaystyle\int_\alpha^{\alpha_h}\dfrac{(1-\cos\beta)^2}{I_z(\beta)}\cdot d\beta \\[4mm]
C^{(n)}_{u_x-f_x,th} = \dfrac{1}{(1+\alpha_{th}\cdot\Delta T)^2}\cdot\left(C^{(n)}_{u_x-f_x}+\Delta C^{(n)}_{u_x-f_x}\right), & \Delta C^{(n)}_{u_x-f_x} = \dfrac{R}{E}\cdot\displaystyle\int_\alpha^{\alpha_h}\dfrac{(\cos\beta)^2}{A(\beta)}\cdot d\beta \\[4mm]
C^{(s)}_{u_x-f_x,th} = \dfrac{1}{(1+\alpha_{th}\cdot\Delta T)^2}\cdot\left(C^{(s)}_{u_x-f_x}+\Delta C^{(s)}_{u_x-f_x}\right), & \Delta C^{(s)}_{u_x-f_x} = \dfrac{\alpha_s\cdot R}{G}\cdot\displaystyle\int_\alpha^{\alpha_h}\dfrac{(\sin\beta)^2}{A(\beta)}\cdot d\beta
\end{cases}
\tag{9.81}
$$

with $C^{(b)}_{u_x-f_x}$, $C^{(n)}_{u_x-f_x}$ and $C^{(s)}_{u_x-f_x}$ being expressed in Eqs. (2.121). Based on Eqs. (2.122), the following altered compliances are obtained:

$$
\begin{cases}
C^{(b)}_{u_x-f_y,th} = \dfrac{1}{(1+\alpha_{th}\cdot\Delta T)^4}\cdot\left(C^{(b)}_{u_x-f_y}+\Delta C^{(b)}_{u_x-f_y}\right), & \Delta C^{(b)}_{u_x-f_y} = -\dfrac{R^3}{E}\cdot\displaystyle\int_\alpha^{\alpha_h}\dfrac{\sin\beta\cdot(1-\cos\beta)}{I_z(\beta)}\cdot d\beta \\[4mm]
C^{(n)}_{u_x-f_y,th} = \dfrac{1}{(1+\alpha_{th}\cdot\Delta T)^2}\cdot\left(C^{(n)}_{u_x-f_y}+\Delta C^{(n)}_{u_x-f_y}\right), & \Delta C^{(n)}_{u_x-f_y} = \dfrac{R}{E}\cdot\displaystyle\int_\alpha^{\alpha_h}\dfrac{\sin\beta\cdot\cos\beta}{A(\beta)}\cdot d\beta
\end{cases}
\tag{9.82}
$$

where $C^{(b)}_{u_x-f_y}$ and $C^{(n)}_{u_x-f_y}$ are given in Eq. (2.122). Equation (2.123), which expresses $C^{(b)}_{u_x-m_z}$, results in:

$$
C_{u_x-m_z,th} = C^{(b)}_{u_x-m_z,th} = \dfrac{1}{(1+\alpha_{th}\cdot\Delta T)^4}\cdot\left(C^{(b)}_{u_x-m_z}+\Delta C^{(b)}_{u_x-m_z}\right),
$$

$$
\Delta C^{(b)}_{u_x-m_z} = \dfrac{R^2}{E}\cdot\displaystyle\int_\alpha^{\alpha_h}\dfrac{1-\cos\beta}{I_z(\beta)}\cdot d\beta.
\tag{9.83}
$$

The original compliances $C_{u_y-f_y}^{(b)}$ and $C_{u_y-f_y}^{(n)}$, which are given in Eqs. (2.124), transform into:

$$
\left\{
\begin{aligned}
C_{u_y-f_y,th}^{(b)} &= \frac{1}{(1+\alpha_{th}\cdot\Delta T)^4}\cdot\left(C_{u_y-f_y}^{(b)}+\Delta C_{u_y-f_y}^{(b)}\right), \quad \Delta C_{u_y-f_y}^{(b)}=\frac{R^3}{E}\cdot\int_{\alpha}^{\alpha_h}\frac{(\sin\beta)^2}{I_z(\beta)}\cdot d\beta \\[2mm]
C_{u_y-f_y,th}^{(n)} &= \frac{1}{(1+\alpha_{th}\cdot\Delta T)^2}\cdot\left(C_{u_y-f_y}^{(n)}+\Delta C_{u_y-f_y}^{(n)}\right), \quad \Delta C_{u_y-f_y}^{(n)}=\frac{R}{E}\cdot\int_{\alpha}^{\alpha_h}\frac{(\sin\beta)^2}{A(\beta)}\cdot d\beta
\end{aligned}
\right.
$$
$$(9.84)$$

Eventually, the compliances $C_{u_y-m_z}^{(b)}$ and $C_{\theta_z-m_z}^{(b)}$, which are defined in Eqs. (2.125), yield the following temperature-defined compliances:

$$
\left\{
\begin{aligned}
C_{u_y-m_z,th}=C_{u_y-m_z,th}^{(b)} &= \frac{1}{(1+\alpha_{th}\cdot\Delta T)^4}\cdot\left(C_{u_y-m_z}^{(b)}+\Delta C_{u_y-m_z}^{(b)}\right), \quad \Delta C_{u_y-m_z}^{(b)}=-\frac{R^2}{E}\cdot\int_{\alpha}^{\alpha_h}\frac{\sin\beta}{I_z(\beta)}\cdot d\beta \\[2mm]
C_{\theta_z-m_z,th}=C_{\theta_z-m_z,th}^{(b)} &= \frac{1}{(1+\alpha_{th}\cdot\Delta T)^4}\cdot\left(C_{\theta_z-m_z}^{(b)}+\Delta C_{\theta_z-m_z}^{(b)}\right), \quad \Delta C_{\theta_z-m_z}^{(b)}=\frac{R}{E}\cdot\int_{\alpha}^{\alpha_h}\frac{d\beta}{I_z(\beta)}
\end{aligned}
\right.
$$
$$(9.85)$$

When shear effects are neglected, Eqs. (9.81) through (9.85) can be written in matrix form as:

$$
\left[C_{ip,th}\right]=\left[C_{ip,th}^{(b)}\right]+\left[C_{ip,th}^{(n)}\right]=[B]\cdot\left(\left[C_{ip}^{(b)}\right]+\left[\Delta C_{ip}^{(b)}\right]\right)+[C]\cdot\left(\left[C_{ip}^{(n)}\right]+\left[\Delta C_{ip}^{(n)}\right]\right),
$$
$$(9.86)$$

with $[B]$ of Eq. (9.74) and

$$
\left[C_{ip,th}^{(b)}\right]=\begin{bmatrix} C_{u_x-f_x,th}^{(b)} & C_{u_x-f_y,th}^{(b)} & C_{u_x-m_z,th}^{(b)} \\ C_{u_x-f_y,th}^{(b)} & C_{u_y-f_y,th}^{(b)} & C_{u_y-m_z,th}^{(b)} \\ C_{u_x-m_z,th}^{(b)} & C_{u_y-m_z,th}^{(b)} & C_{\theta_z-m_z,th}^{(b)} \end{bmatrix};
\left[\Delta C_{ip}^{(b)}\right]=\begin{bmatrix} \Delta C_{u_x-f_x}^{(b)} & \Delta C_{u_x-f_y}^{(b)} & \Delta C_{u_x-m_z}^{(b)} \\ \Delta C_{u_x-f_y}^{(b)} & \Delta C_{u_y-f_y}^{(b)} & \Delta C_{u_y-m_z}^{(b)} \\ \Delta C_{u_x-m_z}^{(b)} & \Delta C_{u_y-m_z}^{(b)} & \Delta C_{\theta_z-m_z}^{(b)} \end{bmatrix};
$$

$$
\left[C_{ip,th}^{(n)}\right]=\begin{bmatrix} C_{u_x-f_x,th}^{(n)} & C_{u_x-f_y,th}^{(n)} & C_{u_x-m_z,th}^{(n)} \\ C_{u_x-f_y,th}^{(n)} & C_{u_y-f_y,th}^{(n)} & C_{u_y-m_z,th}^{(n)} \\ C_{u_x-m_z,th}^{(n)} & C_{u_y-m_z,th}^{(n)} & C_{\theta_z-m_z,th}^{(n)} \end{bmatrix};
\left[\Delta C_{ip}^{(n)}\right]=\begin{bmatrix} \Delta C_{u_x-f_x}^{(n)} & \Delta C_{u_x-f_y}^{(n)} & \Delta C_{u_x-m_z}^{(n)} \\ \Delta C_{u_x-f_y}^{(n)} & \Delta C_{u_y-f_y}^{(n)} & \Delta C_{u_y-m_z}^{(n)} \\ \Delta C_{u_x-m_z}^{(n)} & \Delta C_{u_y-m_z}^{(n)} & \Delta C_{\theta_z-m_z}^{(n)} \end{bmatrix};
$$

$$
[C]=\frac{1}{(1+\alpha_{th}\cdot\Delta T)^2}\cdot\begin{bmatrix} 1 & 0 & 0 \\ 0 & 1 & 0 \\ 0 & 0 & 1 \end{bmatrix}
$$
$$(9.87)$$

In Eq. (9.86) $\left[C_{ip} \right]$ is the in-plane compliance matrix of the circular-axis flexible hinge, which combines bending and axial effects, $\left[C_{ip} \right] = \left[C_{ip}^{(b)} \right] + \left[C_{ip}^{(n)} \right]$. The difference between $[C_{ip,th}]$ and $[C_{ip}]$ (which is not explicitly given here) reflects the error generated by thermal effects.

9.3.2.1.2 Out-of-Plane Compliances

Equations (2.126) provide the compliances $C_{\theta_x-m_x}^{(b)}$ and $C_{\theta_x-m_x}^{(t)}$ – after a temperature increase, they become:

$$
\begin{cases}
C_{\theta_x-m_x,th}^{(b)} = \dfrac{1}{\left(1+\alpha_{th}\cdot\Delta T\right)^4}\cdot\left(C_{\theta_x-m_x}^{(b)}+\Delta C_{\theta_x-m_x}^{(b)}\right), & \Delta C_{\theta_x-m_x}^{(b)} = \dfrac{R}{E}\cdot\displaystyle\int_{\alpha}^{\alpha_h}\dfrac{\left(\sin\beta\right)^2}{I_n(\beta)}\cdot d\beta \\[4mm]
C_{\theta_x-m_x,th}^{(t)} = \dfrac{1}{\left(1+\alpha_{th}\cdot\Delta T\right)^4}\cdot\left(C_{\theta_x-m_x}^{(t)}+\Delta C_{\theta_x-m_x}^{(t)}\right), & \Delta C_{\theta_x-m_x}^{(t)} = \dfrac{R}{G}\cdot\displaystyle\int_{\alpha}^{\alpha_h}\dfrac{\left(\cos\beta\right)^2}{I_t(\beta)}\cdot d\beta
\end{cases}
$$

$$(9.88)$$

The compliances $C_{\theta_x-m_y}^{(b)}$ and $C_{\theta_x-m_y}^{(t)}$ are expressed in Eqs. (2.127), which transform into:

$$
\begin{cases}
C_{\theta_x-m_y,th}^{(b)} = \dfrac{1}{\left(1+\alpha_{th}\cdot\Delta T\right)^4}\cdot\left(C_{\theta_x-m_y}^{(b)}+\Delta C_{\theta_x-m_y}^{(b)}\right), & \Delta C_{\theta_x-m_y}^{(b)} = -\dfrac{R}{E}\cdot\displaystyle\int_{\alpha}^{\alpha_h}\dfrac{\sin\beta\cdot\cos\beta}{I_n(\beta)}\cdot d\beta \\[4mm]
C_{\theta_x-m_y,th}^{(t)} = \dfrac{1}{\left(1+\alpha_{th}\cdot\Delta T\right)^4}\cdot\left(C_{\theta_x-m_y}^{(t)}+\Delta C_{\theta_x-m_y}^{(t)}\right), & \Delta C_{\theta_x-m_y}^{(t)} = \dfrac{R}{G}\cdot\displaystyle\int_{\alpha}^{\alpha_h}\dfrac{\sin\beta\cdot\cos\beta}{I_t(\beta)}\cdot d\beta
\end{cases}
$$

$$(9.89)$$

The following temperature-dependent compliances result from Eqs. (2.128), which define $C_{\theta_x-f_z}^{(b)}$ and $C_{\theta_x-f_z}^{(t)}$:

$$
\begin{cases}
C_{\theta_x-f_z,th}^{(b)} = \dfrac{1}{\left(1+\alpha_{th}\cdot\Delta T\right)^4}\cdot\left(C_{\theta_x-f_z}^{(b)}+\Delta C_{\theta_x-f_z}^{(b)}\right), & \Delta C_{\theta_x-f_z}^{(b)} = -\dfrac{R^2}{E}\cdot\displaystyle\int_{\alpha}^{\alpha_h}\dfrac{\left(\sin\beta\right)^2}{I_n(\beta)}\cdot d\beta \\[4mm]
C_{\theta_x-f_z,th}^{(t)} = \dfrac{1}{\left(1+\alpha_{th}\cdot\Delta T\right)^4}\cdot\left(C_{\theta_x-f_z}^{(t)}+\Delta C_{\theta_x-f_z}^{(t)}\right), & \Delta C_{\theta_x-f_z}^{(t)} = \dfrac{R^2}{G}\cdot\displaystyle\int_{\alpha}^{\alpha_h}\dfrac{\left(1-\cos\beta\right)\cdot\cos\beta}{I_t(\beta)}\cdot d\beta
\end{cases}
$$

$$(9.90)$$

The compliances $C^{(b)}_{\theta_y-m_y}$ and $C^{(t)}_{\theta_y-m_y}$, which are given in Eqs. (2.129), result in:

$$
\left\{
\begin{aligned}
C^{(b)}_{\theta_y-m_y,th} &= \frac{1}{\left(1+\alpha_{th}\cdot\Delta T\right)^4}\cdot\left(C^{(b)}_{\theta_y-m_y}+\Delta C^{(b)}_{\theta_y-m_y}\right), \quad \Delta C^{(b)}_{\theta_y-m_y}=\frac{R}{E}\cdot\int_{\alpha}^{\alpha_h}\frac{(\cos\beta)^2}{I_n(\beta)}\cdot d\beta \\
C^{(t)}_{\theta_y-m_y,th} &= \frac{1}{\left(1+\alpha_{th}\cdot\Delta T\right)^4}\cdot\left(C^{(t)}_{\theta_y-m_y}+\Delta C^{(t)}_{\theta_y-m_y}\right), \quad \Delta C^{(t)}_{\theta_y-m_y}=\frac{R}{G}\cdot\int_{\alpha}^{\alpha_h}\frac{(\sin\beta)^2}{I_t(\beta)}\cdot d\beta
\end{aligned}
\right.
$$

$$(9.91)$$

Equations (2.130), which define $C^{(b)}_{\theta_y-f_z},C^{(t)}_{\theta_y-f_z}$, yield:

$$
\left\{
\begin{aligned}
C^{(b)}_{\theta_y-f_z,th} &= \frac{1}{\left(1+\alpha_{th}\cdot\Delta T\right)^4}\cdot\left(C^{(b)}_{\theta_y-f_z}+\Delta C^{(b)}_{\theta_y-f_z}\right),\ \Delta C^{(b)}_{\theta_y-f_z}=\frac{R^2}{E}\cdot\int_{\alpha}^{\alpha_h}\frac{\sin\beta\cdot\cos\beta}{I_n(\beta)}\cdot d\beta \\
C^{(t)}_{\theta_y-f_z,th} &= \frac{1}{\left(1+\alpha_{th}\cdot\Delta T\right)^4}\cdot\left(C^{(t)}_{\theta_y-f_z}+\Delta C^{(t)}_{\theta_y-f_z}\right),\ \Delta C^{(t)}_{\theta_y-f_z}=\frac{R^2}{G}\cdot\int_{\alpha}^{\alpha_h}\frac{(1-\cos\beta)\cdot\sin\beta}{I_t(\beta)}\cdot d\beta
\end{aligned}
\right.
$$

$$(9.92)$$

The z-axis deflection compliances $C^{(b)}_{u_z-f_z}$ and $C^{(t)}_{u_z-f_z}$ are expressed in Eqs. (2.131) – they change into:

$$
\left\{
\begin{aligned}
C^{(b)}_{u_z-f_z,th} &= \frac{1}{\left(1+\alpha_{th}\cdot\Delta T\right)^4}\cdot\left(C^{(b)}_{u_z-f_z}+\Delta C^{(b)}_{\theta_y-f_z}\right), \quad \Delta C^{(b)}_{u_z-f_z}=\frac{R^3}{E}\cdot\int_{\alpha}^{\alpha_h}\frac{(\sin\beta)^2}{I_n(\beta)}\cdot d\beta \\
C^{(t)}_{u_z-f_z,th} &= \frac{1}{\left(1+\alpha_{th}\cdot\Delta T\right)^4}\cdot\left(C^{(t)}_{u_z-f_z}+\Delta C^{(t)}_{u_z-f_z}\right), \quad \Delta C^{(t)}_{u_z-f_z}=\frac{R^3}{G}\cdot\int_{\alpha}^{\alpha_h}\frac{(1-\cos\beta)^2}{I_t(\beta)}\cdot d\beta
\end{aligned}
\right.
$$

$$(9.93)$$

The cross-section parameters $A(\beta)$, $I_z(\beta)$, $I_n(\beta)$ and $I_t(\beta)$ of Eqs. (9.81) through (9.85) and (9.88) through (9.93) are the ones defined in Eqs. (9.64), where β should be used instead of x, and the subscript n replaces the subscript y of Eq. (9.64).

The out-of-plane, temperature-dependent compliances can be written in matrix form as in Eq. (9.73), where

$$\left[C_{op,th}\right]=\left[C^{(b)}_{op,th}\right]+\left[C^{(t)}_{op,th}\right]; \quad \left[\Delta C_{op}\right]=\left[\Delta C^{(b)}_{op}\right]+\left[\Delta C^{(t)}_{op}\right], \quad (9.94)$$

with

$$
\left[C_{op,th}^{(b)} \right] =
\begin{bmatrix}
C_{\theta_x-m_x,th}^{(b)} & C_{\theta_x-m_y,th}^{(b)} & C_{\theta_x-f_z,th}^{(b)} \\
C_{\theta_x-m_y,th}^{(b)} & C_{\theta_y-m_y,th}^{(b)} & C_{\theta_y-f_z,th}^{(b)} \\
C_{\theta_x-f_z,th}^{(b)} & C_{\theta_y-f_z,th}^{(b)} & C_{u_z-f_z,th}^{(b)}
\end{bmatrix};
$$

$$
\left[\Delta C_{op}^{(b)} \right] =
\begin{bmatrix}
\Delta C_{\theta_x-m_x}^{(b)} & \Delta C_{\theta_x-m_y}^{(b)} & \Delta C_{\theta_x-f_z}^{(b)} \\
\Delta C_{\theta_x-m_y}^{(b)} & \Delta C_{\theta_y-m_y}^{(b)} & \Delta C_{\theta_y-f_z}^{(b)} \\
\Delta C_{\theta_x-f_z}^{(b)} & \Delta C_{\theta_y-f_z}^{(b)} & \Delta C_{u_z-f_z}^{(b)}
\end{bmatrix};
$$

$$(9.95)$$

$$
\left[C_{op,th}^{(t)} \right] =
\begin{bmatrix}
C_{\theta_x-m_x,th}^{(t)} & C_{\theta_x-m_y,th}^{(t)} & C_{\theta_x-f_z,th}^{(t)} \\
C_{\theta_x-m_y,th}^{(t)} & C_{\theta_y-m_y,th}^{(t)} & C_{\theta_y-f_z,th}^{(t)} \\
C_{\theta_x-f_z,th}^{(t)} & C_{\theta_y-f_z,th}^{(t)} & C_{u_z-f_z,th}^{(t)}
\end{bmatrix};
$$

$$
\left[\Delta C_{op}^{(t)} \right] =
\begin{bmatrix}
\Delta C_{\theta_x-m_x}^{(t)} & \Delta C_{\theta_x-m_y}^{(t)} & \Delta C_{\theta_x-f_z}^{(t)} \\
\Delta C_{\theta_x-m_y}^{(t)} & \Delta C_{\theta_y-m_y}^{(t)} & \Delta C_{\theta_y-f_z}^{(t)} \\
\Delta C_{\theta_x-f_z}^{(t)} & \Delta C_{\theta_y-f_z}^{(t)} & \Delta C_{u_z-f_z}^{(t)}
\end{bmatrix}
$$

The thermal-induced errors can be evaluated by analyzing the difference between $[C_{op,th}]$ of Eq. (9.94) and the original $[C_{op}]$.

Example 9.7

A circular-axis flexible hinge has a constant rectangular cross-section defined by the in-plane thickness $t = 0.008$ m, out-of-plane width $w = 0.001$ m, radius $R = 0.04$ m, center angle $\alpha = 60°$, Young's modulus $E = 2 \cdot 10^{11}$ N/m^2 and Poisson's ratio $\mu = 0.33$. Graphically study the relative variation in the out-of-plane bending compliance $C_{u_z-f_z}$ in terms of the temperature variation ΔT. Consider that ΔT varies within the $[50°; 150°]$ interval.

Solution:

The original compliance is expressed based on Eqs. (A2.32) as:

$$
C_{u_z-f_z} = \frac{R^3}{2} \cdot \left[\frac{\alpha - \dfrac{\sin(2\alpha)}{2}}{E \cdot I_n} + \frac{6\alpha - 8\sin\alpha + \sin(2\alpha)}{2G \cdot I_t} \right]
$$

$$(9.96)$$

where $I_n = w^3 \cdot t/12$ and $I_t = w^3 \cdot t/3$. Based on Eq. (9.93) and for $G = E/[2\cdot(1 + \mu)]$, the following relative error function is formed:

$$\text{error}\left(C_{u_z-f_z}\right)=\frac{\left|C_{u_z-f_z,th}-C_{u_z-f_z}\right|}{C_{u_z-f_z}} \quad \text{with} \quad \begin{cases} C_{u_z-f_z,th}=C^{(b)}_{u_z-f_z,th}+C^{(t)}_{u_z-f_z,th} \\ \Delta C_{u_z-f_z}=\Delta C^{(b)}_{u_z-f_z}+\Delta C^{(t)}_{u_z-f_z} \end{cases}. \quad (9.97)$$

The plot of Figure 9.17 shows the variation in the relative compliance formulated in Eq. (9.97) in terms of the temperature variation. As it can be seen, the relative error values are very small.

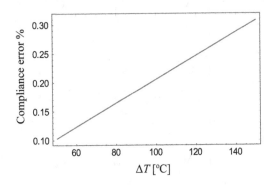

FIGURE 9.17   Plot of percentage relative error of the $C_{u_z-f_z}$ compliance in terms of ΔT.

9.3.2.2   Circular Cross-Section

The temperature-modified compliances of Eqs. (9.81) through (9.93) are also valid here with the geometric parameters of Eqs. (9.77)–(9.79), where β is used instead of x, and n instead of y.

9.4   STRAIGHT-AXIS LAYERED FLEXIBLE HINGES

In order to achieve a wider range of stiffness, inertia and damping properties, flexible hinges can be built of several layers of different materials in a stacked/sandwich manner. The micro-/nanosystems domain, in particular, offers numerous examples of composite cantilevers/bridges that are used in radiofrequency, computers or extraneous particle detection – see Maluf and Williams [36] for a comprehensive description.

This section studies a few straight-axis layered flexible-hinge configurations by using the compliance approach, as well as the finite element method, to model the static, modal and dynamic behaviors of these elastic members. All designs have axial geometric and material symmetry, and are defined by (variable) rectangular cross-sections.

While Chapters 2 through 8 study single-material hinges, this section discusses two categories of multimaterial, layered flexible hinges. One category is formed of hinges with variable in-plane thickness and constant out-of-plane width. In addition to being axially (longitudinally) symmetric, these hinges are also transversely symmetric in terms of both geometry and materials. They are intended to operate through in-plane bending and are modeled by means of the compliance approach. The other category comprises relatively thin hinge configurations where the layers have a constant thickness and variable width generally. These hinges are designed to mainly function through out-of-plane bending and/or torsion, and the finite element method is utilized to model them.

9.4.1 CONSTANT-WIDTH FLEXIBLE HINGES WITH TRANSVERSE GEOMETRY AND MATERIAL SYMMETRY

Consider the sandwich hinge of Figure 9.18a, which is formed of three different layers, of which the middle layer denoted by 1 in Figure 9.18b has a constant thickness and therefore, straight edges along the axial direction x. The two outer layers, denoted by 2 and 3 in Figure 9.18b, are made of the same material, have variable thickness and are geometrically identical being mirrored with respect to the axial plane xz. The entire hinge is also transversely symmetric (mirrored) with respect to the $y_h z_h$ plane. The three layers have the same out-of-plane width w, the same length l and relatively small in-plane thicknesses. The resulting sandwich hinge has a relatively large width w and is intended for in-plane bending motion (in the Oxy plane).

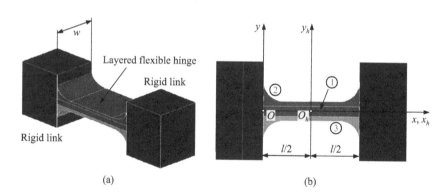

(a)   (b)

FIGURE 9.18 Constant-width three-layer flexible hinge with transverse and longitudinal symmetries: (a) 3D view; (b) front view.

The in-plane compliance and inertia matrices are derived here to enable quasi-static, modal or dynamic analysis to be performed. While the study is limited to only three layers, it can easily extend to five, seven or more odd-number layers as long as longitudinal and transverse symmetry is preserved.

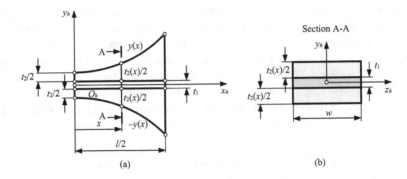

FIGURE 9.19 (a) Side view of three-layer, large-width, small-thickness hinge segment with identical and mirrored external layers; (b) cross-section A-A.

Figure 9.19a is the front view of half a generic three-layer flexible hinge with transverse symmetry. The constant in-plane thickness of the middle segment is t_1, while the minimum in-plane thickness of the identical and mirrored outer layers is $t_2/2$. The variable in-plane thicknesses of these layers are defined by the symmetric curves $y(x)$ and $-y(x)$ as $t_2(x)/2$. Assuming the right end of this flexible hinge is fixed and the left end is free, the three layers of Figure 9.19a act as three springs in parallel; moreover, it is possible to consider that there are only two structural layers coupled in parallel: the constant-thickness central layer 1 and a second layer that combines the mirrored outer layers into a single hollow member denoted by 2 – this is illustrated in the cross-section of Figure 9.19b.

The in-plane compliance matrix of the flexible-hinge segment of Figure 9.19 is calculated based on the rule of parallel spring connection as:

$$\left[C_{O_h}^{(h)}\right]^{-1} = \left[C_{O_h}^{(1h)}\right]^{-1} + \left[C_{O_h}^{(2h)}\right]^{-1}, \tag{9.98}$$

where the superscript "h" denotes the half-hinge. The in-plane compliance matrix of the segments 1 and 2 is arranged as in Eq. (2.13) with the individual compliances being calculated as:

$$\begin{cases} C_{u_x-f_x}^{(1h)} = \dfrac{l}{2E \cdot w \cdot t_1}; \quad C_{u_y-f_y}^{(1h)} = \dfrac{l^3}{2E \cdot w \cdot t_1^3}; \\[3mm] C_{u_y-m_z}^{(1h)} = -\dfrac{3l^2}{2E \cdot w \cdot t_1^3}; \quad C_{\theta_z-m_z}^{(1h)} = \dfrac{6l}{E \cdot w \cdot t_1^3} \\[3mm] C_{u_x-f_x}^{(2h)} = \dfrac{1}{E \cdot w} \cdot \displaystyle\int_0^{l/2} \dfrac{dx}{t_2(x)}; \quad C_{u_y-f_y}^{(2h)} = \dfrac{12}{E \cdot w} \cdot \displaystyle\int_0^{l/2} \dfrac{x^2 dx}{t_2(x)^3}; \\[3mm] C_{u_y-m_z}^{(2h)} = -\dfrac{12}{E \cdot w} \cdot \displaystyle\int_0^{l/2} \dfrac{x dx}{t_2(x)^3}; \quad C_{\theta_z-m_z}^{(2h)} = \dfrac{12}{E \cdot w} \cdot \displaystyle\int_0^{l/2} \dfrac{dx}{t_2(x)^3}, \end{cases} \tag{9.99}$$

where $t_2(x) = 2y(x) - t_1$. The compliances of Eqs. (9.99) are calculated with respect to the $O_h x_h y_h z_h$ reference frame. With the individual compliances of Eq. (9.99), the in-plane compliance matrix of the half-hinge portion is determined as per Eq. (9.98), which is of the form:

$$
\left[C_{O_h}^{(h)} \right] = \begin{bmatrix} C_{O_h, u_x - f_x}^{(h)} & 0 & 0 \\ 0 & C_{O_h, u_y - f_y}^{(h)} & C_{O_h, u_y - m_z}^{(h)} \\ 0 & C_{O_h, u_y - m_z}^{(h)} & C_{O_h, \theta_z - m_z}^{(h)} \end{bmatrix}.
$$
(9.100)

As demonstrated in Section 3.2.2 (which studies single-material, transversely symmetric hinges), the in-plane compliance matrix of the generic full hinge of Figure 9.18 is calculated with respect to the $Oxyz$ reference frame as in Eq. (3.53), namely:

$$
[C_O] = \begin{bmatrix} C_{O, u_x - f_x} & 0 & 0 \\ 0 & C_{O, u_y - f_y} & C_{O, u_y - m_z} \\ 0 & C_{O, u_y - m_z} & C_{O, \theta_z - m_z} \end{bmatrix}
$$

$$
= \begin{bmatrix} 2 C_{O_h, u_x - f_x}^{(h)} & 0 & 0 \\ 0 & 2 C_{O_h, u_y - f_y}^{(h)} + \dfrac{l^2}{2} \cdot C_{O_h, \theta_z - m_z}^{(h)} & -l \cdot C_{O_h, \theta_z - m_z}^{(h)} \\ 0 & -l \cdot C_{O_h, \theta_z - m_z}^{(h)} & 2 C_{O_h, \theta_z - m_z}^{(h)} \end{bmatrix}.
$$
(9.101)

The in-plane inertia matrix of the full symmetric flexible hinge shown in Figure 9.18b whose half is represented in Figure 9.19 is calculated with respect to the $Oxyz$ reference frame directly by adding the corresponding inertia matrices of the two segments 1 and 2, namely:

$$
[M_O] = \left[M_O^{(1)} \right] + \left[M_O^{(2)} \right] \quad \text{with}
$$

$$
\left[M_O^{(i)} \right] = \begin{bmatrix} M_a^{(i)} & 0 & 0 \\ 0 & M_b^{(i)}(1,1) & M_b^{(i)}(1,2) \\ 0 & M_b^{(i)}(1,2) & M_b^{(i)}(2,2) \end{bmatrix}, \quad i = 1, 2,
$$
(9.102)

as per Eq. (7.53). The inertia fractions of the core segment 1 are calculated with the aid of Eqs. (7.30) and (7.53):

$$
M_a^{(1)} = \frac{1}{3} \cdot m^{(1)}; \quad M_b^{(1)}(1,1) = \frac{13}{35} \cdot m^{(1)}; \quad M_b^{(1)}(1,2) = \frac{11}{210} \cdot m^{(1)};
$$

$$
M_b^{(1)}(2,2) = \frac{1}{105} \cdot m^{(1)}; \quad m^{(1)} = \rho_1 \cdot w \cdot t_1 \cdot l.
$$
(9.103)

The axial and bending inertia components of the variable-thickness segment 2 are determined based on Eqs. (7.29) and (7.49):

$$m_a^{(2)} = \rho_2 \cdot w \cdot \int_0^l t_2(x)\left(1-\frac{x}{l}\right)^2 dx; \quad \left[M_b^{(2)}\right] = \begin{bmatrix} M_b^{(2)}(1,1) & M_b^{(2)}(1,2) \\ M_b^{(2)}(1,2) & M_b^{(2)}(2,2) \end{bmatrix}$$

$$= \rho_2 \cdot w \cdot \int_0^l t_2(x)\left[S_{b,1}(x)\right]\left[S_{b,1}(x)\right]^T dx \qquad (9.104)$$

with $\left[S_{b,1}(x)\right] = \begin{bmatrix} 1-3\cdot\dfrac{x^2}{l^2}+2\cdot\dfrac{x^3}{l^3} & x-2\cdot\dfrac{x^2}{l}+\dfrac{x^3}{l^2} \end{bmatrix}^T$.

Equations (9.104) assume a single profile (curve) of the outer layers; in case there are several profiles along the longitudinal axis, the integrals of Eqs. (9.104) need to be expressed as a sum of integrals, each integral being defined for the independent subinterval corresponding to a different curve profile over the x-axis limits defining that subinterval.

Example 9.8

The outer curve of layer 2 in Figures 9.18 and 9.19 is an inverse parabola of the $(0, \alpha_e)$ type.

 i. Compare the compliance C_{0,θ_z-m_z} of this sandwich configuration to the same compliance of a similar hinge having the same outer profile (geometry) and which is built homogeneously of the material of segment 2 when t_1 varies as a fraction of the total minimum thickness $t = t_1 + t_2$. The two materials' Young's moduli are $E_1 = 2.2 \cdot 10^{11}$ N/m² and $E_2 = 0.052 \cdot 10^{11}$ N/m². Known are the minimum total thicknesses $t = 0.004$ m, the out-of-plane width $w = 0.006$ m, the length $l = 0.03$ m and the inverse parabolic constant $c = 0.004$ m;
 ii. Make a similar comparison between the smallest natural frequencies of the layered flexure hinge and of the homogeneous hinge described in part (i) when the mass densities are $\rho_1 = 7,800$ kg/m³ and $\rho_2 = 3,000$ kg/m³. All other geometric and material properties are the ones used at (i).

Solution:

 i. Equations (9.98) through (9.101) are used to evaluate the compliances C_{0,θ_z-m_z} of the two similar configurations. The variable thickness of layer 2 is calculated in the reference frame $O_hx_hy_hz_h$ as in Eq. (2.70):

$$t_2(x) = 2y_2(x) - t_1; \quad y_2(x) = \frac{a}{b^2 - x^2} \quad \text{with} \quad \begin{cases} a = \dfrac{l^2 \cdot t_2 \cdot (t_2 + 2c)}{16c} \\[4mm] b^2 = \dfrac{l^2 \cdot (t_2 + 2c)}{8c} \end{cases}. \tag{9.105}$$

The following compliance ratio is formed to compare the performance of the two geometrically similar flexible hinges:

$$rC_{\theta_z - m_z} = \frac{C^{(c)}_{O,\theta_z - m_z}}{C_{O,\theta_z - m_z}}, \tag{9.106}$$

where $C^{(c)}_{O,\theta_z - m_z}$ represents the compliance of the homogeneous hinge made up of material 2, whereas $C_{O,\theta_z - m_z}$ is the similar compliance of the bi-material hinge. As seen in Figure 9.20, the ratio increases from a value slightly larger than 1 for $t_1 = 0.3 \cdot t$ to a value of approximately 17 when $t_1 = 0.9 \cdot t$, which is an expected outcome as the two-material hinge becomes stiffer relative to the similar homogenous-material hinge when its higher-stiffness core thickness increases.

ii. Equations (9.102)–(9.104) are utilized to obtain the inertia matrix of the three-layer flexure hinge. The inertia matrix of the flexure hinge built of material 2 and having the same geometrical outline with the sandwich beam is:

$$\left[M_O^{(c)} \right] = \begin{bmatrix} M_a^{(c)} & 0 & 0 \\ 0 & M_b^{(c)}(1,1) & M_b^{(c)}(1,2) \\ 0 & M_b^{(c)}(1,2) & M_b^{(c)}(2,2) \end{bmatrix}. \tag{9.107}$$

The inertia components of Eq. (9.107) are calculated by means of Eqs. (9.104) and (9.105) with:

$$t_2(x) = 2y_2(x) - t_1; \quad y_2(x) = \frac{a}{b^2 - x^2}. \tag{9.108}$$

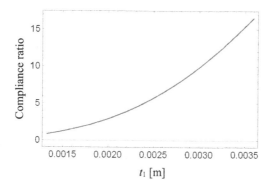

FIGURE 9.20 Plot of compliance ratio $rC_{\theta_z - m_z}$ in terms of the thickness t_1.

The natural frequencies of the following two dynamic matrices can now be evaluated:

$$[D] = [M_O]^{-1}[C_O]^{-1}; \quad [D^{(c)}] = [M_O^{(c)}]^{-1}[C_O^{(c)}]^{-1}. \tag{9.109}$$

The following ratio is formed to compare the first natural frequency of the homogeneous hinge to that of its layered counterpart:

$$r\omega_{n1} = \frac{\omega_{n1,\theta_z-m_z}^{(c)}}{\omega_{n1}}. \tag{9.110}$$

This ratio is graphed in Figure 9.21 in terms of the core layer thickness. The ratio increases mildly nonlinearly from a value of 1.146 for $t_1 = 0.3 \cdot t$ to a value of 2.882 when $t_1 = 0.9 \cdot t$, which indicates that although the sandwich configuration becomes heavier with increasing t_1, it also becomes stiffer (as demonstrated at (i)) in a more substantive manner, such that, overall, the first natural frequency of the layered beam increases relative to the similar frequency of the homogenous-material hinge.

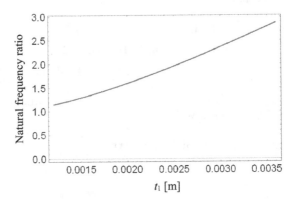

FIGURE 9.21 Plot of the first natural frequency ratio $r\omega_{n1}$ in terms of the thickness t_1.

9.4.2 CONSTANT-THICKNESS FLEXIBLE HINGES

The flexible hinge of Figure 9.22 does not have transverse symmetry and is formed of several layers of constant thickness. A core layer 2 of thickness t_2 extends over the entire length of the hinge, as illustrated in Figure 9.23. Two identical layers 1 of thickness t_1 are placed on the two faces of layer 2 starting at one end (point O) of the composite hinge and extend over a length l_1; at the other end of the hinge are two identical layers 3 of thickness t_3 and extending over a length l_3. The layers 1 and 3 have a variable width, while the central layer has the same variable width as layers 1 over the length l_1 and the same variable width as layers 3 over the length l_3. While layer 2 is shown with a constant width w (see Figure 9.23a) in the middle section of the hinge, it can generally have a variable width. This design reinforces the end portions of the composite hinge while preserving more flexibility in the central part. The composite hinge can mainly be utilized in out-of-plane bending around the z axis.

The finite element model of Figure 9.23c is formed, which consists of three collinear line elements and four nodes, of which node 4 is fixed. When only the bending effects are considered, the out-of-plane reduced, assembled stiffness and inertia matrix can be formed corresponding to the following overall nodal vector:

$$[u] = \begin{bmatrix} u_{1y} & \theta_{1z} & u_{2y} & \theta_{2z} & u_{3y} & \theta_{3z} \end{bmatrix}^T, \qquad (9.111)$$

which includes only the nonzero displacements. The resulting system stiffness and inertia matrix with respect to the origin reference frame are:

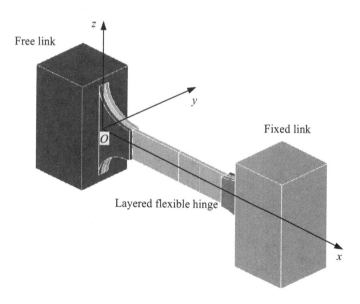

FIGURE 9.22 3D view of multiple-layer flexible hinge without transverse symmetry and layers of constant thickness.

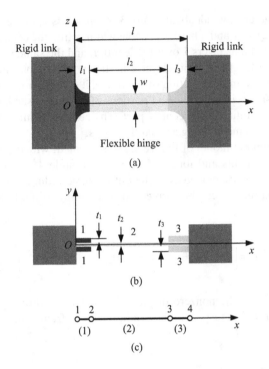

FIGURE 9.23 Flexible hinge with layers of constant thickness: (a) top view; (b) side view; (c) finite element model.

$$[K] = \begin{bmatrix} K^{(1)}(1,1) & K^{(1)}(1,2) & K^{(1)}(1,3) & K^{(1)}(1,4) & 0 & 0 \\ K^{(1)}(1,2) & K^{(1)}(2,2) & K^{(1)}(2,3) & K^{(1)}(2,4) & 0 & 0 \\ K^{(1)}(1,3) & K^{(1)}(2,3) & K^{(1)}(3,3)+K^{(2)}(1,1) & K^{(1)}(3,4)+K^{(2)}(1,2) & K^{(2)}(1,3) & K^{(2)}(1,4) \\ K^{(1)}(1,4) & K^{(1)}(2,4) & K^{(1)}(3,4)+K^{(2)}(1,2) & K^{(1)}(4,4)+K^{(2)}(2,2) & K^{(2)}(2,3) & K^{(2)}(2,4) \\ 0 & 0 & K^{(2)}(1,3) & K^{(2)}(2,3) & K^{(2)}(3,3)+K^{(3)}(1,1) & K^{(2)}(3,4)+K^{(3)}(1,2) \\ 0 & 0 & K^{(2)}(1,4) & K^{(2)}(2,4) & K^{(2)}(3,4)+K^{(3)}(1,2) & K^{(2)}(4,4)+K^{(3)}(2,2) \end{bmatrix}$$

$$[M] = \begin{bmatrix} M^{(1)}(1,1) & M^{(1)}(1,2) & M^{(1)}(1,3) & M^{(1)}(1,4) & 0 & 0 \\ M^{(1)}(1,2) & M^{(1)}(2,2) & M^{(1)}(2,3) & M^{(1)}(2,4) & 0 & 0 \\ M^{(1)}(1,3) & M^{(1)}(2,3) & M^{(1)}(3,3)+M^{(2)}(1,1) & M^{(1)}(3,4)+M^{(2)}(1,2) & M^{(2)}(1,3) & M^{(2)}(1,4) \\ M^{(1)}(1,4) & M^{(1)}(2,4) & M^{(1)}(3,4)+M^{(2)}(1,2) & M^{(1)}(4,4)+M^{(2)}(2,2) & M^{(2)}(2,3) & M^{(2)}(2,4) \\ 0 & 0 & M^{(2)}(1,3) & M^{(2)}(2,3) & M^{(2)}(3,3)+M^{(3)}(1,1) & M^{(2)}(3,4)+M^{(3)}(1,2) \\ 0 & 0 & M^{(2)}(1,4) & M^{(2)}(2,4) & M^{(2)}(3,4)+M^{(3)}(1,2) & M^{(2)}(4,4)+M^{(3)}(2,2) \end{bmatrix}$$

$$(9.112)$$

Because elements 1 and 3 are formed each of three different layers, which behave as three elements in parallel, the following components of the matrices in Eqs. (9.112) are calculated as:

$$K^{(1)}(i,j) = 2K^{(1,1)}(i,j) + K^{(1,2)}(i,j);$$

$$M^{(1)}(i,j) = 2M^{(1,1)}(i,j) + M^{(1,2)}(i,j), \quad i,j = 1,2,3,4$$

$$K^{(3)}(m,n) = K^{(3,2)}(m,n) + 2K^{(3,3)}(m,n);$$

$$M^{(3)}(m,n) = M^{(3,2)}(m,n) + 2M^{(3,3)}(m,n), \quad m,n = 1,2.$$

(9.113)

The second number in the superscripts of Eqs. (9.113) indicates the layer/material. Overall, the following independent 4×4 elemental matrices are needed: two for the interval/element 1 pertaining to layers 1 and 2 and spanning the length l_1, one for element 2 over the length l_2 and defined by the properties of layer 2, and two for the element 3 defined by layers 2 and 3 over the length l_3. These elementary stiffness matrices are determined based on the definition of Eq. (8.35) with the shape functions of Eqs. (8.26); similarly, an equal number of elementary inertia matrices are calculated based on Eq. (7.88) with the same shape functions.

Example 9.9

The generic composite flexible hinge of Figure 9.23 has its element (1) geometrically defined by a quarter circle of radius r_1, such that $l_1 = r_1$; similarly, element (3) is defined by a quarter circle of radius r_3 resulting in $l_3 = r_3$ – see Figure 9.24. Known are the following material and geometric parameters: $E_1 = 2 \cdot 10^{11} \, \text{N/m}^2$, $E_2 = 1 \cdot 10^{11} \, \text{N/m}^2$, $E_3 = 6 \cdot 10^{10} \, \text{N/m}^2$, $\rho_1 = 7{,}500 \, \text{kg/m}^3$, $\rho_2 = 3{,}200 \, \text{kg/m}^3$, $\rho_3 = 2{,}800 \, \text{kg/m}^3$, $t = 0.01 \, \text{m}$, $w = 0.001 \, \text{m}$, $r_1 = 0.01 \, \text{m}$ and $l = 0.04 \, \text{m}$. Study graphically the variation of the first (smallest) natural frequency of this flexible hinge when the length l_2 of the constant cross-section portion varies from 0.1 to 0.9 of the length $l - r_1$.

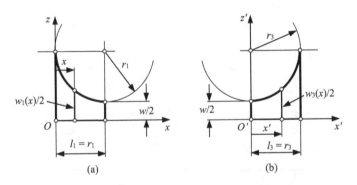

FIGURE 9.24 Geometry of half the end segments: (a) element 1; (b) element 3.

Solution:

The variable widths of the two segments 1 and 3 of Figure 9.24 are expressed as:

$$\begin{cases} w_1(x) = w + 2\left(r_1 - \sqrt{2r_1 \cdot x - x^2}\right) \\ w_3(x') = w + 2\left(r_3 - \sqrt{r_3^2 - x'^2}\right) \end{cases} \quad (9.114)$$

For $w_1(x)$, x is measured from point O and varies from 0 to r_1, while for $w_3(x')$, x' varies from 0 to r_3, as illustrated in Figure 9.24. The variable widths of Eq. (9.114) are used to calculate the cross-section moments of area and areas for the various segments of the composite hinge. The bending shape functions of Eq. (9.104) are used to obtain element stiffness and inertia matrices with a total length l_1 for the first interval of the serial hinge and a length l_3 for the third interval.

The reduced assembled stiffness and inertia matrices are formed as per Eqs. (9.112) and (9.113), which enable formulating the dynamic matrix $[D] = [M]^{-1}[K]$ whose eigenvalues provide the natural frequencies of the composite flexible hinge. The following notations are used:

$$l_2 = f \cdot (l - r_1); \quad r_3 = l - r_1 - l_2 = (1 - f) \cdot (l - r_1); \quad r\omega_{n1} = \frac{\omega_{n1}}{\omega_{n1,min}}, \quad (9.115)$$

where f is the fraction that varies from 0.1 to 0.9. It should be noted that as f increases, the length l_2 of the middle segment 2 increases as well, which results in a reduced radius (and length) r_3 of the third portion. It is therefore expected that the first natural frequency will also decrease since the middle thinner segment increases its length. This trend is, indeed, illustrated in Figure 9.25, which plots the frequency ratio of Eq. (9.115) in terms of f. The minimum natural frequency is $\omega_{n1,min} = 1{,}514$ rad/s and corresponds to $f = 0.9$, while the maximum natural frequency is 3,073 rad/s for $f = 0.1$.

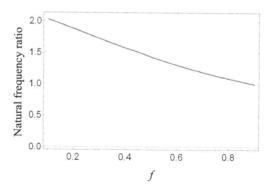

FIGURE 9.25 Plot of the first natural frequency ratio $r\omega_{n1}$ in terms of the length fraction f.

9.5   PIEZOELECTRIC ACTUATION AND SENSING IN FLEXIBLE-HINGE MECHANISMS

This section uses a simplified mathematical model of piezoelectric materials to illustrate transduction (actuation and sensing) by means of layered flexible hinges that incorporate piezoelectric strips and with piezoelectric blocks. Piezoelectric materials display a *direct effect* whereby external mechanical deformations applied to them generate an output voltage; this principle is applied in sensing. The piezoelectric materials are also capable of actuation through a *converse effect* manifested by mechanical deformation (motion) of a member connected to an external voltage source – see Reece [37] and Rupitsch [38], for instance. Natural piezoelectric materials include quartz crystals, tourmaline and the Rochelle salt, whereas a familiar manufactured piezoelectric material is the PZT (an alloy based on lead, zirconium and titanium).

The following equation models three-dimensional actuation produced by a piezoelectric member:

$$[\varepsilon] = [C^E][\sigma] + [d]^T[E]$$
$$[\varepsilon] = [d]^T[E] \quad \text{for} \quad [\sigma] = [0] \tag{9.116}$$

where $[\varepsilon]$ and $[\sigma]$ are the mechanical strain and stress vectors, $[E]$ is the electric field vector, $[C^E]$ is the compliance matrix evaluated at a constant electric field, and $[d]$ is a matrix comprising *piezoelectric coefficients*. The first Eq. (9.116) reflects the fully coupled electrical–mechanical interaction between external mechanical factors (represented by the stress vector) and electrical input (illustrated by the electric field) resulting in mechanical deformation (the strain). In the absence of the external mechanical stress, the second (simplified) Eq. (9.116) results, which directly connects the electric field input to the mechanical strain output. The second Eq. (9.116) simplifies further to the following equation when the problem is unidimensional, i.e., when a single electric field E is applied to a piezoelectric line member to produce mechanical strain aligned with the electric field direction:

$$\varepsilon = d \cdot E \rightarrow \varepsilon = d \cdot \frac{v}{l} \rightarrow \Delta l = \varepsilon \cdot l = d \cdot v, \tag{9.117}$$

where v is the voltage applied across a line member of initial length l and Δl is the member axial elongation aligned with the direction of E.

In sensing, the governing three-dimensional equation is:

$$[\sigma_e] = [d][\sigma] + [\varepsilon^{\sigma_m}][E]$$
$$[\sigma_e] = [d][\sigma] \quad \text{for} \quad [E] = [0] \tag{9.118}$$

where $[\sigma_e]$ is the electric charge density vector and $[\varepsilon^{\sigma_m}]$ is the electric permittivity vector determined for constant mechanical stress. The second Eq. (9.118) connects directly the electric charge density output to the mechanical stress input for zero electric field. While not demonstrated here (details can be found in Lobontiu [39], for instance), the one-direction counterpart of the second Eq. (9.118) is:

$$v = \frac{1}{d} \cdot \Delta l, \qquad (9.119)$$

which also results from Eq. (9.117). Equation (9.119) shows that the output voltage is proportional to the input mechanical deformation of a piezoelectric sensing member.

9.5.1 ACTIVE LAYERED (PIEZOELECTRIC) FLEXIBLE HINGES

Flexible hinges can also be active when provided with means of either actuation resulting in deflection or sensing when external mechanical deformation is converted into (electrical) voltage. Shape memory alloys (SMA), as presented in Bellouard and Clavel [40] and Du et al. [41], for instance, can be used as active connectors of passive hinge segments when a state phase change is initiated in the SMA portion. This section includes a brief study of active multilayer flexible hinges to be used for actuation or sensing purposes and whose active strip is piezoelectric.

9.5.1.1 Actuation

Multiple-layer flexible hinges can be (self-)actuated through mechanical strain that is induced at the interface(s) between active-material layers and passive layers. Impeded expansion/compression of the active layers by the passive layer(s) results in bending of the flexible hinge. The active, strain-induced composite portion can extend over the entire length of the hinge. Alternatively, the active segment can be coupled in series with other passive portions resulting in a composite hinge configuration with longitudinal and (possibly) transverse symmetries. The strain-induced actuation can be realized thermally, such as in bimetallic strips – see Rees [42] and Gere and Goodno [43], for instance – where two layers of distinct metallic materials expand differently and the impeded thermal expansion results in bending of the strip. Strain induction can also be realized by means of piezoelectric, magnetostrictive or shape memory materials – see Maluf and Williams [36], Lobontiu et al. [44] or Garcia and Lobontiu [45]. Analyzed here is an active hinge segment formed of two identical piezoelectric layers that act antisymmetrically (one in extension, while the other in compression), and a passive (base) layer in between. Such an actuation unit with piezoelectrically induced strain is further incorporated into a flexible hinge by serially connecting it to two end passive segments.

9.5.1.1.1 Three-Layer Antisymmetric Strain-Induced Actuation Segment

The composite flexible-hinge segment, which is sketched in Figure 9.26a, is formed of three layers that have the same length and the same out-of-plane width w (it is assumed constant here, but it can also be variable).

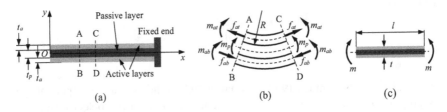

(a)   (b)   (c)

FIGURE 9.26 Layered, strain-induced flexible segment producing antisymmetric bending: (a) side view; (b) enlarged deformed portion; (c) actuation unit of length l with equivalent end bending moments m.

The top and bottom layers are active and made of piezoelectric material; they are identical in terms of both geometry (same in-plane thickness t_a) and material properties. Enclosed between these layers is a passive layer of thickness t_p. Assume that, if free and unattached to the middle layer, one of the external layers, like the bottom one in Figure 9.26a, extends axially, while the other layer compresses axially, both by the same amount. Because the free extension/compression of the two active layers is impeded by the connecting middle layer, the composite beam will bend as illustrated in Figure 9.26b, which is the deformed and magnified view of the segment comprised between the arbitrary sections AB and CD in Figure 9.26a. This type of actuation is known as *antisymmetric* – see Rees [42] and Gere and Goodno [43] – because of the opposite deformations of the two external layers, which actually concur to enhance the bending effect. The aim is to evaluate the two statically equivalent moments m (see Figure 9.26c, which shows an arbitrary free-free actuation segment of length l) that produce the bending of the composite-beam segment.

The tentative compression of the top layer results in the forces f_{at} applied to it by the middle layer opposing that deformation; similarly, the free expansion of the bottom layer is opposed by the two forces f_{ab} produced by the same middle layer – see Figure 9.26b. Force equilibrium shows that $f_{at} = f_{ab} = f_a$. The following strain continuity equations – see Rees [42], Lobontiu et al. [44] or Garcia and Lobontiu [45] for a similar approach – can be formulated for the two layer interfaces:

$$
\begin{cases}
\dfrac{f_a}{E_a \cdot A_a} + \dfrac{m_{at} \cdot t_a}{2 E_a \cdot I_a} - \varepsilon_0 = -\dfrac{m_p \cdot t_p}{2 E_p \cdot I_p} \\[4mm]
-\dfrac{f_a}{E_a \cdot A_a} - \dfrac{m_{ab} \cdot t_a}{2 E_a \cdot I_a} + \varepsilon_0 = \dfrac{m_p \cdot t_p}{2 E_p \cdot I_p}
\end{cases}
\tag{9.120}
$$

The total moment m has been split into three moments: m_{at}, m_{ab} and m_p acting on the three layers. The first terms in the left-hand side of Eqs. (9.120) are the constant axial strains produced by the force f_a applied perpendicularly to the cross-section of area A_a. The second terms in the left-hand side of the same equations and on the right-hand sides represent the axial strains produced through bending, whereas ε_0 is the strain induced externally through actuation of the top and bottom active layers. I_a and I_p are the cross-sectional area moments of the active and passive layers with respect to their own centroidal axes z. Comparing the two Eqs. (9.120) shows that $m_{at} = m_{ab} = m_a$. The following equations can also be written:

$$\begin{cases} R = \dfrac{E_a \cdot I_a}{m_a} = \dfrac{E_p \cdot I_p}{m_p} \\[2mm] 2m_a + m_p = f_a \cdot \left(t_a + t_p\right) \end{cases} \qquad (9.121)$$

The first Eq. (9.121) expresses the curvature radius R at the interface between the top and middle layers – see Figure 9.26b – while the second Eq. (9.121) results from the moment equilibrium on one of the element faces AB or CD. Either of the two Eqs. (9.120) is solved together with the two Eqs. (9.121) for f_a, m_a and m_p. The moments can now be used to find the total moment m on the faces AB and CD as:

$$m = 2m_a + m_p = \frac{t_a \cdot w \cdot \left(t_a + t_p\right) \cdot \left(2t_a^3 \cdot E_a + t_p^3 \cdot E_p\right) \cdot E_a}{2t_a \cdot \left(4t_a^2 + 6t_a \cdot t_p + 3t_p^2\right) \cdot E_a + t_p^3 \cdot E_p} \cdot \varepsilon_0. \qquad (9.122)$$

Equation (9.122) also used the following cross-sectional areas and area moments: $A_a = t_a \cdot w$, $I_a = t_a^3 \cdot w/12$, $I_p = t_p^3 \cdot w/12$. Because the expression of the bending moment m does not depend on the distance between the planes AB and CD, a length l can be selected conveniently for an actuation segment, like the one sketched in Figure 9.26c where the total actuator thickness is $t = 2t_a + t_p$. Equation (9.122) is valid for any type of induced strain ε_0, including the piezoelectric sort, which is utilized here. The strain ε_0 is generated by applying a voltage v to the active layer between its ends and along the length l in Figure 9.26c.

9.5.1.1.2 Active Flexible Hinges with Three-Layer Antisymmetric Strain-Induced Actuation Segment

Figure 9.27 shows the 3D view of a flexible hinge for out-of-plane bending that incorporates a strain-induced actuation unit in its middle section. The two actuation patches are attached to a passive elastic substrate that runs continuously between two rigid links. While the center passive-layer portion is of a constant rectangular cross-section, the two end portions are (or can be) of variable width.

The side (front) view of this hinge is pictured in Figure 9.23a. The top view of this hinge configuration is sketched in Figure 9.28a, which identifies the lengths l_1 and l_3 of the passive sections, the length l_2 of the middle actuation section, as well as the thicknesses t_a and t_p of the active and passive layers.

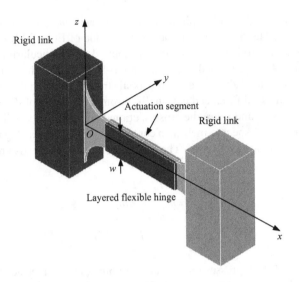

FIGURE 9.27 3D view of multiple-layer flexible hinge without transverse symmetry, layers of constant thickness and middle actuation unit.

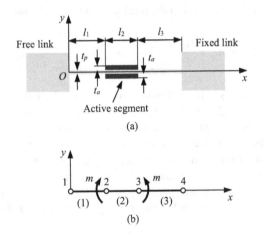

FIGURE 9.28 Layered flexible hinge with active segment used as an actuator: (a) side view; (b) finite element model.

As in Section 9.4.2, the flexible hinge is modeled using the finite element method, based on the line-element sketch shown in Figure 9.28b. When only the bending-related out-of-plane deformations are of interest, the nonzero nodal displacement vector is the one of Eq. (9.111), which is rewritten again:

$$[u] = \begin{bmatrix} u_{1y} & \theta_{1z} & u_{2y} & \theta_{2z} & u_{3y} & \theta_{3z} \end{bmatrix}^T. \qquad (9.123)$$

In a manner similar to the one used in Section 9.4.2, the reduced and assembled stiffness and inertia matrices of the finite element model of Figure 9.28b are formally given in Eqs. (9.112). While the matrices of elements (1) and (3) are calculated based on a single material (of the passive layer), the matrices of element (2) combine in parallel components from the two active layers and the passive layer, namely:

$$K^{(2)}(i,j) = 2K^{(2,a)}(i,j) + K^{(2,p)}(i,j);$$

$$M^{(2)}(i,j) = 2M^{(2,a)}(i,j) + M^{(2,p)}(i,j), \quad i,j = 1,2,3,4. \tag{9.124}$$

In the absence of other external loads, the forcing vector corresponding to the six degrees of freedom (DOF) defined in Eq. (9.123) is:

$$\left[f_{op} \right] = \begin{bmatrix} 0 & 0 & 0 & -m & 0 & m \end{bmatrix}^T. \tag{9.125}$$

The bending moment m is defined in Eq. (9.122) and depends on the active segment geometry and material properties, as well as on the value of the mechanical strain induced on the two interfaces through the converse piezoelectric effect.

Example 9.10

A flexible hinge of the type depicted in Figures 9.27 and 9.28 has its end segments formed of quarter circles, as illustrated in top view in Figure 9.24. Known are the geometric and material parameters of the hinge: $w = 0.008$ m, $r_1 = 0.014$ m, $l_2 = 0.025$ m, $r_3 = 0.01$ m, $t_p = 0.001$ m, $t_a = 0.0008$ m, $E_p = 2 \cdot 10^{10}$ N/m², $E_a = 5 \cdot 10^{10}$ N/m², $\rho_p = 4,000$ kg/m³ and $\rho_a = 5,300$ kg/m³. The two actuation patches are piezoelectric with a constant $d = 4 \cdot 10^{-10}$ m/V; the voltage $v = 150 + 120 \cdot e^{-0.5t} \cdot \sin(12t)$ V is applied to one actuation patch, while a voltage $-v$ is applied to the opposite patch. It is also known that there is overall structural damping defined by the coefficients $\alpha = 0.25$ and $\beta = 0.35$. Plot the time response of the flexible-hinge free-end deflection u_{1y}.

Solution:

In piezoelectric actuation of a straight-line member, such as the two patches are, a voltage difference v applied at their ends along the length results in an axial deformation proportional to the voltage; for the patch of length l_2 of this example and based on Eq. (9.117), the corresponding induced strain is:

$$\varepsilon_0 = \frac{\Delta l_2}{l_2} = \frac{d \cdot v}{l_2}, \tag{9.126}$$

where d is the piezoelectric constant. Combining Eqs. (9.122) and (9.126) shows that the actuation moment m is proportional to the applied voltage, namely:

$$m = c \cdot v \quad \text{with} \quad c = \frac{t_a \cdot w \cdot (t_a + t_p) \cdot (2t_a^3 \cdot E_a + t_p^3 \cdot E_p) \cdot E_a}{2t_a \cdot (4t_a^2 + 6t_a \cdot t_p + 3t_p^2) \cdot E_a + 2t_p^3 \cdot E_p} \cdot \left(\frac{d}{l_2} \right). \tag{9.127}$$

The numerical value of the voltage multiplier in Eq. (9.127) is $c = 7.489 \cdot 10^{-7}$ N·m/V.

To plot the required time response, the transfer function approach is utilized and the following transfer function matrix is formed, as per Eq. (8.215), which accounts for proportional damping:

$$[G(s)] = \left((s^2 + \alpha \cdot s)[M] + (\beta \cdot s + 1)[K] \right)^{-1}, \qquad (9.128)$$

where $[M]$ and $[K]$ are determined as discussed in the previous theoretical portion of this section. The Laplace transform of the required output is calculated based on the following transfer function components of $[G(s)]$ in conjunction with $M(s)$ – the Laplace transform of the actuation moment m of this example:

$$U_{1y}(s) = -G_{1,4}(s) \cdot M(s) + G_{1,6}(s) \cdot M(s) = \left(G_{1,6}(s) - G_{1,4}(s) \right) \cdot M(s). \quad (9.129)$$

The corresponding time-dependent plot of u_{1y} is graphed in Figure 9.29.

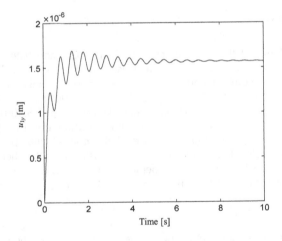

FIGURE 9.29 Time plot of the hinge out-of-plane deflection at its free end.

9.5.1.2 Sensing

Flexible hinges that include active-material layers can also be utilized as sensors since externally generated deformation of these members is converted into electric voltage by means of the active portions. Piezoelectric, piezomagnetic or magnetostrictive materials, to cite just a few sensing options, can be incorporated as active segments into multiple-layer flexible-hinge sensors. The following *Example* studies a flexible hinge with active piezoelectric layers by means of the finite element method.

Example 9.11

A flexible micro hinge of the generic structure depicted in Figures 9.27 and 9.28 has circular-shaped end segments, as in the front view of Figure 9.24. The two identical active layers are relatively thick and made of piezoelectric material, whereas the thinner passive layer is polysilicon. The hinge is used as a sensor to evaluate the external input displacement u_i of its two ends, which are guided – see Figure 9.30a; the input displacement varies as $u_i = U_i \cdot \sin(323,900 \cdot t)$ m. Use finite element modeling to calculate the amplitude U_i, knowing that the steady-state amplitude of the voltage induced through piezoelectric effect is $V = 50$ V. The hinge is defined by $r_1 = r_3 = 3\,\mu m$, $l_2 = 10\,\mu m$, $w = 5\,\mu m$, $t_p = 50\,nm$, $t_a = 100\,nm$, $E_p = 2.9 \cdot 10^{11}\,N/m^2$, $E_a = 5.8 \cdot 10^{10}\,N/m^2$, $\rho_p = 3,000\,kg/m^3$, $\rho_a = 5,600\,kg/m^3$ and $d = 5.6 \cdot 10^{-10}$ m/V. Consider there is no damping.

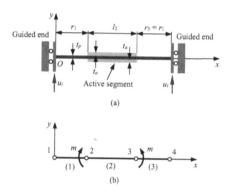

FIGURE 9.30 Layered flexible hinge with active segment used as a sensor: (a) side view with guided end boundary conditions; (b) finite element model.

Solution:

Figure 9.30b is the finite element model of the hinge illustrated in Figure 9.30a. Like the generic model of Figure 9.28b, the current model is formed of three elements accounting for the three different hinge segments. The external translation displacement u_i of the end points/nodes 1 and 4 results in bending deformation of the hinge; the related strain of the active element 2 produces a voltage v through piezoelectric effect – as a consequence, two equivalent moments m can be applied at nodes 2 and 3 that are calculated as in Eqs. (9.126) and (9.127).

The nonzero nodal displacement vector and its corresponding external load vector are:

$$[u] = \begin{bmatrix} u_{1y} & u_{2y} & \theta_{2z} & u_{3y} & \theta_{3z} & u_{4y} \end{bmatrix}^T$$

$$= \begin{bmatrix} u_i & u_{2y} & \theta_{2z} & u_{3y} & \theta_{3z} & u_i \end{bmatrix}^T;$$

$$[f] = \begin{bmatrix} f_{1y} & f_{2y} & m_{2z} & f_{3y} & m_{3z} & f_{4y} \end{bmatrix}^T = \begin{bmatrix} 0 & 0 & -m & 0 & m & 0 \end{bmatrix}^T.$$

$$(9.130)$$

Based on $[u]$ of Eq. (9.130), the assembled stiffness and inertia matrices are:

$$[K] = \begin{bmatrix} K^{(1)}(1,1) & K^{(1)}(1,3) & K^{(1)}(1,4) & 0 & 0 & 0 \\ K^{(1)}(1,3) & K^{(1)}(3,3)+K^{(2)}(1,1) & K^{(1)}(3,4)+K^{(2)}(1,2) & K^{(2)}(1,3) & K^{(2)}(1,4) & 0 \\ K^{(1)}(1,4) & K^{(1)}(3,4)+K^{(2)}(1,2) & K^{(1)}(4,4)+K^{(2)}(2,2) & K^{(2)}(2,3) & K^{(2)}(2,4) & 0 \\ 0 & K^{(2)}(1,3) & K^{(2)}(2,3) & K^{(2)}(3,3)+K^{(3)}(1,1) & K^{(2)}(3,4)+K^{(3)}(1,2) & K^{(3)}(1,3) \\ 0 & K^{(2)}(1,4) & K^{(2)}(2,4) & K^{(2)}(3,4)+K^{(3)}(1,2) & K^{(2)}(4,4)+K^{(3)}(2,2) & K^{(3)}(2,3) \\ 0 & 0 & 0 & K^{(3)}(1,3) & K^{(3)}(2,3) & K^{(3)}(3,3) \end{bmatrix}$$

$$[M] = \begin{bmatrix} M^{(1)}(1,1) & M^{(1)}(1,3) & M^{(1)}(1,4) & 0 & 0 & 0 \\ M^{(1)}(1,3) & M^{(1)}(3,3)+M^{(2)}(1,1) & M^{(1)}(3,4)+M^{(2)}(1,2) & M^{(2)}(1,3) & M^{(2)}(1,4) & 0 \\ M^{(1)}(1,4) & M^{(1)}(3,4)+M^{(2)}(1,2) & M^{(1)}(4,4)+M^{(2)}(2,2) & M^{(2)}(2,3) & M^{(2)}(2,4) & 0 \\ 0 & M^{(2)}(1,3) & M^{(2)}(2,3) & M^{(2)}(3,3)+M^{(3)}(1,1) & M^{(2)}(3,4)+M^{(3)}(1,2) & M^{(3)}(1,3) \\ 0 & M^{(2)}(1,4) & M^{(2)}(2,4) & M^{(2)}(3,4)+M^{(3)}(1,2) & M^{(2)}(4,4)+M^{(3)}(2,2) & M^{(3)}(2,3) \\ 0 & 0 & 0 & M^{(3)}(1,3) & M^{(3)}(2,3) & M^{(3)}(3,3) \end{bmatrix}$$

$$(9.131)$$

As discussed in Section 9.5.1, the element stiffness and inertia matrices of element (2) are calculated by combining the three layers as in Eqs. (9.124). For elements (1) and (2), the element matrices are calculated using a single layer. With the matrices of Eqs. (9.131), the following transfer function matrix $[G(s)]$ is calculated, which connects the Laplace-transformed nodal displacement and load vectors:

$$[U(s)] = [G(s)][F(s)]$$

with $\begin{cases} [U(s)] = \begin{bmatrix} U_i(s) & U_{2y}(s) & \Theta_{2z}(s) & U_{3y}(s) & \Theta_{3z}(s) & U_i(s) \end{bmatrix}^T ; \\ [F(s)] = \begin{bmatrix} 0 & 0 & -M(s) & 0 & M(s) & 0 \end{bmatrix}^T \\ [G(s)] = \left(s^2[M] + [K] \right)^{-1} \end{cases}$

$\qquad$ (9.132)

Taking into account that $M(s) = c \cdot V(s)$, where $V(s)$ is the Laplace transform of the voltage v (see Eq. (9.127)), Eqs. (9.132) enable forming the following scalar transfer function:

$$G(s) = \frac{V(s)}{U_i(s)} = \frac{1}{c \left(G_{15}(s) - G_{13}(s) \right)}, \qquad (9.133)$$

where $G_{13}(s)$ and $G_{15}(s)$ are the elements in the first row and the third and fifth columns, respectively, in the $[G(s)]$ matrix. The steady-state amplitudes of the input and output signals defining the transfer function of Eq. (9.133) are related as:

$$V = |G(\omega \cdot j)| \cdot U_i \rightarrow U_i = \frac{V}{|G(\omega \cdot j)|}. \qquad (9.134)$$

For the provided numerical values, the following value is obtained for the input-displacement amplitude: $U_i = 0.285$ μm.

9.5.2 Block (Piezoelectric) Actuation and Sensing

This section succinctly analyzes actuation and sensing produced by means of block (prismatic) piezoelectric members.

9.5.2.1 Actuation

In the fabrication stage, a block piezoelectric actuator, like the one shown in Figure 9.31a, is subjected to an electro-thermal process named *poling*, which aligns the piezoelectric material electric dipoles along the block longitudinal direction. When an external voltage v_a is applied to the piezoelectric block, which has the same direction as the original poling direction, a mechanical elongation of the block occurs along the axial direction based on Eq. (9.117).

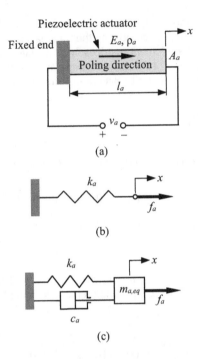

FIGURE 9.31 (a) Fixed-free piezoelectric block actuator with free-end expansion; (b) equivalent lumped-parameter quasi-static model; (c) equivalent lumped-parameter dynamic model.

The lumped-parameter quasi-static model of the fixed-free block actuator consists of a spring (which captures the axial elasticity of the block) of stiffness k_a and a force f_a (also known as block force) applied at the free end – see Figure 9.31b. The maximum, block force is calculated from Eq. (9.117) as:

$$f_a = k_a \cdot \Delta l_a = \left(\frac{E_a \cdot l_a}{A_a} \right) \cdot d \cdot v, \tag{9.135}$$

where E_a is the piezoelectric material Young's modulus along the axial direction, l_a is the length, and A_a is the cross-sectional area – see Figure 9.31a. A similar lumped-parameter dynamic model of the piezoelectric block is represented in Figure 9.31c, which adds a point mass $m_{a,eq}$ at the free end of the block and a viscous damper of coefficient c_a. The point mass can be calculated by one of the methods described in Chapters 7 and 8 when using a consistent or a lumped-parameter approach.

Displacement-amplification mechanism with two symmetry axes and block piezoelectric actuators:

Of the many flexible-hinge mechanisms that utilize piezoelectric actuation, this section analyzes the double-symmetric displacement-amplification

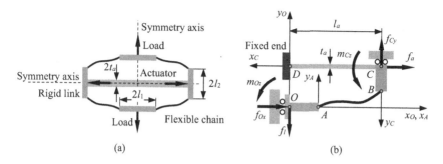

FIGURE 9.32 Double-symmetry displacement-amplification device with four identical flexible chains and piezoelectric block actuator: (a) skeleton representation; (b) quarter mechanism.

device of Figure 5.23, which is also studied in Section 5.1.3.1 and is illustrated in Figure 9.32a – it adds the block piezoelectric actuator, which is aligned with the horizontal symmetry axis. Due to structure and load symmetry, the quarter mechanism of Figure 9.32b can be analyzed instead of the full device. Note that due to two-axis symmetry, the actuator preserves its original horizontal position during deformation; as a consequence, the actuator left end in the quarter mechanism model is fixed; also, note that l_a and t_a are the actuator half-length and half-thickness, respectively. The quasi-static response of the mechanism under the actuation force f_a and external (load) force f_l involves evaluating the performance criteria defined in Section 5.1.3.1, namely, displacement amplification, input stiffness, output stiffness, cross-stiffness, block load, as well as the actual input displacement u_{Cx} or output displacement u_{Oy}.

The reactions f_{Ox}, m_{Oz}, at O and f_{Cy}, m_{Cz} at C need to be determined first. The four unknown reactions are found using the following compatibility (zero boundary conditions) equations:

$$u_{C_y} = 0; \quad \theta_{C_z} = 0; \quad u_{O_x} = 0; \quad \theta_{O_z} = 0. \tag{9.136}$$

The following in-plane displacement and load vectors correspond to points C and O of Figure 9.32b:

$$
\begin{cases}
[u_O] = \begin{bmatrix} 0 & u_{Oy} & 0 \end{bmatrix}^T; \quad [f_O] = \begin{bmatrix} f_{Ox} & -f_l & m_{Oz} \end{bmatrix}^T; \\[2ex]
[u_C] = \begin{bmatrix} u_{Cx} & 0 & 0 \end{bmatrix}^T; \quad [f_C] = \begin{bmatrix} f_a & f_{Cy} & m_{Cz} \end{bmatrix}^T.
\end{cases}
\tag{9.137}
$$

The displacements and loads of Eqs. (9.137) are expressed in terms of the in-plane compliance matrices related to the flexible-hinge chain and actuator as:

$$\begin{cases} [u_O] = [C_O][f_O] + [C_{OC}][f_C] \\ [u_C] = [C_{CO}][f_O] + [C_C][f_C] \end{cases}. \tag{9.138}$$

The compliance matrices of Eqs. (9.138) are:

$$[C_O] = \left[C_O^{(h)}\right] + \left[C_O^{(a)}\right]; \quad [C_{OC}] = \left[C_{OC}^{(a)}\right]; \quad [C_{CO}] = [C_{OC}]^T; \quad [C_C] = \left[C_C^{(a)}\right]. \tag{9.139}$$

In Eqs. (9.139), the superscript "h" refers to the flexible-hinge chain AB in Figure 9.32b, while the superscript "a" denotes the actuator. The compliance matrices in the right-hand sides of Eqs. (9.139) are calculated as:

$$\begin{cases} \left[C_O^{(h)}\right] = [T_{OA}]^T \left[C_A^{(h)}\right][T_{OA}]; \\ \left[C_O^{(a)}\right] = [T_{OC}]^T \left[R^{(a)}\right]^T \left[C_C^{(a)}\right]\left[R^{(a)}\right][T_{OC}]; \\ \left[C_{OC}^{(a)}\right] = [T_{OC}]^T \left[R^{(a)}\right]^T \left[C_C^{(a)}\right]\left[R^{(a)}\right]; \\ \left[C_C^{(a)}\right] = \left[R^{(a)}\right]^T \left[C_C^{(a)}\right]\left[R^{(a)}\right] \end{cases} \tag{9.140}$$

$\left[C_A^{(h)}\right]$ is the compliance matrix of the hinge chain in its local frame $Ax_A y_A z_A$, while $\left[C_C^{(a)}\right]$ is the compliance matrix of the actuator in its local frame $Cx_C y_C z_C$. The global frame is $Ox_O y_O z_O$, which is identified in Figure 9.32b. The in-plane translation matrices of Eqs. (9.140) are evaluated based on the necessary x and y offsets between points O, A and C, which are defined by a specific application, while the rotation matrix of the same equation is evaluated for an angle $\delta = \pi -$ this is due to the angular offset between the axes Ox_O and Cx_C.

The zero boundary conditions of Eqs. (9.136) are used in conjunction with Eqs. (9.137) and (9.138) to solve for the four unknown reaction components as:

$$[f_r]_{4\times1} = [E]_{4\times2}[f]_{2\times1} \quad \text{with} \quad \begin{cases} [f_r] = \begin{bmatrix} f_{O_x} & m_{O_z} & f_{C_y} & m_{C_z} \end{bmatrix}^T \\ [f] = \begin{bmatrix} f_a & f_l \end{bmatrix}^T \\ [E] = [A]_{4\times4}^{-1}[B]_{4\times2} \end{cases}, \tag{9.141}$$

where

$$[A] = \begin{bmatrix} C_O(1,1) & C_O(1,3) & C_{OC}(1,2) & C_{OC}(1,3) \\ C_O(3,1) & C_O(3,3) & C_{OC}(3,2) & C_{OC}(3,3) \\ C_{CO}(2,1) & C_{CO}(2,3) & C_C(2,2) & C_C(2,3) \\ C_{CO}(3,1) & C_{CO}(3,3) & C_C(3,2) & C_C(3,3) \end{bmatrix};$$

$$[B] = \begin{bmatrix} -C_{OC}(1,2) & C_O(1,2) \\ -C_{OC}(3,1) & C_O(3,2) \\ -C_C(2,1) & C_{CO}(2,2) \\ -C_C(3,1) & C_{CO}(3,2) \end{bmatrix}.$$

$$(9.142)$$

The $[A]$ and $[B]$ matrices of Eq. (9.142) are formed with components of the matrices introduced in Eqs. (9.138). The pair (i, j) indicates the element located on the row i and column j of a related matrix.

The two load vectors defined in Eqs. (9.137) can now be expressed as:

$$[f_O] = [E_1][f]; \quad [f_C] = [E_2][f]$$

$$\text{with}: \quad [E_1] = \begin{bmatrix} E(1,1) & E(1,2) \\ 0 & -1 \\ E(2,1) & E(1,1) \end{bmatrix}; \quad [E_2] = \begin{bmatrix} 1 & 0 \\ E(3,1) & E(3,2) \\ E(4,1) & E(4,2) \end{bmatrix}, \quad (9.143)$$

where the elements of the $[E_1]$, $[E_2]$ matrices are components of the matrix $[E]$ of Eqs. (9.141) and (9.142). The following displacement vectors are found on substituting the force vectors of Eqs. (9.143) into Eqs. (9.138):

$$[u_C] = [F]_{3\times2}[f]; \quad [u_O] = [H]_{3\times2}[f] \quad \text{with} \quad \begin{cases} [F] = [C_{CO}][E_1] + [C_C][E_2] \\ [H] = [C_O][E_1] + [C_{OC}][E_2] \end{cases}.$$

$$(9.144)$$

The input and output displacements are expressed from Eqs. (9.144) as:

$$\begin{cases} u_{C_x} = a_{11} \cdot f_a + a_{12} \cdot f_l \\ u_{O_y} = a_{21} \cdot f_a + a_{22} \cdot f_l \end{cases} \quad \text{with} \quad \begin{cases} a_{11} = F(1,1); \quad a_{12} = F(1,2) \\ a_{21} = H(2,1); \quad a_{22} = H(2,2) \end{cases}. \quad (9.145)$$

The performance qualifiers, which are introduced in Section 5.1.3.1, can now be formulated from Eqs. (9.145). For no external load, $f_l = 0$, the displacement amplification $d.a.$, input stiffness k_{in} and cross-stiffness k_c are:

$$d.a = \left|\frac{u_{O_y}}{u_{C_x}}\right| = \left|\frac{a_{21}}{a_{11}}\right|; \quad k_{in} = \frac{f_a}{u_{C_x}} = \frac{1}{|a_{11}|}; \quad k_c = \frac{f_a}{u_{O_y}} = \frac{1}{|a_{21}|}. \qquad (9.146)$$

When there is no actuation, which is $f_a = 0$, Eqs. (9.145) enable to evaluate the output stiffness k_{out} and the cross-stiffness k_c as:

$$k_{out} = \left|\frac{f_l}{u_{O_y}}\right| = \frac{1}{|a_{22}|}; \quad k_c = \frac{|f_l|}{u_{C_x}} = \frac{1}{|a_{12}|}. \qquad (9.147)$$

Since the cross-stiffness k_c is unique, Eqs. (9.146) and (9.147) show that $a_{21} = a_{12}$. The block load $f_{l,max}$ is found by setting the output displacement to 0, i.e., $u_{O_y} = 0$ in the second Eq. (9.145), namely:

$$f_{l,max} = -\left|\frac{a_{21}}{a_{22}}\right| \cdot f_a. \qquad (9.148)$$

Example 9.12

Consider the mechanism of *Example 5.10* whose quarter is depicted in Figure 5.25a, which is also shown in Figure 9.33 with the actuator block. The piezoelectric block is defined by a constant $d = 4 \cdot 10^{-10}$ m/V, Young's modulus $E_a = 2.9 \cdot 10^{11}$ N/m^2, semi-length $l_a = 0.02$ m and side length of square cross-section $2t_a = 0.005$ m. Known are also $l_1 = 0.0063$ m and $l_2 = 0.005$ m. Use all the numerical values of the hinge geometry and material parameters provided in *Example 5.10* to:

  i. Calculate the displacement amplification, relevant stiffnesses, and the block load;
  ii. Calculate the output displacement u_{O_y} for an input voltage to the piezoelectric actuator $v = 500$ V and no external load;
  iii. Determine the piezoelectric voltage that will produce an output displacement $u_{O_y} = -2.5$ µm when the load force is $f_l = -5$ N.

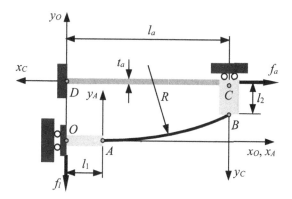

FIGURE 9.33 Quarter model of double-symmetry displacement-amplification device with one curvilinear-axis flexible hinge and block actuator.

Solution:

i. The in-plane compliance matrix $\left[C_A^{(h)}\right]$ of the circular-axis flexible hinge is evaluated in *Example 5.10* by means of Eq. (5.172), while the straight-axis, in-plane compliance matrix of the actuator $\left[C_C^{(a)}\right]$ is calculated as:

$$\left[C_C^{(a)}\right] = \begin{bmatrix} C_{u_x-f_x}^{(a)} & 0 & 0 \\ 0 & C_{u_y-f_y}^{(a)} & C_{u_y-m_z}^{(a)} \\ 0 & C_{u_y-m_z}^{(a)} & C_{\theta_z-m_z}^{(a)} \end{bmatrix} \quad \text{with}$$

$$\begin{cases} C_{u_x-f_x}^{(a)} = \dfrac{l_a}{E_a \cdot A_a}; \quad C_{u_y-f_y}^{(a)} = \dfrac{l_a^3}{3E_a \cdot I_{za}}; \quad C_{u_y-m_z}^{(a)} = -\dfrac{l_a^2}{2E_a \cdot I_{za}}; \quad C_{\theta_z-m_z}^{(a)} = \dfrac{l_a}{E_a \cdot I_{za}}; \\ A_a = (2t_a) \cdot t_a = 2t_a^2; \quad I_{za} = \dfrac{t_a^3 \cdot (2t_a)}{12} = \dfrac{t_a^4}{6} \end{cases}$$

$$(9.149)$$

The cross-section of the actuator shown in Figure 9.33 is rectangular with an out-of-plane width $2t_a$. The translation matrices and the rotation matrix of Eqs. (9.140) are:

$$[T_{OA}] = \begin{bmatrix} 1 & 0 & 0 \\ 0 & 1 & 0 \\ 0 & -l_1 & 1 \end{bmatrix};$$

$$[T_{OC}] = \begin{bmatrix} 1 & 0 & 0 \\ 0 & 1 & 0 \\ l_2 + R \cdot (1 - \cos\alpha) & -(l_1 + R \cdot \sin\alpha) & 1 \end{bmatrix}; \qquad (9.150)$$

$$[R^{(a)}] = \begin{bmatrix} \cos(\pi) & \sin(\pi) & 0 \\ -\sin(\pi) & \cos(\pi) & 0 \\ 0 & 0 & 1 \end{bmatrix}.$$

Equations (9.141) through (9.148) are combined and result in the following performance qualifiers: $d.a. = 5.312$, $k_{in} = 186{,}082{,}010\,\text{N/m}$, $k_c = 35{,}029{,}407\,\text{N/m}$, $k_{out} = 1{,}787{,}134\,\text{N/m}$, $f_{l,max} = 0.051 \cdot f_a$. Note that, as expected, the displacement-amplification value is identical to that obtained in *Example 5.10*. The inclusion of the (stiff) actuator block results in considerably larger stiffnesses than those of *Example 5.10*, which is, again, a realistic evaluation.

ii. Combining the second Eq. (9.145) for $f_l = 0$ with the piezoelectric force of Eq. (9.135) results in:

$$u_{O_y} = a_{21} \cdot f_a = a_{21} \cdot \frac{E_a \cdot A_a}{l_a} \cdot d \cdot v, \qquad (9.151)$$

which is $u_{O_y} = 1.03\,\mu\text{m}$. The output displacement can also be calculated from Eq. (9.146) as $u_{O_y} = f_a / k_c$, which produces the same numerical value.

iii. The second Eq. (9.145) is written as:

$$f_a = \frac{1}{a_{21}} \cdot \left(u_{O_y} - a_{22} \cdot f_l \right). \qquad (9.152)$$

The actuation force of Eq. (9.151) is substituted into Eq. (9.152), which yields:

$$v = \frac{l_a}{E_a \cdot A_a \cdot d \cdot a_{21}} \cdot \left(u_{O_y} - a_{22} \cdot f_l \right), \qquad (9.153)$$

and the numerical value is $v = 143.874\,\text{V}$.

9.5.2.2 Sensing

A block piezoelectric member can also be used to sense external mechanical input, which it converts into electrical voltage by means of the direct effect and based on Eq. (9.119). The following example illustrates a piezoelectric sensing application.

Example 9.13

A block piezoelectric sensor is attached to the rigid link of the mechanism shown in Figure 9.34a, which also consists of two identical right elliptically corner-filleted flexible hinges. The mechanism uses all geometric and material features provided in *Example 5.1*, which studies the same mechanism. Known are also the constant rectangular cross-section dimensions of the rigid link: $t_r = 0.025$ m and $w_r = 0.024$ m, the mass density of the rigid link and hinges $\rho = 7,800$ kg/m³, and the properties of the sensor: $l_s = 0.03$ m, square cross-section side length $t_s = 0.005$ m, Young's modulus $E_s = 2.9 \cdot 10^{11}$ N/m² and mass density $\rho_s = 7,200$ kg/m³. A small initial displacement y_0 is applied to the mechanism-sensor system, which results in a sensor voltage whose time variation is qualitatively plotted in Figure 9.34b and indicates that the free response is underdamped. Propose a single-DOF, lumped-parameter system to model the system's free vibrations along the y-direction at the center point C and evaluate the equivalent viscous damping coefficient of the full system knowing that the ratio of any two consecutive voltage amplitudes, such as V_1 and V_2 in Figure 9.34b, is $V_1/V_2 = 60$.

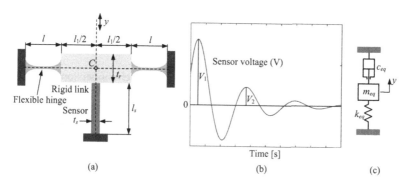

(a) (b) (c)

FIGURE 9.34 (a) Symmetric mechanism with two identical straight-axis, right elliptically corner-filleted flexible hinges, one rigid link and block piezoelectric actuator; (b) sensed output voltage time variation; (c) equivalent, lumped-parameter model of mechanism-sensor system.

Solution:

The single-DOF, lumped-parameter mechanical system modeling the y-axis vibrations of the actual mechanism-sensor system is depicted in Figure 9.34c. It consists of a body of equivalent mass m_{eq}, a viscous damper of unknown damping coefficient c_{eq} and a spring of stiffness k_{eq}.

The mass m_{eq} is calculated as:

$$m_{eq} = m_r + 2m_{h,eq} + m_{s,eq}, \qquad (9.154)$$

where $m_r = \rho \cdot w_r \cdot t_r \cdot l_1$ is the mass of the rigid prismatic link, $m_{h,eq}$ is the equivalent mass of a flexible hinge, and $m_{s,eq}$ is the equivalent mass of the block sensor, all expressed with respect to the center point C in Figure 9.34a. The latter masses are calculated as in Eqs. (7.55) and (7.56) by the lumped-parameter approach as:

$$\begin{cases} m_{h,eq} = \dfrac{33}{140} \cdot m_h = \dfrac{33}{140} \cdot \rho \cdot w \cdot \int\limits_0^l t(x)\,dx \\[4mm] m_{s,eq} = \dfrac{1}{3} \cdot m_s = \dfrac{1}{3} \cdot \rho_s \cdot t_s^2 \cdot l_s \end{cases} \qquad (9.155)$$

In Eqs. (9.155), m_h and m_s are the actual masses of the hinge and the sensor, while the integral in the same Eq. (9.155) is calculated based on the thickness of the right elliptically filleted flexible hinge; the integral is expressed with the aid of Eq. (2.50) as:

$$\int\limits_0^l t(x)\,dx = a \cdot \int\limits_0^{\pi/2} t_1(\varphi) \cdot \sin(\varphi)\,d\varphi + t \cdot (l - 2a) + a \cdot \int\limits_0^{\pi/2} t_2(\varphi) \cdot \cos(\varphi)\,d\varphi \qquad (9.156)$$

$$\text{with} \quad \begin{cases} t_1(\varphi) = t + 2b \cdot (1 - \sin(\varphi)) \\[2mm] t_2(\varphi) = t + 2b \cdot (1 - \cos(\varphi)) \end{cases}$$

Combining Eqs. (9.154)–(9.156) and using the given numerical values yields $m_{eq} = 0.0507\,\text{kg}$.

The equivalent stiffness results from the two parallel-connection hinges of stiffness k_h and the sensor stiffness k_s as:

$$k_{eq} = k_h + k_s; \quad k_s = \frac{E_s \cdot A_s}{l_s} = \frac{E_s \cdot t_s^2}{l_s}. \qquad (9.157)$$

The stiffness of the parallel-connection flexible hinges is calculated in *Example 5.1* as $k_y = k_h = 10.268{,}549\,\text{N/m}$ and k_s is the sensor stiffness, which is related to the moving point C. The numerical value of the equivalent stiffness is $k_{eq} = 251{,}935{,}216\,\text{N/m}$.

The free damped mathematical model of the lumped-parameter mechanical system of Figure 9.34c can be written in terms of the variable y or in terms of the sensed voltage v – see Rao [46], for instance – as follows:

$$\left.\begin{aligned} m_{eq} \cdot \ddot{y} + c_{eq} \cdot \dot{y} + k_{eq} \cdot y &= 0 \\[2mm] y &= d \cdot v \end{aligned}\right\}$$

$$\rightarrow \begin{cases} m_{eq} \cdot \ddot{v} + c_{eq} \cdot \dot{v} + k_{eq} \cdot v = 0 \\[2mm] \text{or} \\[2mm] \ddot{v} + 2\xi_{eq} \cdot \omega_n \cdot \dot{v} + \omega_n^2 \cdot v = 0 \quad \text{with:} \quad 2\xi_{eq} \cdot \omega_n = \dfrac{c_{eq}}{m_{eq}}; \quad \omega_n^2 = \dfrac{k_{eq}}{m_{eq}}, \end{cases} \qquad (9.158)$$

where d is the piezoelectric constant, ξ_{eq} is the equivalent damping ratio and ω_n is the natural frequency. The free underdamped response plotted in Figure 9.34b allows expressing ξ_{eq} as:

$$\xi_{eq} = \frac{\delta}{\sqrt{4\pi^2 + \delta^2}}; \quad \delta = \ln\left(\frac{V_1}{V_2}\right), \tag{9.159}$$

with δ being the logarithmic decrement – see Rao [46]. Combining Eqs. (9.158) and (9.159) determines the damping coefficient:

$$c_{eq} = 2\xi_{eq} \cdot \sqrt{m_{eq} \cdot k_{eq}}, \tag{9.160}$$

and its numerical value is $c_{eq} = 3{,}933.09$ Ns/m.

REFERENCES

1. Peterson, R.E., *Stress Concentration Factors*, John Wiley & Sons, Hoboken, NJ, 1974.
2. Pilkey, W.D. and Pilkey, D.F., *Peterson's Stress Concentration Factors*, 3rd ed., John Wiley & Sons, Inc., Hoboken, NJ, 2008.
3. Young, W.C. and Budynas, R.G., *Roark's Formulas for Stress and Strain*, 7th ed., McGraw Hill, New York, 2002.
4. Budynas, R.G. and Nisbett, J.K., *Shigley's Mechanical Engineering Design*, 9th ed., McGraw Hill, New York, 2011.
5. Isida, M., On the tension of the strip with semi-circular notches, *Transactions of Japan Society of Mechanical Engineers*, 19, 5, 1953.
6. Ling, C.B., On stress concentration at semicircular notch, *Transactions of ASME: Applied Mechanical Section*, 89, 522, 1967.
7. Noda, N.A., Sera, M., and Takase, Y., Stress concentration factors for round and flat test specimens with notches, *International Journal of Fatigue*, 17 (3), 163, 1995.
8. Gray, T.G.F. et al., Closed-form functions for elastic stress concentration factors in notched bars, *Journal of Strain Analysis for Engineering Design*, 30 (2), 143, 1995.
9. Durelli, A.J., *Stress concentrations*, U.M. Project SF-CARS, School of Engineering, University of Maryland, Office of Naval Research, Washington, DC, 1982.
10. Zappalorto, M., Berto, F., and Lazzarin, P., Practical expressions for the notch stress concentration factors of round bars under torsion, *International Journal of Fatigue*, 33, 382, 2011.
11. Zheng, M. and Niemi, E., Analysis of the stress concentration factor for a shallow notch by the slip-line field method, *International Journal of Fatigue*, 19 (3), 191, 1997.
12. Panasyuk, V.V., Savruk, M.P., and Kazberuk, A., Stress concentration near sharp and rounded V-notches, *Materials Science*, 49 (6), 711, 2014.
13. Taylor, D. et al., The variable-radius notch: two methods for reducing stress concentration, *Engineering Failure Analysis*, 18, 1009, 2011.
14. Radaj, D., State-of-the-art review of extended stress intensity factor concepts, *Fatigue and Fracture of Engineering Materials and Structures*, 37, 1, 2014.
15. Gomez, F.J. et al., A generalized notch stress intensity factor for U-notched components loaded under mixed mode, *Engineering Fracture Mechanics*, 75, 4819, 2008.

16. Neuber, H., Theory of stress concentration for shear-strained prismatic bodies with arbitrary non-linear stress-strain law, *Journal of Applied Mechanics*, 28, 545, 1961.
17. Zappalorto, M., Lazzarin, P., and Yates, J.R., Elastic stress distributions for hyperbolic and parabolic notches in round shafts under torsion and uniform antiplane shear loadings, *International Journal of Solids and Structures*, 45, 4879, 2008.
18. Guimin, C., Wangu, J., and Liu, X., Generalized equations for estimating stress concentration factors of various notch flexure hinges, *Transactions of ASME: Journal of Mechanical Design*, 136, 031009–1, 2014.
19. Ugural, A.C. and Fenster, S.K., *Advanced Strength and Applied Elasticity*, Prentice Hall, Englewood Cliffs, NJ, 1995.
20. Timoshenko, S.P., *History of Strength of Materials*, Dover, New York, 1983.
21. Solecki, R. and Conant, R.J., *Advanced Mechanics of Materials*, Oxford University Press, Oxford, 2003.
22. Volterra, E. and Gaines, J.H., *Advanced Strength of Materials*, Prentice Hall, Englewood Cliffs, NJ, 1971.
23. Cook, R.D. and Young, W.C., *Advanced Mechanics of Materials*, Macmillan, New York, 1985.
24. Kobayashi, A.S. (editor), *Handbook of Experimental Mechanics*, 2nd ed., VCH Publishers, Weinheim/New York, 1993.
25. Lobontiu, N. et al., Corner-filleted flexure hinges, *ASME Journal of Mechanical Design*, 123, 346, 2001.
26. Lobontiu, N., Planar compliances of thin circular-axis notch flexure hinges with midpoint radial symmetry, *Mechanics Based Design of Structures and Machines*, 41, 202, 2013.
27. Lobontiu, N. and Cullin, M., In-plane elastic response of two-segment circular-axis symmetric notch hinges: the right circular design, *Precision Engineering*, 37, 542, 2013.
28. Nicholson, D.W., *Finite Element Analysis: Thermomechanics of Solids*, 2nd ed., CRC Press, Boca Raton, FL, 2008.
29. Hetnarski, R.B. and Eslami, M.R., *Thermal Stresses-Advanced Theory and Applications*, 2nd ed., Springer, Berlin, 2019.
30. Li, S.R. and Zhou, Y.H., Geometrically nonlinear analysis of Timoshenko beams under thermomechanical loadings, *Journal of Thermal Stresses*, 26 (9), 861, 2003.
31. Manoach, E. and Ribeiro, P., Coupled, thermoelastic, large amplitude vibrations of Timoshenko beams, *International Journal of Mechanical Sciences*, 46 (11), 1589, 2004.
32. Li, S.R. and Xi, S., Large thermal deflections of Timoshenko beams under transversely non-uniform temperature rise, *Mechanics Research Communications*, 33 (1), 84, 2006.
33. Hou, W.F. and Zhang, X.M., Dynamic analysis of flexible linkage mechanism under uniform temperature change, *Journal of Sound and Vibration*, 319 (1–2), 570, 2009.
34. Zhang, X. and Hou, W., Dynamic analysis of the precision compliant mechanisms considering thermal effect, *Precision Engineering*, 34 (13), 592, 2010.
35. Burgreen, D., *Elements of Thermal Stress Analysis*, C.P. Press, New York, 1971.
36. Maluf, N. and Williams, K., *An Introduction to Microelectromechanical Systems Engineering*, 2nd ed., Artech House, Boston, MA, 2004.
37. Reece, P.L. (editor), *Smart Materials and Structures*, Nova Science Publishers, New York, 2006.
38. Rupitsch, S.J., *Piezoelectric Sensors and Actuators: Fundamentals and Applications*, Springer, Berlin, 2018.

39. Lobontiu, N., *System Dynamics for Engineering Students*, 1st ed., Academic Press, Burlington, MA, 2010.

40. Bellouard, Y. and Clavel, R., Shape memory alloy flexure hinges, *Materials Science and Engineering A*, 378, 210, 2004.

41. Du, Z. et al., Static deformation modeling and analysis of flexure hinges made of shape memory alloy, *Smart Materials and Structures*, 25 (11), 115029, 2016.

42. Rees, D.W.A., *Basic Solid Mechanics*, McMillan Press, London, 1997.

43. Gere, J.M. and Goodno, B.J., *Mechanics of Materials*, 9th ed., CENGAGE Learning, Boston, MA, 2019.

44. Lobontiu, N., Goldfarb, M., and Garcia, E., Achieving maximum tip displacement during resonant excitation of piezoelectrically actuated beams, *Journal of Intelligent Materials and Structures*, 10, 900, 1999.

45. Garcia, E. and Lobontiu, N., Induced-strain multimorphs for microscale sensory actuation design, *Smart Materials and Structures*, 13, 725, 2004.

46. Rao, S.S., *Mechanical Vibrations*, 5th ed., Prentice Hall, Englewood Cliffs, NJ, 2011.

Index

Note: **Bold** page numbers refer to tables and *italic* page numbers refer to figures.

Printed in the United States
by Baker & Taylor Publisher Services